COMPARATIVE ANATOMY OF VERTEBRATES

COMPARATIVE ANATOMY
OF
VERTEBRATES

SECOND REVISED & ENLARGED EDITION

S.K. KULSHRESTHA
M.Sc., Ph.D., F.A.Z., F.A.E.B., F.Z.S.I., F.N.E.S.A., F.N.C.
Former Professor of Zoology, Principal
Additional Director, Higher Education, Govt. of M.P.

ANMOL PUBLICATIONS PVT. LTD.
NEW DELHI - 110 002 (INDIA)

ANMOL PUBLICATIONS PVT. LTD.
4374/4B, Ansari Road, Daryaganj
New Delhi - 110 002
Ph.: 23261597, 23278000
Visit us at: www.anmolpublications.com

Comparative Anatomy of Vertebrates

First Edition, 1992

Reprint 1999

Second Revised and Enlarged Edition, 2004

PRINTED IN INDIA

Published by J.L. Kumar for Anmol Publications Pvt. Ltd., New Delhi and Printed at Mehra Offset Press, Delhi.

Contents

Preface to the Second Edition

For the second edition much new information has been added where necessary. All diagrams have been redrawn to make them easily understandable. The glossary of technical terms has been included alongwith bibliography.

I wish to express thanks to M/s Anmol Publications Pvt. Ltd. for bringing out this revised edition in short time.

Author

Preface to the First Edition

The present title "Comparative Anatomy of Vertebrates" is designed for undergraduates, but hope other people will read it as well. It is abut the vertebrates and the few other animals, which are included with them in the phylum chordata. It will be obvious that this is too small a book to contain a complete account of modern knowledge of the chordates.

Most of the book is about animal structure and how they work. In this volume, unlike most textbooks on anatomy, every efforts has been made to include discussions of the function of body structure along with anatomical description. Order concepts are delineating various scientific disciplines are falling by the wayside.

The subject matter is presented in twelve chapters starting form the Phylum Chordata to Receptor Organs through the various systems of the body. Much of this portion is presented according to the most generally used scheme of classification of chordates. Emphasis is placed for the most part, upon more familiar living chordates. Each organ system is covered in a separate chapter. Country to the conviction of certain biologists, the endocrine glands are treated as a separate system because of their profound effect upon the development and sense organs are for convenience, treated independently despite their obviously close connection. At the end of each chapter is a summary of the main parts emphasised in that chapter.

Every effort has been made to present the subject matter clearly and accurately. The author will appreciate it if readers will call to his attention any inaccuracies and inconsistencies which may have escaped his notice.

It has been my good fortune to receive assistance form many generous friends and colleagues, without whose help I could not have assembled this material. They are too many to mention. I extend my warmest thinks for their time, case and tolerance. Without their help this book would be a different one and an inferior one.

Few words of appreciation are also due to M/s Anmol Publications for bringing out a handsome print of book in a comparatively short time.

Author

1

Phylum Chordata

1.0. Origin of Chordata
- 1.1 Concept of Protochordata
  - 1.1.1 *Protochordata Phylogeny*

2.0 The Nature of Vertebrates Morphology
- 2.1 Approach to Comparative Anatomy
  - 2.1.1 Definition, Scope and Relation to Other Disciplines
  - 2.1.2 Importance of the Study of Vertebrate Morphology

3.0. Origin of Vertebrates
- 3.1 *Branchiotoma Ancestry*
- 3.2 *Balanroglossus Ancestry*
- 3.3 Urochordata Ancestry
- 3.4 Probable Ancestry of Vertebrates
- 3.5 Time and Place of Origin

4.0 General, Characteristics and Classification
- 4.1 Subphylum Tunicata (Urochordata)
- 4.2 Subphylum Cephalochordata (Leptocardi)
- 4.3 Subphylum Agnatha
  - 4.3.1 *Class Ostracodermi*
  - 4.3.2. *Class Cyclostomata*
- 4.4 Subphylum Gnathastomata
  - 4.4.1 *Superclass Pisces*
  - 4.4.2 *Superclass Tetrapoda*
    - *Class Amphibia*
    - *Class Reptilia*
    - *Class Aves*
    - *Class Mammalia*

1.0. ORIGIN OF CHORDATES

Chordates are organisms with notochord, dorsal hollow nerve cord and pharyngeal gill slits. They share common features such an bilateral symmetry, axial organization, triploblastic condition and segmentation etc. with non chordates. Considerable controversy exists regarding the origin of chordates. According the geological records the chordates originated prior to Chordates period as evident by fossils of some lower chordates. There are a few theories regarding the origin the chordates from the nonchordates groups. Most of the theories, however, suffer from serious defects. Garstang suggested that chordates have involves some free-swimming echinoderm larvae, possible auricularian larvae by paedogenesis. This view is not accepted because later view is that difference between the vertebrates and nonchordates are artificial in nature. Inclusion of the echinoderm, pogonophores and chordates in one group Deuterostomia, where the anus develops from the blastopore and the mouth is formed a new is more acceptable. The protochordates including hemichordates, urochordates and cephalochordates are considered as the invertebrate members of the phylum Chordata. The protochordates provide connecting link between the deuterostomes and vertebrates. According to Barrington (1965) the deuterostomes have evolved from sessile or semisessile ancestors having bylaterally symmetrical and tripartite body with coelom. The echinoderms have departed a long way from the ancestors, while the hemichordates remained closer. The hemichordates have developed pharyngeal wall opening associated with ciliary mode of feeding. In course of time a group with internal food collection mechanism by elaborate and complicated pharynx gave rise to the urochordates. cephalochordates and vertebrates.

1.1. Concept of Protochordata

Diverse opinions exist on the taxonomical subdivisions of the phylum chordata. According to a compromised scheme the phylum chordata is divided into four subphyla including Cephalochordata or Acrania. Hemichordata. Urochordata or Tunicata and Vertebrata or Craniata. This association of the first three subphyla with the vertebrates suggests the possibility of intergradation between these groups and the sequences for evolutionary transitions from one to another. The current view is in favour of removal of the Hemichordata from the phylum Chordata leaving the Urochordata. Cephalochordata and Vertebrata as the subphyla of the phylum Chordata. The Hemichordata is now given status of a separate phylum. It is however, a common practice to designate the hemichordates, urochordates and the cephalochordates as lower chordates or protochordates or invertebrate chordates. The naming of these groups collectively as Protochordata or Invertebrate chordates appears to be most convincing. The invertebrate chordates include a heterogenous group of marine coelomates having a notochord, which may persist throughout life or may be confined only to the larval stage. They have a dorsal tubular nerve chord present throughout life or restricted in larval stage and permanent gill slits in the pharyngeal region. The skull jaws and paired appendages are absent.

1.1.1. Protochordate Phylogeny

The relationship of the protochordates to the vertebrates is well documented on embryological and structural aspects of larvae and adult.

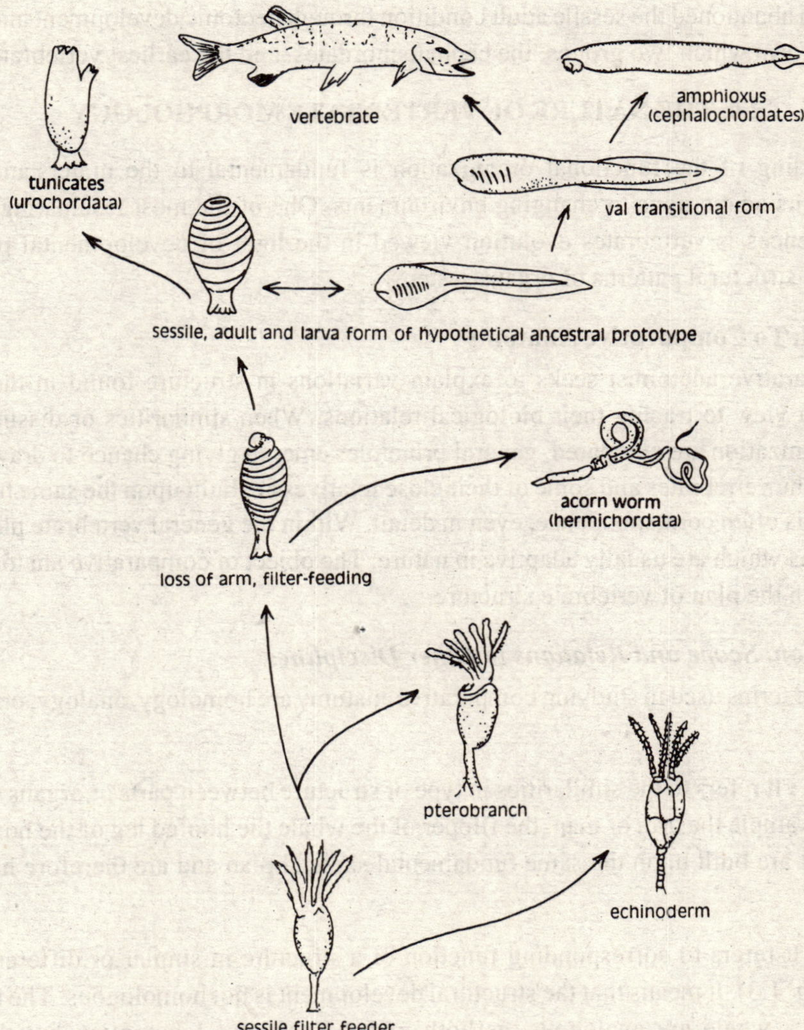

Fig. 1.1 Diagrammatic representation of the major steps in vertebrate evolution from a primitive echinoderm like ancestor

The placement of the hemichordata within the phylum Chordata or in its own phylum reflects the lack of agreement by workers on the precise relationship of this group. It is however, agreed that the Hemichordata are important in understanding vertebrate origin. The three protochordata groups are represented by highly specialized forms which tend to obscure their relationships with other animal groups. It is difficult to find an ancestor from which the protochordates may have arisen. However, the echinoderms present the most striking evidence of a relationship to chordate ancestry. The most generally accepted suggestion is that the protochordates represent specialized offshoot stages along the main line of evolution of the vertebrates from a shared ancestor with echinoderms.

(Fig 1.1) The earliest sessile filter-feeding forms, represented today by hemichordates, gave rise to a group, which abandoned the sessile adult condition through neotonic development and represents the ancestors form which two groups, the cephalochordates, and the earliest vertebrates arose.

2.0. THE NATURE OF VERTEBRATE MORPHOLOGY

An understanding of the functional organization is fundamental to the understanding of the organism and its adaptations to changing environments. One of the most fascinating area of the biological sciences is vertebrates evolution viewed in the light of developmental patterns and functional and structural patterns of organs systems.

2.1. Approach To Comparative Anatomy

The comparative anatomist seeks to explain variations in structure found in the bodies of animals with a view to tracing their biological relations. When similarities or dissimilarities in structural organization are compared, general principles emerge giving chance to draw deductive conclusions. The vertebrates and some of their close relatives are built upon the same fundamental plan and there is often correspondence, even in detail. Within the general vertebrate plan there are many variations which are usually adaptive in nature. The object of comparative anatomy is to get acquainted with the plan of vertebrate structure.

2.1.1. Definition, Scope and Relations to Other Disciplines

Some of the terms used in studying comparative anatomy are homology, analogy, ontogeny and phylogeny.

Homology : It refers to the similarities in type of structure between parts or organs of different animals. For example the arm of man, the flipper of the whale the hoofed leg of the horse, and the wing of the bat are built upon the same fundamental skeletal plan and are therefore homologous (Fig. 1.2).

Analogy : It refers to corresponding function of a structure in similar or different organ or organ parts (Fig. 1.3). It means that the structural development is not homologous. The fin of a fish and flipper of a whale are analogous, as both may be used for locomotion but the two are structurally different.

Ontogeny And Phytogeny : Early embryos of all vertebrates show remarkable similarities in their structural development (Fig. 1.4). In order to explain the structures such as pharyngeal pouches and aortic arches making their appearance in the embryos of higher vertebrates recapitulation theory or Law Biogenesis was furnished. This theory accounts for the fact the early embryonic stages of all vertebrates are remarkably similar up to the point where each group diverges in its own characteristic manner. This concept has been useful to biologists in establishing homologies and making certain phylogenetic relationships, and in helping to give a picture of probable structure of ancestral forms, and stages in vertebrate evolutionary history.

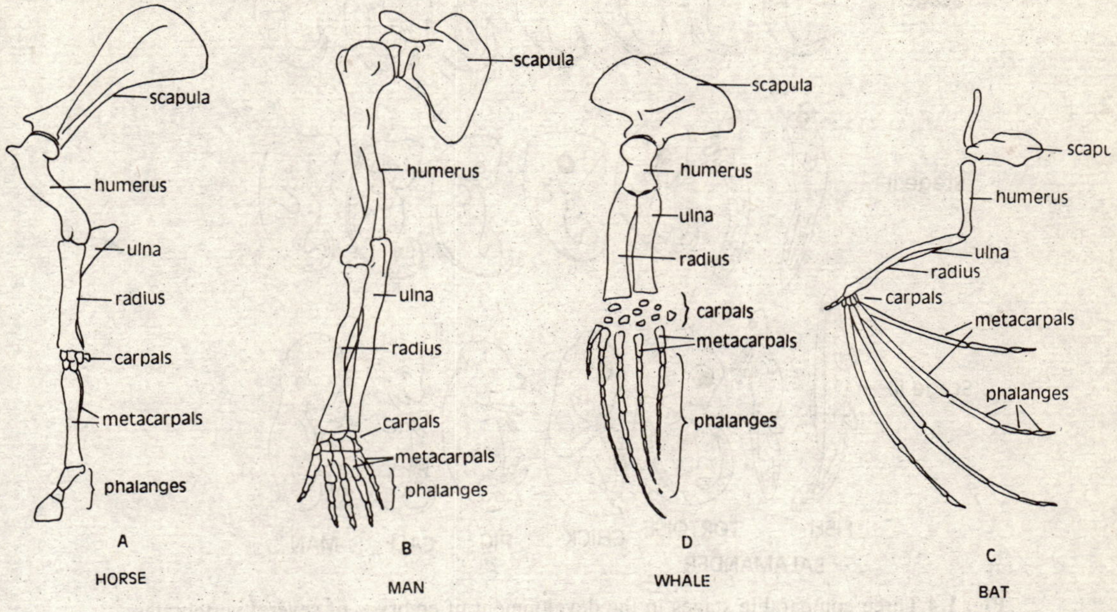

Fig. 1.2 Skeleton of forelimb showing modification of structure (Homologous structured)

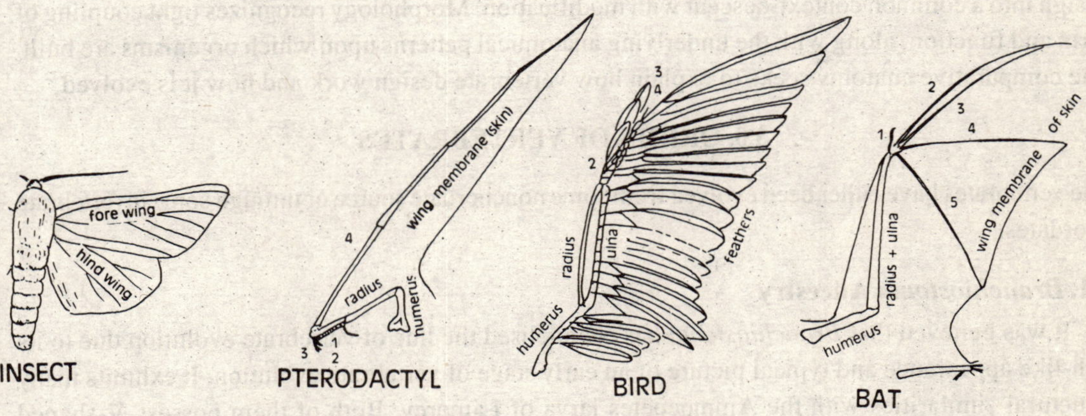

Fig. 1.3 Analogy between wings of insects with no internal skeleton and of vertebrates with skeleton which have a like function hut different origins. Homology in the wing bones of vertebrates all derived from the common pattern of the forelimb in land vertebrate but variously modified. Pterodactyl (extinct reptile) with very long fourth finger: bird with first and fifth lacking, third and fourth partly fused: bat with second to fifth linger long.

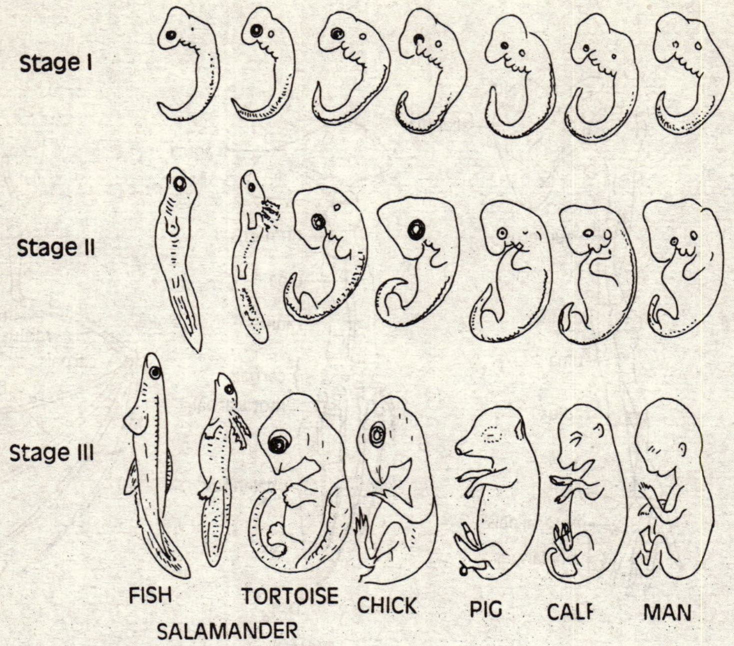

Fig. 1.4 Three comparable stages in the development of embryos of several vertebrates

2.1.2. Importance of the Study of Vertebrate Morphology

Anatomy and its significance are parts of comparative morphology. This provides insight into how organisms work and how they are evolved. Charles Darwin added united issue of biological design into a common context-descent with modification. Morphology recognizes tight coupling of form and function, along with the underlying anatomical patterns upon which organisms are built. The comparative anatomy seeks to explain how vertebrate design work and how it is evolved.

3.0. ORIGIN OF VERTEBRATES

The vertebrates have either been evolved from some nonchordate source or through some invertebrate chordates.

3.1. *Branchiostoma* Ancestry

It was believed that *Branchiostoma* has channelised the line of vertebrate evolution due to its fish-like appearance and typical picture of an early stage of vertebrate evolution. It exhibits many structural similarities with the Ammocoetes larva of Lamprey. Both of them possess V-shaped myotomes, oral hood, continuous dorsal fin. velum, velar tentacles and endostyle. The Ammocoetes larva has been considered just above *Branshiostoma* in the evolution of vertebrates. *Branchiostoma* stands nearest to the true ancestral vertebrates but its sedentary habit offers the strongest objection in assuming the biological changes of the vertebrates from sessile animals.

3.2. *Balanoglossus* Ancestry

Balanoglossus is regarded as ancestor of vertebrates. The tornaria larva shows remarkable resemblance with echinoderm larvae. *Balanoglossus* itself does not fulfil the standard criterian of a standard chordata. Its notochord is very doubtful structure. Its nervous system is more of invertebrate type than that of a chordata. The similarities of *Balanoglossus* with the vertebrates are due to remote phylogenetic relationship. The presence of gill slits brings it in chordata organization.

3.3. Urochordate Ancestry

According to Berril the sedentary habit of the urochordates shows close conformity with other sedentary invertebrate chordates and the free swimming tadpole larva of urochordates has come from the sessile life of the vertebrates. The free swimming tadpole larva through developmental paedomorphosis has avoided the sedentary condition and maintains a perenial free-living habit. Hence, the dyanamic tadpole larva of urochordates is the actual source from which vertebrates have been evolved in time and space. The probable line of evolution of vertebrates according to Berril is shown in figure 1.5.

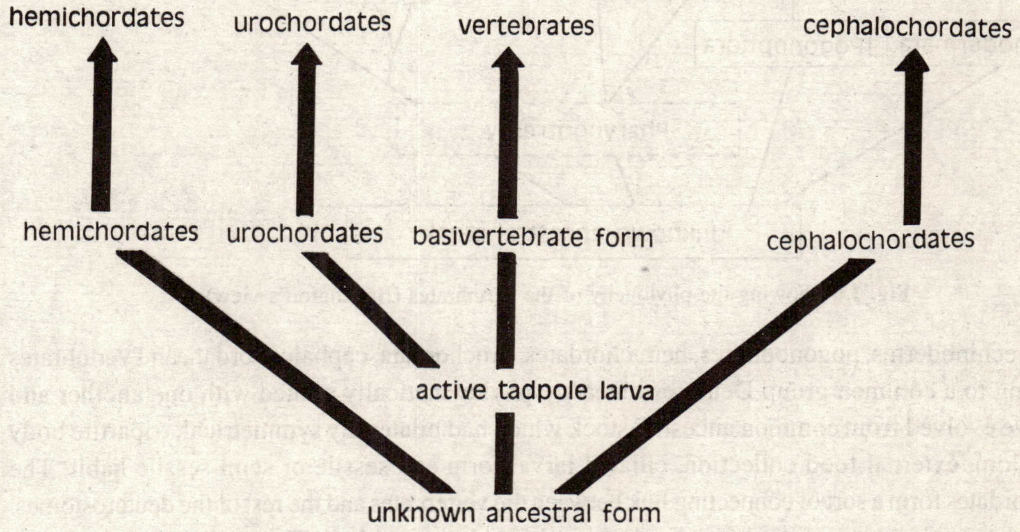

Fig. 1.5 Showing Berrill's view on the probable line of vertebrate origin.

3.4. Probable Ancestry of Vertebrates

Baldwin, Ennor and Morrisson and many others have shown phylogenetic relationship on the distribution of phosphagens, the phosphate carrier which participate in energy production in muscular activities. They have established that phosphagen in the form of phosphocreatine is present in *Amphioxus* and Vertebrates whereas phosphoarginine is present in majority of the nonchordates and urochordates. Both these compounds are found in echinoderms and hemichordates.

Form this evidence it is suggested that the nonchordates and vertebrates are related to echinoderms and hemichordates but urochordates come in anamolous situation. This suggestion justified the notion that the urochordates are situated far from the main line of chordates. The phylogenic value of the existence of phosphagens in evaluating the relationship between different groups is doubtful. Presence of common metabolic pathways in different groups may be interpreted as instances of independent evolution of similar biochemical adaptations. But the distribution of phosphagens fits well with the current views regarding the relationships of the deuterostomia as shown in figure 1.6.

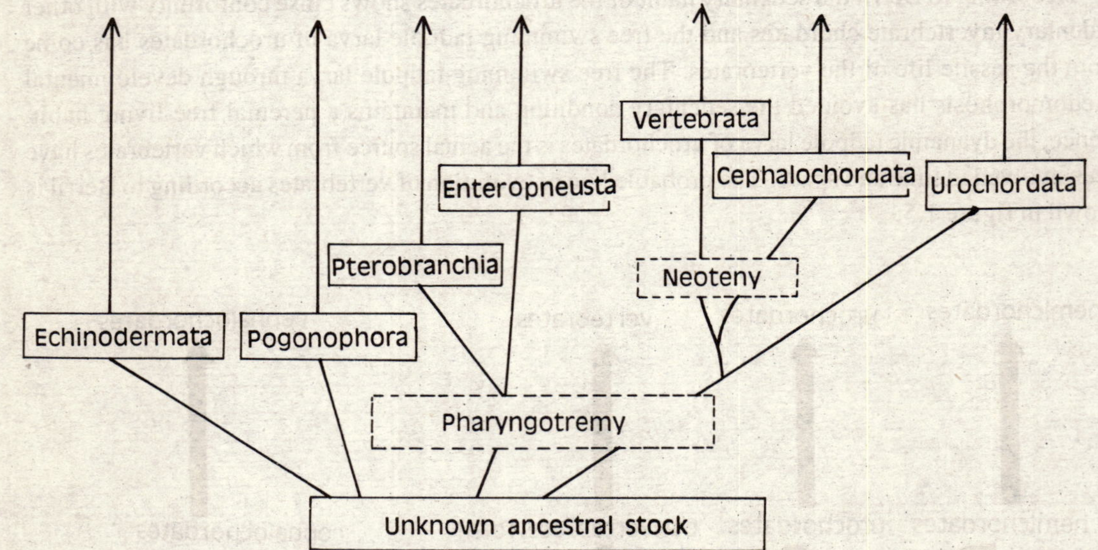

Fig. 1.6 Showing the phylogeny of the vertebrates (Barrington's view)

The echinoderms, pogonophores, hemichordates, urochordata, cephalochordata and vertebrates belonging to a common group Deuterostomea are phylogenetically related with one another and they have evolved from common ancestral stock which had bilaterally symmetrical, tripartite body and coelom, external food collection, ciliated larval form and sessile or semi-sessile habit. The protochordates form a sort of connecting link between the vertebrates and the rest of the deuterostomes. The phylogenetic status of the hemichordates is difficult to establish. The cephalochordates and urochordates are true chordate but their relationships with each other as well as with vertebrates are difficult to ascertain. The concept of neoteny as a factor of evolution is quite convincing.

3.5. Time and Place of Origin

Fragmentary remains of the invertebrate chordates have been recorded in the Cambrian strata and that of ostracoderms in the middle Ordovician period. But the relics discovered in the strata are those of creatures who had already travelled a long evolutionary way. It may be inferred that the time of origin was not later than the beginning of the Ordovician, if not long before. The place of origin is doubtful. The place of origin of early chordates is probably the sea but the place of origin of vertebrates is not known. There are evidences that the vertebrates first appeared in sea and migrated into the river system during early phase of vertebrate evolution.

4.0. GENERAL CHARACTERISTICS AND CLASSIFICATION

The Phylum chordata includes tunicates, lancelets, lampreys, sharks, rays, bony fishers, amphibians, reptiles, birds, and mammals. The chordates exhibit great diversity of structure, function and habit. The animals belonging to the phylum chordata are bilaterally symmetrical, triplobalstic with segmented body, well developed coelom, and complete digestive tract. Three fundamental characteristics of chordates which distinguish them from other animals are notochord, dorsal tubular nerve cord, and pharyngeal gill slits, present at some stage of the life history.

The *notochord* is found in the embryos of most chordates as long and flexible rod of vacuolated cells extending from the anterior to the posterior part of the body on mid dorsal line. It is the first supporting structure of the chordate body. In lower chordates like lancelets and lampreys, it exists throughout life as main axial support, but in fishes, amphibians, reptiles, birds, and mammals it is replaced by the vertebral column.

The *nerve cord* is developed after the gastrula stage on the dorsal surface of embryo. It lies above the notochord and assumes hollow tubular shape by the infolding of the ectoderm. In lancelets and tunicate larvae its anterior end becomes enlarged to form cerebral vesicle, while in vertebrates it thickens to form the brain. In higher chordates the nerve cord is surrounded by the neural arches of the vertebrae and the brain is surrounded by the cranium.

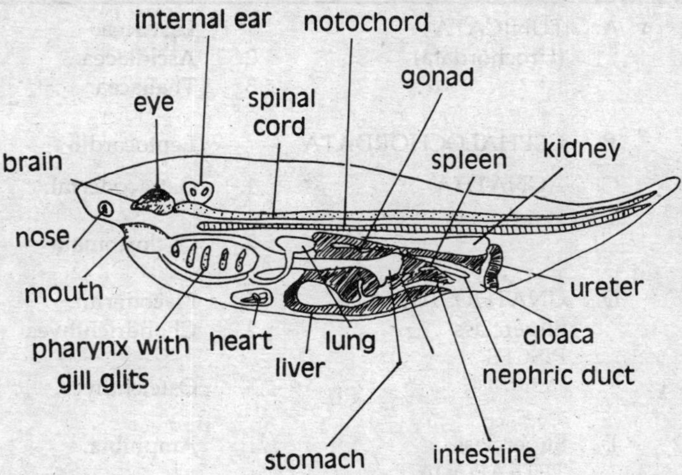

Fig. 1.7 A diagram showing the anatomical features of a chordate.

The *pharyngeal gill slits* are paired structures developed on the sides of the embryonic pharynx. Each gill slit is formed by an outgrowth of endoderm in the pharynx, which joins a corresponding inpushing of ectoderm on the outer side of the body. The wall present in between breaks down to form a gill slit. These paired series of pores leading from the anterior section of the alimentary canal to the lateral surface of the body constitute the universal feature of the chordata. Within these pores the gill of the aquatic animals are situated which are used in respiration. In amphibians the gills are lost during metamorphosis. In reptiles, birds and mammals several gill slits are formed during the early embryonic life but are nonfunctional and the respiration is effected by the newly acquired lungs.

The classification of Phylum Chordata has been described under different schemes by several authors. Parker and Haswell revised by Marshall (1962) have divided phylum Chordata into four subphyla including Hemichordata, Tunicata. Acrania (Cephalochordata) and Craniata (Vertebrata). Romer (1959, 1962) have also used similar subphyla with minor difference in naming certain superclasses as classes. The group Hemichordata has recently been given the status of a separate phylum under nonchordata or invertebrate chordates (Hyman, 1959; Johnson et al. 1967). The classification used in the present text is based on the scheme given by Storer and Usinger (1965) with minor changes to make it simplified. The classification of class Mammalia has been modified according to Young (1962).

The outline classification of Chordata is given in table 1.1.

TABLE 1.1
Phylum Chordata

Group	*Subphyla*	*Classes*	*Examples*
I. ACRANIA	A. TUNICATA (Urochordata)	1. Larvacea.	*Appendicularia*
		2. Ascidiacea.	Ascidians.
		3. Thaliacea.	*Salpa* *Doliolum.*
	B. CEPHALOCHORDATA.	Leptocardii.	Lancelets
II. CRANIATA	C. AGNATHA	1. Ostracodermi.	*Ancient armoured fishes
		2. Cyclostomata	Petromyzon, *Myxine.*
	D. GNATHOSTOMATA Superclass PISCES	1. Placodermi.	*Climatius. *Bothriolepis.*
		2. Chondrichthyes	sharks, Rays.
		3. Osteichthyes.	Bony fishes. lung fishes.
	E. Superclass TETRAPODA	1. Amphibia.	Salamanders, frogs, Caecilians.
		2. Reptilia	Lizards, snakes, turtles crocodiles.
		3. Aves	Birds.
		4. Mammalia	Mammals.

Phylum Chordata

Characters: The notochord, dorsal nerve cord, and paired gill slits in some stage or throughout life; coelom and segmentation usually present; tail present behind anus.

Examples: Chordates.

Group I. Acrania

Characters: Cranium, jaws, vertebrae and paired appendages absent.

Examples: Lower Chordates.

4.1. Subphylum Tunicata (Urochordata)

Characters:
i. Larva is minute, freeliving, tadpole-like with gill slits, and notochord in the tail.
ii. Adults tubular, globe-like or irregular, covered usually with transparent test or tunic.
iii. Notochord absent in adult, nerve cord reduced to a ganglion; nephridia and segmentation absent.
iv. The larva changes into adult by retrogressive metamorphosis. The adult is hardly recognizable as a chordate as a result of degenerative change due to retrogressive metamorphosis.

Class 1. Larvacea (Appendicularia)

Characters:
i. Free swimming or pelagic forms.
ii. Larva-like structure and appearance.
iii. Notochord, nerve cord, and brain persistent.
iv. Test not persistent: two gill slits present.

Examples: *Oikopleura* (Fig 1.8. *Appendicularia*; *Fritilluria*; *Kowalevskia.*

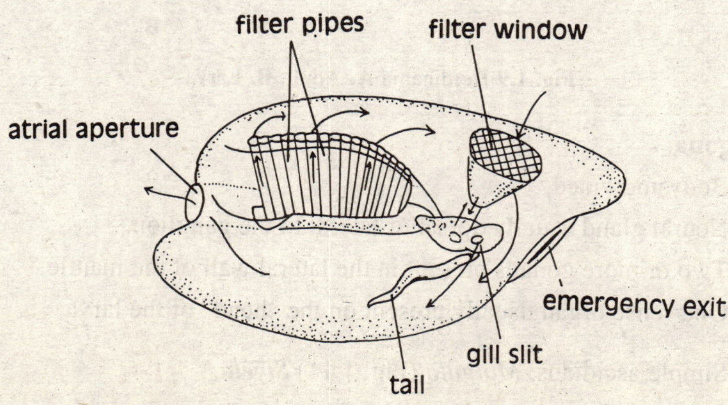

Fig. 1.8 Oikopleuro Side view.

Class 2. Ascidiacea

Characters:
i. Solitary, colonial or compound ascidians, usually sessile in the adult condition.
ii. Larva contains notochord, nerve cord, and tail which are lost during the retrogressive metamorphosis. The brain is reduced to a ganglion after metamorphosis.
iii. Many persistent gill slits present.
iv. The test is permanent and well developed.
v. The atrium opens dorsally.
vi. The stolon simple or absent

Order 1. Enterogona

Characters:
i. Body sometimes divided into thorax and abdomen.
ii. Neural glands are usually ventral to ganglion.
iii. One gonad present behind or within the intestinal loop.
iv. Two sense organs present on the 'head' of larva.

Examples: *Herdmania.* (Fig. 1.9 A.B): *Clavelina; Ciona: Ascidia.*

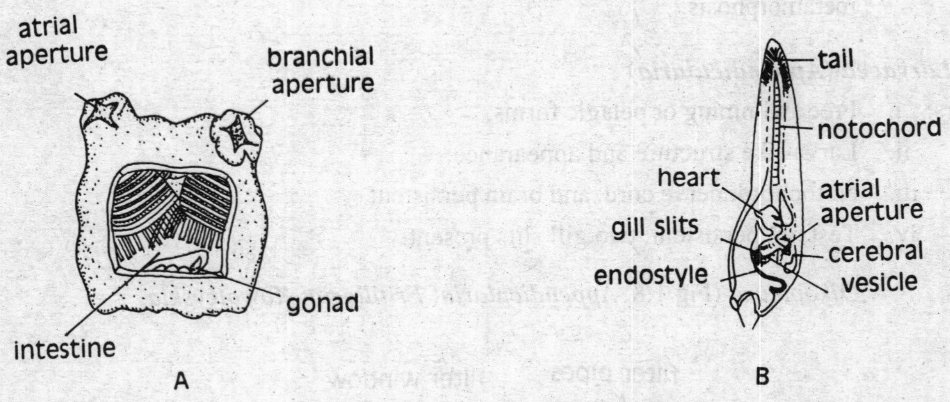

Fig. 1.9 Herdmania **A.** Adult, **B.** Larva

Order 2. Pleurogona

Characters:
i. Body undivided,
ii. Neural gland usually dorsal or lateral to the ganglion.
iii. Two or more gonads present in the lateral wall of the mantle.
iv. One sense organ usually present on the "head" of the larva

Examples: Simple ascidians; *Molgula* (Fig. 1.11) *Styela;*
Compound ascidians: *Botryllus* (Fig 1. 10); *Amarowcium;*

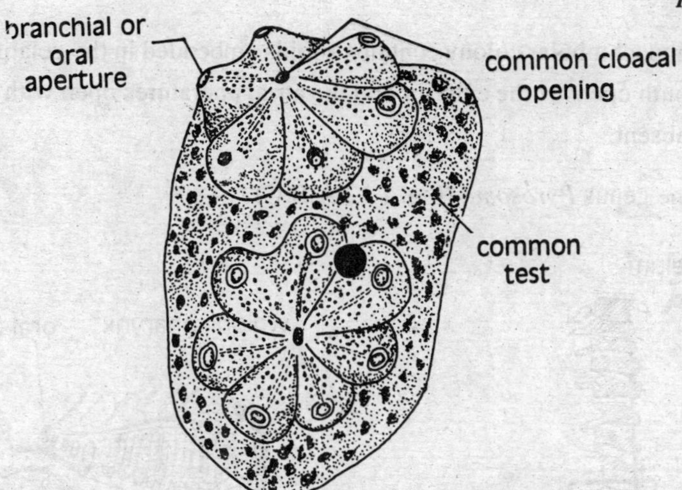

Fig. 1.10 Botryllus

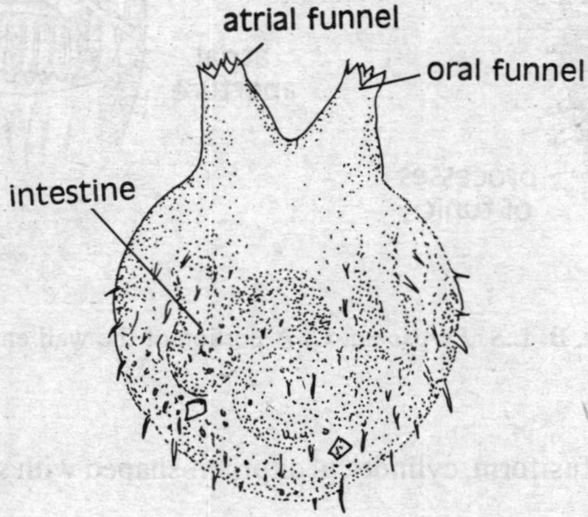

Fig. 1.11 Molgula

Class 3. Thaliacea

Characters:
- i. Chain tunicates of different sizes, free swimming, pelagic forms, simple or colonial.
- ii. Notochord and tail absent in the adult.
- iii. Gill-slits present.
- iv. Test permanent with circular muscle bands.
- v. Atrium opens posteriorly.
- vi. Stolon complex.

Order 1. Pyrosomida

Characters:
- i. Free swimming forms reproduced by budding to form hollow cylindrical colony open at one or both ends.

ii. The compact tubular colony contains zooids embedded in the gelatinous wall.

iii. The mouth opens at the exterior and the atrial aperatures open with cylinder.

iv. Larva absent.

Examples: Only one genus *Pyrosoma* (Fig. 1.12 A.B.C)

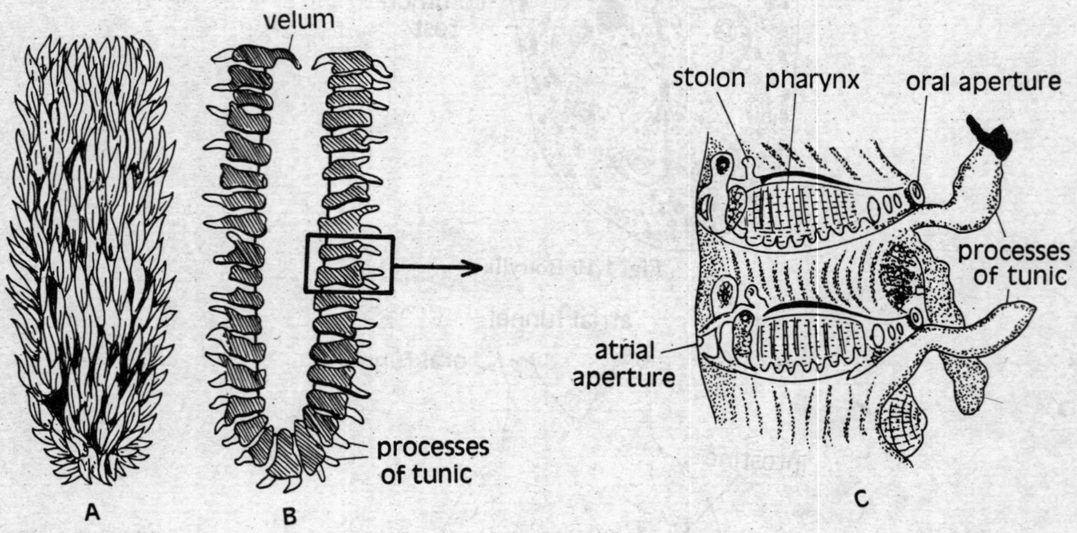

Fig. 1.12 *Pyrosoma*. **A.** Colony, **B.** L.S. *Pyrosoma*. **C.** A portion of the wall enlarged to show two zooids

Order 2. Salpida

Characters: i. The body is fusiform, cylindrical or prism-shaped with subterminal oral and atrial apertures.

ii. Muscle bands are incomplete below and convergent above.

iii. First gill in the adult in a single large opening.

Examples: *Salpa* (Fig. 1.13.A. & B.); *Octacnemus*.

Order 3. Doliolida

Characters : i. Barrel-shaped body with oral and atrial apertures at opposite ends.

ii. Muscle bands form 8 regular complete rings.

iii. Gill slits few or many.

iv. A tailed larva is present.

Example: *Doliolum* (Fig. 1.14)

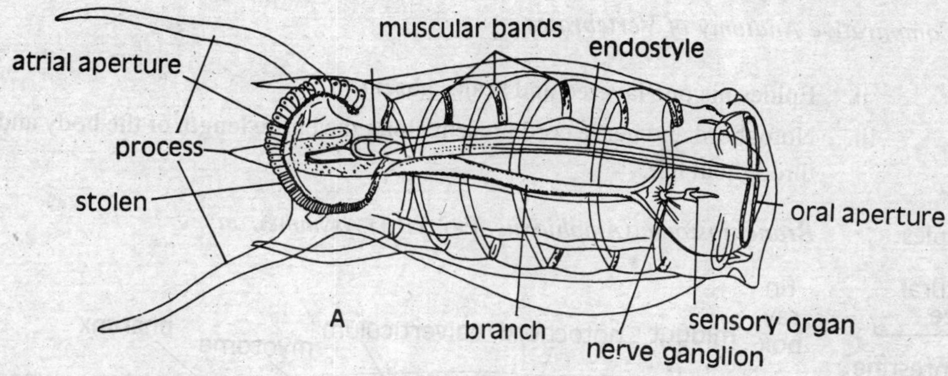

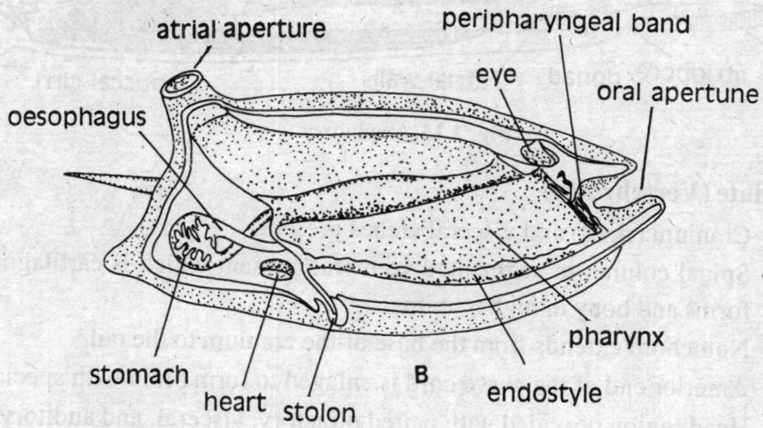

Fig. 1.13 Salpa : Oozooid. **A.** Asexual form of *S. democratica* in ventral view. B.L.S. View.

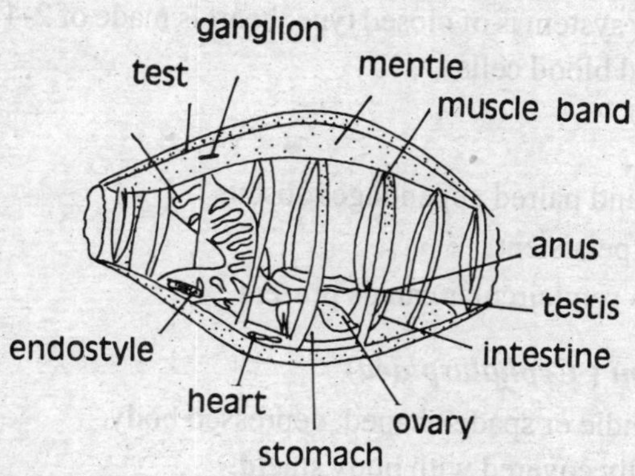

Fig. 1.14 Structure of Doliolum-Sexual Stage

4.2. Subphylum Cephalochordata (Leptocadii)

Characters: i. Slender, fish-like, segmented animals popularly called lancelets, found in shallow waters.

ii. Epidermis one layered and scales absent.

iii. Notochord and nerve cord present along the entire length of the body and persist throughout life.

Examples: *Branchiostoma* (*Amphioxus*, FIG. 1.15) *Asymmetron*

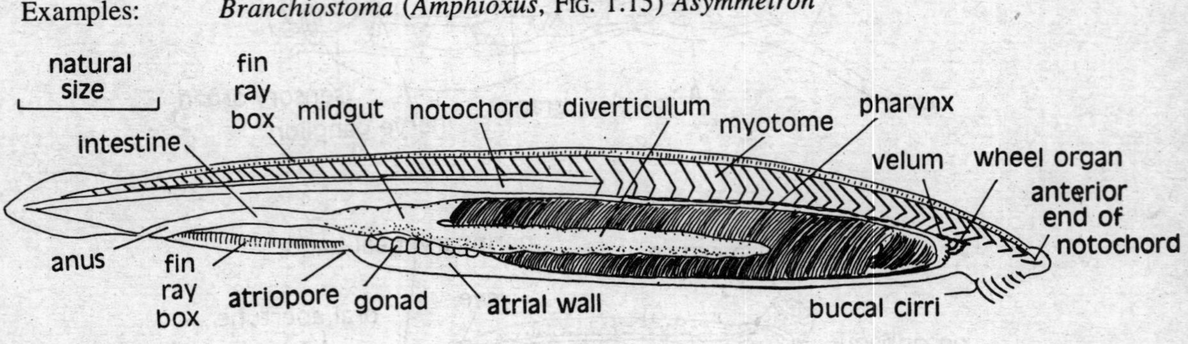

Fig. 1.15 Amphioxus.

Group II. Craniate (Vertebrates)

Characters: i. Cranium (skull) and visceral arches present.

ii. Spinal column of segmented vertebrae present which is cartilaginous in lower forms and bony in higher animals.

iii. Notochord extends from the base of the cranium to the tail.

iv. Anterior end of the nerve cord is enlaged to form brain with specialized parts.

v. Head region provided with paired olfactory, visceral, and auditory organs.

vi. Semicircular canals present for equilibrium.

vii. Circulatory system is of closed type: heart is made of 2-4 chambers and the blood contains red blood cells.

4.3. Subphylum Agantha

Characters : i. True jaws and paired appendages absent

ii. Notochord persistent.

iii. One or two semicircular canals present.

4.3.1. Class 1. Ostracodermi (Cephalaspidea)

Characters: i. Small, Spindle or spade-shaped, depressed body.

ii. Head usually covered with bony shield.

iii. Body is covered by plates or spines.

iv. Nasal openings one or two.

v. Two eyes and one pineal eye present.

vi. Two or ten pairs of gill pouches.

Examples: All fossil forms of Silurian to Devonian.
**Cephalaspis, * Thelodus:* Birkenia:*Pteraspis.*

4.3.2. Class 2. Cyclostomata (Marsipobranchii)

Characters: i. Body long, cylindrical, and eel-like.

ii. Skin smooth, scales absent.

iii. Suctorial mouth with horny teeth,

iv. One nasal Opening,

v. Paired Fins, genital ducts absent,

vi. Skeleton cartilaginous.

vii. Gills situated in pouches having 1 to 16 opening.

viii. Heart two chambered.

Order 1. Petromyzoniformes. *LAMPREYS*

Characters: i. A large, ventral suctorial funnel with horny teeth surrounding the mouth.

ii. Nasal sac is dorsally situated without connection with mouth.

iii. Seven pairs of gill pouches open posteriorly.

iv. Well developed branchial basket. Examples: *Petromyzon* (Fig. 1.16) *Lampetra Entosphenus.*

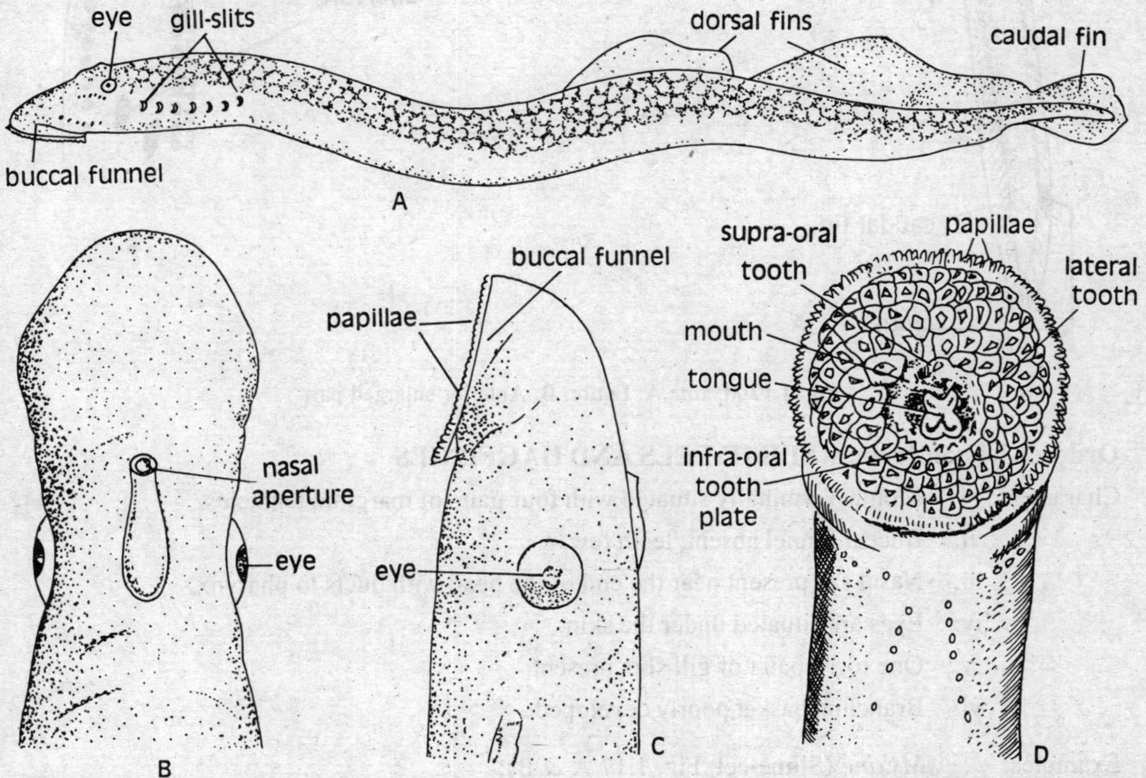

Fig. 1.16 External structure of *Petromyzon* **A.** An entire animal (lateral view). **B.** Dorsal view of head showing the position of the nasal aperture. **C.** Lateral view of the head. 1). Ventral view of the head showing the details of buccal funnel and the position of the mouth and 'tongue'

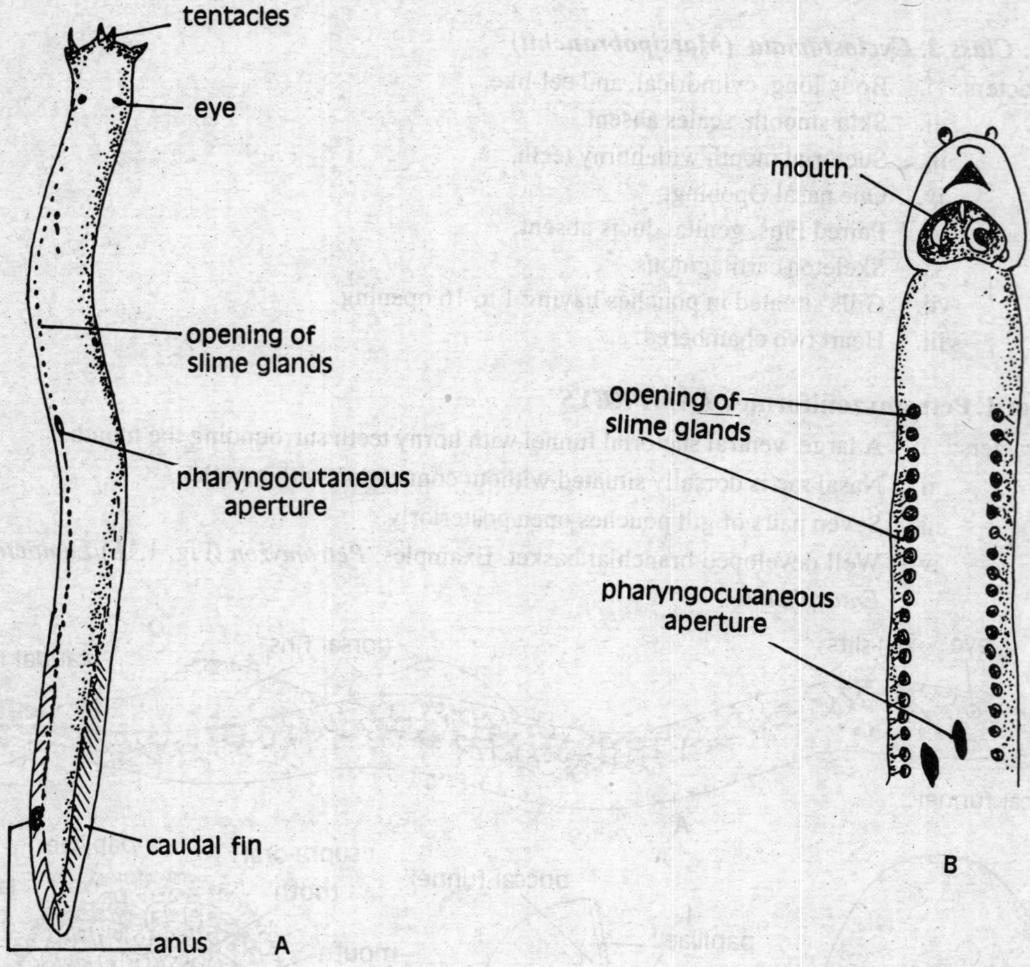

tentacles
eye
opening of slime glands
pharyngocutaneous aperture
caudal fin
anus
A

mouth
opening of slime glands
pharyngocutaneous aperture
B

Fig. 1.17 *Myxine* **A**. Entire, **B.** Anterior enlarged part

Order 2. Myxiniformes SLIME-EELS AND HAGFISHES

Characters:
- i. Mouth Terminally situated with four pairs of marginal tentacles.
- ii. Buccal funnel absent; teeth few.
- iii. Nasal sac present near the end of the head with ducts to pharynx.
- iv. Eyes are situated under the skin.
- v. One to 16 pairs of gill slits present.
- vi. Branchial basket poorly developed.

Examples: *Myxine* (Slime-eel. Fig. 1.17 A & B);

4.4. Subphylum Gnathomata

Characters: i. Paired appendages usually present.

ii. One pair of visceral arches modified as biting jaws.

iii. Two nasal tubes or capsules.

iv. Sexes separate.

4.4.1. Superclass Pisces. Fishes

Characters: i. Paired fins usually present median fins supported by fin rays.

ii. Skin usually provided with scales.

iii. Nasal capsule not connected to mouth cavity.

iv. Heart with one auricle.

v. Aquatic animals having respiration by gills.

Class 1. Placodermi (Aphetohyoidea)

Characters: i. A fully developed functional gill arch is present anterior to the gill arch.

Examples: Extinct ancient fishes of Silurian, Devonian, and Permian periods.

Subclass 1. *Pterichthyes*, (Antiarchi).

Examples: , Devonian **Pterichthys*: **Bothriolepis.*

Subclass 2. *Coccostei.* ARTHRODIRES.

Examples: Devonian **Dinichthys* (Marine up to 20 feet);

Coccosteus (in rivers)

Subclass 3. *Gemuedinia.*

Examples: Devonian **Gemuedina* (ray-like); *Jagorina* (shark-like)

Subclass 4 *Acanthodii.*

Examples: Devonian to Lower Permian, freshwater or marine. *Climatius: *Acanthodius.

Class 2. Chondrichthyes (Elasmobranchii): CARTILAGINOUS FISHES

Characters: i. Living fishes with cartilaginous skeleton.

ii. Operculum absent.

iii. First gill-slit is modified into spiracle.

iv. Skin covered over with minute placoid scales.

v. Mouth and two nostrils situated on ventral side.

vi. Males have claspers.

vii. Mostly marine.

viii. Lack swim bladder.

Subclass 1. Xenacanthi (Pleurocanthii).

Characters: i. Dorsal fin present at the posteriors part of the back.

ii. Notochord in the tail region.

Examples: Extinct ancient fishes of Late Devonian to Triassic freshwaters. **Xenacanthus*

Subclass 2. Cladoselachii

Characters: i. Shark-like fishes with pectoral fin broad at the base; two dorsal fins; claspers absent.

Examples: Upper Devonian to Upper Carboniferous **Cladoselache*..

Subclass 3. Selachii. SHARKS AND RAYS

Characters: i. Mostly marine forms.

ii. Gills in separate gill clefts lateral to pharynx.

iii. Cloaca present.

Order I. Squaliformes (Pleurotremata) SHARKS

Characters: i. Usually spindle-shaped body with hetrocercal tail used for swimming.

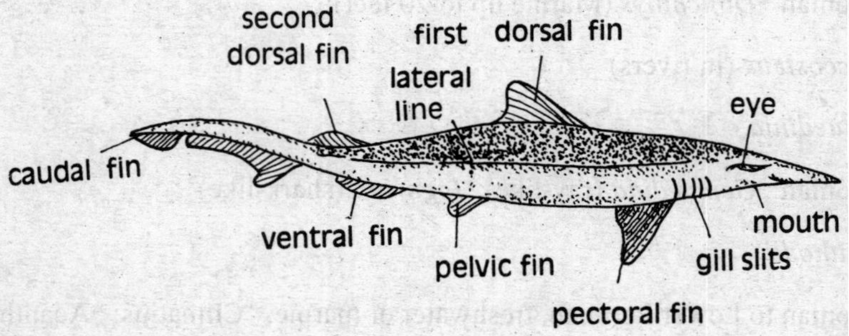

Fig. 1.18 Scoliodon

ii. 5-7 pairs of lateral gill slits present.

iii. Pectoral fin moderate and constricted at base.

Examples: *Scoliodon* (Fig. 1.18.): *Squalus* (*Acanthias*): *Stegostoma*: *Sphyrna* (*Zygaena*): *Hexacanthus*.

Order 2. Rajiformes (Hypotremata, Batoidea), RAYS.

Characters: i. Body dorsoventrally compressed.

ii. Pectoral fins enlarged and joined to sides of the head and body, and used for swimming by flapping them.

iii. Five pairs of ventral gill -slits present.

iv. Spiracle functional.

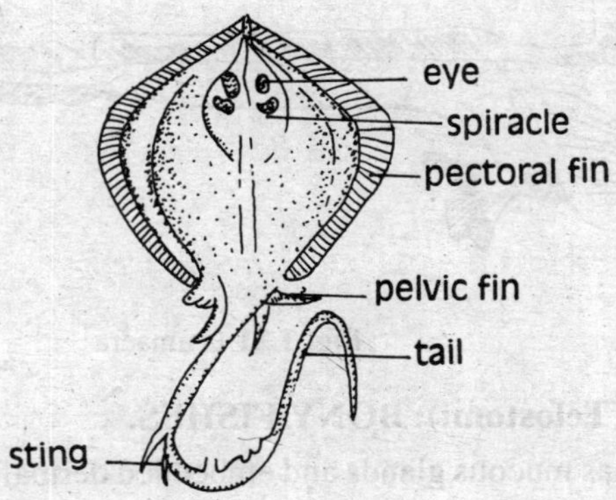

Fig. 1.19 Trygon - Sting ray

Examples: *Trygon* (Sling ray. Fig. 1.19.) *Raja* (Common Ray); *Torpedo* (Electric ray): Sting rays *Dasyatis*. *Urobatis*, and *Aetobatus*: *Pristis* (Saw fish. Fig. 1.20): *Rhinobatis* (Banjoray): *Myliobatis* (Eagle ray).

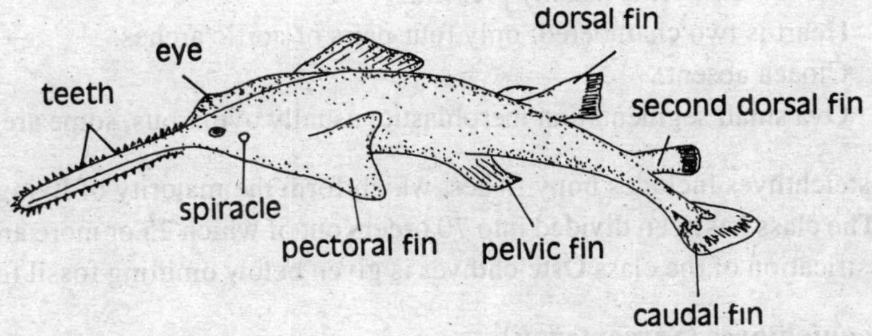

Fig. 1.20 Pristis-Saw Fish

Subclass 4. Holocephali.

Order 3: Chimaeriformes. CHIMAERAS

Characters: i. Gill slits are covered by operculum.

ii. Spiracle, cloaca absent.

iii. First dorsal fin with a strong spine.

iv. Whip-like slender tail.

v. Each jaw is provided with a large tooth plate,

vi. Adults have no scales.

Examples: *Hydrologus* (*Chimaera.* Fig. 1.21): *Harriotta.*

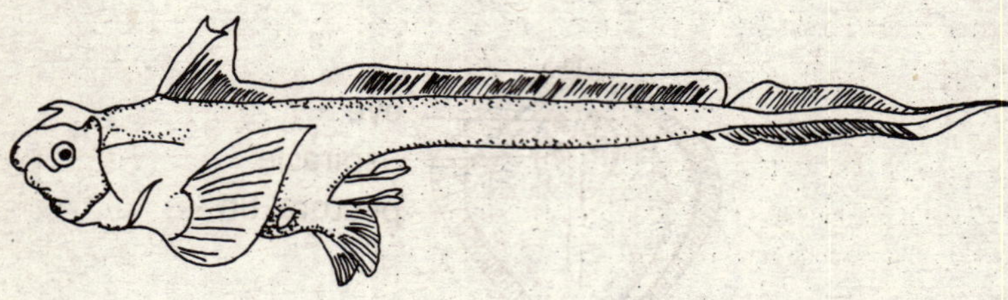

Fig. 1.21 Chimaera

Class 3. Osteichthyes (Telostomi): BONY FISHES.

Characters: i. Skin has mucous glands and embedded dermal or bony scales of cycloid, ctenoid, or ganoid type: few fishes have naked skin.

ii. Skeleton is usually bony with many vertebrae.

iii. Paired lateral fins and median fins supported by fin rays usually present.

iv. Mouth usually terminal.

v. Gills usually present on bony gill arches in a common chamber at each side of the pharynx, covered by an operculum.

vi. A swimbladder is usually present.

vii. Heart is two chambered; only four pairs of aortic arches.

viii. Cloaca absent.

ix. Ova small segmentation meroblastic; usually oviparous, some are viviparous.

The class Osteichthyes includes bony fishes, which form the majority of living fishes; about 15000 species. The class has been divided into 70 orders out of which 25 or more are fossil forms. The outline classification of the class Osteichthyes is given below omitting fossil forms

Subclass 1. Choanichthyes (Sarcoptengii).

Characters: i. Mouth Cavity connected with nostrils.

ii. Paired fins with dermal rays and provided with jointed skeleton and muscles.

iii. Caudal fin with epichordal lobe.

iv. Usually two dorsal fins present.

Super order 1. Dipnoi (Dipneusti) : LUNG FISHES.

Character: i. Body elongated and cylindrical.

ii. Premaxillae and maxillae absent: three pairs of hard teeth plates present on the palate and margins of the lower jaws.

iii. Paired fins narrow or reduced.

iv. Air bladder functional and lung-like.

TABLE 1.2
Class Osteichthyes

Subclass	*Super Order*	*Order*	*Examples*
1. Choanichthyes (Sarcopterygii)	I. Dipnoi	1. Ceratodontiformes	* *Ceratodus; Neoceratodus.*
		2. Lepidosireniformes	*Lepidosiren Protopterus.*
	II. Crossopterygii.	3. Osteolepiformes	* *Osteolepis*
		4. Coelacanthiformes	**Eusthenopteron Latimeria*
2. Actinopterygii.		1. Polypteriformes.	*Polypterus*
		2. Acipensoriformes.	*Acipenser*
		3. Amiiformes	Amia
		4. Semionotiformes (Ginglymodi)	*Lepidosteus*
		5. Clupeiformes	*Salmo,* herring.
		6. Myetophiformes	*Harpadon* (Bombay Duck)
		7. Saccopharyn-	*Saccopharynx* giforms
		8. Cypriniformes.	*Cyprinus: Labeo: Catla; Clarias Hetropneustes*
		9. Anguilliformes.	*Anguilla; Muraena*
		10. Notacanthiformes	*Notacanthus*
		11. Beloniformes	*Exocoetus; Xenentodon.*
		12. Cypinodontiformes	*Gambusia*
		13. Gasterosteiformes	*Gasterosteus.*
		14. Gadiformes	*Godus.*
		15. Lamprediformes	*Lampris.*
		16. Perciformes	*Anabas; Tipapia; Lates.*
		17. Pleuronectiformes	*Solea Psettodes; Cynogolossus.*
		18. Echeniformes	*Echenesis Remora.*
		19. Tetradontiformes	*Tetradon; Diodon; Ostracion.*
		20. Mastacembeli-formes	*Mastacembelus; Macrognathus.*
		21. Channiformes.	*Channa*

**Extinct.*

Order 1. Ceratodontiformes

Characters: i. Scales cycloid, large, and overlapping.

ii. Fins leaf-like.

iii. Lung with single lobe.

Examples : *Neoceratodus forsteri* (Fig. 1.22): *Ceratodus*

Order 2. Lepidosireniformes

Characters: i. Body eel-like.

ii. Scales small; fins thread-like.

iii. Lungs bilobed.

Examples: *Lepidosiren paradoxa* (Fig. 1.22.)
Protopterus annectens (Fig. 1.22.)

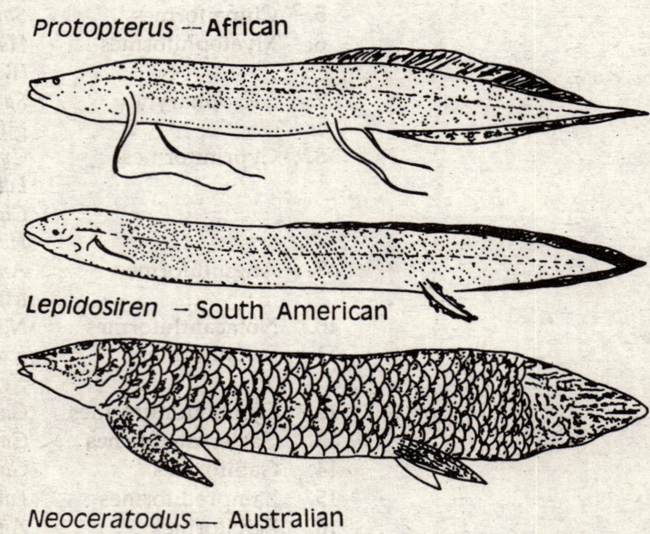

Fig 1.22 Lung Fishes

Super Order II. Crossopterygii. ***LOBE FINNED FISHES.***

Character: i. Premaxillae and maxillae present.

Order 3. Osteolepiformes

Characters: i. Head normal, skull bony; pineal eye present.

ii. Tail heterocercal.

Example: **Osteolepis.* Devonian to Carboniferous.

Order 4. Coelacanthiformes (Actinistia)

Characters: i. Head small and deep; skull mostly cartilaginous.

ii. Paired fins, lobed.

iii. Pineal eye absent,

iv. Tail three lobed.

Examples: **Eusthenopteron:* **Macropoma:* Devonian to Cretaceous. Recent *Latimeria chalumanae* (Fig 1.23).

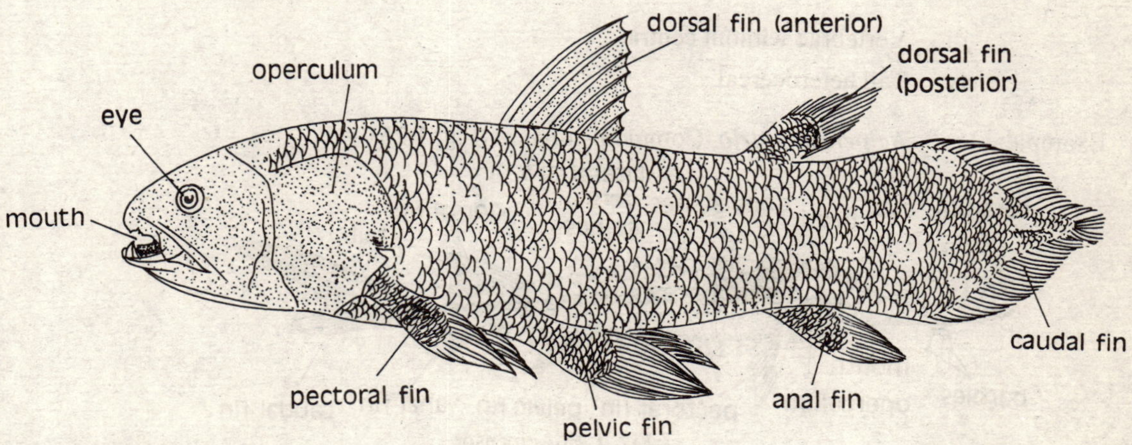

Fig. 1.23 Latimeria chalumnae

Subclass 2. Actinopterygii. ***RAY FINNED FISHES.***

Characters: i. Nostrils and mouth cavity not connected.

ii. Paired fins without muscles or skeletal parts.

iii. Caudal fin without epichordal lobe.

Order 1. Polyptery formes

Characters: i. Body slender with thick rhomboid ganoid scales.

ii. Dorsal fin is divided into 8 or more finlets. each with a spine.

iii. Caudal fin pointed, arrow-like.

iv. Air bladder is probably used for respiration, and is lung-like.

Examples: *Polypterus* (Fig 1.24)

Order 2. Acipenseriformes (Chondrostei)

Characters: i. Long snout.

ii. Skeleton cartilaginous.

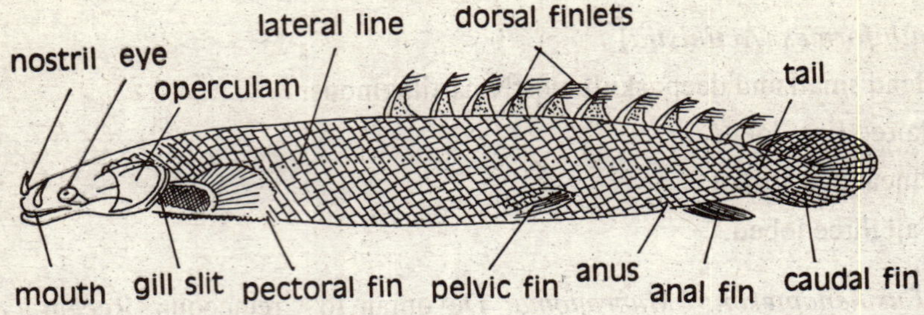

Fig 1.24 Polypterus.

iii. Vertebrae without centra.

iv. Tail heterocercal.

Example: *Acipensor sturio*, Common sturgeon (Fig. 1.25)

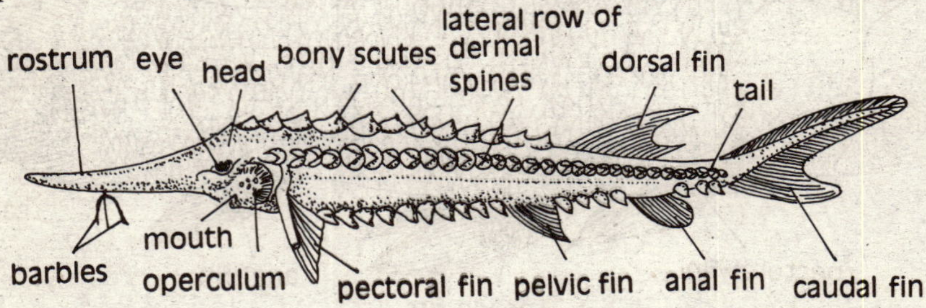

Fig. 1.25 Acipensor

Order 3. Amiiformes (Halecomorphi)

Characters:
i. Long dorsal fin; caudal fin short.
ii. Scales cycloid, overlapping, and thin.
iii. Snout normal.

Example: *Amia calva*. bowfin (fig 1.26)

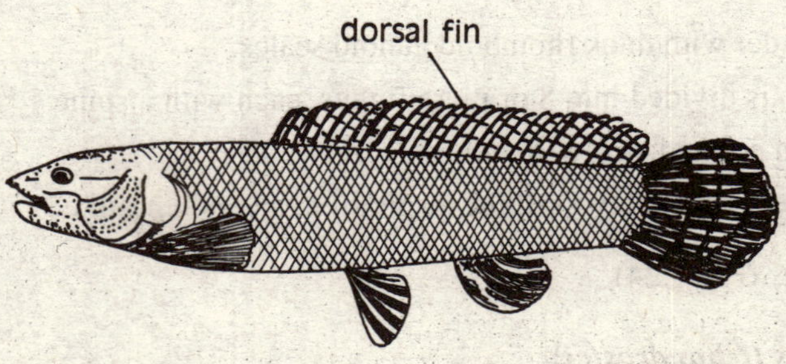

Fig 1.26 Amia.

Order 4. Semionotiformes (Ginglymodi)

Characters: i. Body and snout long.

ii. Ganoid scales in oblique rows.

iii. Caudal fin heterocercal and short.

iv. Teeth conical and strong.

Examples: *Lepidosteus osseus.* Garpike. Alligator gar (fig. 1.27)

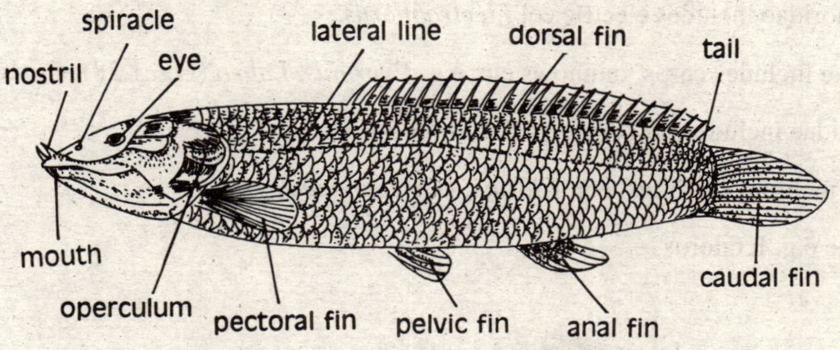

Fig. 1.27 Lepidosteus

Order 5. Clupeiformes (Isospondyli+ Haplomi)

Characters: i. Fins without spiny rays: caudal fin homocercal; pelvic fins abnormal.

ii. Air bladder connected with pharynx by open ducts.

Examples: Herrings. *Clupea;* Salmons, *Salmo* (trout); *Salmo gairdnerii* (Indian trout introduced from America); *chitala* (*Chital* fish); *Notopterus notopterus* (Faloi fish); *Elops saurus: Hilsa ilisha*; *Ilisha clongata; Megalops Cyprinoides; Albula vulpes.*

Order 6. Myctophiformes (Iniomi) LANTERN FISHES

Characters: i. Premaxillae long and form jaw.

ii. Air bladder reduced, absent, or filled with fat.

Examples: "Bombay duck". *Harpadon;* Lizard fish *Synodus; Myctophum.*

Order 7. Saccopharygiformes (Lyomeri.). GULPERS

Characters: i. Body elongated, eel-shaped and naked.

ii. Mouth very large.

iii. Soft fin rays: ribs, caudal fin and pelvic fins absent.

Examples: *Saccopharynx.*

Order 8. Cyriniformes (Ostariophysi). LAOCHES, CARPS, CATFISHES

Characters: i. Anterior vertebrate and lateral elements of first four vertebrae modified to form Weberian apparatus between air bladder and internal car.

This order has been divided into 6 families given below.

Family Characinidae includes Characins e.g. *Hydrocynus.*

Family Electrophoridae includes electric eel *Electrophorus.*

Family Cyprinidae includes carps, minnows etc. e.g. *Cyprinus, Labeo* (Fig. 1.28); *Catla.*

Family Catostomidae includes suckers, e.g. *Catostomus.*

Family Siluridae includes catfishes. e.g. *Clarias*, *Heteronpneustes*

Family Ictaluridae e.g. Ictalurus.

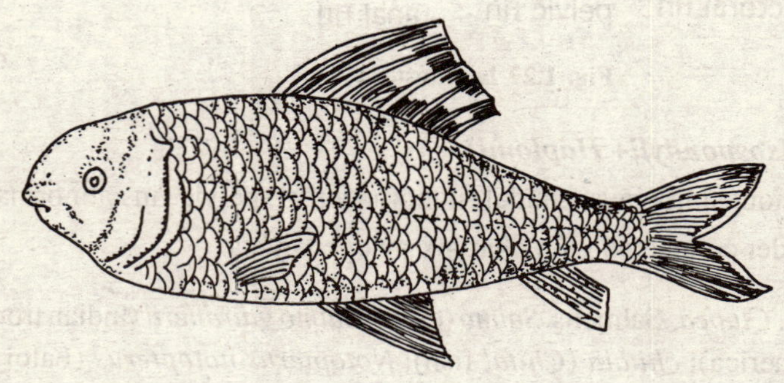

Fig. 1.28 Labeo

Order 9. Anguilliformes (Apodes). EELS.

Characters: i. Body elongated and cylindrical.

ii. Pelvic fin usually absent ; caudal, dorsal and anal fins usually continuous.

iii. Scales reduced or absent.

Examples: Freshwater eel *Anguilla* (Fig 1.29): marine eel *Muraena.*

Order 10. Notacanthiformes (Heteromi). SPINY EELS.

Characters: i. Body cylindrical; fins with spines; tail long caudal fin absent.

Examples: Deep sea form *Notacanthus.*

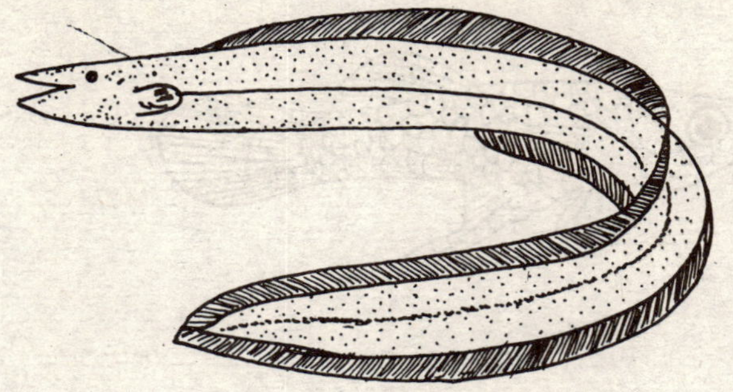

Fig. 1.29 Anguilla

Order 11. Beloniformes (Synentognathi) FLYING FISHES, GARFISHES.

Characters: i. Pectoral fins enlarged and shifted to the dorsal side of the body.

ii. Ventral fin shifted to abdominal side.

Examples: Garfishes- *Strongylura Xenodon* (*Belone*). Flying fishes -*Exocoetus* (Fig. 1.30) *Cypselurus*. Halfbeak-*Hemiramphus*

Fig. 1.30 Exocoetus

Order 12. Cyprinodontiformes (Microcyprini). TOP MINNOWS, CAVE FISHES.

Characters: i. Single dorsal fin; pelvic fin abdominal or absent; Fins spineless.

ii. Mouth protractile.

Examples: Top minnows - *Cyprinodon. Gambusia* (Fig. 1.31) *Cavefish -Typhlichtyes.*

Order 13. Gasterosteiformes.

Characters: i. Dorsal fin provided with spiny rays.

ii. Two free spines may be present before dorsal fin.

iii. Air bladder closed.

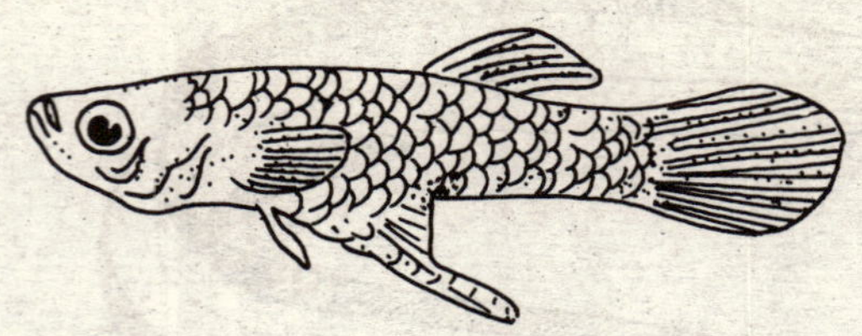

Fig. 1.31 Gambusia

Examples: Family Gasterosteidae : Stickleback Gasterosteus. Family Syngnathidae; Pipefish - *Syngnathus; Trachyrhampus -serratus* (Indian species of Pipefish): Sea Horse -*Hippocampus* (Fig. 1.32).

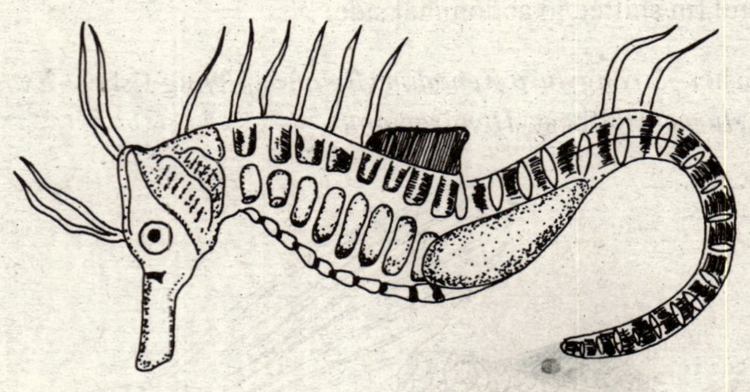

Fig. 1.32 Hippocampus

Order 14. Gadiformes (Anacanthini). CODS.

Characters:
- i. Soft rayed fins; usually many fin rays in the thoracic part.
- ii. Caudal fin formed without broad hypural and consists of dorsal and anal rays only.
- iii. Ducts are absent in air bladder.

Examples: Godfish - *Gadus morhua.*

Order 15. Lampridiformes (Allotriognathi).

Characters:
- i. Mostly marine fishes with soft rayed fins or dorsal fin with 1-2 spines.

Examples: Lampris; Ribbonfish - *Trachypterus.* Oarfish - *Regaleus.*

Order 16 Perciformes (Percomorphi, Acanthopteri). PERCHES.

Characters:
- i. Dorsal and anal fins with spines.

ii. Pectoral fins present on lateral sides on upper position.

iii. Ducts absent in air bladder.

Examples. Freshwater bass - *Micropterus Perch-Perca*: Mackerel-*Scomber*; *Lates; Tilapia: Petropus*; Indian climbing perch-*Anabas* (Fig 1.33).

Fig. 1.33 Anabas

Order 17. Pleuronectiformes. FLATFISHES.

Characters: i. Asymmetrical fishes with both movable eyes shifted to one side of the head.

ii. Body extremely compressed, fishes adapted for bottom living.

iii. Dorsal and anal fins fringe the body.

Example: *Solea ovata Psettodes: Cynoglossus.*

Order 18. Echeniformes (Fiscocephali). REMORAS.

Characters: i. Dorsal surface of head provided with flat. Oval, adhesive disc by which the fish adheres to a shark or other object in water.

Example: Sucking fish - *Echeneis* (Fig 1.34). Remora.

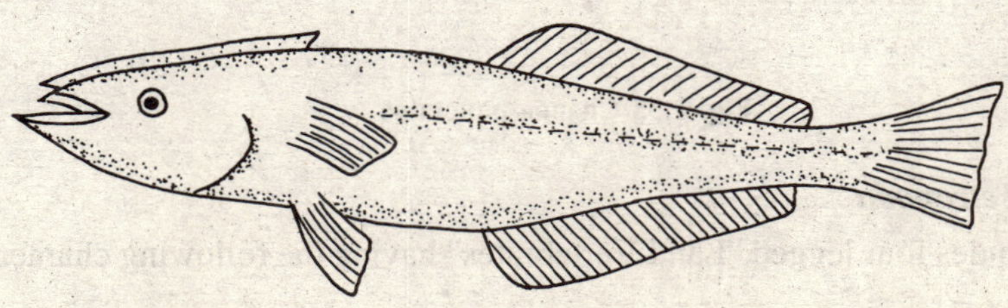

Fig. 1.34 Echeneis

Order 19. Tetradontiformes (Plectognathi)

Characters: i. Jaws short and powerful.

ii. Scales bony and modified into spines.

Examples: Porcupine fish - *Diodon hystrix;* Sunfish-*Molamola:* Trunkfish - *Ostracion gibbossus* (Fig. 1.28); Globefish -*Tetradon cutcutia.*

Order 20. Mastacembelifromes. (Opisthomi). SPINY EELS.

Characters: i. Body eel-shaped with tubular nostrils situated at the end of the snout on fleshy tentacles.

ii. Dorsal, anal, and caudal fins united to form a continuous fin.

Examples: *Macrognathus: Mastacembelus.*

Order 21. Channiformes (Ophiocephaliformes)

Characters: i. Head is depressed and covered over by large plate-like scales.

ii. Spines absent in fins.

iii. Air bladder without ducts.

iv. Accessory respiratory organs present.

Examples: Lata fish - *Channa, (Ophiocephalus punctatus* (Fig. 1.35).

Sholafish - *Channa striatus*; Chanfish - *Channa gachua; Shalfish - Channa marulius.*

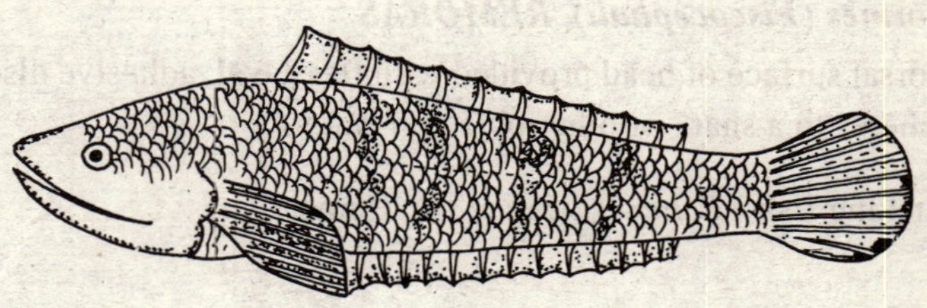

Fig. 1.35 Channa punctatus

4.4.2. Superclass Tetrapoda

This superclass includes four legged "Land Vertebrates" having the following characteristics:

i. Two pairs of pentadactyle limbs, modified, reduced or absent.

ii. Double circulation of blood; two chambered heart.

ii. Endoskeleton bony.

iv. Nasal capsule connected to mouth cavity.

Class 1. Amphibia

The class Amphibia which represents the primitive group of land vertebrates includes many extinct and living forms. About 2,500 living species of Amphibia are known and the living Amphibians show the following features:

i. Body is bilaterally symmetrical.

ii. Skin moist and glandular without external scales.

iii. Cold blooded animals.

iv Two pairs of limbs usually present: forelimbs with 4 digits, hindlimbs with 5 digits: paired fins absent.

v. Two nostrils connect the mouth cavity.

vi. The heart is three chambered with 2 auricles one undivided ventricle, one sinous venosus and one conus arteriosus. Red blood cells nucleated and oval.

vii. Skeleton mostly bony; skull has two occipital condyles; ribs, when present, are not attached to the sternum.

viii. Tympanum present and connected to the inner ear by rod-shaped columella.

ix. Respiration by lungs, skin, or inner lining of oral cavity.

x. Fertilization external or internal: oviparous usually; eggs laid in water with gelatinous covering.

xi. Cleavage unequal but holoblastic: larva called tadpole is aquatic; adults aquatic or amphibious.

TABLE 1.3

Outline Classification of Amphibia

Subclass	*Super order*	*Order*	*Suborder/ Family*	*Examples*
Aspidospondyli	Labryntho-dontia (Stegocephalia)	1. *Temnos-pondyli	—	**Branchiosaurus; *lchthyostega * Mastodonosaurus.*
		2. *Anthraco pondyli	—	**Eogyrinus; *Seymauria*
	Salientia	1. *Proanura -		**Protobatrachus*
	(Anura)	2. Amphicoela	Fam. Leiopel-midae (Ascaphidae)	*Leiopelma; Ascaphus truci.*

cont...

Subclass	*Super order*	*Order*	*Suborder/ Family*	*Examples*
		3. Opisthocoela	Fam. Discoglossidae	*Alytes.*
			Fam. Pipedae	*Pipa; Xenopus.*
		4. Anomocoela	Fam. Pelobatidae	*Pelobates.*
			Fam. Pelodytidae	*Pelodytes.*
		5. Procoela	Fam. Palaeobatrachidae	**Palaecobatrachus*
			Fam. Bufonidae	*Bufo*
			Fam. Brachycephalidae	*Brachycephalus*
			Fam. Hylodae	*Hyla.*
		6. Diplasiocoela	Fam. Ranidae	*Rana*
			Fam. Polypedatidae	*Polypedates Rhacophorous)*
			Fam. Mierochylidae (Brevicipitidae)	*Gastrophryne*
Lepospondyli		1. * Microsauria	—	**Euryodus*
		2. *Aistopoda	—	** Ophiderpeton.*
		3. *Nectridia	—	**Sauropleura.*
		4. Caudata (Urodela)	Sub. O. Cryptobranchoidea	
			Fam Cryptobranchidae	*Cryptobranchus*
			Fam. Hynobidae	*Hynobius; Randon.*
			S.O. Ambystomatoidea	
			Fam. Ambystomidea	*Ambystoma*
			S.O. Salamandroidea	*Salamandra*
			Fam. Salamandridae	
			Fam. Amphiumidae	*Amphiuma*
			Fam. Plethodontidae	*Pleuthodon*
			S.O. Proteida	
			Fam. Proteidae	*Necturus.*
			S.O. Meantes	
			Fam. Siorenidae	*Siren.*
		5. Gymnophiona (Apoda)	Fam. Caeciliidae	*Ichthyophis*

Subclass 1. Aspidospondyli.

Characters: i. Amphibians having the centra formed from cartilage block in units of two which ossify as intercentra and pleurocentra.

**Superorder 1. Labrynthodontia (Stegocephalia).*

Characters: i. Salmander-like tailed amphibians of Lower Carboniferous to Triassic periods.

ii. Usually long tail and legs of uniform size present.

iii. Bony plates making complete covering over cranium and checks.

iv. Overlapping scales usually present on the ventral side and sometimes on the dorsal side making armour

*Order 1. *Temnospondyli*

Characters: i. Vertebral pleurocentra incomplete discs on laterodorsal side, vomers broad.

Examples: Late Devonian to Triassic **Ichthyostega;* * *Eryops* (upto 5 feet long); **Branchiosaurus;* **Mastodonosaurus* (Skull 3 feet long)

*Order 2 *Anthracosauria.*

Characters: i. Pleurocentra forming complete discs enclosing notochord: intercentra small.

Examples: Upper Carboniferous to Triassic **Eogyrinus* (upto 15 feet long); *Seymouria* (Small 20 inches long)

Superorder 2 Salientia (Anura). FROGS. TOADS

Characters: i. Tail absent.

ii. Skull without roof, reduced, and thin with few bones.

iii. Fewer vertebrae terminated by a slender urostyle.

iv. Ribs absent or reduced.

v. Hindlimbs usually larger than forelimbs.

vi. Fertilization mostly external.

vii. Tadpole larva has ovoid head and long tail with median fins, without true teeth; larval metamorphosis prominent.

vii. About 2,200 species.

Order 1. *Proanura.

Characters: i. Hind legs not larger than the forelegs.

ii. Urostyle absent; 16 precaudal and 3 or 4 caudal vertebrae pesent.

Example: Triassic **Protobranchus.*

Order 2. Amphicoela.

Characters: i. Amphicoelous vertebrae.
ii. 9 presacral vertebrae.
iii. Free ribs in adults.

Example: Family Leiopelmidae (Ascaphidae) is represented by only two genera of bell loads - *Leiopelma* (New Zealand), *Ascaphus* (American).

Order 3. Opisthocoela.

Characters: i. Opisthocoelous vertebrae.
ii. Free ribs found either in larva or adult.

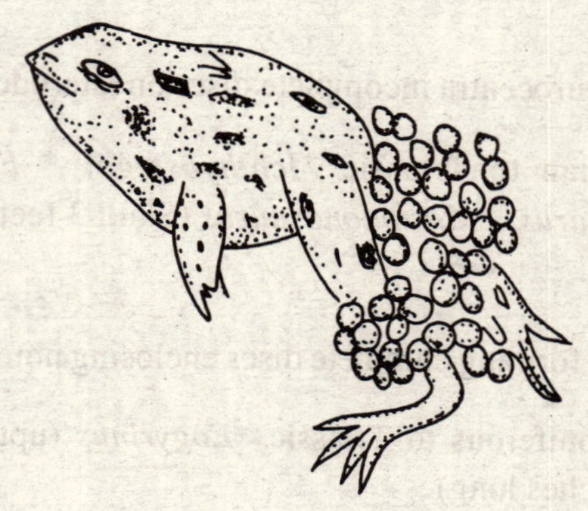

Fig. 1.36 Alytes obstetricians

Family: Discogossidae includes *Discoglossus* (European); *Bombina; Alytes obstetricians* (midwife toad of Europe (Fig. 1.36).

Family: Pipidae includes *Pipa* (Surinam toad of North America, (Fig. 137); *Xenopus* (Clawed frog of Africa).

Order 4. Anomocoela.

Characters: i. Procoelous or amphicoelous vertebrae.
ii. Ribs absent,
iii. Teeth present in upper jaw only.

Family: Pelobatidae includes Spadefoot toads *Pelobates* of Europe and North Africa, and *Scaphiopus* of Amrica.

Family: Pelodytidae includes *Pelodytes* of Europe.

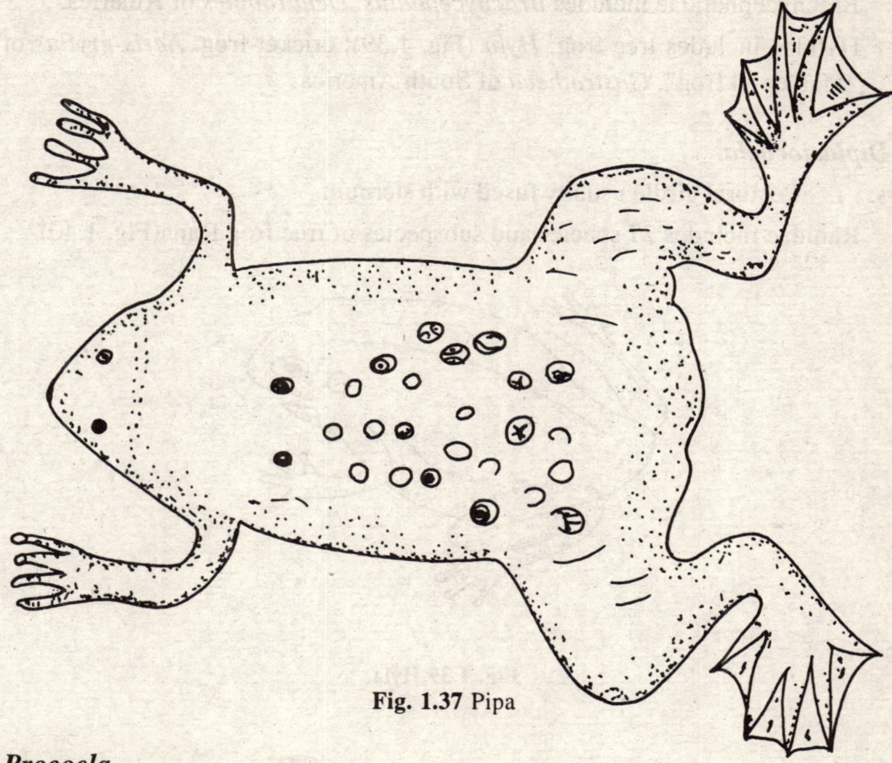

Fig. 1.37 Pipa

Order 5. Procoela.

Characters: i. Vertebrae usually precoelous.

ii. Urostyle has two condyles

Family: *Palaeobatrachidae includes extinct **Palaeobatrachus* of Miocene period.

Family: Bufonidae includes true toads having worldwide genus *Bufo* (Fig. 1.38): *Leptodactylus* (South Amrican) *Rhinoderma* (Chile); *Eleurotherodactylus* (Central American).

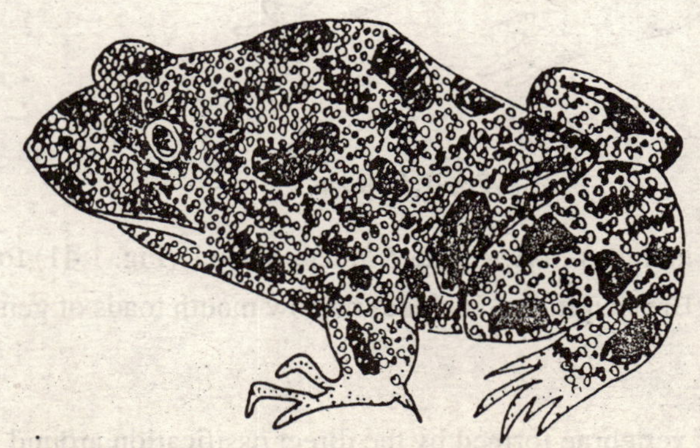

Fig. 1.38 Bufo

Family: Brachycephalidae includes *Brachycephalus*: *Dendrobates* of America.

Family: Hylidae includes tree frog: *Hyla* (Fig. 1.39); cricket-frog. *Acris gryllus* of Canada: "Marsupial frog". *Gastrotheca* of South America.

Order 6. Diplasiocoela.

Characters: i. Pectoral girdle usually fused with sternum.

Family: Ranidae includes 27 species and subspecies of true frog Rana(Fig. 1.40).

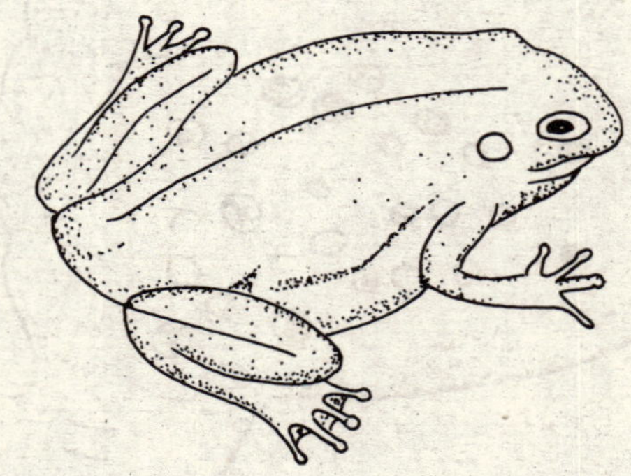

Fig. 1.39 Hyla

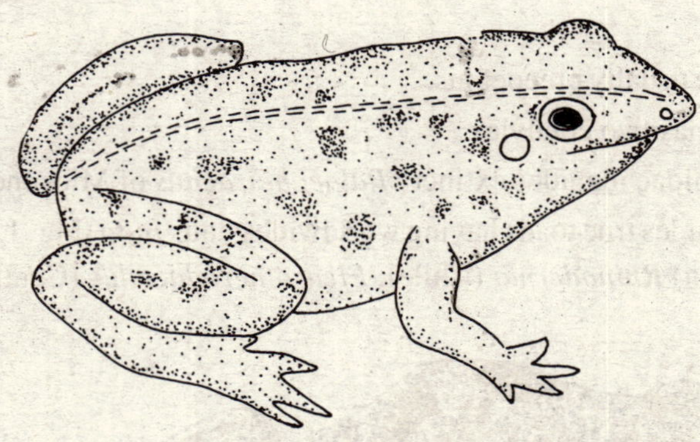

Fig. 1.40 Rana

Family: Polypedatidae includes *Polypedates (Rhacophorus)*, (Fig. 1.41) found in old world.

Family: Microhylidae (Brevicipitidae) includes narrow mouth toads of genus *Breviceps.*

Subclass 2. Lepospondyli.

Characters: i. Centra of vertebrae formed by the direct ossification around notochord without intermediate cartilaginous stage.

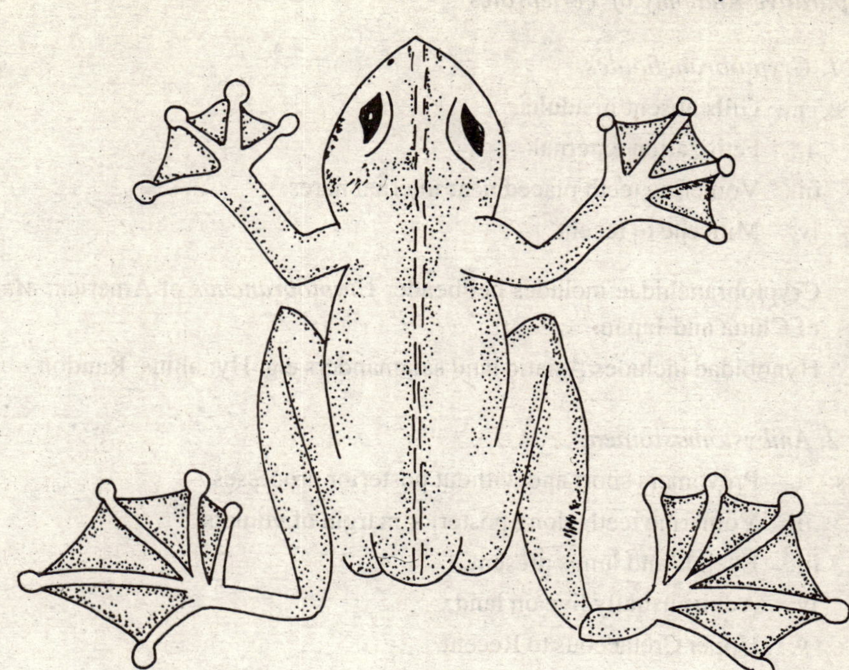

Fig. 1.41 Rhacophorus

Order 1. *Microsauria.

Characters: i. Body elongated, small with short limbs.

Examples: Carboniferous to Permian- **Euryodus; *Microbrachis: *Lysorophus.*

Order 2. *Aistopoda.

Characters: i. Limbs absent.

ii. Body elongated, snake-like with 100 ribs.

Examples: Carboniferous **Ophiderpeton.*

Order 3. *Nectridia.

Characters: i. Expanded neural and haemal processes present in tail.

Examples: Carboniferous to Lower Permian. **Sauropleura: Diplocaulus.*

Order 4. Caudata (Urodela). SALAMANDERS, NEWSTS ETC.

Characters: i. Body usually distinguishable into head, trunk, and tail.

ii. Limbs usually of equal size.

iii. Girdles mainly cartilaginous; dermal pectoral girdle absent.

iv. Larvae look like adults in appearance and have teeth in both jaws.

v. Cretaceous to Recent: more than 250 species

Suborder 1. Cryptobranchoides.

Characters: i. Gills absent in adults.
ii. Fetilization external.
iii. Vomerine teeth placed near internal nares.
iv. Miocene to recent.

Family: Cryptobranchidae includes hellbender *Cryptobranchus* of America: *Megabatrachus* of China and Japan.

Family: Hynobidae includes Asiatic land salamanders e.g. Hynobius: Randon.

Suborder 2. Ambystomastoidea

Characters: i. Prevomers short and without posterior processes,
ii. Vomerine teeth along posterior margin of vomers.
iii. Eyelids and lungs present,
iv. Adults usually live on land,
v. Upper Cretaceous to Recent.

Family: Ambystomidae includes 13 species of Ambystoma (Fig. 1.42) and its larva is called Axolotl.

Suborder 3. Salamandroidea

Characters: i. Premolars extended into tooth bearing extensions along parasphenoids.
ii. Eocene to Recent.

Family: *Salamandridae* (salamanders, includes *Triturus* (American); *Triton* (European); *Tylotriton verracosus* (Indian in Eastern Himalayas); *Salamandra* (European fire salamander), (Fig. 1.43).

Family: *Amphiumidae* includes "Congo eel". *Amphiuma* (American, Fig. 1.44).

Family: *Plethodontidae* (American lungless salamanders) includes *Desmognathus: Spelerpes.*

Suborder 4. Proteida.

Characters: i. Permanent larval forms with depressed body and tail with fin.
ii Three pairs of gills and two pairs of gill slits present in the adult.
iii. Lungs present.
iv. Eyelids absent.
v. Two rows of teeth present in upper jaw and one row in lower jaw.
vi. Aquatic and bottom dwellers.
vii. Eocene to Recent.

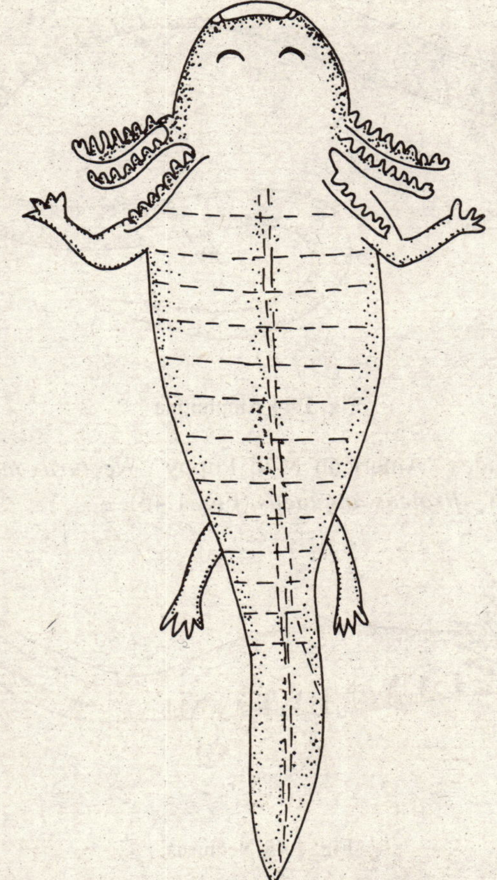

Fig. 1.42 Axolotl larva

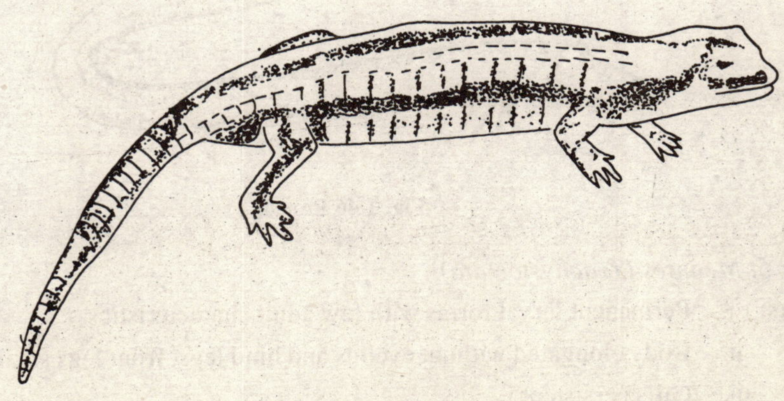

Fig. 1.43 Salamandra

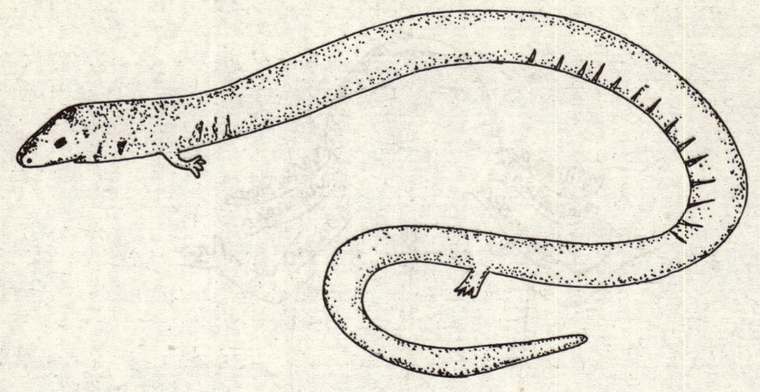

Fig. 1.44 Amphiuma

Family: *Proteidae* includes "American Mud Puppy". *Necturus-maculosus* (Fig. 1.45) and European "Olm". *Proteus anguneus* (Fig. 1.46).

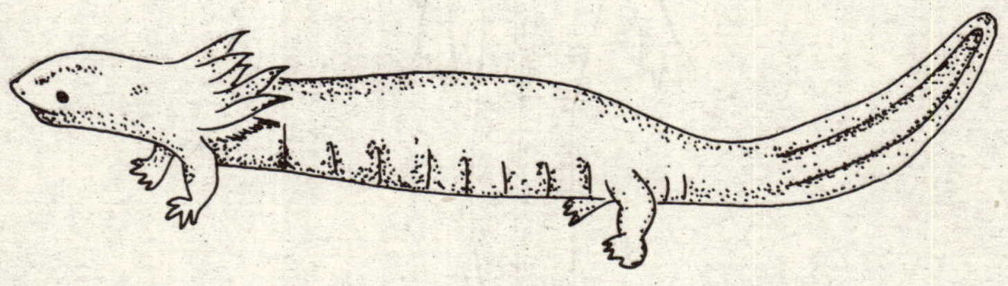

Fig. 1.45 Necturus

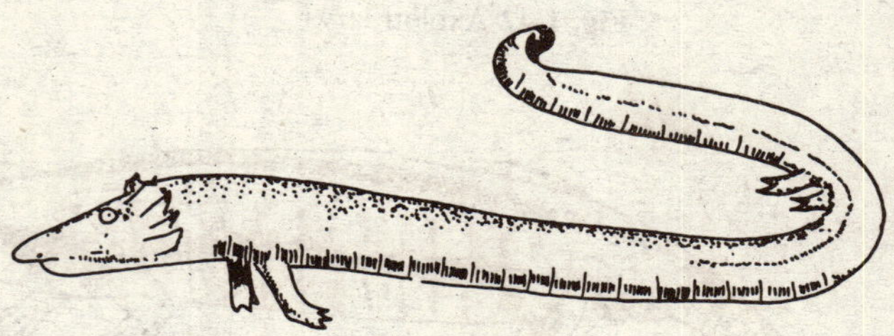

Fig. 1.46 Proteus

Suborder 5. Meantes (Trachystomata).

Characters: i. Permanent larval forms with few adult characteristics.

ii. Body elongated without eyelids and hind legs; front legs small.

iii. Gills persistent.

iv. Horny covering present over jaws.

v. Pleistocene to Recent.

Family: Sirendidae includes 30 inches long "mid eel" of America. *Siren lacertina Pseudobranchus striatus.*

Order 5. Gymnophiona (Apoda). CAECILIANS

Characters: i. Body elongated, worm-like, without limbs or girdles.

ii. Skull roofed with bone and compact.

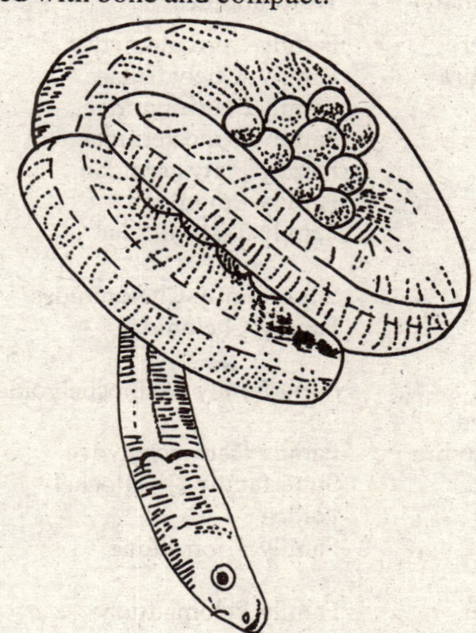

Fig. 1.47 Ichthyophis

iii. Vertebrae many and ribs long.

iv. Smooth skin with slims glands and squirt glands: sometimes embedded with mesodermal scales.

v. One small protractile tentacle present between nostril and eye.

vi. Eyes without lids, often under skin or maxillary bone

vii. Tail short.

viii. Anus near body end.

ix. Protractile copulatory organ present in males.

x. Fossils unknown: about 55 recent species.

Family: Caeciliidae includes tropical forms like *Ichthyophis* (Fig. 1.47): *Gymnophis Typhlonectes* (Northern part of south America: *Uraeotyphlus* (India)

TABLE 1.4

Outline Classification of Reptilia

	Subclass	*Order/Suborder*	*Super Family/Family*	*Examples*
		Order 1		
1.	Anapsida	*Cotylosauria		**Seymouria* **Scutosaurus*
		Order 2 Chelonia (Testudinate)	Superfamily Testudinoidea	
		Suborder- Cryptodira	Family-Dermatemydea Family -Chelydridae Family-Staurotypidae Family -Kinosternidae Family-Platysternidae Family-Emydidae Family-Testudinidae	*Dermatemys* *Chelydra* *Staurotypus* *Kinosternon* *Platysternon.* *Terrapine;* *Testudo;* *Gopherus.*
			Superfamily-Chelonioides Family Chelonidae	*Cholonia;* *Caretta;*
			Superfamily Dennochelyoidea	
		Suborder 1. Cryptodira	Family Dermochelydee Superfamily Cerettochel- yoidea Family Trionchidae	Dernochelys. *Trionyx* *Lissemys.*
		Suborder 2. Pleurodira	Family Pelomedudae Family Chelydae	*Pelomedusa;* *Podocnemis.* *Chelys.*
2.	Synapside	Order 1. *Mesosaurus. Order 2. *Pelycosauria Order 3.		**Mesosaurus.* **Varmiosaurus* **Dimetrodon.* **Moschops;* **Dicynodon;* **Cynognathus.*
3.	Euryapsida (Synaptosauria. Parapsida)	Order 1. *Protosauria Order 2. *Sauropterygia		**Araeoscelis.* **Pleiosaurus;* **Elasmosaurus*
4.	Ichthyoptery- gia (Parapsida)	Order **Ichthyosauria*		**Ichthyosaurus:* **Ophthalmosaurus*

	Subclass	*Order/Suborder*	*Super Family/Family*	*Examples*
5.	Lepidosauria (Diapsida)	Order 1. *Eosuchia		*Youngina*
		Order 2. Rhynchocephalia		*Sphenodon.*
		Order 3. Squamata	Suborder 1. Lacertilia (Sauria)	*Lizards- Uromastix: Draco*
			Suborder 2. Ophidia (Serpentes)	Snakes.
6.	Archosauria (Diapsida)	Order 1. * Thecodontia	*Phylosaurus	
		Order 2. Crocodilia (Loricata)		*Crocodilus; Gavialis,*
		Order 3. *Saurischia	Suborder 1. * Theropoda	**Tyramnosaurus*
			Suborder 2. *Sauropoda	**Brontosaurus.*
		Order 4 Ornithischia	Suborder 1. *Ornithopoda	**Iguanodon.*
			Suborder 2. *Stegosauria	**Stegosaurus*
			Suborder 3. Ankylosauria	**Ankilosaurus*
			Suborder 4. Ceratopsia	**Triceratops.*
		Order 5. Pterosauria		Rhamphorhy- nchus; *Pteanodon.

CLASS REPTILIA

The class Reptilia includes turtles, tortoises, crocodiles, lizards, and snakes. Most of the reptilies are extinct and the living forms exhibit divergent features which are given below.

i. The skin is dry with horny epidermal scales, scutes, of shields: skin glands few

ii. Usually two pairs of limbs with five clawed toes present; limbs modified into paddles, reduced or absent.

iii. Skeleton fully ossified and skull has only one occipital condyle.

iv. Cervical, thoracic, sacral, and caudal vertebrae usually present in the vertebral column.

v. Heart has two auricles and a partly divided ventricle which is completely divided in crocodilies.

vi. Three aortic trunks present; red blood cells nucleated, biconvex and oval.

vii Respiration by lungs.

viii. Twelve pairs of cranial nerves present.

ix. Cold clouded animals.

x. Fertilization usually internal.

xi. Cleavage usually meroblastic; embroyonic membranes including amnion, chorion, yolk sac, and allantois present during development: youngs resemble adults; metamorphosis absent.

xii. Permain to Recent; terrestrial or aquatic in tropics and temperate areas

xiii. About 6.000 species known.

Subclass 1. Anapsida.

Characters: i. Reptiles with solid skull roof without temporal fossa behind the eye.

Order 1. *Cotylosauria. STEM REPTILIES.

This order includes stem reptilians resembling labyrinthodont amphibians. e.g. Permain **Seymouria* (Fig. 1.48); Triassic **Scutosaurus.*

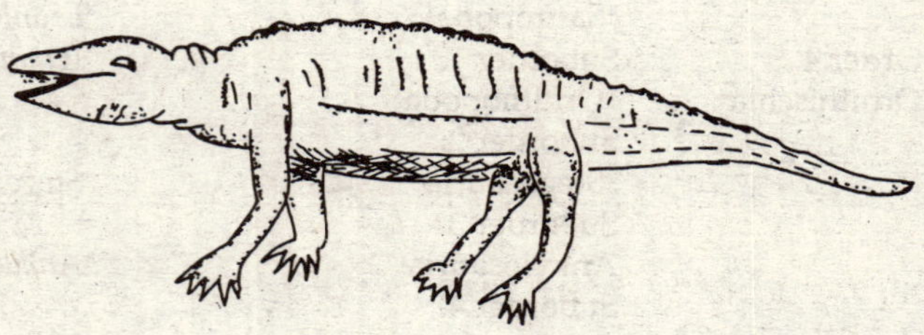

Fig. 1.48 Seymouria

Order 2. Chelonia (Testudinata). TORTOISES, TURTLES, TERRAPINS.

Characters: i. Body enclosed in a shell, made up of dorsal carapace and ventral plastron joined at sides and covered by leathery skin or epidermal scutes; thoracic vertebrae and ribs fused with carapace.

ii. Quadrate is immovable.

iii. Anus longitudinally split.

iv. Single copulatory organ.

v. Teeth absent and jaws provided with horny sheaths.

vi. Egg laying; eggs are laid in holes in ground and covered by females.

vii. Jurassic to Recent: 335 species known.

Suborder l. Cryptodira includes turtles with usually retractile neck which bends in S curve and pelvic girdle not fused with shell.

Examples: Tortoise and turtles - *Dermatemys: Chelydra* (Fig. 1.49A): *Macrochelys: Testudo elegans* (Fig. 1.49B) Starred tortoise *Chelonia Trionyx* (Found in fresh water rivers of India. Fig. 1.49C, Dia 1½-2ft.) *Desmochelyscoriacea* (leathery turtle upto pounds found in south India).

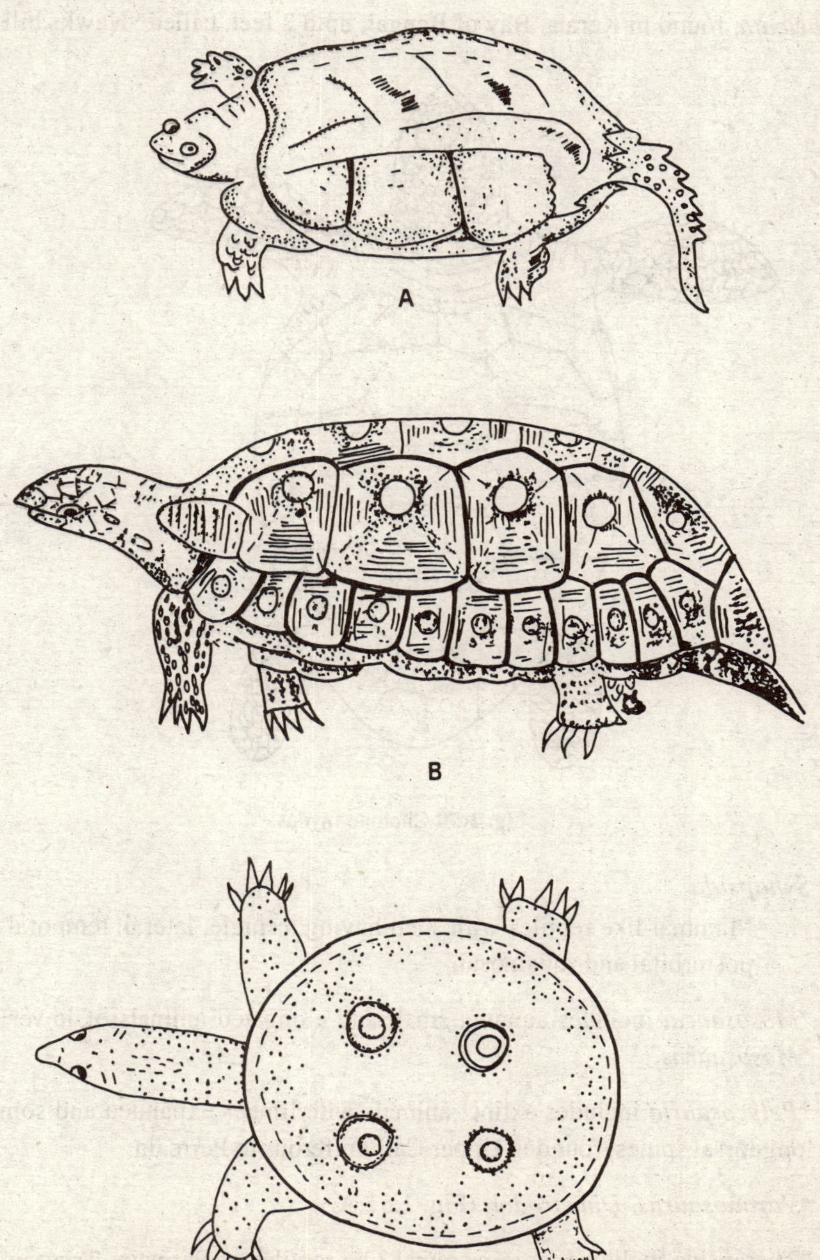

Fig. 1.49 A. Chelydra, **B.** Testudo, **C.** Trionyx

Suborder 2. Pleurodira includes turtles with neck fold sideways under the shell and pelvic girdle fused with the shell.

Examples: *Pelomedusa; Chelys: Chelone mydas* (Fig. 1.50); *Eretomochelys imbricata. Chelone imbricata,* found in Kerala. Bay of Bengal, upto 2 feet, called "Hawks bill turtle".

Fig. 1.50 Chelone mydas

Subclass 2. Synapsida

Characters: i. Mammal-like reptiles with shell having a single, lateral, temporal cavity below postorbital and squamosal.

Order 1. **Mososauria* includes aquatic, small and elongated animals of lower Permain e.g. **Mesosaurus.*

Order 2. **Pelycosauria* includes extinct animals with limbes expanded and some forms with longdorsal spines, found in Upper Carboniferous to Permain.

Examples: **Varanosaurus, Dimetrodon* (Fig. 1.51.A.).

Order 3. *Therapsida included large mammal-like reptiles of Permain. Triassic, and Jurassic periods.

Examples: **Moschops;* **Dicynodon:* **Cynognathus* (Fig. 1.51.B.).

Fig. 1.51 A. Dimetrodon, **B.** Cynognathus

*Subclass 3. *Euryapsida (Synaptosauria, Parapsida).*

Characters: i. Roof of the skull provided with upper opening behind eye lined below by postorbital and squamosal.

Order 1. **Protosauria* includes small lizard-like reptiles of Permain to Jurassic periods e.g. *Araeoscelis*

Order 2. **Sauropterygia* includes Plesiosaures of Triassic to Cretaceous periods, e.g. **Plesiosaurus; *Elasmosaurus* (Fig. 1.52 upto 50 feet).

Subclass 4. Ichthyopterygia (Parapsida):

Characters: i. Roof of the skull with upper opening behind eye bound by postorbital and supratemporal below.

Order 1. Ichthyosauria included Ichthyosaures of medium size with limbs as paddles; porpoise-like appearance.

Examples: * *Ichthyosaurus* (Fig- 1.53. A.); *Ophthalmosaurus* (Fig. 153.B)

Fig. 1.52 C. Elasmosauws

Subclass 5. Lepidosauria (Diapsida).

Characters: i. Roof of the skull with two openings behind eye separated by postorbiral and squamosal some specialized forms have reduced openings.

Order 1. *Eosuchia included two arched reptiles of Permian Period, e.g. **Youngina*

Order 2. Rhynchocephalia is represented by a surviving representative *Sphenodon punctatum* (Fig. 1.56) having following characteristics:

Fig. 1.53 A. Ichthyosarus **B.** Opthalmosarus

i. Lizard-like, with granular scales.
ii. A mid dorsal row of low spines present.
iii. Quadrate fixed and mandibles joined by a ligament.
iv. Amphicentrous vertebrae.
v. Abdominal ribs present.
vi. Transverse anal opening.
vii. Copulatory organs absent in males.

Order 3. *Squamata* is represented by lizards and snakes having the following characteristics:

i. Horny scales or shields are present over the skin forming an armour.
ii. Movable quadrate bone.
iii. Usually procoelous vertebrae.
iv. Copulatory organ is double and eversible.
v. Transverse anal opening.

Fig 1.54 Restoration of some forms of reptiles. **A.** Brontosaurus. **B.** Triceratops **C.** Stegosaurus. **D.** Iguanodon.

Fig 1.55 Restorations of Mesozoic reptiles Brontosaurus and Stegosaurus Jurassic, all others Cretaceous four at bottom are marine. Numbers are length in feet.

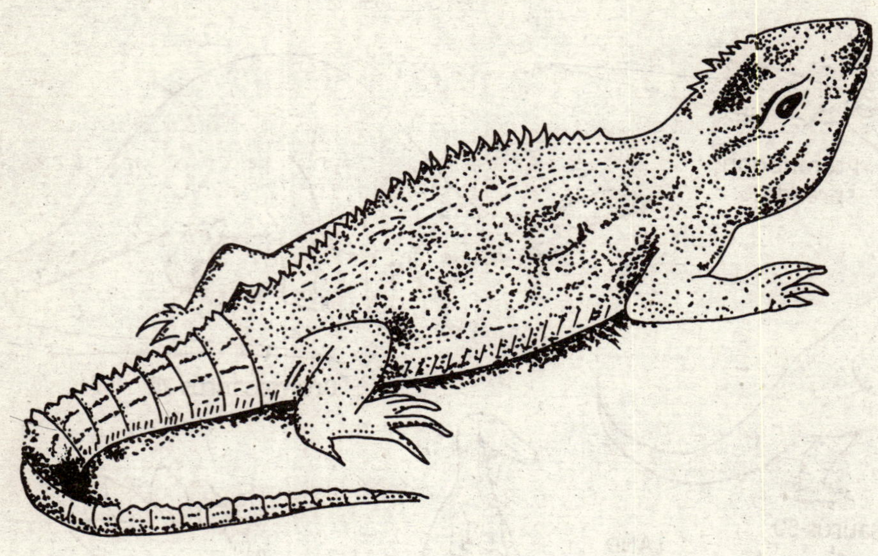

Fig. 1.56 Sphenodon

Suborder 1. Lacertilia (Sauria) includes lizards having movable eyelids: ear openings and mouth not expansible; girdles usually present; about 3,140 species.

Examples: *Uromastix* (Sand lizard. Fig. 1.57.A.); *Varanus* (Monitor lizard. Fig 1.57. B.): *Iguana* (Fig. 1.57. C); *Calotes* (Garden lizard. Fig. 1.57. D.): *Heloderma* (Gila monster. Fig. 1.57.E.); *Hemidactylus* (Wall lizard. Fig. 1.57.F.); *Chamaelon* (Fig. 1.57. G.): *Draco* (Flying lizard. Fig. 1.57. H.); *Ophiosauras* (Glass snake. Fig. 1.57.1).

Suborder 2. Ophidia (Serpantes) includes snakes having no limbs, feet, sternum, ear openings, and urinary bladder, (ii). Eyes immovable without lids. (iii). Tongue bifid and protrusible. (iv). About 2,500 species.

Examples: *Vipera russeli* (Viper. Fig. 1.58): *Bungarus* (Krait. Fig. 1.59); *Naia* (Cobra. Fig. 1.60); *Eryx. Johni; Python; Typhlops.*

Subclass 6. Arcosauria (Diapsida).

Characters: Skull of diapsid type without interparietal and parietal foramina.

Order 1. **Thecodontia* included Dinosaura with teeth in sockets, e.g. **Saltoposuchus* *Phytosaurus; resembling crocodile.

Order 2. *Crocodilia* (Loricata) includes crocodiles, alligators, and gavials having the following features;

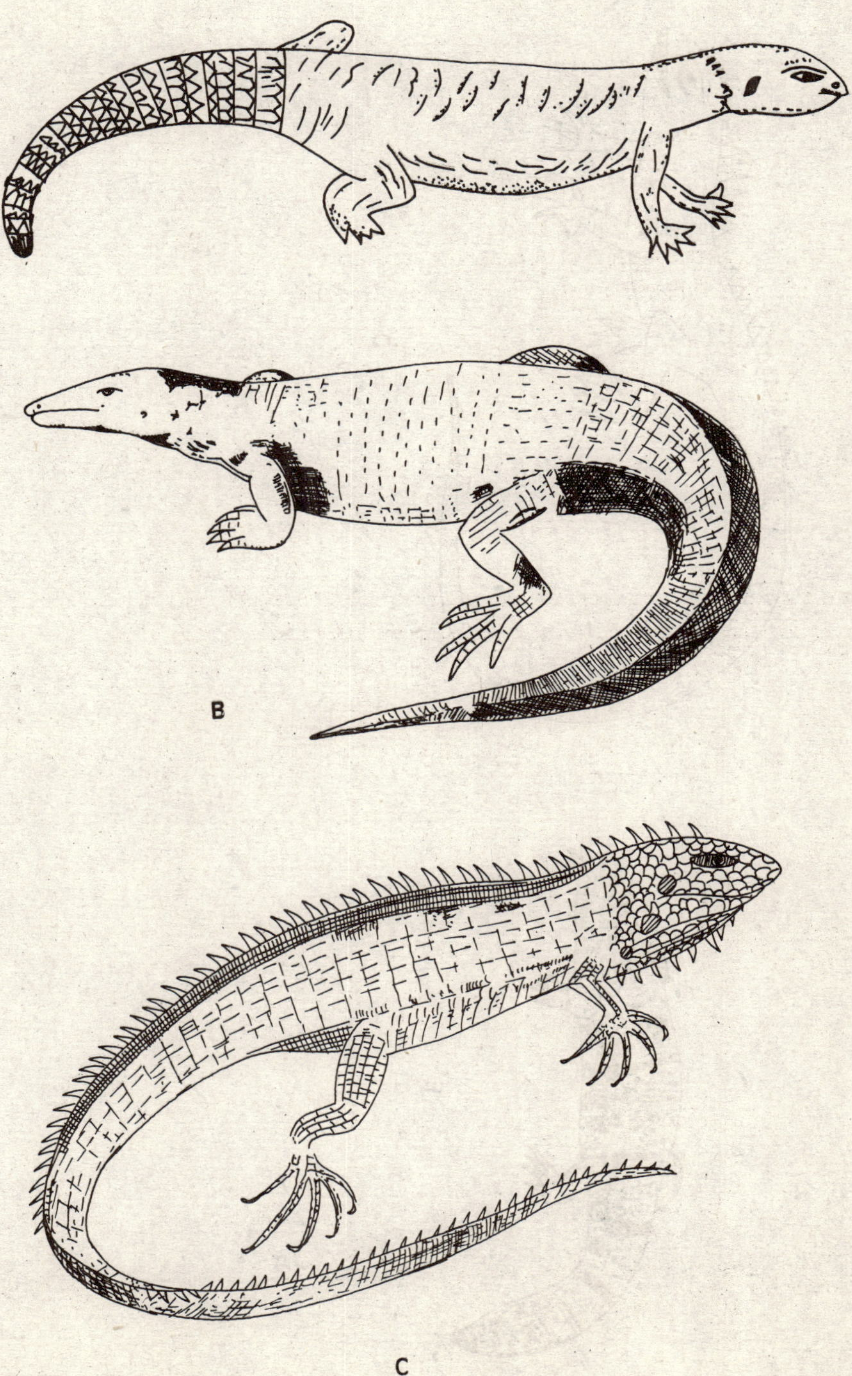

Fig. 1.57 A. Uromastix, **B.** Varanus, **C.** Iguana

Fig. 1.57 D. Calotes, **E.** *Heloderma*

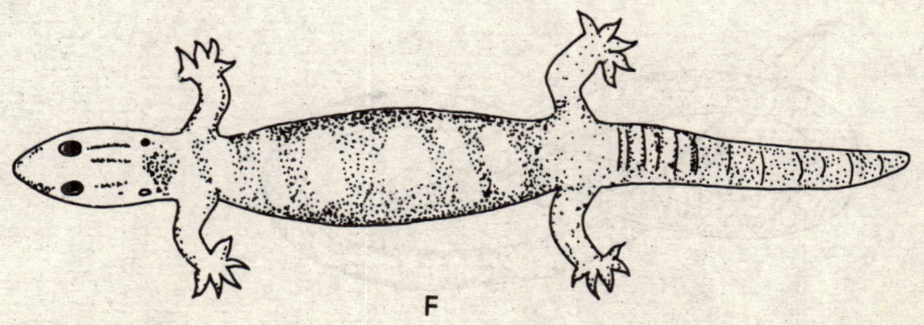

Fig. 1.57 F. *Hemidactylus,* **G.** Chamaelon

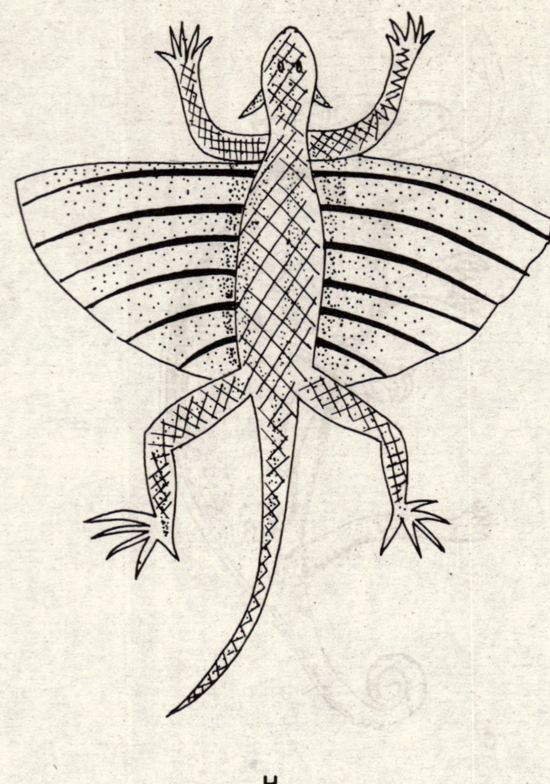

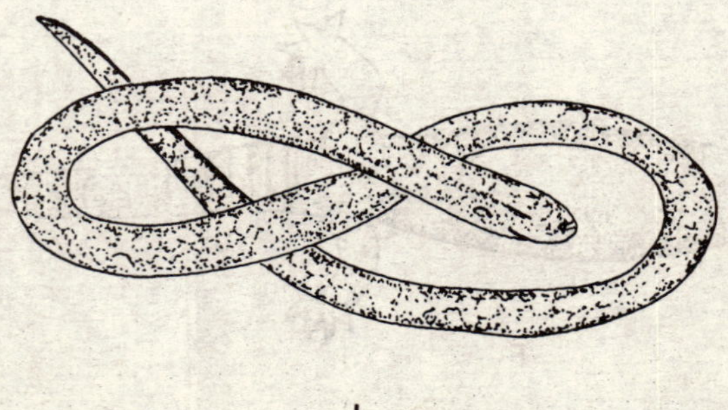

Fig. 1.57 H. Draco, **I.** Ophiosaurus

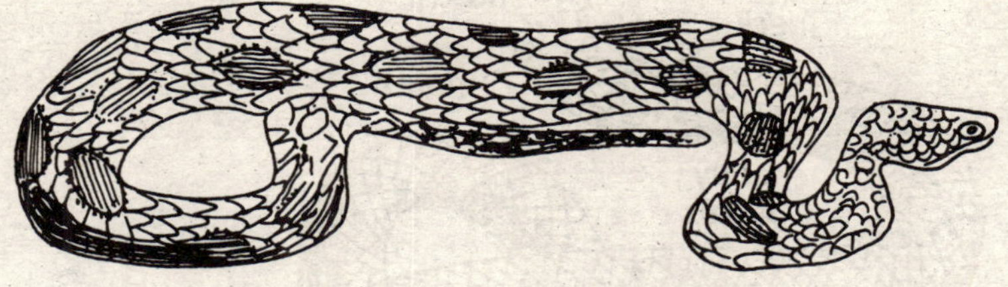

Fig. 1.58 Vipera russeli

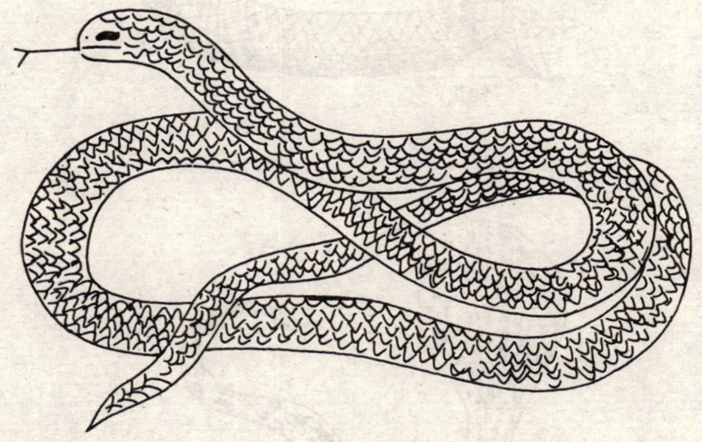

Fig. 1.59 Bungarus

i. The heart is four chambered with complete interventricular septum.

ii. Abdominal ribs present.

iii. A bony secondary palate is formed by the fusion of shelves from the maxillae, palatines and pterygoids.

iv. Snout elongated with external nares at the tip.

Examples: *Crocodilus* (Fig. 1.61.A.): *Alligator* (Fig. 1.61.B.): *Gavialis* (Fig. 1.6 l.C.)

Order 3. **Saurischia,* includes Dinosaurs having diverging ishium and pubis.

Suborder 1. **Theropoda* includes bipedal, carnivorous Dinosaurs extending from Traissic to Cretaceous

Examples: Strythiomimus (Ostrich-like): Tyrannosaurus (huge, upto 50 feet Fig. 1.62).

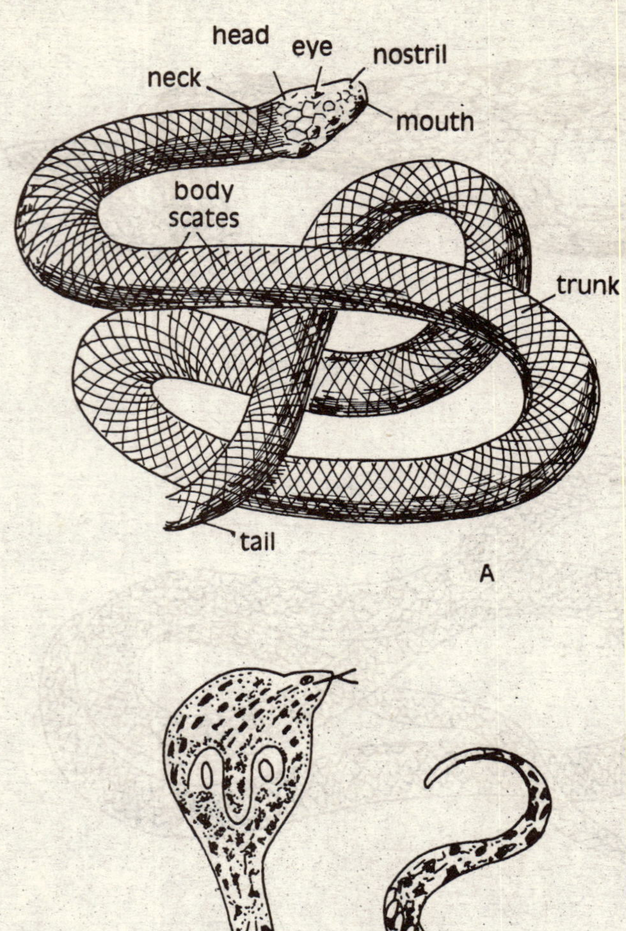

Fig. 1.60 A. Eryx Johni, **B.** Cobra

Suborder 2. **Sauropoda* includes huge, quadripedal, herbivorous Dinosaurs of Jurassic e.g. **Brontosaurus* (upto 70 feet. Fig. 1.63); **Diplodocus.*

Order 4. **Ornithischia* includes Dinosaurs with bird-like pelvis; ischium and publis fused.

Fig. 1.61 A. Crocodilus, **B.** Alligator, **C.** Gavia lis

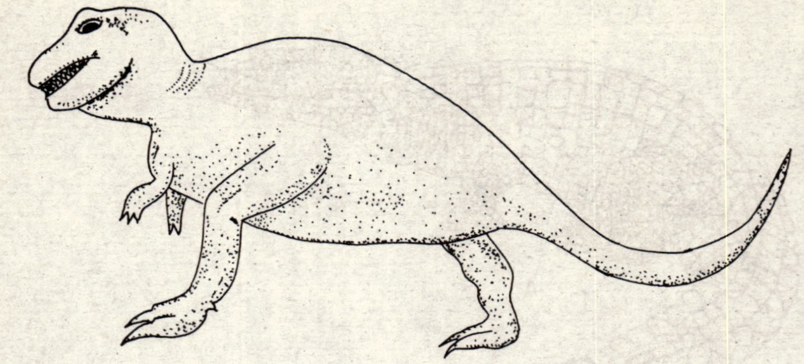

Fig. 1.62 Tyannosaurus

Fig. 1.63 Diplodocus

Suborder 1. **Ornithopoda* includes duck billed Dinosaurs, bipedal or quadripedal of Jurassic to Cretaceous, e.g. **Camptosaurus;* **Iguanodon* (30 feet long Fig. 1.64).

Suborder 2. **Stegosauria* includes armoured Dinosaurs of Jurassic, e.g. **Stegosaurus* (20 feet. Fig. 1.65).

Suborder 3. **Ankylosauria* includes armoured Dinosaurs of Cretaceous e.g. **Ankylosaurus* (25 feet. Fig. 1.66).

Suborder 4. Ceratopsia includes horned Dinosaurs of Cretaceous e.g. **Triceratops* (20 feet. Fig. 1.67).

Fig. 1.64 Iguanodon

Order 5. *Pterosauria* includes flying reptiles having wing membranes in fore limb. e.g. *Rhamphorhychus* (small 18 inches wing span, tail long. Jurassic. Fig. 1.68); *Pteranodon* (without tail wing span 27 feet. Cretaceous. Fig. 1.69.).

Fig. 1.65 Stegosaurus

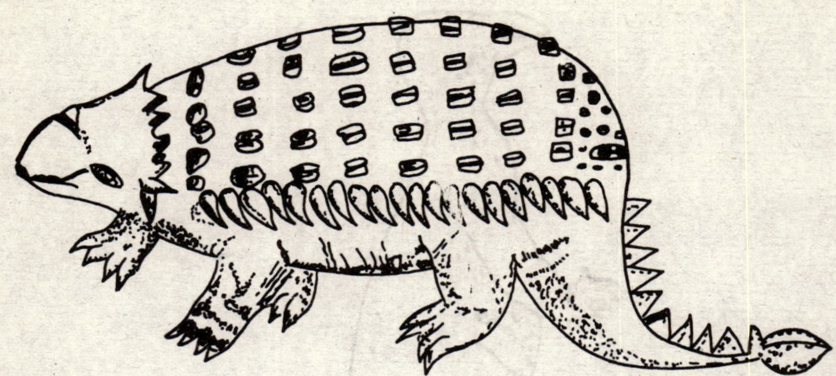

Fig. 1.66 Ankylosaurus

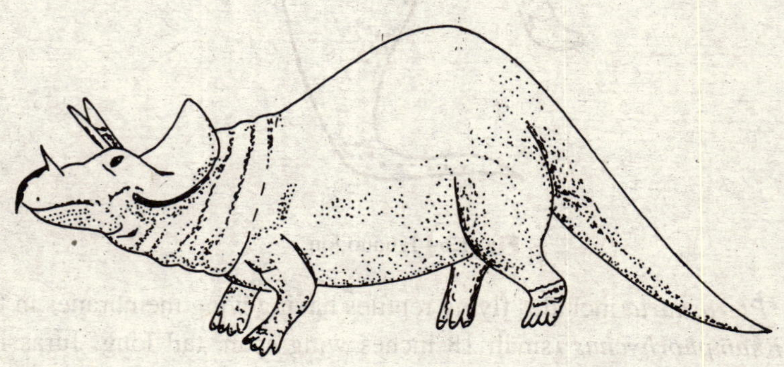

Fig. 1.67 Triceratops

CLASS AVES

The class Aves is represented by about 8,600 species of birds having the following characteristics:

i. Body covered over by feathers.
ii. The fore limbs are modified into wings adapted for flying, the hind limbs with four toes usually adapted for walking, perching or swimming.
iii. Mouth is produced into a beak with horny sheaths.
iv. Teeth absent in living forms.
v. Skull with one occipital condyle; skeleton fully ossified, delicate, spongy, with number of air cavities.
vi. Cervical and free thoracic vertebrae usually heterocoelous sometimes procoelous or amphicoelous. Some thoracic, sacral, and a few caudal vertebrae fuse to form a synsacrum supporting the ilia of the pelvic girdle.

Fig. 1.68 Rhamphorhynchus

Fig. 1.69 Pteranodon

vii. The Posterior caudal vertebrae are usually fused to form pygostyle.

viii. Heart is four chambered with two auricles and two ventricles.

ix. Lungs are compact with air sacs: sound box is present at the base of the trachea.

x. Urinary bladder absent.

xi. Warm blooded animals; red blood cells usually nucleated.

xii. Egg laying; eggs covered with large yolk and hard limy shell; segmentation meroblastic; embryonic membranes present.

xiii. Upper Jurassic to Recent.

The outline classification of the Class Aves is given in table 1.5

TABLE 1.5

Outline Classification of Aves

Subclass	*Superorder*	*Order*	*Example*
1. Neornithes			**Archaeopteryx; *Archaeornis.*
2. Neornithes	1. *Odontognathi		**Hesperonis.*
	2. *Ichthyornuthes		**Ichthyornis*
	3. Impennes	Sphenisciformes	*Aptenodytes*, Penguin
	4. Palaeognathi (Walking birds)	1. Struthioniformes	*Struthio.* Ostrich
		2. Rheiformes	*Rhea*
		3. Casuariformes	*Casuarius*
		4. Aepyornithiformes	*Aepyornis,* Elephant bird
		5. Dinornithiformes	*Dinornis,* Moas
		6. Apterygiformes	*Apteryx,* Kiwi
		7. Tinamiformes	*Tinamus*
	5. Neognathi	1. Gaviiformes	*Ga via.* Divers
		2. Podicipediformes	*Podiceps* (Colymbus). *Grebes*
		3. Procellariformes	Petrels
		4. Pelecaniformes	Pelicans, Cormorants
		5. Ciconiiformes	Herons, Storks
		6. Anseriformes	Ducks, Geese, Swan
		7. Falconiformes	Vultures, Kites. Eagles
		8. Galliformes	Pheasants, Quail
		9. Gruiformes	Cranes, Coot
		10. *Diatrymiformes	*Diatryma.
		11. Charadriiformes	Glulls, Auks
		12. Columbiformes	Pigeons
		13. Psittaciformes	Parrots
		14. Cuculiformes	Cuckoos
		15. Strigiformes	Owls
		16. Caprimulgiformes	Nighthawks
		17. Apodiformes	Hummingbirds
		18. Coliiformes	Colies
		19. Trogoniformes	Trogons
		20. Coraciiformes	Kingfisher, Hornbills
		21. Piciformes	Woodpecker
		22. Passeriformes	Perching birds

*Subclass 1. *Archacornithes.*

This subclass includes ancestral birds having the following characteristics:

i. Teeth present in sockets in both jaws.

ii. Wing feathered broadly.

iii. Three clawed digits present in the hand.

iv. Tail long with 20 vertebrae with paired lateral feathers; Pygostyle absent.

v. Poor flyers.

Examples: Only two genera **Archaeopteryx* (Fig. 1.70) and **Archaeornis*

Fig. 1.70 Archaeopteryx

Subclass 2. Neornithes

This subclass includes true birds having the following characteristics:

i. Metacarples fused; second digit finger longest.

ii. Tail vertebrae compressed and less than 13 in number.

iii. Well developed sternum bears a distinct keel or carina.

iv. Cretaceous to Recent.

Superorder 1. *Odontognathi includes extinct New World toothed birds having following characters.

i. Sternum without keel.

ii. Clavicle free.

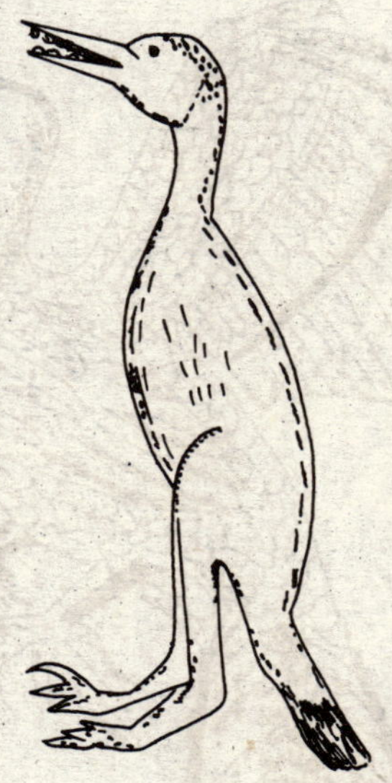

Fig. 1.71 Hesperornis

iii. Wing present on vestigial humerus only.

iv. Teeth present on both jaws.

v. Found in Upper Cretaceous represented by one extinct genus **Hesperornis* (Fig. 1.71) which was flightless bird of about 5 feet length, adapted for swimming.

Superorder 2. **Ichthyornithes* includes extinct gull-like birds of Upper Cretaceous having the following features:

i. Sternum with keel.

ii. Clavicle fused.

iii. Wings well developed.

iv. Jaws absent e.g. * *Ichthyornis.*

Superorder 3. *Impennes: Order Sphenisciformes* includes Penguins having the following characteristics:

i. Flightless birds with paddle-like wings.

ii. Teeth absent.

iii. Fusion of metacarpals incomplete.

iv. Four toes, first small feet with webs.

v. Small scale-like feathers over body without barbules.

vi. Stand erect on metatarsi and may dive easily and swim with the help of forelimbs. e.g. *Aptenodytes.* Penguins with 17 species (Fig. 1.72).

Fig. 1.72 Aptenodytes

Superorder 4. *Palaeognathi* includes Walking birds without teeth and with reduced wings incapable of flight. The tail vertebrae are usually free.

Order 1. *Struthioniformes* includes Ostriches with following Characteristics:

i. Landliving flightless birds.

ii. Sternum without keel.

iii. Public Symphysis present.

iv. Pygostyle absent.

v. Only third and fourth toes present in each foot.

vi. Few feathers on head, neck, and legs.

vii. Miocene to Recent: found in Africa and Arabis. e.g. *Struthio camelus* (Fig. 1.73A), Largest surviving bird about 300 pound weight and 7 feet height.

Fig. 1.73 A. Struthio camelus

Fig. 1.73 B. Rhea

Order 2. *Rheiformes* includes Rheas of South America, with following characteristics:

i. Landliving flightless birds.

ii. Keel absent in the sternum.

iii. Each foot has three front toes.

iv. Aftershaft absent in feathers.

v. Few feathers on head, neck, and legs.

vi. Miocene to Recent, e.g. *Rhea* (Fig. 1.73B.). 4 feet tall.

Order 3. *Casuariiformes* includes Cassowaries and Emus of Australia,

i. Neek and body thickly feathered.

ii. Wings small and feathers with equal sized aftershafts.

iii. Sternum without keel.

iv. Pliocene to Recent e.g. *Casuarius* (Fig. 1.74A.), Cassowary of Australia, 5 feet; *Dromiceius* (Fig. 1.74B.) Emu, about 5 feet tall.

Fig. 1.74 A. Casuarius

Order 4 *Acpyornithiformes* includes Elephant birds of Medagascar, about 10 feet tall with largest known eggs of about 12 inches.

i. Landliving, flightless birds with unkeeled. short and broad sternum.

ii. Wings vestigial.

iii. Four toes.

iv. Recent but extinct for several centuaries. e.g. **Aepyornis.*

Order 5. *Dinornitheformes* includes Moas of New Zealand.

Characters: i. Flightless birds with reduced, unkeeled sternum.

ii. Wingbones. Scapula, and coracioid reduced or absent.

iii. Massive hindlimbs with 3-4 toes.

Fig. 1.74 B. Dromiceius

iv. Large aftershaft present with feathers.

v. Recent but extinct for several centuaries. e.g. **Dinornis* many species, one 8 feet tail.

Order 6. *Apterygiformes* includes kiwis of New Zealand.

Characters: i. landliving, flightless birds.

ii. Nostrils stituated at the tip of long and slender bill.

iii. Wings degenerate and unkeeled sternum.

iv. Four toes.

v. Body covered over by hairy feathers without aftershafts. e.g. Apteryx (Fig. 1.75). A).

Fig. 1.75 A. Apteryx

Order 7. *Tinamiformes* includes Tinamus of south America.

Characters: i. Landliving; flying birds.

ii. Wings short, rounded, and adapted for flying.

iii. Keel present in the sternum.

iv. Pygostyle short, tail short e.g. *Tinamus* (Fig. 1.75.B.); *Rhynchotus.*

Fig. 1.75 B. Tinamus

Super Order 3. *Neognathi* includes modern birds with following

Characters: i. Teeth absent in jaws.

ii. Sternum is provided with keel.

iii. Wings fully developed, forelimbs with jointed metacarpals and fingers.

iv. Tail vertebrae upto 6 and pygostyle present.

v. Ecocene to Recent.

Order 1. *Gaviiformes* includes Divers with short legs at the body end; toes fully webbed, ii. Tail with 18-20 stiff and short feathers. iii. Adapted for diving and feed on fishes, e.g. *Gavia* (loon)

Order 2. *Podicipediformes* includes Grebes with following.

Characters: i. Legs situated far back on the body.

ii. A tuft of down feathers present over the tail.

iii. Tarsus compressed and feet provided with lobes.

iv. Food consists of small aquatic animals and plant materials, e.g. Podiceps (Colymbus). Grebes.

Order 3. *Procellariiformes* includes Petrels with the following

Characters: i. Nostrils tubular and skull with large nasal glands.

ii. Feathers compact with oily texture:

iii. Wings narrow and long.

iv. Hind toes vetigeal or absent.

v. Oceanic birds with nests on islands, e.g. *Albatrosses* Fulmans; Petrels.

Order 4. *Pelecaniformes* includes pelicans. Cormorants. Bobbies. Gannets etc. having the following characterstics;

i. Foot web includes all the four toes.

ii. Nostril vestigial or may be absent.

iii. A gular pouch is usually present on the throat.

iv. Aquatic birds e.g. *Pelicanus; Sula* (Bobby). *Morus* (Garnet).

Order 5. *Ciconiiformes* includes Herons. Storks, and Flamingoes with the following characteristics:

i. Long legged and necked birds.

ii. Usually web is reduced or absent between toes.

iii. Stay near water and nesting on trees.

iv. Food mainly of fishes and other aquatic animals e.g. pond Herons; *Egrets* (Egrets): *Ibis* (Stork): *Platella* (Spoonbill); *Phoenicopterus* (Flamingo).

Order 6. *Anseriformes* includes Ducks, Geese. Swans having the following characteristics:

i. Bill wide with covering of soft cornified eqidermis containing sense pits; hard nail or cap at the tip.

ii. Legs short and webs present in feet.

iii. Many feathers present on the til which is short.

iv. Stay near water; about 200 species, e.g. *Anseranus* (Magfrie goose.); *Cygnus* (Swan); *Anser* (True geese); *Anas* (Dabbing ducks).

Order 7. *Falconiformes* includes vultures, kites, hawks, eagles etc. having the following characteristics:

i. Bills sturdy with soft naked skin at the base and hooks at the tips.

ii. Jaws sharp edged.

iii. Predaceous with usually rapid and active flight.

iv. Feet With sharp curved claws adapted for grasping.

v. Food consists of smaller animals, e.g. *Vultures-Aegypius; Gypa; Neophron.* Kites-*Milvus*: *Haliastur;* Hawks - *Circus; Accipiter;* Fulcoons-*Faleo.*

Order 8. *Galliformes* includes Quail, Pheasants. Fowl. Turkeys etc. with the following characteristics:

i. Short bills.

ii. Feathers have aftershafts.

iii. Feet adapted usually for running and scratching.

iv. Several species domesticated.

v. Food mainly of plant materials e.g. *Lagopus*, (Grouse): *Coturnix* (Quail): *Phasianus* (Pheasants); *Gallus* (Fowl): *Meleneogris*. (Turkeys).

Order 9. *Gruiformes* includes Cranes. Coots etc. having the following characteristics:

i. Aftershafts present on feathers.

ii. Size varies from small with weak flight to larger with strong flight, e.g. *Grus* (Crane); *Fulca: Rallus.* (Rails); *Otis* (Bustard).

Order 10. * *Diatrymiformes* includes extinct flightless birds of Eocene of America with reduced wings, four toes on each foot, and large bill. e.g. **Diatryma.*

Order 11. *Charadriiformes* includes Waders or shore birds. Gulls, and Auks etc. having the following characteristics:

i. Toes are usually weobed at the base.

ii. Feathers firm and dense.

iii. Long-legged, strong-winged or with three toes and legs far situated on the body. e.g. *Charadrius* (Shore birds); *Larus* (Full.); *Plautus* (Awks).

Order 12. *Columbiformes* includes Pigeons and Doves having the following characteristics:

i. Bill short and long usually with thick soft skin at the base.

ii. Tarsus shorter than toes usually, e.g. *Columba: Streptopelia* (Dove).

Order 13. *Psittaciformes* includes Parrots with following feathers;

i. Beak sturdy with sharp edged hoks at the tip.

ii. Upper jaw movable on frontal bone.

iii. Bill with soft skin and usually with feathers.

iv. Toes two infront and two behind; feet adapted for grasping; outer hind toes not reversible.

v. Plumage brilliantly coloured, e.g. *Psittacus* (parrots.); *Ara* (Macaw); *Kakatoe* (Cocatoo).

Order 14. *Cuculiformes* includes Cuckoos. Road runners, and Anis etc. characters:

i. Toes two infront and behind; outer toes reversible.

ii. Feet adapted for grasping.

iii. Tail long e.g. *Cuculus* (Cuckoo.); *Eudynamus* (Koel); *Crotophaga* (Anis): *Geococcyx* (Roadrunner).

Order 15. *Stringiformes* includes Owls. Characters:

i. Head large rounded with large eyes directed, forwards. each in a disc of radial feathers.

ii. Large ear openings usually with flap as cover.

iii. Beak small.

iv. Feet adapted for grasping with sharp claws.

v. Food consists of small land vertebrates and arthropods. e.g. *Tyco* (Barn-Own); *Otus* (Screech-Owl.); *Bubo* (Eagle Owl).

Order 16. *Caprimulgiformes* includes Nighthawks etc.

Characters: i. Wide mouth with long bristle-like feathers usually margining the mouth.

ii. Bill small.

iii. Legs and feet not adapted for grasping and weak.

iv. Active mostly at dusk and night.

v. Food consists of night flying insects captured in air. e.g. *Chordeiles* (Nighthawk); Goatsuckers.

Order 17. *Apodiformes (Micropodiformes)* includes Humming birds and Swifts. Characters:

i. Usually small size.

ii. Legs small with very small feet.

iii. Wings pointed.

iv. Bill elongated with tubular tongue, or mouth weak and small.

v. Food captured on wings, e.g. *Archilochus* (Humming bird.): *Chaitura*, *Apus* (Swifts).

Order 18. *Coliiformes* includes Colics or Mouse birds.

Characters: i. Passer-like, small birds with first and fourth toes reversible.

ii. Tail very long. e.g. *Colies.*

Order 19. *Trogoniformes* includes Trogons.

Characters: i. Bill sturdy and short with bristles at base.

ii. Feet weak and small.

iii. Plumage usually brilliant green, e.g. *Harpactes: Trogon* (Trogons).

Order 20. *Coraciiformes* includes Kingfishers. Hornbills etc.

Characters: i Third and fourth toes fused at the base.

ii. Bill very strong, e.g. *Alcedo* (King-fisher): *Buceros* (Hornbill.) *Coracias* (Roller).

Order 21. *Piciformes* includes Woodpeckers. Puffbirds. Honey-guides. Barbets etc. Characters:

i. Tail feathers stiff with pointed tips.

ii. Bill stout.

iii. Tongue protractile with bards near tip or roughened.

iv. Toes two infront and usually two, non reversible behind.

v. Digout insects and larvae out of wood by clinging on trees, e.g. *Picus: Dendrocopos* (Woodpeckers); *Indicator* (Honeyguide).

Order 22. *Passeriformes* includes Perching birds.

Characters: i. Toes adapted for perching, three infront and one behind

ii. Largest order of birds including four. Suborders, 69 families and about 5.100 species, e.g. *Hirundo* (Swallows,); *Menura* (Lyrebird); *Motacile* (Wagtail); *Pycnonotus* (Bulbul); *Minus* (Mocking bird); *Orthotomus* (Tailer bird); *Luscinia* (Nightingale), *Aethopyga* (Sunbird); *Gracula* (Talking Mynah); *Corvus* (Crow) *Passer domesticus* (House sparrow).

Some interesting birds have been shown is Fig. 1.76 and 1.77.

CLASS MAMMALIA

This class mammalia is represented by about 4000 species of mammals having the following characteristics:

i. The body is covered over by epidermal hairs.

ii. The skin is provided with sweat glands, which help in the regulation of body temperature.

iii. Sebaceous glands are found in the skin which secrete oily secretion to keep the hair smooth; the sebaceous glands are modified in some parts into mammary glands which secrete milk to nourish the young.

iv. The coelmic cavity is divided into thoracic and abdominal cavities by disphragm which helps in respiration.

v. Fleshy lips are present on edges of jaws.

vi. Teeth are thecodont. heterodont, and diphyodont.

vii. The vertebral centrum bears disc-like bone, epiphysis, at each end.

viii. The vertebrae are acoelous: neck has 7 cervical vertebrae.

ix. The half of the lower jaws consists of only one bone dentary which articulates with the squamosal of the skull.

x. The skull bears two occipital condyles.

Fig. 1.76 Birds are unique by having their body covered by feathers and fly by wings. They have conquered the air by making significant modifications in their body form. They are also noted for their less diversity in their structure.

Fig. 1.77 Some interesting birds. **A.** Little blue kingfisher. **B.** Baya with a hanging nest behind. **C.** Openbilled stock. **D.** Barheaded goose. **E.** Grey heron. **F.** Flamingo. **G.** Spoonbill.

xii. The nasal bone bears scroll bone or turbinals.

xiii. Heart 4-chambered. only left aortic arch present.

xiv. The red blood cells are non nucleated.

xv. Eggs small usually retained in uterus for development.

xvi. Placenta is present in Eutherian mammals.

The outline classification of Mammalia is given in Table 1.6 where some extinct orders have been omitted.

TABLE 1.6

Outline Clarification and Mammals

Subclass	*Infraclass*	*Order*	*Suborder*	*Examples*
1. Prototheria	Monotremata			*Ornithorhynchus* *Tachyglossus*
2. *Allotheria		*Multituberculata		**Ctenodon.*
3. Theria	1. *Triconodonta	*Troconodonta		**Triconodon*
	2. *Pantotheria	*Pantotheria		* *Amphitherium*
		*Symmetrodontia		**Eurylambda*
	3. Metatheria	Marsupialia		*Dideiphys; Dasyurus*
	4. Eutheria	1. Insectivora Mole.		Hedgehogs,
		2. Dermoptera		Flying lemurs
		3. Chiroptera	i. Megachiro-ptera	Flying Foxes
			ii. Microchir-optera	Bats.
	Cohort 1: Unguiculata			
		4. Primates	i. Lemuroidea	Lemurs.
			ii. Anthropoidea	Monkeys Man
		5. Edentata		Sloths.
		6. Pholidota		Scaly ant eater.
		7. Lagomorpha		Rabit. Hares
	Cohort 2: Gkures			
		8. Rodentia		Rat
	Cohort 3: Mutica	9. Cetacea	i. *Archaeoceti	*Zeuglodonts
			ii. Odontoceti	Toothed whales,
			iii. Mysteceti	Toothless whales
		10. Carnivora	i. "Creodonta	*Eocene to Mieccae.
			ii. Fissipedia	Cats; dogs, bears
			iii. Pinnipedia	Seals; Sealions.

Cohort 4: Ferungulata	11 Tubulidentata		Aard-vark
	12 Proboscidea		Elephants.
	13. Hyracoidea		*Hyrax*
	14. Sirenia		Seacows or Manatees
	15. Perrissodactyla		Horse
	16. Artiodactyla	i. Suiformes	Pigs.
		ii. Tylopoda	Camels.
		iii. Ruminanta	Ruminants.

Subclass 1. Prototheria. EGG LAYING MAMMALS.

Order Monotremata

Characters: i. Coracoid and precoracoid present.
ii. External ear or pinna absent,
iii. Adults have beak without teeth,
iv. Cloaca present.
v. Testes abdominal; penis present which conducts sperms only.
vi. Oviducts open into cloaca separately.
vii. Uterus and vagina absent.
viii. Mammary gland simple, without nipples.
ix. Females egg laying, e.g. *Ornithorhynchus* (Duch billed platypus. Fig. 1.78.B); *Tachyglossus* (Spiny ant eater. Fig. 1.78. A)

Fig. 1.78 A. *Tachyglossus* (Echidna). **B.** *Ornithorhynchus* (Platypus).

Subclass 2. **Allotheria. Order Multituberculate* includes extinct animals of Upper Jurassic to Eocene, e.g. * *Ctenodon.*

Subclass 3. *Theria* includes Marsupials and plascental mammals.

Characters:
- i. External ear with pinna present.
- ii. Teeth usually present.
- iii. Cloaca usually absent.
- iv. Tests usually in scrotal sacs.
- v. Oviducts differentiated into fallopian tubes and uteri, both opening into distinct vagina.
- vi. Mammary glands with nipples.
- vii. Females oviparous

Infraclass and *Order. *Triconodonta:* includes carnivorous cat-sized mammals of Middle and Upper Jurassic, e.g. *Triconodon.*

Infraclass and *Order *Pantotheria* (Trituberculata) includes Jurassic **Amphitherium.*

Order 2. **Symmetrodontia* includes Upper Jurassic **Eurylambda.*

Infraclass 3. *Metatheria* Order 1. *Marsupialia* includes marsupials or animals with pouches. Characters:

- i. Epipubic bone usually present.
- ii. Female usually provided with marsupial pouch or marsupial folds on abdomen surrounding nipples.
- iii. Uterus and vagina double.
- iv. Placenta usually absent.
- v. Early development of eggs takes place in uterus but premature young ones crawl out to marsupium and attach themselves to mammary nipples until full development. The marsupuim also serves to provide shelter . It includes American Opossums. Didelphis (Fig. 1.79); *Chironectes* (Water Opossum) Family Dasyuridae Australian tiger cat. *Dasyurus;* marsupial wolf, *Thylacimus.* Family Notoryctidao - Marsupial mole. Notoryctus Family *Peramelidae* includes bandicoot e.g. Perameles. Family Caenolestidae includes *Caenolestes* of South America. Family Phalangeridae includes Koala or "Teddy bear" *Phascolarctus* (Fig. 1.80); Cuscus. Phalanger; flying phalanger. *Petaurus.* Family Phascolomidae includes wombat. *Phascolomis* Family Macropodidae includes Kangaroo *Macropus* (Fig. 1.81) rock wallaby, *Petrogale*, hare wallaby, *Lagorchestes;* tree Kangaroo. *Dendrolagus.*

Fig. 1.79 Didelphis

Infraclass 4. *Eutheria* includes placental mammals.

Characters: i. Marsupial pouch or marsupial bone absent,

ii. Vagina single.

iii. Full development of foetus inside the uterus of the female

iv. Placenta present.

Fig. 1.80 Phasolaretus

COHORT 4. UNGUICULATA

This division comprises order in which archaic mammalian characters are retained.

Fig. 1.81 Macropus- Kangaroo

***Order 1. Insectivora.* MOLE. SHREWS**

Characters: i. Small size.

ii. Snout usually long and tapering.

iii. Feet usually five toed and clawed, inner toes not opposable.

iv. Teeth pointed and sharp, e.g. Mole-*Scalopus; Scapanus.* Elephant Shrews—*Macroscelides*; Shrews—*Sorex* Hedgehogs—*Hemiechinus*.

***Order 2. Dermoptera.* FLYING LEMURS.**

i. Four limbs equal and included in a "parachute" with tail e.g. *Galeopithecus* Flying lemur of Southeast Asia.

***Order 3. Chiroptera.* BATS. FLYING FOXES.**

Characters: i. Flying mammals.

ii. Second to fifth digits long and support wing or flight membrane.

iii. Teeth sharp.

Suborder 1 *Megachiroptera* includes Fruit bats or Flying foxes e.g. Pteropus

Suborder 2 *Microchiroptera* includes insectivorous bats e.g. Vampire bats, ***Desmodus;*** *Rhinopoma.*

***Order 4. Primates.* LEMURS, MONKEYS, APES, MAN.**

Characters:
- i. Completely hairy and generally arboreal.
- ii. The pollex or thumb and hallux or toe are shorter than other digits and are opposable for holding except hallux in man.
- iii. Hand and feet often enlarged and provided with 5 distinct digits and flat nails instead of the claws.

Suborder 1 *Lemuroidea (Prosimii)* includes small Primates having head with snout; long tail; fur usually thick and wooly. e.g. Lemur (Lemurs); *Tupia* (Tree shrews): *Nycticebus* (Loris); *Galago* (Bush baby); *Tarsius* (Trasier)

Suborder 2. *Anthropoidea* includes monkeys, apes, man, etc.

Characters:
- i. Cranium enlarged.
- ii. Eyes directed forwards.
- iii. Facial muscles allow emotional expression.
- iv. Posture more or less upright.
- v. Forelimbs usually longer than hind limbs.
- vi. Digits usually with flat nails.
- vii. Often social animals, e.g. New World Monkeys; *Cebus* (Capuschin); *Saimiris* (Squirrel monkey); Ateles (Spinder monkeys); *Callithrix* (Marmosets). Old world monkeys; *Maccaca* (Rhesus monkey); *Cynocephalus* (Baboon) *Mandrillus* (Mandrill); *Semnopithecus* (Langour); *Nasalis* (Proboscis monkey); *Hylobates* (Gibbon); *Pongo* (Orangutan) *Gorilla. Anthropopithecus* (Pan; Chimpanzee) *Home Sapiens* (Man).

***Order 5. Edentata* (Xenarthra). SLOTHS.**

Characters:
- i. Teeth absent or reduced to molar in the anterior part of jaws.
- ii. Toes clawed e.g. giant Anteater of South American. Myrmecoohaga: Armadillos. *Dasypus*; Arboreal Sloths, *Bradipus*.

***Order 6. Pholidota* SCALY ANTEATER.**

Characters:
- i. Body covered over by large overlapping horny plates with few hairs in between.
- ii. Teeth absent, tongue elongated and used for capturing insects, e.g. Pangolin, or scaly anteater *-Manis.*

COHORT 2. CLIRES

This division includes two orders *Lagomoroha* and *Rodentia* formerly included in a single order *Rodentia.* They probably represent early deviations from the primitive insectvore stem.

Order 7. Lagomorpha. RABBITS. HARES

Characters: i. Size moderate or small.
ii. Toes with claws.
iii. Incisors chisel-like. 4 in number; canines absent.
iv. Jaw motion lateral only: palate broad.
v. Tail short, e.g. Hare. *Lepus;* Rabbit. *Oryctolagus.*

Order 8. Rodentia. RODENTS

Characters: i. Usually small animals.
ii. Limbs with five toes and claws.
iii. Incisor chisel - like and 2 in number; canines absent; a gap between incisors and premolars presents called diastemma.
iv. Palate narrow.
v. Jaw motion lateral and anteroposterior.
vi. Tail long. e.g. Squirrel (*Sciurus*); Rat (*Rattus*) Porcupine (*Erethizon*): Guienea pig (*Çavia*)

COHORT 3. MUTICA

This division includes highly specialized and widely divergent mammals with obscure phylogenetic history. They are perfectly adapted to marine life.

Order 9. Cetacea. WHALES. DOLPHINS. PORPOISES.

Characters: i. Aquatic mammals of small to very large size.
ii. Body usually spindle shaped.
iii. Head joined directly to the body; no neck.
iv. Digits embedded; claws and hind limbs absent.
v. Tail long with two broad transverse fleshy lobes.
vi. The notochord lies in the middle line of the tail.
vii. Hairs usually absent excepting on muzzle.
viii. The forelimbs are modified as paddles and digits are enclosed in a common integument.

Suborder 1. *Odontoceti* including toothed whales including sperm whale (*Physeter*) of size upto 60 feet: Indian Dolphin (*Platanista*) of 6 to 9 feet; Common Dolphin (*Delphinus*); Killer whale (*Orcinus*) of about 30 feet.

Suborder 2. Mysticeti includes whales without teeth, e.g. Blue whale (*Balaenoptera*) of 105 feet length, the largest animal in the world; Porpoise (*Phocaena*) of smaller size.

COHORT 4. FERUNGULATA.

This division includes all mammalian groups not dealt with previously.

Order 10. Carnivora.

Characters: i. Usually five toes with claws.

ii. Incisors small; canines strong.

Suborder 1. Fissipedia. DOGS, CATS, BEARS.

Characters: i. Terrestrial animals with separate digits bearing sharp claws.

Family Ursidae-Bear (*Ursus*).

Family Procyonidae-Racoon (*Procyon*); Panda (*Ailurus*).

Family Mustelidae-fur bears. River Otters (*Lutra*).

Family Viverridae-Mangoose (*Herpestes*).

Family Hyaenidae-Hyaenas (*Hyaena*).

Family Felodae-Lion (*Felis leo*); Tiger (*Panthra tigris*)

Felis tigiris Cat (*Felis cattus*); Wild cat (*Lynx rwfus*).

Suborder 2. Pinnipedia includes Seals and Sea Lions.

Character: i. Body spindle shaped and limbs modified into paddles or flippers.

Family Otaridae-eared seals. *Callorhinus.*

Family Odobenidae- Walruses. *Odobenus*

Family Phocidae-Earless seals. *Phoca*

Order 11. Tubulidentata AARD VARKS.

Characters: i. Body pig-like, stout, and scanty haired.

ii. Ear and snout long.

iii. Tubular mouth with slender, protrusible tongue.

iv. Milk teeth many, few permanent teeth; canines and incisors absent.

v. 4-5 toes with heavy claws, e.g. Earthpig Aard-Vark (*Oryctyropus*).

Order 12. Proboscidea ELEPHANTS.

Characters: i. Huge, massive animals with nose and upper lip produced into a long, flexible, muscular proboscis.

ii. Legs piller-like, neck short, skin thick,

iii. Two upper incisors elongated as tusks.

iv. Feet club-shaped with 5, 3 or 4 hoofed toes. e.g. Indian Elephant; *Elephas maximus indica* African Elephant, *Laxodonta africana.*

Order 13. Hyracoidea CONEYS.

Characters: i. Small Animals.

ii. Fore limb has 4 toes and hind limb 3.

iii. Digits, excepting 2nd which is clawed, have small hoofs.

iv. Resemble guinea pigs superficially but related to hoofed animals, e.g. *Pròcavia* (Hyrax): *Dendrohyrax.*

Order 14. Sirenia MANATEES.

Characters: i. Large, spindle shaped body.

ii. Paddle-like forelimbs; hindlimbs absent.

iii. Tail without notochord and has lateral flukes.

iv. Mouth small and lips fleshy.

v. External cars absent.

vi. Hairs few and scattered.

vii. Herbivores e.g. *Dugong (Helicore dugong).*

Order 15. Perrissodactyla ODD-TOED HOOFED ANIMALS

Characters: i. Hoofed animals, large size. long legs.

ii. Foot with odd number of toes each with cornified hoof.

Family Equidae-Horses (*Equs calallus*): Asses (*Equs asinus*) zebra (*Equs zebra)*

Family Tapiridae-Tapirs (*Tapirus*).

Family Rhinocerotidae-Rhinoceros (*Rhinoceros unicornis*)

Order 16. Artiodactyla EVEN-TOED HOOFED ANIMALS.

Characters: i. Size varying usually long legs.

ii. Two or rarely four toes in each foot with cornified hoof.

iii. Antlers or horns usually present.

iv. Dentition reduced excepting in pigs.

Suborder 1. Suiformes (Bunodontia). Includes pigs.

Characters: i. Horns absent: canines enlarged into curved tusks e.g. Pigs (*Sus scrofa):* European wild Boar (*Phacochoerus*); Wart hog of Africa; *Hippopotomus* (*Hipopotamus amphibious*)

Suborder 2. Tylopoda includes Camels.

Characters: i. Soft and broad feet, without hoofs, e.g. Camel, Camelus lama of South America. *Auchenia*

Suborder 3. Ruminantia includes Ruminants having hoofs.

Family Bovidae includes about, 50 species of Hollow-horned Ruminants e.g., Oxen (*Bos*); Goat (*Capra*); Sheep (*Ovis*): Black buck (*Antelope cervicapra*): Neelgai or Blue bull (*Boselephus*).

CHAPTER SUMMARY

1. The chordates are organisms with notochord, dorsal hollow nerve cord and pharyngeal gill slits. A more acceptable theory for the origin of chordates is from the group Deuterostomia, which comprises of echinoderms. pogonophores and chordates, where the anus develops from the blastopore and the mouth is formed a new. The deuterostomes have evolved from sessile or semi-sessile ancestors having bilateral symmetry and tripartite body with coelom.
2. The protochordata or lower chordata or invertebrate chordata comprise of these subphyla Hemichordata, Urochordata and Cephalochordata.
3. The phytogeny of protochordata is controversial. The most accepted view is that protochordata represent specialized offshoot stages along the main line of evolution of the vertebrate from a shared ancestor with echinoderms.
4. An understanding of the functional organization is fundamental to the understanding of the organism and its adaptations to the changing environment.
5. The comparative anatomy seeks to explain variations in structure found in the bodies of animals with a view of tracing their biological relations.
6. Homology refers to similarities in the type of structure between parts or organs of different animals. Analogy refers corresponding function of a structure in similar or different organs or parts not homologous to each other.
7. The concept of ontogeny has been useful in establishing homologies and making certain phylogenetic relationships. The vertebrate morphology provides insight into how organisms work and how they are evolved.
8. The Vertebrates have either evolved form some nonchordata source or through some invertebrate chordata. The protochordata form a sort of connecting link between the vertebrate and rest of the deuterstomes.
9. The time of origin of vertebrate is not later than the beginning of Ordovician period. The likely place of their origin is sea form where they migrated into the river system during early phase of vertebrate evolution.
10. Phylum Chordata is divisible into two groups Acrania and Craniata. The Acrania include two sulphyla Tunicata and Cephalochordata. The Craniata includes two subphyla Agnatha and Gnathostomata.

11. Outline classification of Phylum Chordata is given below:

TABLE 1.7
Outline Classification of Chordata

Group	*Subphyla*	*Classes*	*Examples*
I. Acrania	1. Tunicata (Urochordata)	1. Larvacea	*Appendicularia*
		2. Ascidiacea	*Ascidians*
		3. Thalacea	*Salpa Doliolum*
	2. Cephalochordata	1. Leptocardii	Lancelots, fishes
II. Craniata	1. Agnatha		
	2. Cyclostomata	Petromyzon.	Myxine
	2. Gnathostomata		
	Superclass Pisces	1. Placodermi	Extinct *Climatius*
		2. Chondrichthyes	Sharks. Rays
		3. Osteichthyes	Bony fishes
	Superclass Tetrapoda	1. Amphibia	Salamanders. Frogs, Caecilians
		2. Reptilia	Lizzards, Snakes. Turtles, Crocodile
		3. Aves	Birds
		4. Mammalia	Mammals

2

The Integument

2.1. GENERAL NATURE OF THE INTEGUMENT

The integument comprising of skin and associated structures forms external covering of the animal. The skin may be regarded as a boundary tissue that sets the individual apart as a separate entity among other organisms and in the environment. It is one of the largest of the organs making up some 16% of the body weight in mammals and upto 10.7% in fishes. Being in continuous contact with the environment, the integument performs many important functions, chiefly of protective nature and produces different end products. The skin is in continuation with the mucous membrane lining of mouth, rectum and openings of urinogenital ducts.

The vertebrate skin has two layers, an outer *epidermis* and an inner *dermis* or *chorium*. The epidermis is derived from the ectoderm and dermis from the mesoderm. The epidermis consists of two layers in all vertebrates excepting mammals where it has four layers of cells. The innermost layer of epidermis is *stratum malpighii*, which consists of a single layer of cells which keep on dividing and providing new cells to upper layers. The cells of this layer move upwards and get accumulation of keratin and replace the lost cells of upper layers. The outer layer, *stratum corneum* consists of several layers of hard, flattened and dead cells made mostly by the keratin.

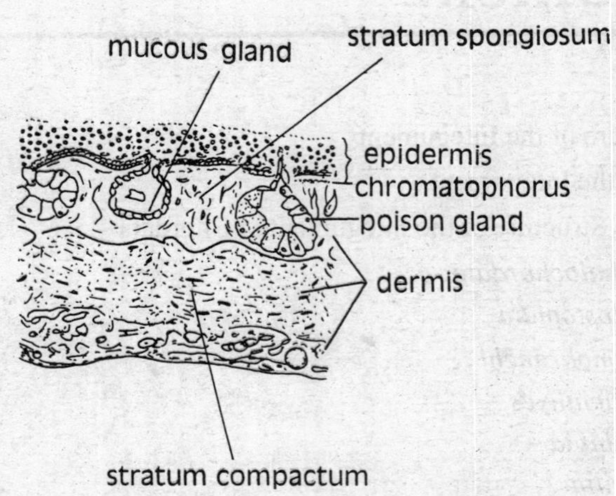

Fig. 2.1 Cross section of a generalized skin

In mammals two more layers are distinguishable in epidermis called *stratum granulosum and stratum lucidum.* In Amphibia and Reptilia the *stratum cornenum* is shed off periodically (ecdysis); in birds and mammals, feathers and furs etc. are seasonally changed (moulting). In fishes the horny epidermal layer is usually absent as their skin is usually protected by scales or *mucous glands* or both.

The *dermis* lies below the epidermis. It has an outer loose layer *stratum laxum* and an inner dense *stratum compactum.* It is made up of dense connective tissue having muscles, blood vessels, lymph vessels, collagen fibres and nerves. The dermis allows the diffusion of substances from dermis to the epidermis.

2.2. FUNCTIONS OF THE INTEGUMENT

The integument has several important functions including protection, storage of food, temperature control, secretion, water and electrolyte balance, exteroreception, respiration, locomotion, sexual selection and skeleton formation.

Protection: The exoskelstal elements produced by the integument including armour of scales and plates and skin proper protect soft parts of the body from foreign particles like dust, bacteria, excessive loss of moisture, mechanical injury and injurious contact. In some animals the skin shows protective colouration. Some skin glands produce bitter substances, which tend to ward off enemies. Presence of claws, nails, spines, barbs, needles etc. give additional protection to animals.

Food Storage: In the deeper subcutaneous layers of the skin reserve food is stored as fat to be drawn upon in times of need. In whales and seals the subcutaneous fat forms thick layer called *blubber,* which is used as food reserve and also as nonconductor.

Temperature Control: The heat produced by the oxidation of tissues is distributed evenly by the circulation of blood within the body. The heat is lost from the animal body through expired breath, excrete and urine and from skin surface. The skin effects the regulation of the loss of heat in two ways, *physiological* and *physical*. The *physiological regulation* is brought about by the expansion or relaxation and contraction of the skin and walls of the capillaries contained therein, when exposed to heat and cold. *Physical regulation* is accomplished by the evaporation of sweat which is constantly being excreted from the mammalian skin. Hairs and feathers provide insulation against cold or warm climates. The *blubber* of whale, film of oil in mammals, feathers of birds etc. also help in checking the loss of heat.

Secretion and Nourishment of young Ones: The skin acts as an organ of secretion. In aquatic animals mucus is secreted for protection. Sebaceous glands of mammals secret oil for the lubrication of skin and hairs. The mammary glands of mammals produce milk for the nourishment of young ones. The uropygial glands secrete oil for preening the feathers in birds.

Water and Electrolyte Balance: Sweat glands of mammals share the function of elimination of salts and water with kidneys. The gills of marine fishes have chloride secreting cells.

Exteroreception: Different type of sensory tactile cells and sensory corpuscles in the skin serve as tactile organs for the detection of temperature variation, external environment, pressure and pain.

Respiration: In amphibians the respiratory function of the moist skin which serves for the exchange of gases is more important than that of lungs. The gills of aquatic animals serve for respiration.

Locomotion: Some integumental derivatives such as fins of fishes, webs of skin in the feet of frog and aquatic birds, feathers of the wings, tails of birds and extension of integument to form wings in flyng lizards, flying squirrels and bats help in locomotion of animals.

Skeleton Formation: The integument participates in the formation of the skeleton by forming dermal bones of vertebrates. It also forms some parts of the teeth.

2.3. COMPARATIVE STRUCTURE OF THE INTEGUMENT IN VERTEBRATES

2.3.1. Cephalochordata

In *Branshiostoma* the epidermis consists of a single layer of columnar cells which are ciliated in the larval stage and later produce a thin, noncellular cuticle. The dermis is thin layer of soft connective tissue overlying the musculature. The skin is transparent without glands and pigments. (Fig. 2.2)

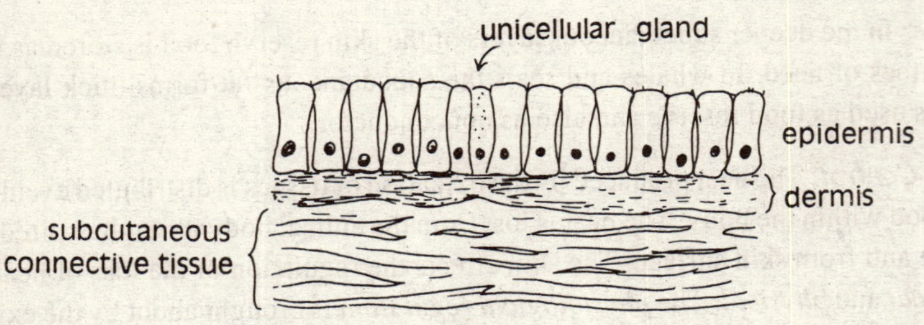

Fig. 2.2 V.S. Integmment of Branchiostoma

2.3.2. *Cyclostomata*

The integument of lampreys and hagfishes is highly glandular and slippery. The epidermis is many layered and outermost cells retain their cytoplasmic character and secrete a cuticle over their exposed surface. A large number of gland cells of varying shapes and size are scattered among the epidermal cells Fig. 2.3. The corneal cells are not formed. The dermis is thinner than epidermis and contains interwoven network of vertical and horizontal connective tissue fibres practically undifferentiated into strata. The chrompatophores are present in the dermis. The skin is without scales in living froms but ancestral forms contained scales. The fleshy tongue has horny teeth which are corneal modifications of the epidermis and renewed periodically.

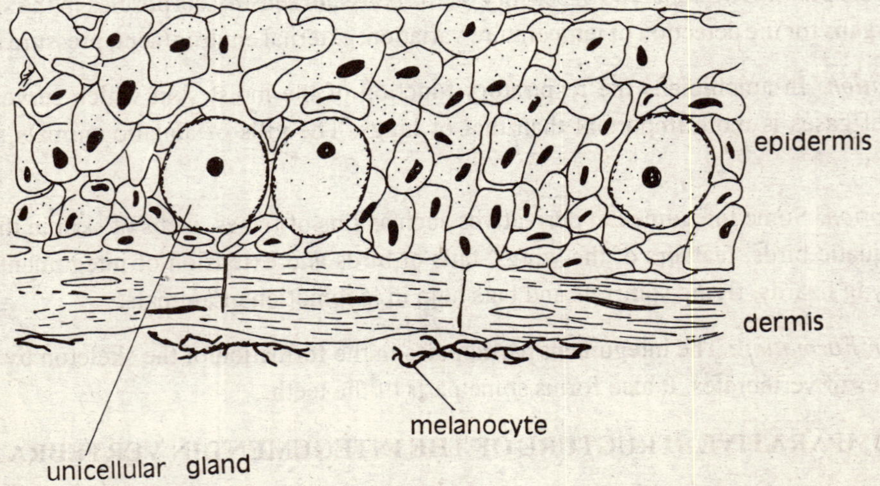

Fig. 2.3 Skin of a larval lamprey

2.3.3. Elasmobranchi

In Elasmobranchi the skin is tough in texture and rough to touch due to the presence of minute placoid scales which resemble the structure of a tooth in all important characteristics. The placoid scales are embedded in skill as broad base of bone-like substance and have projecting spine consisting of dentine covered by enamel layer. The enamel is product of epidermis and dentine of dermis. The points of placoid scales extend out through the epidermis. The dermis has three layers called *stratum laxum. stratum compactum* and a *subcutaneous layer* of fine fibres. The outermost layer *stratum laxum* around the neck of scales is free from fibres. The *stratum compactum* is laminated and fibrous layer. (Fig. 2.4)

Development of placoid scales: The placoid scales develop by the accumulation of mesenchyme cells in the dermis forming dermal papillae. The papilla increase in size and projects outwards pushing the malpighian layer of the epidermis outwards. The cells of the outer margin of the papilla develop odontoblasts which secrete a hard substance resembling dentine of the higher animals. The cells of malpighian layer overlying dentine form *enamel-organ* made of columnar cells called *ameloblasts* which produce hard enamel-like coating over the outer portion of the conical mass of dentine. As the scale grows in size it pushes the epidermis and projects above the surface of the skin as palcoid scale. (Fig. 2.5).

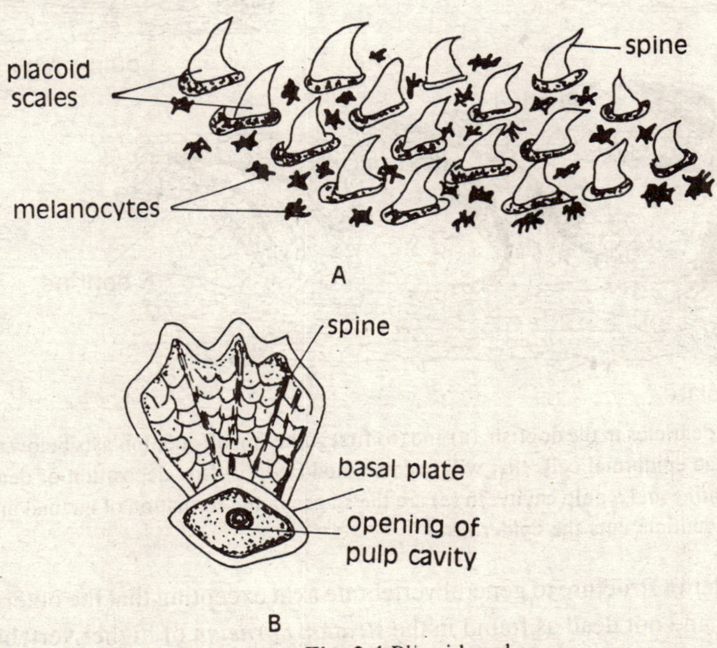

Fig. 2.4 Placoid scales

2.3.4. Osteichthyes

The skin in bony fishes is rich in mucous glands which make protective covering preventing entrance of foreign materials and growth of fungi and make the skin smooth so as to reduce the friction in swimming. The skin of bony fishes differs from that of cartilaginous fishes in the possession of bony scales deeply buried in it. The scales are dermal in origin in most fishes.

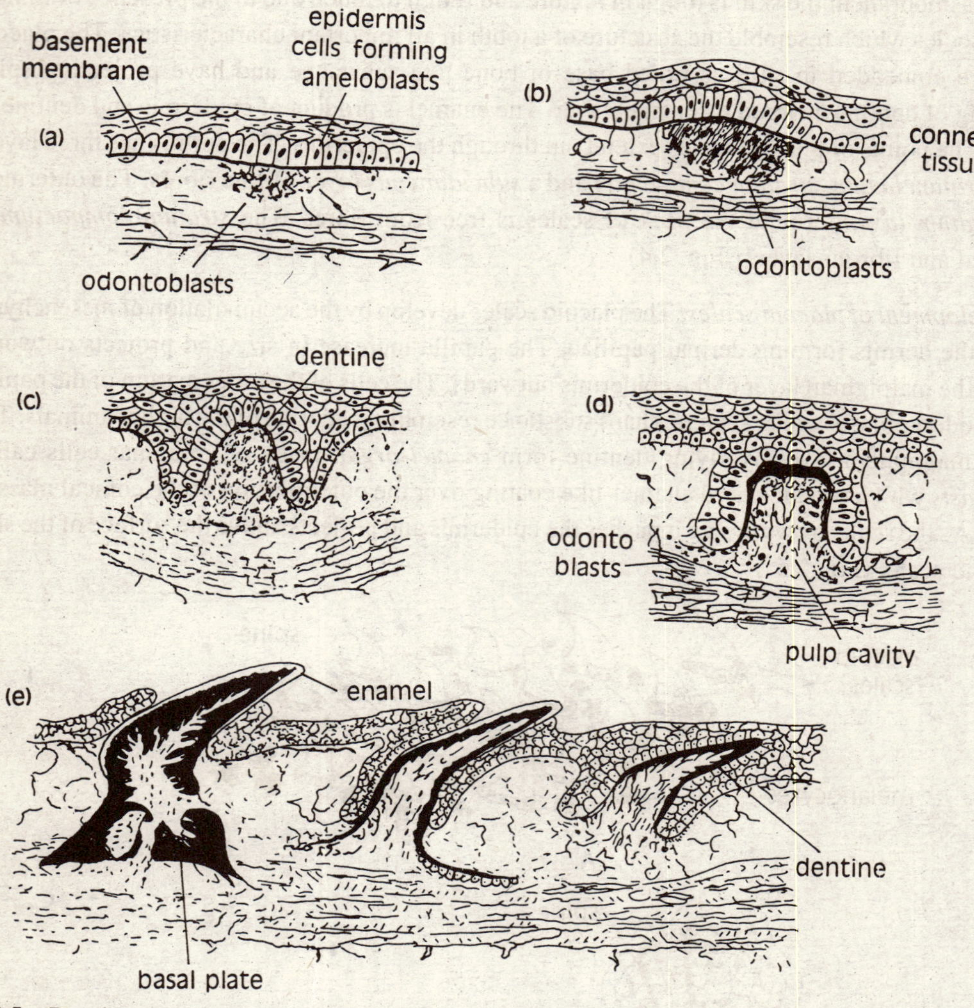

Fig. 2.5 Development of denticles in the dogfish, (a) and (b) first gathering of odontoblasts below the basment membrane and the epidermal cells that will become modified, (c) first deposition of dentine. In (d) there is more dentine and a pulp cavity. In (e) are the stages in the formation of enamel and the basal plate, while the denticle cuts the epidermis.

The fish skin is similar in structure to general vertebrate skin excepting that the outermost cells are nucleated and living and not dead as found in the *stratum corneum* of higher vertebrates. The *stratum malpighii* keeps on producing outer cells with some keratin. The *epidermis* has goblet vessels, nerve, lymph vessels, smooth muscles and collagen fibres with some stratification. The connective tissue fibres run almost parallel to the surface. (Fig. 2.6. A).

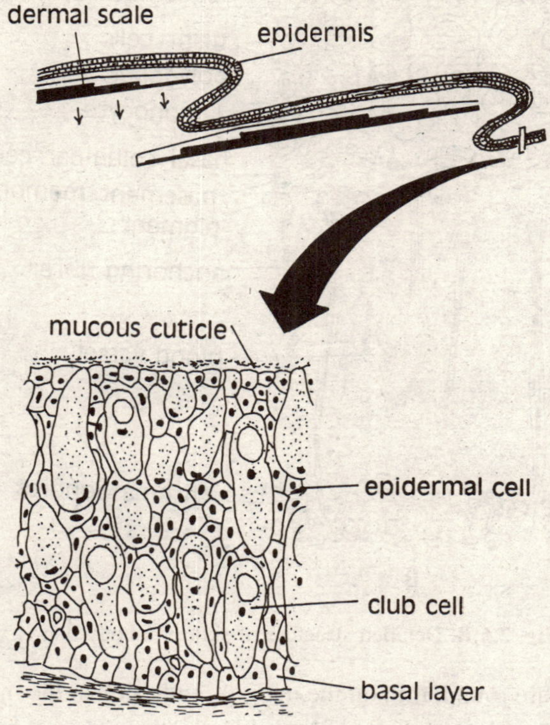

Fig. 2.6 **A.** Bony fish skin, (a) Dermal scales in the skin of a teleost fish (arrows indicate the growth), (b) Enlargement of epidermis.

In some catfishes scales are absent and epidermis consists of many layers of cells distinguishable into several morphologically distinct cells types. These are *basal columnar cells, giants, cells, mucous cells, polygonal cells, squamous cells and lymphocytes.* (Fig. 2.6 B.).

The fish scales have been classified into *ganoid, cycloid, ctenoid* and *cosmoid* types. *Placoid seales* are found in elasmobranches only.

Ganoid scales: These scales are found in Acanthodians and Actinopterygii. These are thick, rhomboid, non overlapping scales with little or no spongy bone between the basal and superficial layers. The outer layer is of hard shiny material ganoin with few or no pulp cavities. Two types of ganoid scales, the *palaeoniscoid* and *lepidoteoid* type are found in *Polypterus* and *Lepidosteus* respectively. *Lepidosteoid scales* do not contain cosmine layer. (Fig. 2.7)

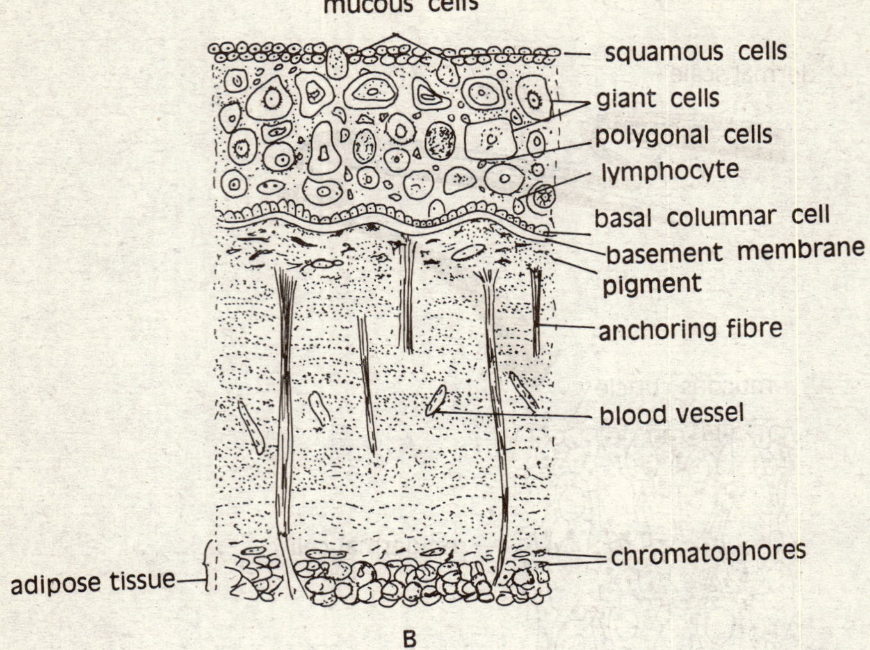

Fig. 2.6 B. Detailed structure of skin of a bony fish

Cycloid scales: These are round, thin on the margins and thick in the middle with lower layer of fibrous connective tissue and upper layer of bone like dentine. They exhibit concentric lines of growth indicating the age of the fish. The scales remain embedded in the dermis by anterior parts and posterior part of each scale overlaps the anterior part of the scale behind. The exposed part of the scale has spiny or rough margin. Cycloid scales are found in teleosts.

Ctenoid scales: These scales have structure and concentric lines like cycloid scales but differ in the possession of teeth or *cteni* on their free posterior side (Fig. 2.7).

The *cycloid* and *ctenoid scales* form dermal exoskeleton of most of the bony fishes. Scales covering lateral line cannals are perforated by a vertical tube of the lateral line opening. In some flat fishes both cycloid and ctenoid scales are found.

Cosmoid scales: These scales were present in the primitive members of the subclass sarcopterygii. They are not found in any living form today. The cosmoid scale consists of four layers. The outermost layer is thin made of hard substance *vitrodentine* followed by that of dentine-like *cosmine layer* with many pulp cavities. The third layer is *vascular* and *bony* and the lowermost *isopedine layer* consists of several laminae of bone tranversed by vascular canals.

The Relationship of Scales with Age of Fish

The cycloid and ctenoid scales are helpful in calculating the age of fishes. These scales grow alongwith growth of the fish. The scales appear first at a very young stage. During summer when nutrients are abundant the fish grows very fast and the scales also grow alongwith the rapid growth of the fish. In winter the growth rate is slow and the scales grow correspondingly. The difference between the growth of scales during summer and winter is marked distinctly by concentric *annual rings* on the scale. By counting the lines of growth or annual rings the age of the fish can be determined with considerable accuracy.

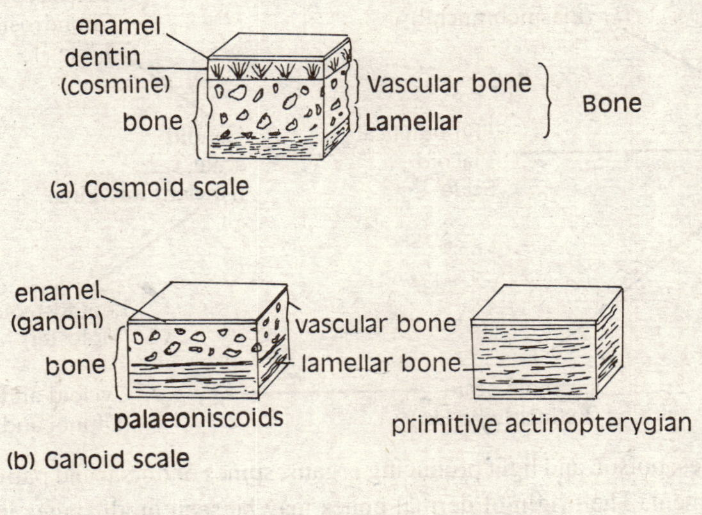

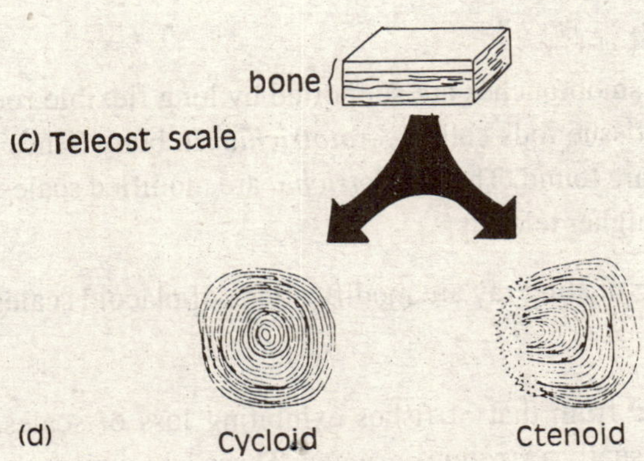

Fig. 2.7 Scale types of long fishes. (a) cross section of a cosmoid scale, (b) a ganoid scale and (c) a teleost scale. (d) Surface views of the two types of teleost scale, cycloid and ctenoid scales.

Evolution of fish scales: The fish scales have probably originated from the ancient bony dermal armour of ostracoderms and placoderms. The cosmoid scales found in extinct Choanichthyes exhibited small bony plates. The ganoid scales appear to be structurally modified ctenoid scales with a hard substance *ganoin.* The palaeoniscoid type of ganoid scales lack spongy bone found in cosmoid scales, while in lepidosteoid type one more layer is absent. Cosmoid and ganoid type of scales existed together in lobe-finned fishes and ray finned fishes respectively. The cycloid and ctenoid scales probably derived from ganoid type. The dentine layer is thin without covering of ganoin in these scales. The placoid scales have been regarded as equivalent of dentine part of cosmoid scales. The dentine is covered over by a hard and glossy material referred as enamel. The evolutionary trends in fish scales may be summerised as follows:

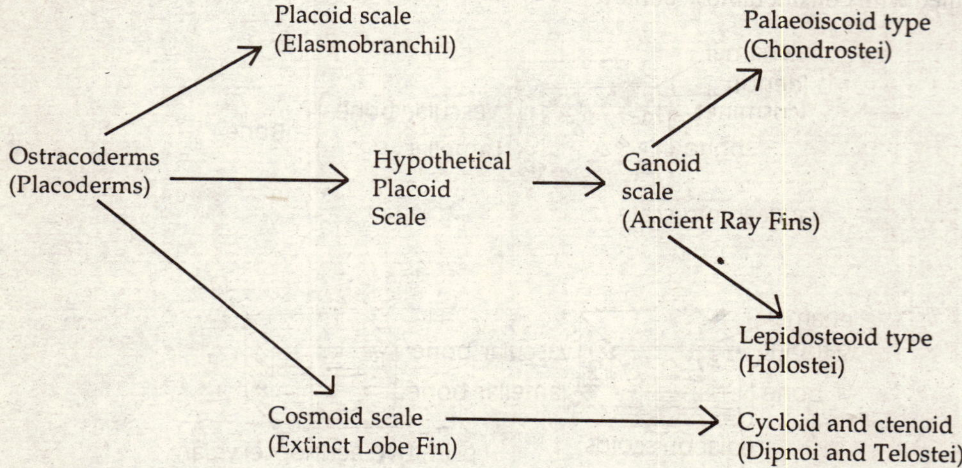

In addition to scales, poison and light producing organs, spines of dorsal and paired fins may be present in the integument. The origin of dermal bones may be seen in all stages in lower fishes where dermal plates are added to the chondrocranium and to other parts of the skeleton and thus function in the building of the bones.

Dermal fin rays: The fins of elasmobranches are supported by long flexible rod-like fin rays having hair like fibrous connective tissue rods called *ceratotrichia.* In bony fishes branched and jointed fin rays called *lepidotrichia* are found. The *lepidotrichia* are modified scales joined end to end and get modified into spines in higher teleosts.

Fin spines of spiny dogfish, sting of sling ray are modifications of placoid scales.

2.3.5. Amphibia

The amphibian skin has evolved from that of fishes exhibiting loss of scales, presence of multicellular epidermal glands and usually a *stratum corneum.* These integumental modifications are adaptations in relation to semiterrestrial existence. The amphibian skin lacks dermal epidermal exoskeleton excepting in Gymnophiona and some extinct forms where dermal scales are present. The dermal scales of caecilians lie between ring like folds of skin with several scales confined to one pocket. The skin of Ladyrinthodonts, the first amphibians was heavily covered over by scales. The amphibian skin is thin with large lymph spaces separating it from the muscle layer lying below. In aquatic forms the skin is richly glandular with mucous secretion. In terrestrial forms it is less glandular and hardened to prevent water loss.

The skin of frog has two distinct regions *epidermis* and *dermis.* (Fig. 2.8). The epidermis consists of *stratum corneum* and *stratum malpighii.* The *stratum corneum* consists of flattened and highly keratinized dead cells, which protect excessive loss of moisture. The *mucous glands* are epidermal but invade the *dermis*. Another type of epidermal glands are *serous glands* which secrete acrid or toxic substance to ward off the enemies. The stratum corneum is absent in water dweling urodeles but well developed in landforms such a toads. The dermis is a relatively thin layer consisting of upper loose *stratum spongiosum* and lower *stratum compactum.* Connective tissue, muscle fibres, nerve endings and blood vessels are found in dermis. The upper part of the dermis contains *chromatophores* having black *melanophores* and yellow *lophophores* which produce colour of the skin. The colour change in frog is slow and the chromatophores are under the control of pituitary gland and not under the direct influence of the nervous system. The frog takes up dark colour in moist environment and yellow colour in dry environment. The amphibian skin serves as an important organ of respiration.

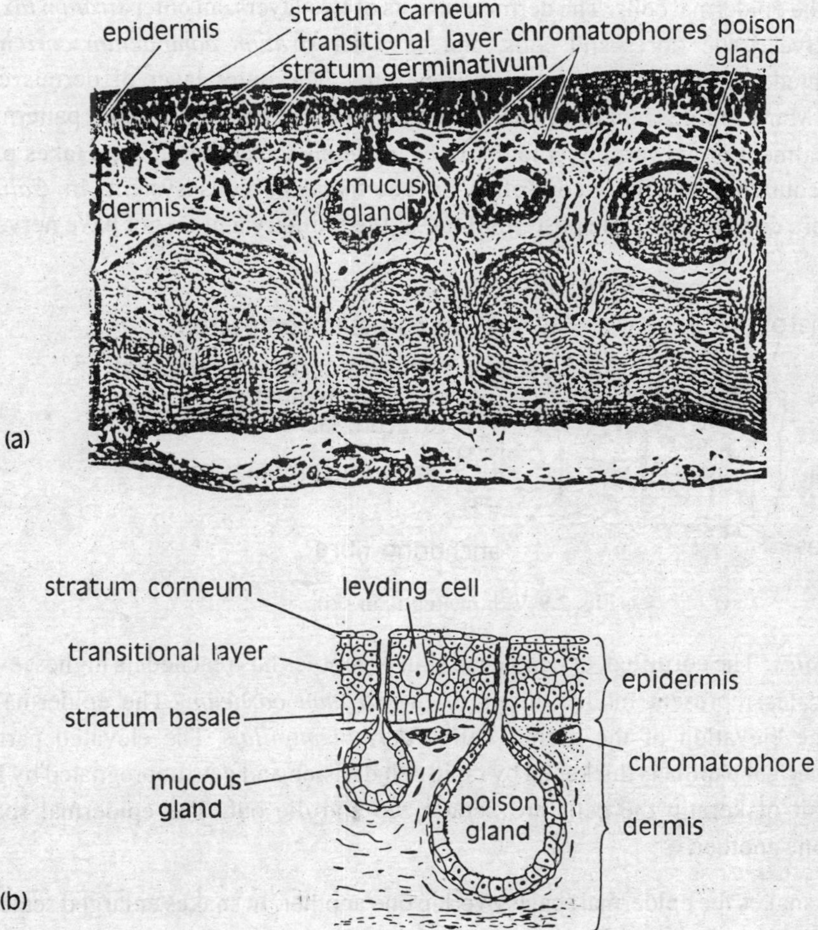

Fig. 2.8 Amphibian skin (a) Section through an adult frog skin, (b) Diagrammatic view of amphibian skin showing ducts of mucous and Poisonous glands.

2.3.6. Reptilia

Reptiles are usually adapted to a terrestrial life. The chief integumental adaptations of reptiles are presence of a thick stratum corneum, evolution of modifications such as claws, horns, spikes, rattles, beaks and epidermal scales and a general scarcity of skin glands. Most of the modern reptiles are equipped with exoskeleton of horny epidermal scales or in some cases with bony plates. Unlike fishes and amphibians the integument of reptiles is dry to prevent water loss. The scales are horny epidermal structures and often form spines. In tortoise, crocodiles and some lizards (*Heloderma*) dermal bony plates are found below the epidermal scales.

The *epidermis* has three distinct layers including upper *stratum* corneum, middle *stratum intermedium* and lower *stratum malpighii* (Fig. 2.9). The stratum corneum is further divisible into outer *stratum. corneum compactum,* a noncellular layer and inner cellular *stratum corneum relaxatum.* The *stratum malpighii* consists of many layers of cells. Some hair-like processes take their origin form the epidermal cells. The dermis consists of two layers, an outer *stratum laxum cori* made of connective tissue fibres and cells, and an inner *stratum compactum cori* made of horizontally directed bundles of connective tissue fibres. The outer layer of dermis contains *chromatophores*. Many reptiles including snakes and lizards have elaborate colour patterns which may be for concealment or as warning colours. In chameleons the colour change takes place for concealment or courtship which is controlled by automatic nervous system. In *Calotes* the chromatophores are controlled by the posterior lobe of the pituitary and do not have nerve fibres.

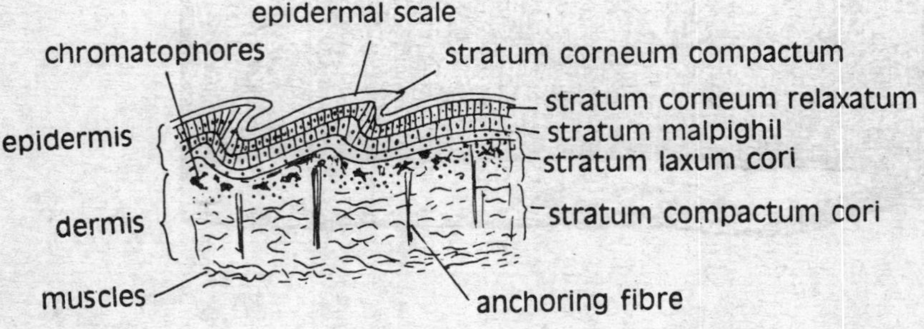

Fig. 2.9 V.S. of Reptilian skin

Epidermal Scales: The cornified, dry and dead skin of reptiles has reached its highest evolution and epidermal scales represent modifications of the *stratum corneum.* The epidermal scales develop above the elevation of the dermis called *dermal papillae.* The elevated part of the epidermis above dermal papilla is thickened by rapid cell division and gets impregnated by Keratin. Due to the deposit of keratin the cells are keratinized and die out. The epidermal scales are continuous with one another.

In lizards and snakes the epidermal scales overlap one another. In snakes enlarged scales on the head form head shields and on the anal region anal shields. The horns of *Phrynosoma* (horned toad) are modifications of the stratum corneum. Snakes get friction, necessary for locomotion through, transversely arranged scales. The rattle of rattle snake is made of a series of dried, old scales attached to one another. The stratum corneum of lizards and snakes is periodically discarded and left behind attached to a rock or log as cast.

In turtles the epidermal scales, usually called *scutes,* cover the *plastron* and *carapace* and each scale develops separately. The scutes lie in groups and have definite names. The arrangement of scutes does not conform with that of underlying dermal plates. The limbs, tall and neck of turtle have smaller epidermal scales.

In crocodiles the epidermal scales cover the entire body and form continuous sheet of *stratum corneum.* They are large rectangular on the trunk and tail and smaller on limbs. Moulting of entire stratum corneum does not take place at one time and patches of a few scales are occasionally sloughed.

The *claws* and *beaks* of reptiles are modification of the stratum corneum at the tip of the digit. The beaks of turtles are also modifications of the stratum corneum.

Dermal Scales: Reptiles retain traces of bony dermal scales inherited from their ancestors. The dermal armour of many extinct reptiles is large and living reptiles still possess its remnants.

In crocodiles and alligators the dermis is thick and contains many plates embedded in it on dorsal side and on the throat. Small dermal plates are also present in some snakes and reptiles. The dermal "ribs" or gastralia of crocodiles are located in the ventral abdominal region. Gastralia are also found in *Sphenodon.* They are not homologous to true ribs.

In turtles the dermal scales are highly developed. The bony dermal plates get fused intimately with the endoskeleton. The *carapace* is composed of bony costal plates fused to the neural arches of vertebrae and laterelly to the ribs present below. The *plastron* is made up of nine dermal plates covered over by epidermal scales in most of the turtles. The median row contains one nuchal, eight vertebral and two precaudal plates. Lateral rows of plates include eight costal plates, each fused with a rib. The peripheral ring of plates is made by marginal plates and the last marginal plate called pygal, is unpaired. The *plastron* consists of one pair each of *epiplastrons, hyaoplastrons. hypoplastrons* and *xiphiplastrons* and an unpaired *entoplastron.*

2.3.7. Aves

The integument in birds is thin, loose, dry and almost without glands. There is a *uropygial gland* at the base of the tail; its oil is used for preening the feathers. The epidermis and dermis are delicate membranes loosely attached to underlying muscles excepting on head and neck where they are intimately attached and thick. The epidermis is delicate structure excepting in areas not covered by feathers like foot or beak. It is made of *stratum corneum* and *stratum malpighii.* The skin glands are few in number and *stratum corneum* is modified into scales, claws, horny covering of the beak, spurs and comb. The *stratum corneum* consists of loosely packed cells covered over by horny layer in the beak and foot. The *stratum malpighii* consists of two rows of cells usually and sometimes ine row only. The epidermal cells have pigment granules. The *dermis* consists of connective tissue cells and fibres and smooth muscle fibres. The dermis may have pigment cells in its upper part below the epidermis. (Fig 2.10)

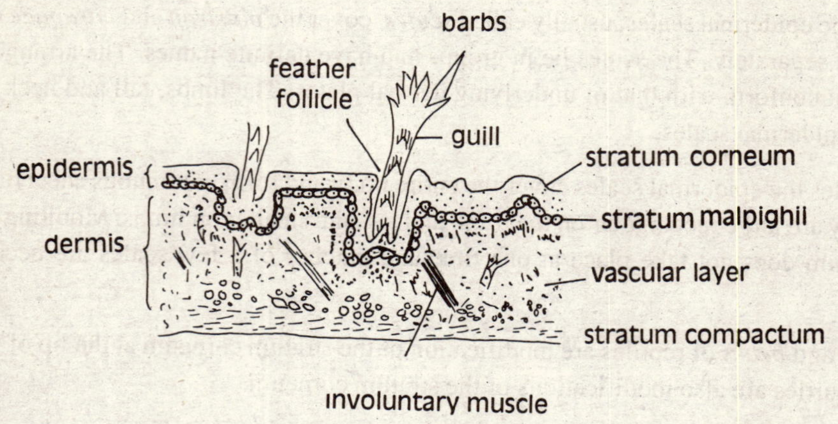

Fig. 2.10 Avian skin in V.S.

The skin of birds is characterized by the presence of feathers formed by the stratum corneum. The feathers are of these types including *contours, down feathers* and *filoplumes.*

Contour feathers or *pennae* are found all over the body and provide the main bulk of the body covering and wing plumage. They are of two types: (1) *Quill feathers* or *flight feathers* on wings called *remiges* and on tail called *rectrices:* (2) *Coverts* which are smaller in size and cover the wings. These are smaller contour features and their lower barbules lack hooklets.

A *contour feather* has three main parts called *calamus, rachis* and *aftershaft.* The *calamus* or the *basal part* is embedded in the skin. It is hollow and has a *pith* which opens below by an *inferior umbilicus* and above by a *superior umbilicus*. The *rachis* is the stalk above the calamus with an umbilical groove on its under side. The *rachis* has an expanded *vane* or *vaxillum* made of several parallel barbs on both sides of the rachis. Each barb is thin, plate-like with distal barbules on one side with many curved hooks and each barb of the other side has proximal barbule having grooved edges. Hooks of distal barbules hook over the proximal barbules over the grooved edges. In this way the barbs are interlocked together so that the entire vane acts as one flat piece for resisting the air. The *aftershaft* or *hyporachis* is situated between calamus and rachis and has tufts of barbs and barbules. Down feathers have very small quill with barbs and barbules at their tips without hooks. They remain concealed in contour feathers and make a dense layer so that loss of body heat is prevented. In very young birds the down feathers cover the body and contour feathers are lacking. *Filoplumes* are hair-like delicate feathers having long and slender shaft and a vestigial vane consisting of a few loose terminal barbs without hooks. They lie among contour feathers. (Fig 2.11)

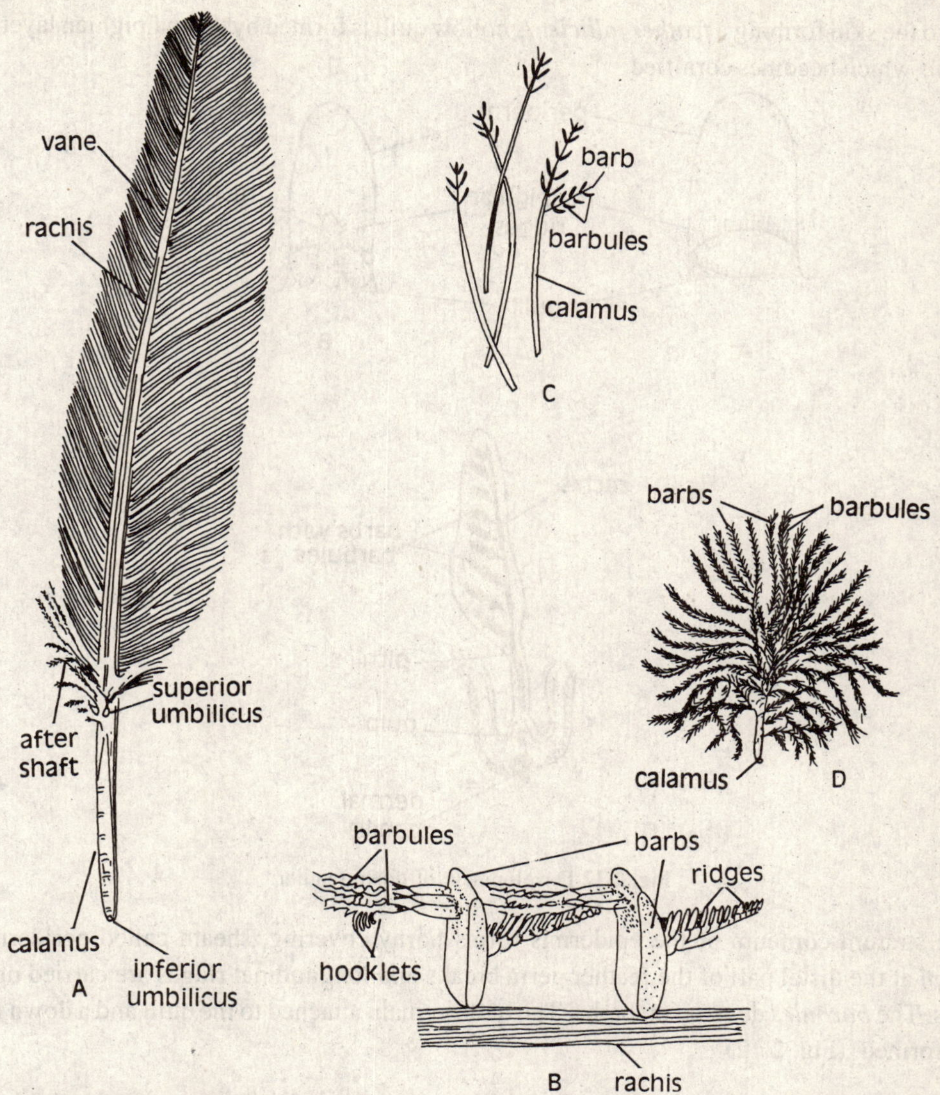

Fig. 2.11 Feathers in birds **A.** Typical flight feather. **B.** Sectional view of two barbs showing the disposition of the barbules and the locking mechanism. **C.** Filoplume. **D.** Down feather.

The nestlings down feathers are produced first and are replaced by adult feathers. The adult feathers are moulted once or twice every year. In pigeon there are 23 wing feathers or remiges and 12 rectrices which act as rudder. Down feathers are not found in pigeon.

Development of down feather: The development starts as a *germ papilla* formed by the mesodermal tissue of the dermis including blood vessels. The epidermis above the papilla becomes cone shaped. The feather papilla is nutritive and called *feather pulp.* The feather pulp and overlying epidermis elongate into a cylinder called *feather germ.* The epidermis surrounding the feather germ sinks into the skin forming a *feather follicle.* A hollow quill is formed by the malpighian layer of the epidermis which becomes cornified.

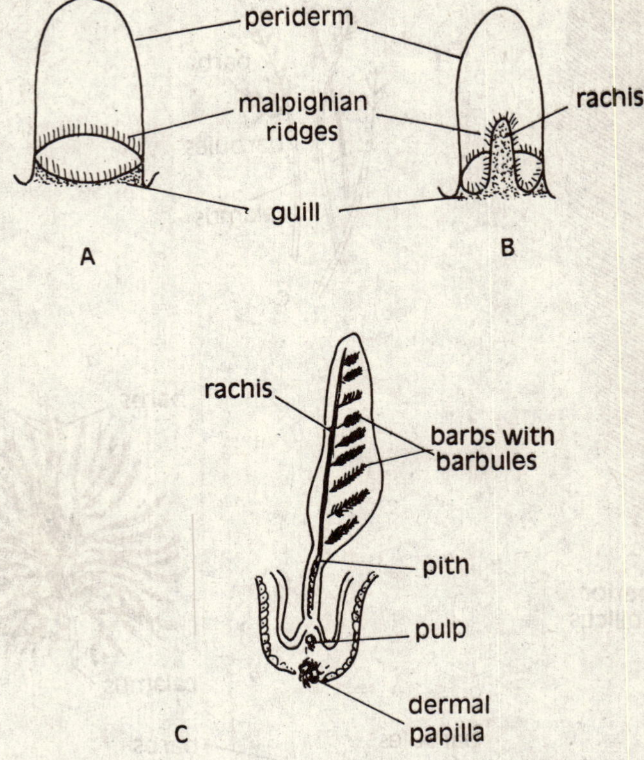

Fig. 2.12 Development of down feather

The stratum corneum of the epidermis forms horny covering, sheath called periderm. The *periderm* at the distal part of the feather germ breaks and longitudinal ridges are carried upwards as barbs. The *barbules* develop on barbs. The barbs remain attached to the quill and a down feather is thus formed. (Fig. 2.12)

Development of contour feather: The early development of the contour feather is similar to that of down feather. After the formation of periderm and longitudinal ridges the two middorsal malpighian ridges thicken and fuse together forming a solid *rachis.* The rachis grows outwards and remaining longitudinal ridges migrate on the lateral sides of the rachis to become barbs. New barbs appear below and barbules arise from the barbs. The periderm dries and splits along the midventral line. The barbs form the vane. The quill is converted into calamus and its feather pulp dries to form a pith. The original proximal and distal openings of the quill become inferior and superior umbilicus respectively. (Fig. 2.13)

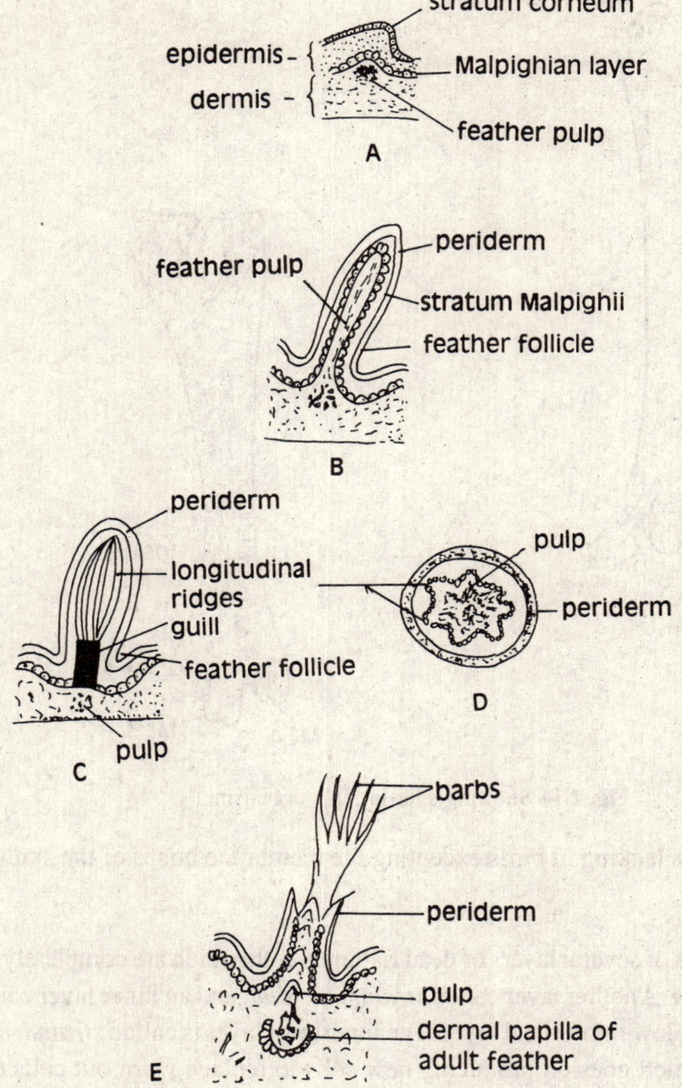

Fig. 2.13 Development of contour feather.

Epidermal Scales: In birds the epidermal scales of reptilian type overlapping one another are confined to the base of the beak and lower parts of the legs and feet.

The *horny beak* of birds is identical with that of turtles but much elaborated and diversified.

The *claws* of birds, comparable to reptilian ones are adapted to many different habits.

The *spurs* of birds are dermal bones covered over by thickening of the stratum corneum. (Fig. 2.14)

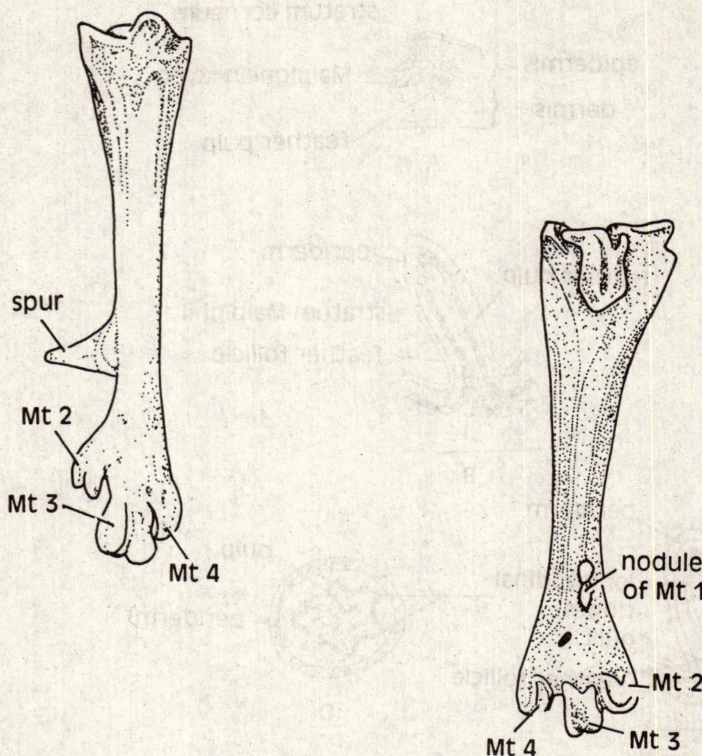

Fig. 2.14 Showing spur in male organism

The dermal skeleton is lacking in birds excepting the membrane bones of the skull.

2.3.8. Mammalia

The epidermis consists of several layers of dead flattened cells which are completely keratinized and form *stratum corneum.* Another layer called *stratum lucidum* and an inner layer called *stratum granulosum* are present below the corneal layer. The innermost layer is called *stratum malpighii* or *stratum germinativum* which goes on producing new cells to replace worn out cells of the outer layer of the epidermis. The *dermis* is situated below the epidermis. It is tough, flexible and elastic layer made of connective tissue fibres, blood capillaries, nerve ending, fat cells and unstrainted muscle fibres. The dermis also contains lymph vessels and small papillae projecting into the epidermis called tactile corpuscles. The lower part of the dermis contains adipose cells serving as heat insulators. The dermis provide flexibility and firmness to the skin, support to the body and facility to carry blood to the general surface. (Fig 2.15)

The *hairs* are epidermal structures characteristic of mammals. Each hair is divided into a *basal part* or *root* lying in the dermis and the upper part called *shaft* passing through the dermis and projecting from the epidermis. Hair is situated in a tube-like invagination of the epidermis formed by the stratum malphigii called *hair follicle.* The dermal tissue near the base of each *hair follicle* forms a thickening called hair papilla rich in nerves and blood vessels. The shaft of the hair is covered by a cuticle. Below cuticle there is a fibrous layer of cortex made of horny elongated cells and its central protion medulla is made of round cells. The hair follicle are situated somewhat slanting in the skin and a small *erector muscle,* called *errector pillis* is responsible for raising the hair due to cold, anger, fear or nerveousness.

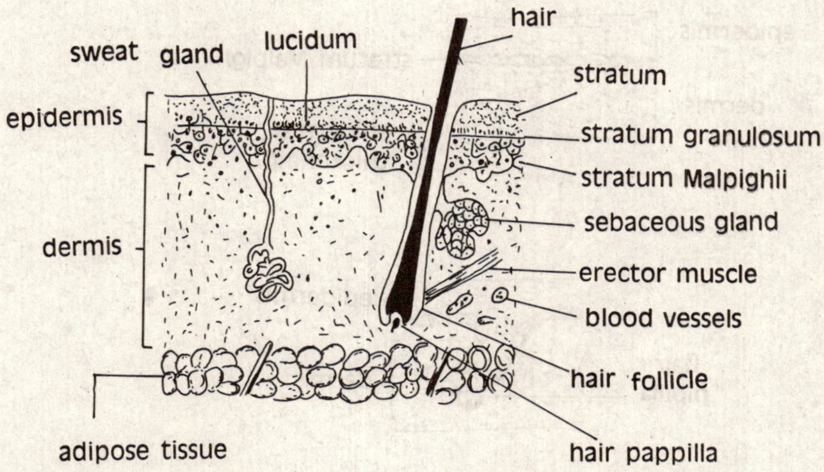

Fig. 2.15 V.S. of Mammalian skin.

The skin has two types of glands called *sweat* and *sebaceous glands.* The *sweat glands* are long, slender, coiled tubes formed by the invagination of the stratum malpighii. Each *sebaceous gland* opens by a duct at skin surface by a pore. The *sweat gland* cells are longer and columnar and remove some metabolic wastes. The *sweat glands* also serve to regulate body temperature by the evaporation of sweat. The sweat glands are absent in rat. The *sebaceous glands* are alveolar made of small branching alveoli sacs opening at the upper part of the hair follicle by ducts. These glands are outpushings of the wall of the hair follicle and produce an oily secretion for keeping the skin and hair greasy and waterproof.

The *mamnary glands* are modification of sebaceous glands made of alveoli and branching system of tubes. In females they produce milk. *Meibomian glands* are found in the eyelaid margins.

Development of hair **:** The cells of malpighii layer of the epidermis start multiplying at a place where the hair is scheduled to appear. A concentration of epidermis cells is produced and increase in size due to faster growth and sinks into the dermis producing *hair germ.* The tip of the hair germ enlarges and a bulb is produced where some dermal cells also get accumulated producing *dermal papilla.* The dermal papilla provides norishment to the developing hair. The *shaft* of the hair is produced by the rapid multiplication of *germinative layer* in contact with the dermal papilla. The hair germ becomes differentiated intι inner and outer parts. The inner part cornifies to make *medulla, cortex, cuticle* and *inner root sheath* of the hair shaft. The outer part of the hair germ is developed into *outer root sheath* made up of several rows of cells. The cells of the hair bulb retain capacity for multiplication and help in the growth of the hair. The hair shaft eventually breaks through the epidermis projecting freely over the skin surface.

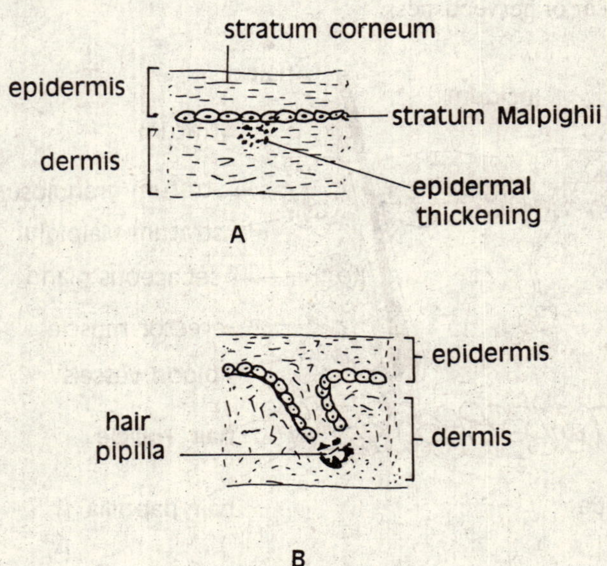

Fig. 2.16 A & B Development of hair

Pigment bearing *chromatophores* are absent in mammals but pigment granules are present in the lowest layer of the epidermis. In man the pigment bearing *melanoblasts* or *branching dendritic cells* are present between the epidermis and the dermis.

Most of the mammals lack dermal skeleton excepting the membrane bones of the skull. In armadillos the bony plates are reinforced be epidermal scales. There has been a reduction of dermal structure and elaboration of epidermal structure in mammals also seen in birds.

Epidermal scales are generally confined to tails and paws excepting armadillows and scaly anteraters.

2.4. DERIVATIVES OF THE INTEGUMENT

Various types of derivatives are produced by the two layers of the skin. The epidermis gives rise to skin glands, epidermal scales, horns, claws, nails, feathers and hairs. The dermis produces dermal scales in fishes and reptiles.

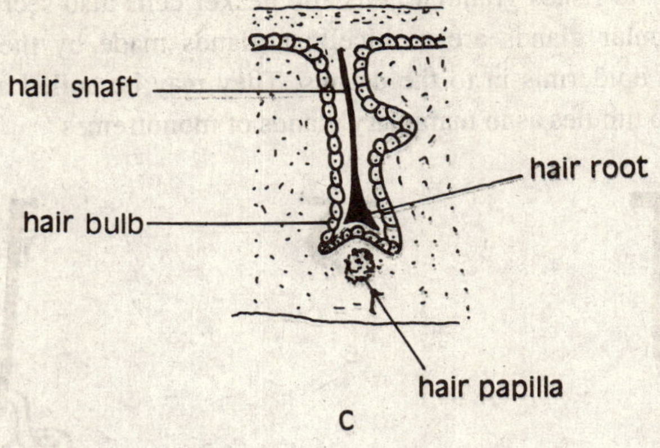

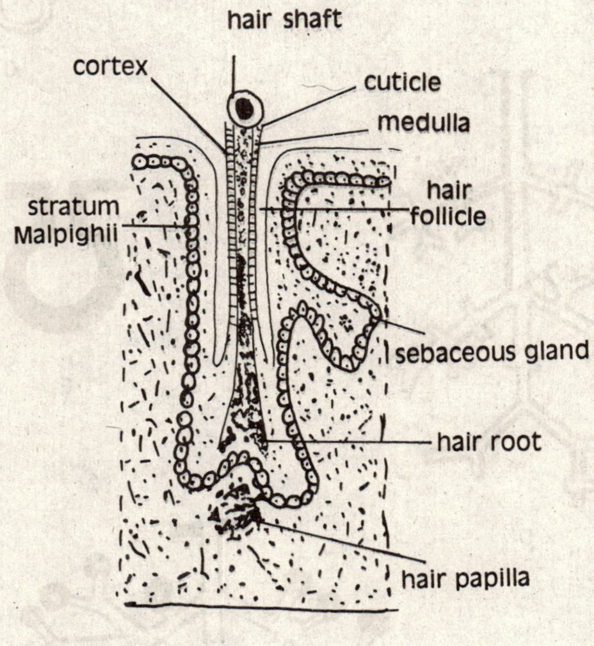

Fig. 2.16 C, D. Development of hair

2. 4.1. Epidermal derivatives

Glands: Three types of glands unicellular, tubular and alveolar or saccular are found in the skin. Unicellular glands are found in *Branchiostoma,* cyclostomes, fishes and amphibian larvae. They consist of single modified cells of the epidermis made by the malpigian layer. They are called mucous *glands* or *goblet cells* and secrete mucin which combines with water or form mucus. In cyclostomes and fishes granular cells and beaker cells also secrete mucus and called *unicellular glands*. Tubular glands are multicelluar glands made by the invagination of the malpighian layer of the epidermis in to the dermis. They may be coiled as in sweat glands of mammals or divided into tubules as in mammary glands of monotremes.

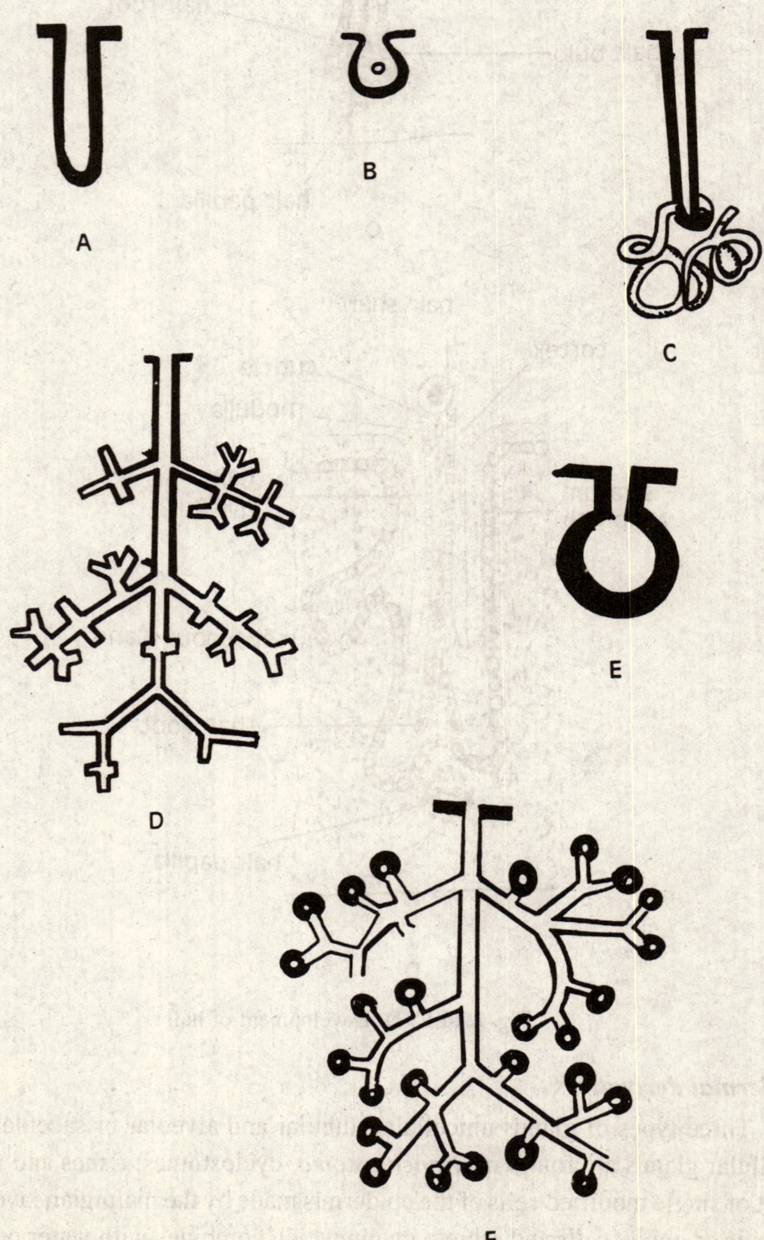

Fig. 2.17 Glands associated with integument

Alveolar or *saccular glands* are complex glands having the multicelluar malpighian layer extended into the dermis as tubular duct with flask-shaped terminal parts. The mucous glands and poisonous glands are alveolar glands.

The *alveolar glands* may be compound with branched tubules which open into a common duct as seen in the mammary glands of eutherians.

Poison glands of fishes: In many fishes poison glands are developed in association with their spines. These structures act as organs of offence and defence. The poison glands are modified skin glands envolved to provide protection from enemies. Among the clasmobranches e.g. in string rays, the serrated spines on the tail can inflict painful and fatal wounds. The spines are not connected with any special poison glands, but the secretory products of the mucous glands of the skin are possibly poisonous in nature. Among teleosts, well formed multicellular poison glands are present in association with the spines. In common cat-fish *Heteropneutes fossils,* each petoral fin bears a spine with a poison gland at its base. In scorpion fishes, the dorsal spines bear deep groves and the duct from the basal poison gland is prolonged up to the tips to open there.

The *epidermal glands* of vertebrates have different structures and functions according to the need of the animal concerned. The mucous glands secrete mucus for keeping the skin moist and slippery in aquatic animals. In some deep sea fishes *photophores* or *luminescent organs* are formed which produce light. In amphibians and some fishes *multicellular mucous glands* are found which produce mucus for lubrication and help in respiration. The poison glands of some amphibians are larger than mucous glands. Reptiles and birds have a few epidermal glands. *Femoral glands* of male lizards help in copulation. The *uropygeal glands* of pigeon secrete an oil which is odoriferous during sexual activity and also helps in preening of feathers. The mammals have the largest number and varieties of tubular and alvelolar glands. The sweat glands, sebaceous glands and meibomian glands are such examples.

The *mammary glands* of mammals secrete milk in females for the nourishment of the young. The raised outgrowth of breast forms nipple in which *teat* is situated. The ducts of the mammary glands open on treats. The number and location of teats differ in different group of mammals (Fig. 2.18).

In monotremes the tubular mammary glands have no nipples and open directly on the surface of the depressed skin area. They young ones lick milk from these hairy areas. In metatherians, man and carnivores the mammary glands open by their ducts at the tip of a nipple and breasts consists of mammary glands and their surrounding tissue. In ungulates a cistern is present at the base of an elongated false nipple or teat which receives milk by a secondary duct. In lower eutheria the two series of mammae are present with nipples. The *mammae* of higher mammals are thoracic situated in the pectoral region (bats, elephant and primates) or inguinal region (ungulates, whales). In many mammals the mammae are thoracic, abdominal, inguinal or axillary. In lemurs a single pair of mammae is situated in arm pits.

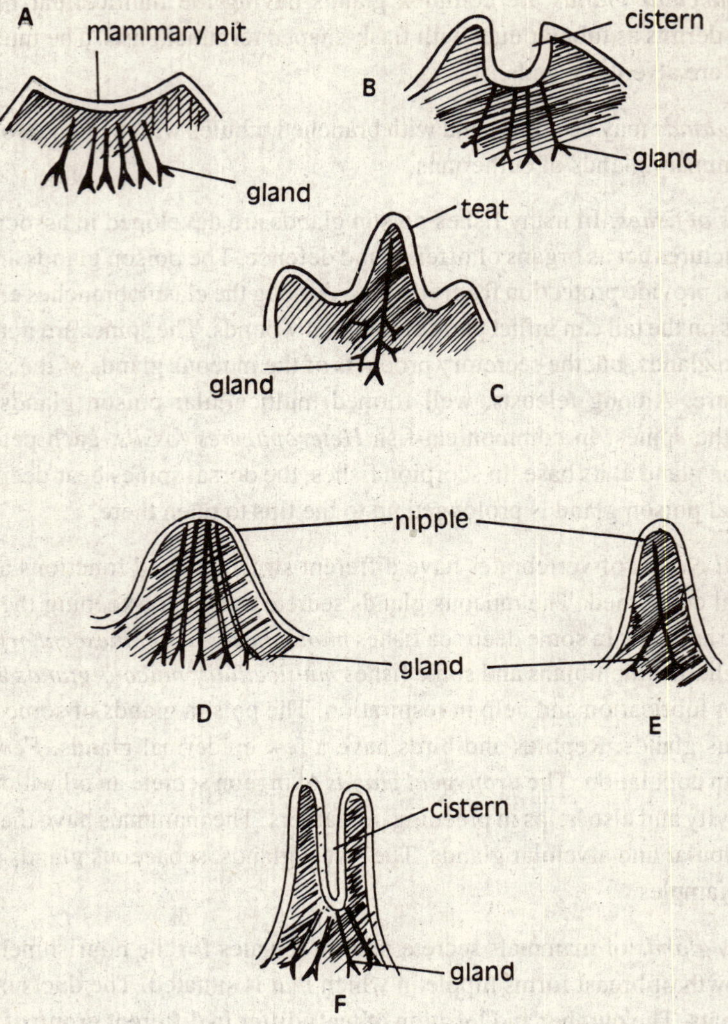

Fig. 2.18. A, B, C, D, E, F Mammary Glands.

Epidermal Scales: The epidermal scales are absent in fishes and amphibia. In reptiles they appear as confired derivatives of the malgihian layer and are replaceable. The scales of snakes undergo ecdysis. The scales of Chelonia and Crocodilia do not overlap and do not undergo ecdysis but are worn out and replaced. The scales present on the shell of turtles are called *scutes.* The scales of birds are restricted of the shanks, feet and base of the beak. The scales of mammals are found on the tail of rats, shrews and beavers and hairs project from beneath the scales. In scaly anteaters the scales are large covering the entire body excepting the ventral side.

Horns: The horns are products of the stractum corneum in ungulate mammals. They may be different types such as *hair horn, prong horn, antlers, hollow horns* and *giraffe horns.* The Hair horn, also called Rhinoceros horn is permanent epidermal structure made of keratinized epidermal cells having keratin fibres held together and perched on the roughened area of nasal bones. Its fibres are not true hairs. These horns occur in both sexes of *Rhinoceros.* The Indian rhino has one horn and African rhino has two horns. These horns are not shed.

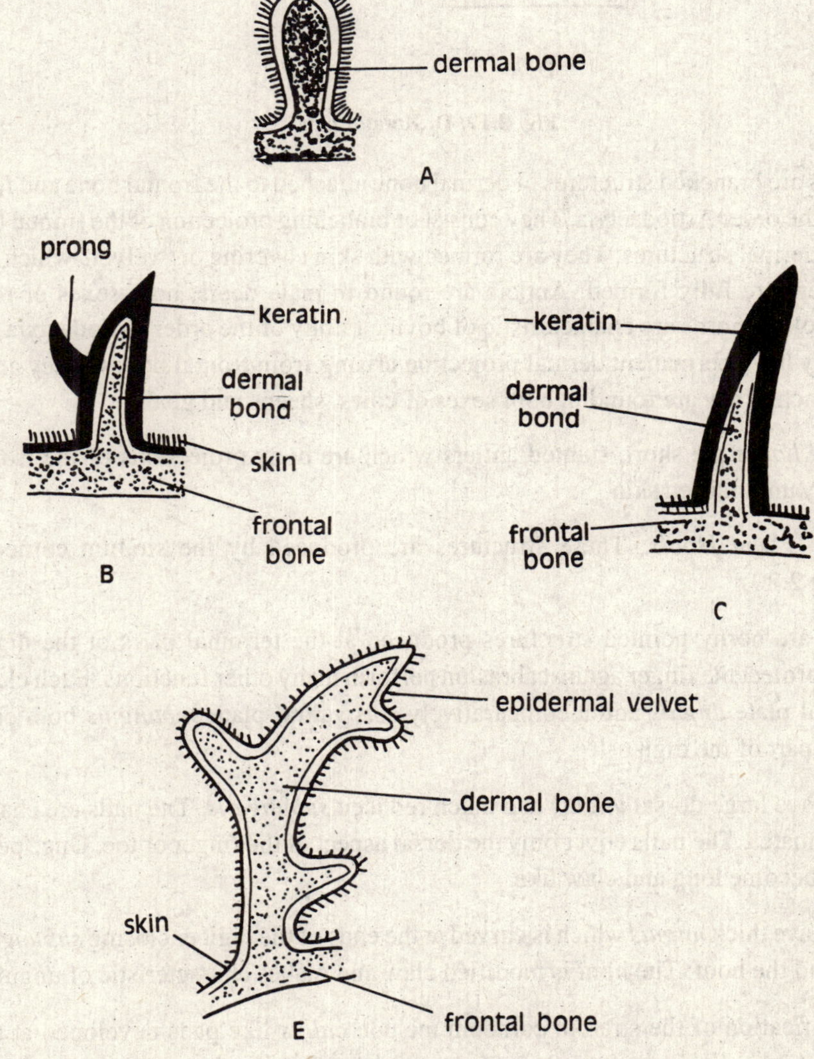

Fig. 2.19 A, B, C, E. Mammalian horn

The *pronghorns* are true horns found in anteleopes. They are permanent projections of the frontal bone of the skull, covered over by a horny epidermal sheath which is forked with 1-3 prongs made of heavy sheath. Prongorns are found in Russian antelopes and differ from hollow horns in branching and their annual shedding.

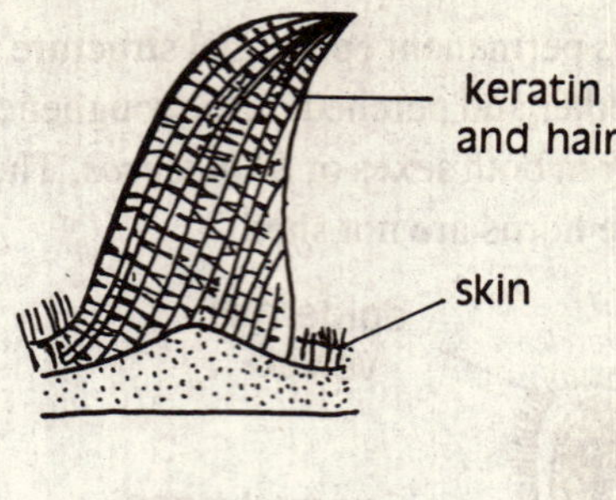

Fig. 2.19. D. Keratin hair

The *antlers* are branched structures of dermal bone attached to the frontal bone and found in the deer family of the order Artiodactyla. They consist of branching projection of the frontal bone. They are solid mesodermal structures. They are formed with skin covering or "velvet" which is shed off when the antlers are fully formed. Antlers are found in male deers, both sexes of reinder and caribou. The hollow horns are characteristic of bovine family of the order Artiodactyla. These are true horns They have permanent dermal projection arising from frontal bones. They are not shed and do not branch. They are found in both sexes of cows, sheeps and goals.

The *giraffe horns* are short, stunted antlers which are bony projections of the frontal bones covered over by unmodified skin.

Claws. Nails and Hoofs: These structures are produced by the stratum corneum of the epidermis. (Fig 2.20)

The *claws* are horny pointed structures produced at the terminal parts of the digits by the epidermis and protect the finger against abrasion perform many other functions. Each claw is made of a hard dorsal plate *unguis* and a comparatively soft vaitral plate *subunguis* both covering to cover terminal part of the digit.

The *nails* have large dorsal *unguis* and much reduced *subunguits.* The nails are characteristic of the order Primates. The nails cover only the dorsal aspect of the finger or toe. Once permitted to grow the nails become long and claw like.

The *hoofs* have thick *unguis* which is curved at the ends of the digit enclosing *subunguis.* A pad is present behind the hoof. The hoof is modified claw and makes characteristic of ungulates.

Other modification of the stratum corneum include *callus* like pads developed at the site of friction. The *ischial callosities* of monkeys, *kneepads* of camels, and *interidigital pads, corns* and *callus* are other such structures.

2.4.2. Dermal Derivatives

These are dermal scales found in fishes, some reptiles and a few mammals. They are mesodermal in origin derived from the bony dermal armour of ostractoderm and placoderm fishes. The scales of primitive placoderms were complex and similar to cosmoid scales. In chondrichthyes, the incomplete armour is formed by placoid scales. In bony fishes cosmoid, cycloid and ganoid scales are found. The details of fish scales have been given in section 2.3.

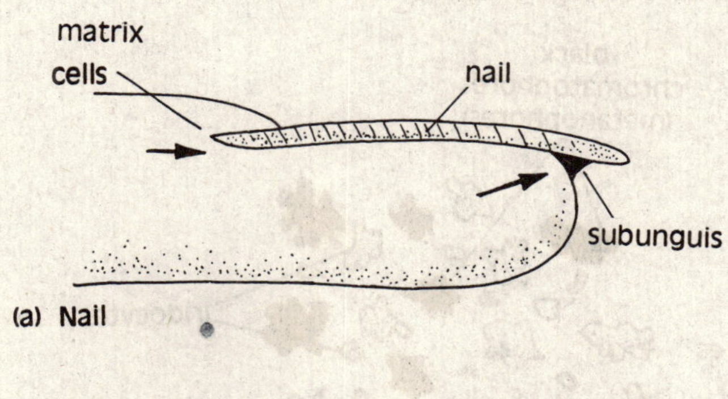

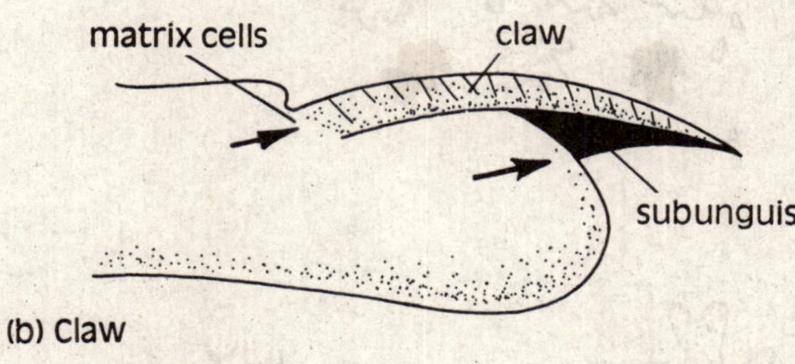

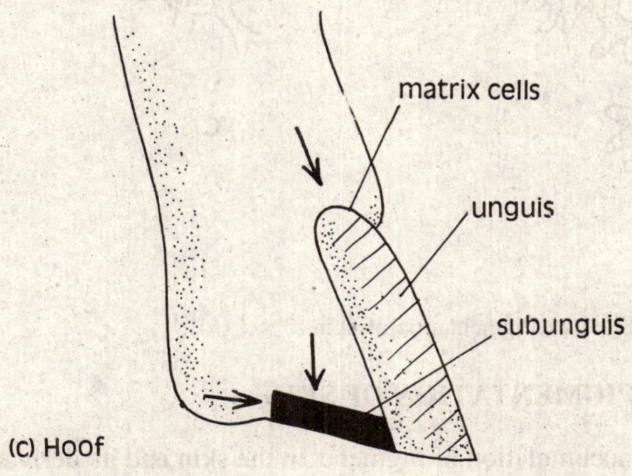

Fig. 2.20 Epidermal derivatives: **A.** The nail is a plate of cornified epithelium growing outward (arrow) from proliferating matrix cells at its base and from the subunguis **B.** Claw, **C.** Hoof.

The dermal scales of tetrapods are called bony plates or *osteoderms.* In apoda (Amphibia) vestiges of osteoderms are found. In Chelonia continuous osteoderms lie below the epidermal scales of the shell and form a rigid dermal skeleton connected with the endoskeleton. In crocodiles, the osteoderms lie below the epidermal scales on the back and the throat. In mammals the osteoderms are found in armadillos below the epidermal scales and in some whales on the back and dorsal fin. Details of the dermal skeleton have been given in section 2.3.

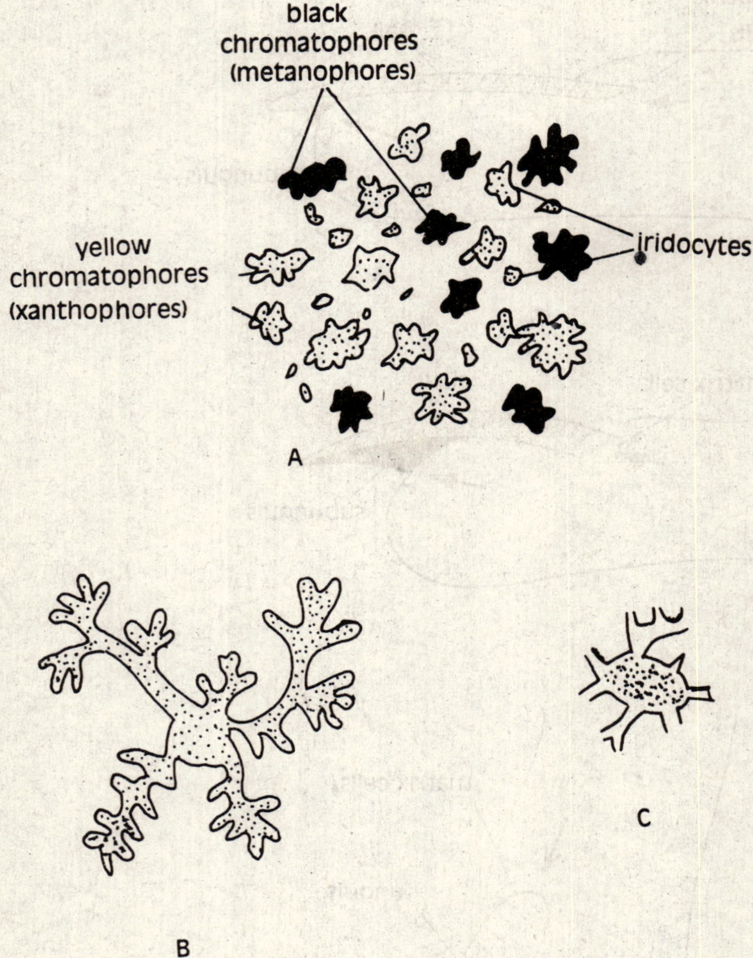

Fig 2.21 Pigmentation of skin

2.5. PIGMENTATION OF SKIN

The colour of animals is due to accumulation of pigments in the skin and its derivatives. The colour is not distributed uniformly in most cases and scattered colour pattern forms characteristics of a species. The colour pattern has some adaptive relationship with the environment in a good number of species.

Epidermal colouration: The colour of skin epidermis results due to pigments in diffused or grannuule state and diffraction or both. The colour of hairs in mammals, feathers in birds, beaks of birds and shell of turtile is due to epidermal pigments.

Dermal colouration: The colour in dermis is produced by pigments within specialized cells called *chromatophores.* In lower vertebrates where the epidermis is transparent such colouration is conspicuous. Chromatophores branch irregularly and pigment is flows out from the center of a cell along its branches resulting into intensification of the colour or its withdrawl to the centre which permits appearance of different colour in adjoining cells. The colour changes take place due to reflexes caused by visual stimulation humidity and other factors, which induce release of *neurohumours* from nerve endings. The intermediate lobe of the pituitary gland also induces colour changes. The chromatophores are capable of amoeboid movements. They have been distinguished into three main types, *melanophores, lippophores* and *iridocyctes.* The melanophores have dark brown pigment *melanin.* The *lippophors* or *xanthophores* have red or yellow cartenoid pigments. The *iridocytes* or *guanophores* are without pigments but have crystals of *guanine* which many alter the effect of pigmented material by refraction.

CHAPTER SUMMARY

1. The integument comprising of the skin and associated structures forms external covering of the animals.
2. The Vertebrate skin consists of two layers, an outer *epidermis* and an inner *dermis* or *chorium.*
3. The epidermis is ectodermal in origin and consists of two layers in all vertebrates excepting in mammals where it contains four layers of cells.
4. The *dermis* is divisible into an outer loose *stratum* laxum and an inner dense *stratum compactum.* The *dermis* is made of dense connective tissue having muscles, blood vessels lymph vessels, collagen fibres and nerves.
5. The skin performs several functions such as protection, storage of food, temperature control, secretion, water and electrolyte balance, exteroception, respiration locomotion sexual selection and skeleton formation.
6. The evolution of integument in vertebrates is related to transition from aquatic to terrestrial life.
7. Both epidermis and dermis produce various types of protective or adaptive integumentary structures. The epidermal derivatives include skin glands (mucous, sweat, poison, sebaceous mammary, scent, ceruminous). scales (epidermal scales of reptiles, birds and mammals), feathers. spurs, beaks, bills, hairs, claws, nails, hoofs and true horns.
8. The dermal derivative are bony scales (*ganoid, placoid, ctenoid* and *cycloid*). bony plates in the skin of some reptiles and mammals and dermal bones of the skull.
9. The colour of animals is due to accumulation of pigments in the skin and its derivatives. The colour of the epidermis is due to pigments in diffused or granule state. The colour in the dermis is produced by pigments within specialised cells called *chromatophores.* The colour pattern has same adaptive relationship with the environment in a good number of species.

3

The Skeleton

The skeleton comprises of harder parts of body developed from mesoderm and helps the body for support, attachment of muscle and protection of soft parts. The skeleton present outside the body is called *exoskeleton* and the one present inside is called *endoskeleton.* It forms the farmeword of the body in vertebrates and is made of cartilage or bone or both. Some vertebrates have a dermal exoskeleton comprising dermal scales and fin rays in fishes and bony armour in turtles crocodiles and some mammals. The dermal bones also share in the formation of some parts of the exoskeleton.

3.1. SKELETON TISSUES

The vertebrates have two characteristic skeletal tissues, *cartilage* and *bone.* Both of them arise from mesenchyme and are specialised derivatives of the connective tissue. They usually differ in their position, mode of origin, and characteristics. One more skeletal tissue is *notochord*, which makes one of the fundamental characters of Chordata. The skeletal tissues have been described in details in the following description.

3.1.1. The Notochord (Fig. 3.1)

It is semi stiff flexible rod situated along the median axis of the body in chordatws ventral to the nerve cord and dorsal to the alimentary cannal giving partial support to the body in embryos. It has large closely packed cells with big cells with big vacuoles filled with a fluid which keeps the cells rigid. The vacuolated cells are surrounded by a single layer of peripheral cells making *notochordal, epithelium.* The notochordal epithelium secrets two layers, the outer *elastica externa* and inner usually fiberous sheath which is thicker. In higher tetrapoda, the notochord is a transitory structure.

The notochord is unique tissue, neither cartilage, nor bone, without intercellular matrix.

3.1.2. Cartilage

The Cartilage is found in abundance in embryos and young vertebrates. In Cyclostomes and Chondrichthyes it constitutes major skeletal structure.

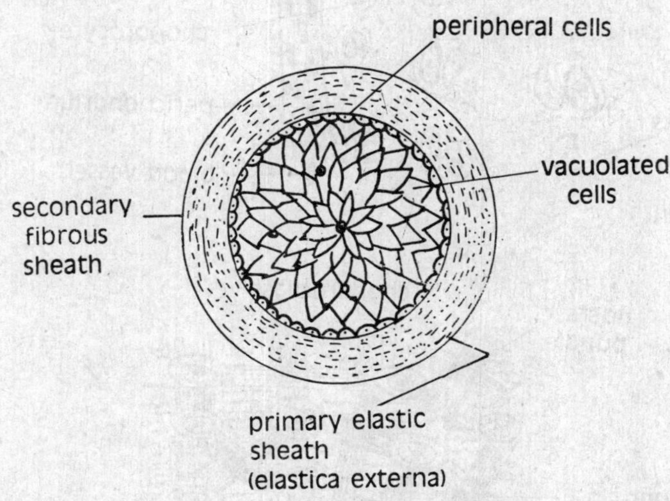

Fig. 3.1 Notochord in a cross section

Three main types of cartilages are *hyaline cartilage, elastic cartilage* and *fibro cartilage.* The cartilage has abundance of intercellular matrix which is made of protein called *chondrin* having delicate network of collagen fibres. The cartilage cells are called *chondroblases* and lie in fluid filled lacunae.The cartilage is surrounded by a connective tissue layer called *perichondrium*

The *hyaline cartilage* (Fig. 3.2.A) has translucent, homogenous, bluish white firm matrix. It is found in trachea, lower ends of ribs, ends of long bones and hyloid apparatus.

The *elastic cartilage* is elastic and flexible. It is found in ear pinna, epiglottis and tips of the nose in mammals. The matrix is limited and many yellow clastic fibres are present.

The *fibrous cartilage* (Fig. 3.2.B) also has limited matrix with many nonelastic collagen fibres running paraller to each other. It is found in the vertebral column and intervertebral discs of mammals.

Formation of cartilage: The irregularly shaped mesenchyme cells round up and make intestitial ground substance with fibres between themselves. The cells are closely aggregated in the beginning but get separated in later stages when more material is gathered from the outer surface formed by the cells of the perichondrium. Further divisions of the cartilages cells result in the formation of cell grouped in pairs or quarters. The newly formed cartilage may grow in size by internal expansion.

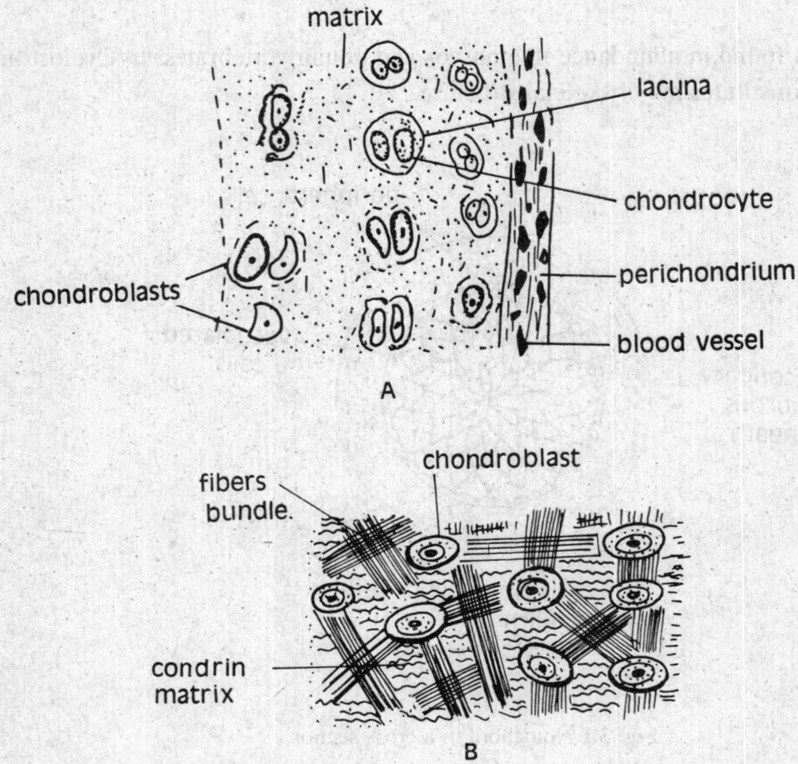

Fig. 3.2 A. Hyaline cartilage in a cross section **B.** A part of librous cartilage.

3.13. Bone (Fig. 3.3): The bone is a dominant structure in most of the adult vertebrates. The bone has intercellular matrix made of protein, some collagen fibres and impregnated by phosphates and carbonates of calcium and magnesium. The bone is a derivative of mesenchyme and consists of mesenchymes cells enclosed in a matrix, which is harder, opaque, and cornified material. The matrix is in concentric circles called lamellae. The bone cells are called *osteocytes* and lie between lamellae and have numerous photoplasmic branching processes, joining process of neighbouring osteocytes. The osteocytes lie in fluid filled space lacunae from where thin tubular canaliculi arise and join other lacunae. A thick layer of fibrous connective tissue periosteum surrounds the bone. The protoplasmic processes of osteocytes pass through canaliculi and provide nourishment to bone cells. The long bones have hollow cavity in the centre where bone marrow is filled in. The innermost layer of bone around bone marrow is called *endosteum.* The bones also serve for the storage of calcium for body needs. The long bones have a system of a series of *Haversian canals* which run parallel to the long axis of the bone and are joined to each other. The blood vessels pass through Haversian canals. The bones may be spongy or compact. The spongy bones called *cancellous bones* consists of interconnecting bars with spaces between them containing bony marrow. The compact bone is solid, hard structure without intercellular spaces. In long limb bones, the expanded extremities are made of spongy bones and shaft is made of compact bone.

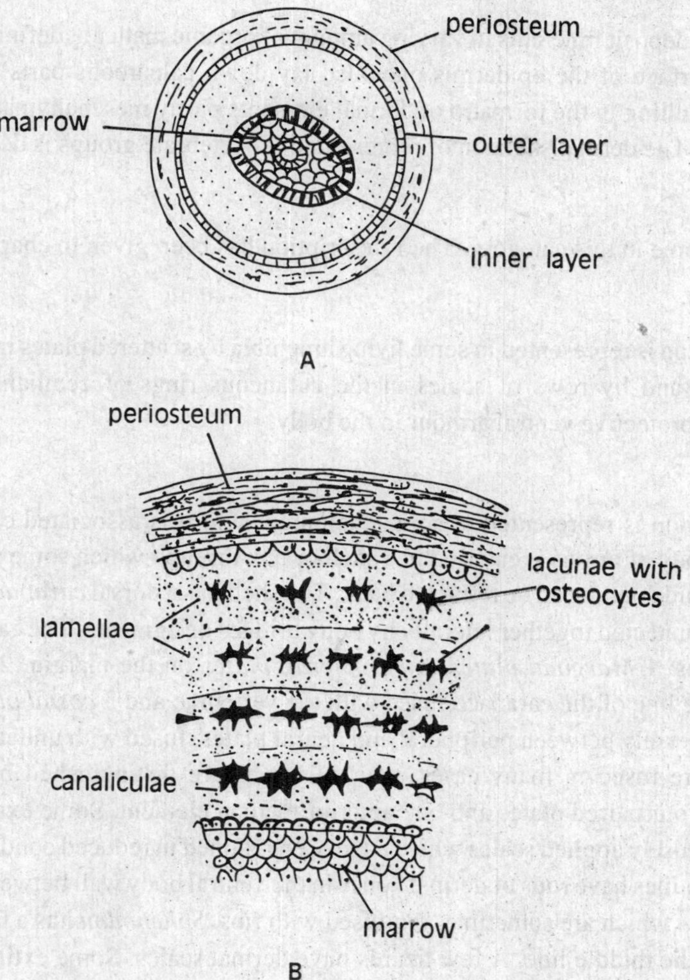

Fig. 3.3 Sectoral view of a bone, **A.** under low power, **B.** under high magnification

Formation of bones:'The bone is formed usually by two ways:

1. By replacement of preexisting cartilage, giving rise to cartilage bones (replacing bone).
2. By the direct ossification of messenchyme or connective tissue giving rise to dermal bones or membrane bone. Some bones called mixed bones may be formed by the joint contribution of cartilage and skin. Some animals have heterotropic bones which do not belong to either dermal or cartilage character.

3.2. DERMAL SKELETON

The dermal seleton arises in some animal by specialised cells called *scleroblasts, odontoblasts,* or *osteoblasts*, which deposit lime salts in varying amount of organic matter at definite points in the dermis. The basal surface of the epidermis·may also lay down calcareous parts on limy plates formed in dermis, resulting in the formatio of dermal elements partly mesenchymatous and partly ectodermal in origin. The dermal skeleton of representative vertebrate groups is described below.

3.2.1. Fishes

The fish scales come in this category. Their description has been given in chapter 2.

3.2.2. Amphibia

The dermal skeleton is represented in some living amphibia by scattered plates in the skin of the back of some toads and by rows of scales in the cutaneous rings of cecilians. Some fossil stegocephalians had protective ventral armour in the body.

3.2.3. Reptiles

The dermal skeleton is represented in some chelonia where it is associated closely with the endoskeleton. The body of turtles is encased in a shell or box through which some parts like head, tail, or legs may protrude out or can be retracted back. The shell has a dorsal *carapace* and a ventral *plastron* which are connected together laterally by bony bridges or ligments. The carapace is made of three sets of plates: 1 *Marginal plates* or *peripheral, plates* on the margin. 2 *Neural plates* situated in the middle line of the carapace fused with the vertebrae and 3 *costal plates* or *pleural plates* situated transversely between peripheral and neural plates, fused with underlying ribs. The carapace elements are fused in many cases, their identities are distinguished by sutures. The *plastron* consists of four paired plates and one anterior median element. Some extinct crocodiles were provided with closely applied scales which have been retained in reduced condition in modern crocodiles. The crocodiles have rods of dermal bones in the ventral body wall between true ribs and pelvis called *gastralia* which are sometimes confused with ribs. *Sphenodon* has a large number of gastralia meeting in the middle line. A few lizards have dermal scales. Some extinct Stegosauria had dermal scales, sometimes of great size in the dorsal and caudal regions. Dermal ossifications are not found in birds.

3.2.4. Mammalia

Dermal bones are of rare occurrence in mammals. In living mammals, only cetacea and armadillos have dermal skeleton on the back. In armadillos, the dermal bone makes a complete armour on the dorsal side having plates arranged in transverse rows some of which are movable on one another.

3.3. ENDOSKELETON

The following classification of endoskeleton will be adopted in this text for terrestrial animals:

3.3.1. Axial skeleton

The axial or median skeleton includes skull, vertebral column, ribs and sternum in most land vertebrates.

TABLE 3.1
Classification of Endoskeleton

Axial Skeleton or Median unpaired skeleton	Skull	Cranium or brain box Sense capsule; Eye, Visceral arches;	 Nose, ear. Jaws, Hyoid visceral arches.
	Vertebral Column	Cervical vertebrae (neck) Thoracic vertebrae (In chest) Lumber vertebrae (In lower back) Sacral vertebrae (In hip) Caudal vertebrae (In tail)	
	Ribs & Sternum	Ribs paired, bony or cartilaginous Sternum or brest bone.	
Appendicular Skeleton or Lateral paired Skeleton	Pectoral (Anterior)	Pectoral or shoulder girdle	Scapula (Dorsal) Clavicle (Anterior) Coracoid (posterior)
		Forelimb	Humerous (Upper arm) Radius, Ulna Carpals (Wrist) Metacarpals (palm) Phalanges (fingers)
	Pelvic (posterior)	Pelvic or Hip girdle	lium (Dorsal) Publis (Anterior) Ischium (Posterior)
		Hindlimb	Femur (Thigh) Tibia, Fibula (Shank) Tarals (Ankle) Metatarals (Sole) Phalanges (Toes)

SKULL

The head skeleton of vertebrates is called *skull,* which is complex structure formed of different components. The skull may be divided into *cranium,* containing a box for the brain, sense capsules enclosing sense organs and *visceral skeleton* related to the anterior part of the digestive tract. In its simplest forms the skull is made of *chondrocranium,* and is joined by *osteocranium* later on during its evolution into higher types. The osteocranium is having two structures, *splanchnocranium* and *dermatocranium* The chondrocranium is made of cartilage and is usually replaced by bones in vertebrates. The splanchnocranium is formed usually by the ossification of visceral skeleton in higher vertebrates. The dermatocranium is formed by dermal bones which become attached to the chondrocranium of bony fishes and terapoda; it is absent in cyclostomata, clasmobranches and a few higher fishes having cartilaginous skull.

Development of the skull: The skull passes through three stages in its development; membranous, cartilaginous, and bony. The outlines of skull development are as follows (Fig. 3.4):

i. Two paired cartilaginous *parachordal plates* takes their origin form mesenchyme below the posterior part of the brain on the lateral sides of the notochord after the formation of central nervous system.
ii. The *trabeculae* take their origin as curved cartilaginous rods infront of parachordals below the anterior part of the brain from the mesenchyme.
iii. Paired *sense capsules, olfactory* around the organ of smell, *optic* around eyes and auditory or otic around *auditory* organ take their origin simultaneously with the formation of trabecuale and parachordals.
iv. An opening called *basicranial fenestra* is formed by the growth and fusion of the parachordals in the middle line and a plate thus formed is called ***basilar plate.***

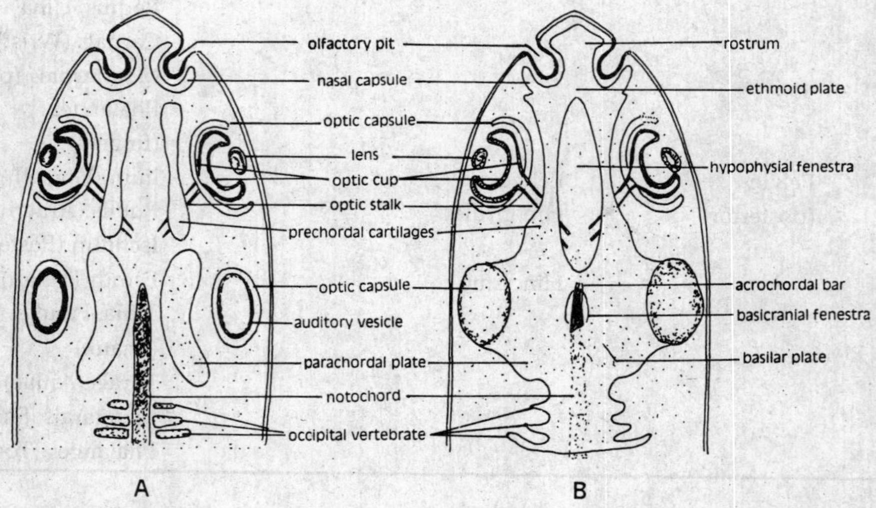

Fig. 3.4 Diagrams illustrating development of the chondrocranium. A. separate chondrocranium cartilage surrounding the special sense organs and flanking the notochord and brain. B. all the cartilage except the optic capsules have fused to form the early chondrocranium.

v. An *ethmoid plate* is formed by the fusion of the anterior ends of the trabeculae enclosing a large space called *hypophyseal fenestra* for the accommodation of pituitary glands.

vi. The floor of the brain is formed by the fusion of the ethmoid plate with the posterior ends of trabeculae.

vii. The chondrocranium is formed by the joining of olfactory and auditory capsules. The basal plate grows upwards and laterally. In elasmobranches, the lateral walls meets above the brain to make a brain box or *neurocranium*, but on other vertebrates it is formed by dermal bones.

viii. The *splanchnocranium* (Fig. 3.5) is made of paired visceral cartilaginous bars united with one another on ventral side by an unpaired cartilage; these bars constitute horse hoe shaped visceral arches encircling the pharynx from the lateral and ventral sides. The fishes have typically 7 visceral arches. The first visceral arch is mandibular, second is hyoid, and remaining are branchial arches which support gills.

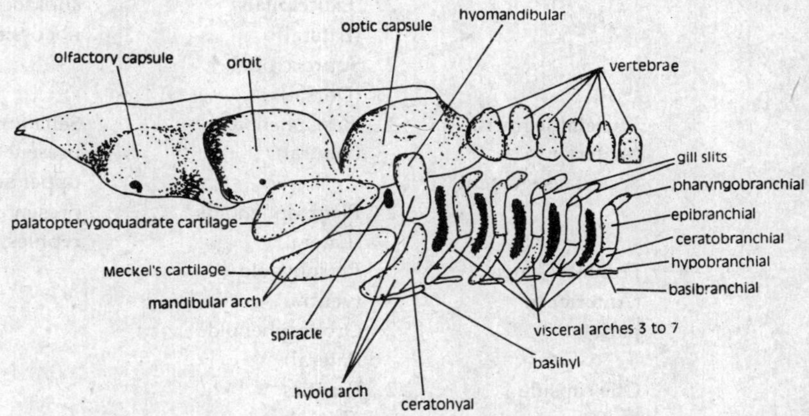

Fig. 3.5 Diagram showing relation of visceral arches of an x elasmobranch fish to the chondrocranium, vertebral column, and gill slits.

ix. The *mandibular arch* is absent in Agnatha and forms jaws in all vertebrates for encircling the mouth. It is divided into a dorsal *palatoquadrate cartiage* forming the lower jaw. The *hyoid arch* supports gills in placoderms but forms a *hyomandibular cartilage* in other animals to provide attachment of jaws to the chondrocranium. Remaining portions of hyoid arch provide support to gills or tongue.

x. The *Branchial arches* support gills in fishes and do not contribute to the formation of skull. In higher animals, they are reduced and form part of the hyoid apperatus and cartilage of the larynx. Modification of visceral arches has been given in table 3.4.

xi. The *dermatocranium* is formed by the dermal scales or bones of exoskeleton produced inwords to join the condrocranium for making a protective covering around the brain. The dermal bones and some cartilage bone get associated with the skull in vertebrate excepting cyclostomes and clasmobranches. The dermal bones may remain independent of remaining skull but sometimes get fused with cartilage bones making distinction between different types of bones.

xii. The cartilage bones are formed in fishes and tetropodes by the replacement of cartilages of condrocranium and splanchnocranium by ossification into bones.

The principal skull bones have been given in table 3.2 according to there origin.

TABLE 3.2

Skull Bones

Type of bone	*Part of Skull*	*Segment of skull and position*	*Bones of the segment*	*Special features*
Cartilage Bones	Chandro-cranium	Occipital segment (Posterior)	1. Bassioccipital (Ventral) 2. Exoccipitals (Lateral) 3. Suproccipital (Dorsal)	In modern amphibia basioccipital and supraoccipital do not ossify.
		Parietal segment (Anterior to occipital segement)	1. Basisphenoid (ventral)	Sella tursica present on its upper surface
			2. Pleurosphenoids (Lateral)	present in fishes, reptiles, birds
		Frontal segment (Anterior to parietal	1. Presphenoid (ventral) 2. Orbitosphenoids (lateral)	
		Otic capsule	2. Prootics (Anterior) 2. Opisthotics (Posterior) 2. Epiotics (Dorsal)	
			2. Sphenotics and pterotics	Present in fishes
		Olfactory capsule	1. Mesthmoid	Forms turbunals In mammals
	Splanchno-cranium		1- Ethmoid 2- Lateral ethmoids	In fishes only
		Optic capsule	2- Sclerotics	Make sclerotic bones in reptiles and birds.
		Palatopterygo Quadrate bar	2- Metapterygoids 2- Quadrates	Alisphenoids of mammals.
			2- Epipterygoids	In tetrapods only unossified
		Meckel's cartilage	2- Articulars (Posterior)	
			2- Mentomeckelians	In Anura only
		Hyoid arch	Paired Hyomandibulars,	

Type of bone	*Part of Skull*	*Segment of skull and position*	*Bones of the segment*	*Special features*
	Splandi- nocranium		Symplectics Epihyals, Ceratohyals, Hypohysis and Interhyals.	
			Median-Basihyal	In fish only
			Colmella of Stapes	of middle ear.
		Branchial Arch	Paired pharyngo-, epi, cerato-, and hypobranchials	In fishes only
Dermal Bones	Dermato cranium	Partietal segment	Paired parietals, Interparietal, (Upaired) Paired-Postparietals	
		Frontal segment	Frontals, Post frontals, and Lacrimals	All paired
		Olfactory capsule	Nasals, Vomers,	All paired
			Septomaxillaries	All paired
		Otic capsule	Squamosals, Supratemporals	All paired
		Palate	Parasphenoid (Upaired)	Vomer of mammals
			Endopterygoids or Pterygoids, Ectopterygoids)	Both paired
	Splanch- nocranium	Upper jaw	Premaxillae, Maxillae, jugals, Quadratogugals	All paired
	Splanch nocranium	Lower jaw	Dentaries (Mandibles), Coronoids, Splenials, Angulars, Super angulars	All paired
		Gill cover and Gill arch	Preoperculars, Operculars Suboperculars, Interopoerculars	All paired
			Unpaired Basibranchials	
Mixed bones	Frontal capsule		prefrontals	Paired
		Olfactory capsule	Sphenotics, Pterotics	Paired
		Plate	Palatines	Paired

Jaw Suspension (Suspensorium)

The attachment of splanchnocranium to the condrocranium (neurocranium) is called jaw *suspension* or *suspensorium*. Different types of jaw suspensions (Fig 3.6) have been described in table 3.3.

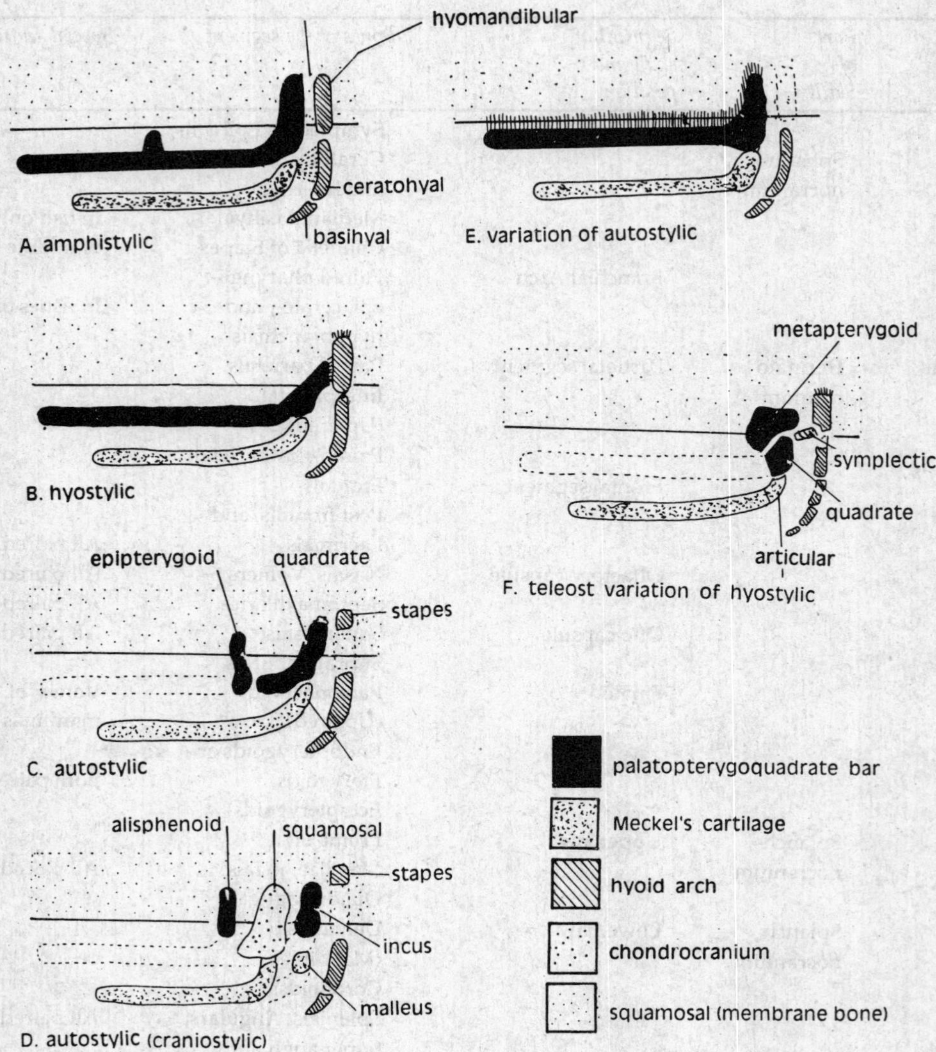

Fig. 3.6 Diagram indicating various methods of jaw suspension. A. Amphistylic. found in a few primitive elasmobranches. B. hyostylic, found in sharks and sturgeons; C. autostylic, found in tetrapods other than mammals; D. craniostylic, typical of mammals; E. variation of autosylic found in Polypterus, holocephalians and lungfish; F. teleost variation of hyostylic, found in Amia, Lepisosteus and teleosts. The membrane bones investing Meckels cartilages are not indicated.

Skull in Vertebrates Series

Skull in class Chondricthyes: Characteristics:

1. The skull of elasmobranch fishes is.made of simple brain case persisting without the addition of bone. It may be called specialised skull or degenerate skull without bone.

TABLE 3.3
Jaw Suspensions

S. No.	*Type*		*Characters of the type*	*Found in*
1.	Autodiastylic		Splanchnocranium attached to chondrocranium by ligaments without participations of hyoid arch.	Early bony fishes (acanthodians)
2.	Amphistylic		Basal and otic processes of the upper jaw attached to chondrocranium by ligaments and hyomandibular of hyoid arch attached to the chondrocranium at one end; both jaws suspended from its other end.	Primitive shark *Hepatanchus* (Fig. 3.6. A)
3.	Hyostylic		Both jaws suspended from hyomandibular attached to otic region of the chondrocranium.	Most of elasmobranchs and bony fishes (Fig. 3.6 B)
4.	Autostylic		Upper jaw fused completely with bony skull by its process and lower jaw is suspended from upper jaw.	
		i)	*Holostylic:* The upper jaw is attached to skull and lower jaw is suspended from it; Hyoid arch is not attached to it.	Holocephali (Fig. 3.6. E)
		ii)	*Monimostylic:* Articular of lower jaw Articulates with Immovable quadrate of Upper jaw.	Tetrapoda excepting mammals. (Fig. 3.6, C)
		iii)	*Streptostylic:* When articulation if between quadrate and articular; the quadrate is movable.	Lizards snakes and birds
		iv)	*Craniostylic:* The squamosal of the upper jaw articulates with the dentary of the lower jaw. (The quadrates and articular are modified into Malleus and Incus of the middle ear.)	Mammals (Fig. 3.6, D)

TABLE 3.4

Modification of Visceral Arches in Vertebrates

	Arch	*Elasmobranch (Dogfish)*	*Teleost*	*Amphibians Frog*	*Reptiles and birds*	*Mammals*
1.	Mandibular Arch	Palatopterygo-quadrate carti-lage	Quadrate epipterygoid	Quadrate Annulus-tyanicus	Quadrate Epiptery-goid	Incus Alisp-henoid
		Meckel's cartilage	Metapterygoid Articular	Articular Mentomeck-elian	Articular	Malleus
II.	Hyoid Arch	Hyomandibula	Hyomandibular Symplectic	Columella	Stapes	Stapes
		Ceratohyal	Interhyal Epihyal Ceratohyal	Anterior horn	Anterior horn	Anterior horn Styloid process
		Basihyal	Hypohyal Entoglossal	Body of Hyobranchial apparatus	Body of hyobranchial apparatus Entoglossas	Body of hyoid
III.	First Branchial Arch	Pharyngobran-chial	Pharyngobranchial			
		Epibranchial	Epibranchial		Second horn	Poster-ior
		Ceratobranchial	Ceratobranchial	Body of hyobran-chial app-aratus	Body of hyobranchial apparatus	Body of hyoid
		Hypobranchial	Hypobranchial			
IV.	Second Branchial Arch	Branchial elements (same as III)	Branchial elements (same as III)	Posterior horn	Posterior horn Body of hyobranchial apparatus	Thyroid cartilage
V.	Third Branchal Arch	Branchial elements	Branchial elements	They may contribute to cartilages of larynx but homologies between laryngeal cartilages of lower vertebrate and mammals are doubtful		Thyroid cartilage crocoid cartilage oid Car-tilage
VI.	Fourth Branchial Arch	Branchial elements	Branchial elements	Missing or not clearly delineated		
VII.	Fifth Branchial Arch	Branchial elements	Reduced	Missing		

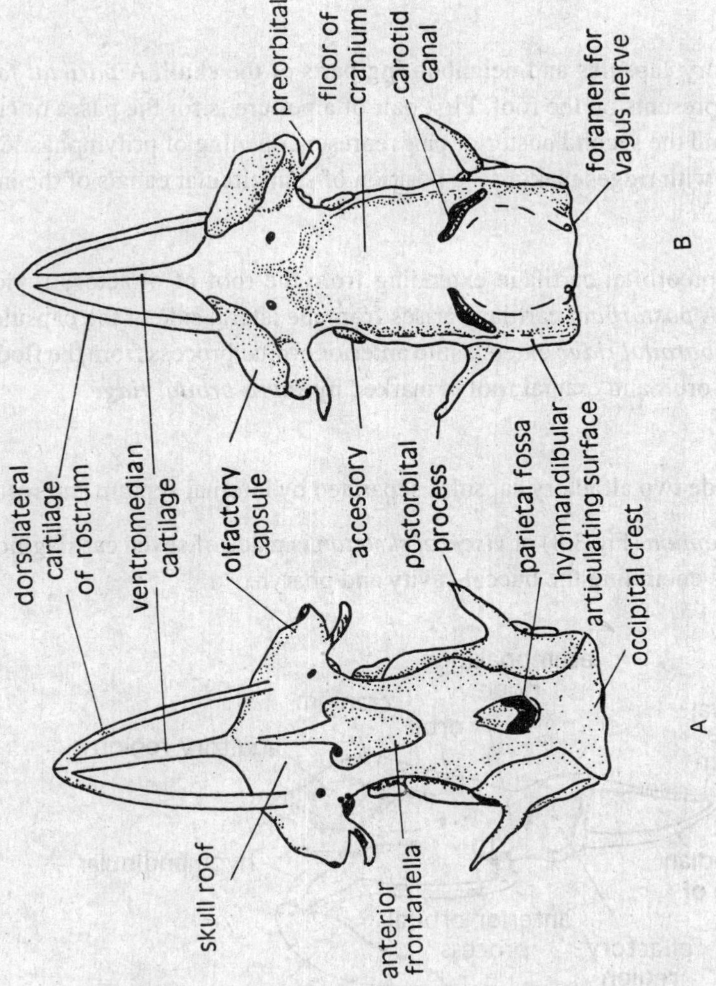

Fig 3.7 skull of scoliodon **A.** Dorsal view **B.** Ventral view

2. The jaw remains closely articulated with the cranium at the basal process with upper part of hyoid arch constituting *hyostylic jaw suspension.*
3. The skull consists of intimately fused cranium, four sense capsules and visceral skeleton forming jaw and supporting pharynx with gills (chondocranium).

Skull of Scoliodon

The chondocranium has a large neurocranium with intimately fused paired olfactory and auditory capsules. It is divided into occipital, auditory, orbital and ethmoid reasons (Fig. 3.7).

Occipital Region

It forms posterior part of the skull with a large median *foramen magnum* from where the brain continues to spinal cord. To occipital chondyles articulate with the first vertebra.

Auditory Region

It includes auditory capsules and neighbouring parts of the skull. A *parietal fossa* with two pairs of apertures is presents on the roof. First pair of a perture is for the pases of endolymphatic duct to internal ear and the second posterior pair represent opening of prilymphastic spaces. Each capsules is provided with ridges making the position of semicircular canals of the internal ear.

Orbital Region

This region has preorbital cartilage extending from the roof of olfactory region and partly encircling the orbit. A *postorbital* cartilage arises from the lateral side of the capsules and curves upwards. A lateral *suborbital ridge* extends into anterior orbital process from the floor of the orbit. The margin between orbit and cranial roof is marked by *supra orbital ridge.*

Ethmoid Region

This region include two alfactory capsules separated by internal septum and rostrum.

The *splanchnocranium* (Fig 3.8) or *visceral skeleton* is made of seven cartilaginous structures called visceral arches encircling the buccal cavity and pharynx.

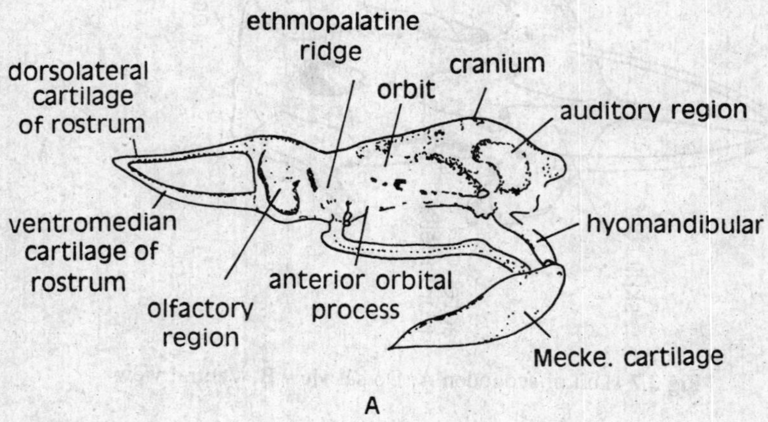

Fig 3.8 A. Skull of scoliodon in side view

Mandibular arch is the first pair of visceral arches which formed the jaws. Each half is divided into upper *palatopterygoquadrate* forming upper jaw and lower *Meckcl's cartilage* forming the lower jaw.

Hyoid arch is the second pair of visceral arches. Each hyoid arch is made of three parts, a ventral *basihyal* a lateral *ceratohyal* and a *dorsal hyomandibular.* The jaws are suspended to the cranium through hyomandibular and make hyostylic type of jaw suspension.

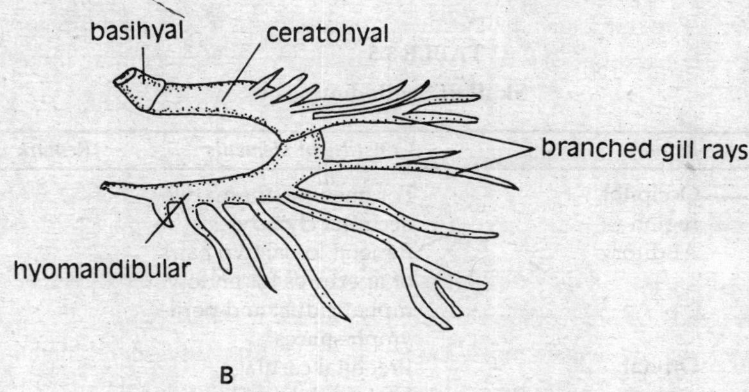

Fig 3.8 B. Hyoid apparatus

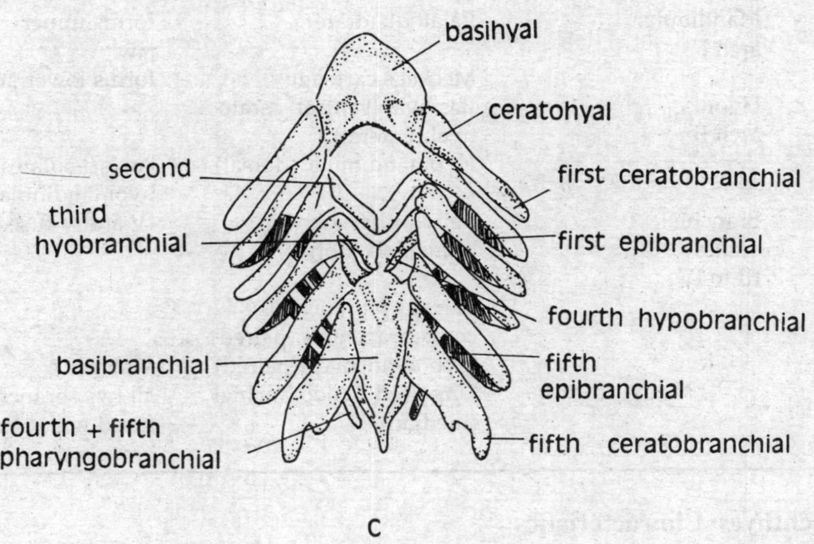

Fig. 3.8 C. Branchial arches

Branchaial arches are third to seventh pairs of cartilaginous rods which support pharynx and gills. Each arch is made of four pieces, a *pharyngobranchial* on dorsal side and an *epibranchial*, a *ceratobranchial* and a *hypobranchial* on the ventral side. The pharyngobranchials of fourth and fifth arches are fused to form one piece. The fifth *hypobranchial* is absent and remaning hypobranchials of the two sides meet below in a median *basibranchial.* The fifth pair of ceratobranchials directly articulate with the basibranchials as fifth hypobranchial is absent.

The structure of the cartilaginous skull of *Scoliodon* may be summarized by the table given below:

TABLE 3.5

Skull of Scoliodon

Part of	*Region*	*Constituent elements*	*Remark*
Neuro Cranium	Occipital region	Foramen magnum; two occipital chondyles	
	Auditory	Perietal fossa; two pairs of apertures for endolymphatic duct and perilymph spaces	
	Orbital region	Preobital cartilage; post orbital cartilage; sub orbital ridge; supra orbital ridge.	
	Ethmoid region	Two olfactory capsules internasal septum, rostrum	
Splanchno-Cranium	Mandibular arch I	Palatoquadrate	forms upper jaw
		Meckel's cartilage	forms lower jaw
	Hyoid arch II	Basihyal (ventral) cerato hyal (Lateral)	
		Hyomandibular (dorsal)	jaw articulates by hyomandibular
	Branchial arches III to IV	Paired pharyngo-branchials (dorsal)	IV and V fused
		Paired epibranchials; ceratobranchials (paired)	
		Hypobranchials (paired)	V absent
		Unpaired basiobranchial (median)	all hypobranchials fused with basi-branchial

Skull in class Osteichthyes: Characteristics

The skull is mostly bony with diversity of structure in different subgroups. (1) The *cranium* is usually covered over by numerous thin dermal bones, some of them having numerous lateral line canals. (2) The *skull* is composed of cranium with immovably united sense capsules and lossely attached visceral arches. (3) The skull is derived from the chondrocranium and visceral arches as in elasmobranches but acquires bones.

Skull of Labeo: The skull of *Labeo* has a very complicated elongated structure with a large number of cartilage bones and dermal bones participating in its composition (Fig. 3.9). The skull is fully ossified and includes cranium, sense capsules and visceral arches. The cranium and sense capsules are intimately fused together to form one compact skull. The visceral skeleton is loosely attached to the cranium.

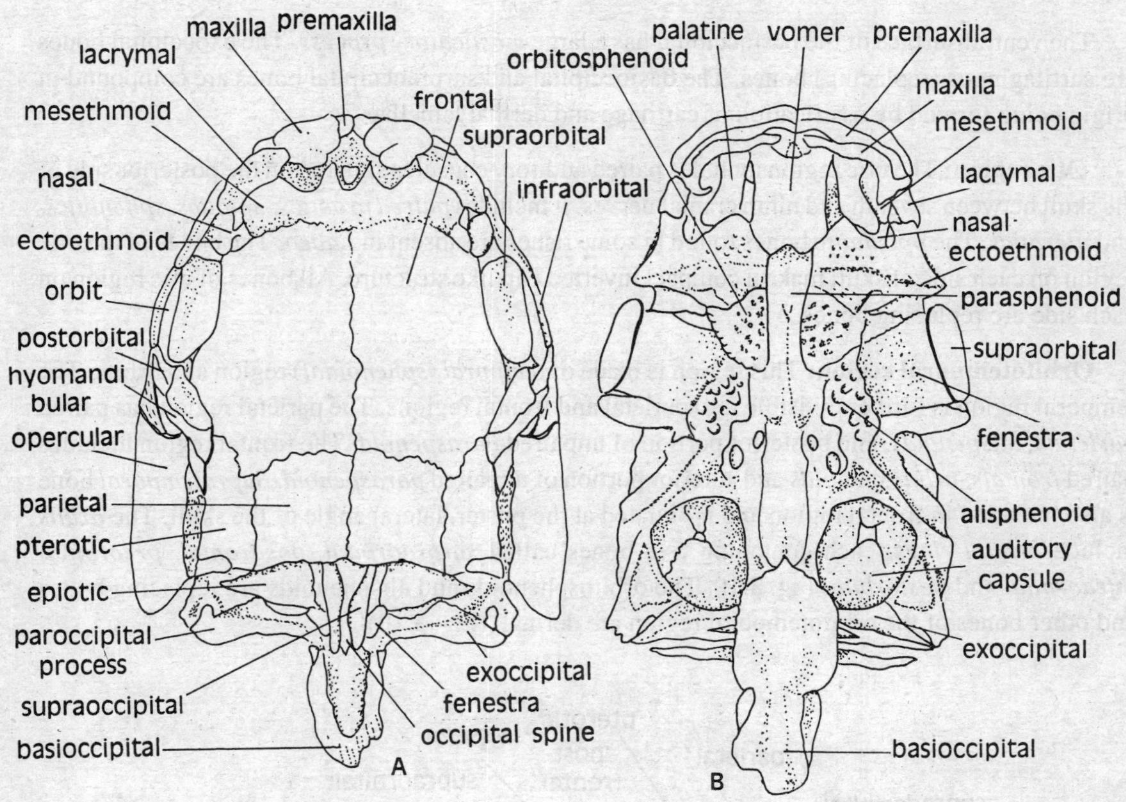

Fig. 3.9 Skull of Labeo **A.** Dorsal view, **B.** Ventral view

Cranium

Occipital region: This region is made of a dorsal *supraoccipital* a ventral *basioccipital* and lateral paired *exoccipitals.* This region forms the posterior part of the skull, articulates with first vertebra and surrounds the foramen magnum for the passage of nerve cord. Two *oval fenestrae* pierce each exoccipital bone. The *supraoccipital* bone forms the main part of the skull and is divided into dorsal and posterior inferior parts. It articulates anteriorly with partietal bones. The dorsal part of the supraoccipital has a vertical *occipital spine* or *keel.* Two small crests form median wedge shaped processes and constitude postero-inferior portion of the supraoccipital. The *exoccipitals* are large bones having three parts each. The first part is *basal plate* forming the part of the floor of cranial cavity; the second part *paroccipital process* forms posterior boundary of the auditory capsule and side walls of the cranial cavity; the third dorsal part encloses *foramen magnum.* The *basioccipital* is a large bone roofed by occipital condyle. The posterior surface of the occipital condyle has a deep depression. The ventral surface of the occipital condyle has a deep depression.

The ventral surface of the basioccipital has a large *masticatory process.* The exoccipital bones are cartilaginous (replacing) bones. The basioccipital and surpraoccipital bones are compound in origin being formed by ossification of cartilage and dermal lamellae.

Otic region: The otic region includes paired auditory capsules situated on the posterior side of the skull between seventh and ninth cranial nerves. It includes *paired prootics; epiotics, sphenotics,* and *pterotics.* The *opisthotic* bones found in some fishes are absent in *Labeo.* The bones of the otic region on each side of skull make a compact inverted cup like structure. All bones of otic region on each side are replacing bones.

Orbitotemporal region: This region is made of *temporal* (*sphenoidal*) region and orbits. The temporal region is further divisible into parietal and frontal regions. The parietal region has paired *parietals, alispenoids,* and posterior portion of unpaired *paraspenoid.* The frontal region includes paired *frontals, orbitospenoids* and anterior portion of unpaired *paraspenoid. Supratemporal* bone is also included in frontal region and is situated at the poster, lateral angle of the skull. The *orbits* include *orbital rings,* each contaning five bones called *surpraorbital, postfrontal, preorbital, infraorbital* and *postorbital* Fig. 3.10. The orbitosphenoids and alisphenoids are replacing bones and other bones of the orbitotemporal region are dermal. (Fig. 3.10)

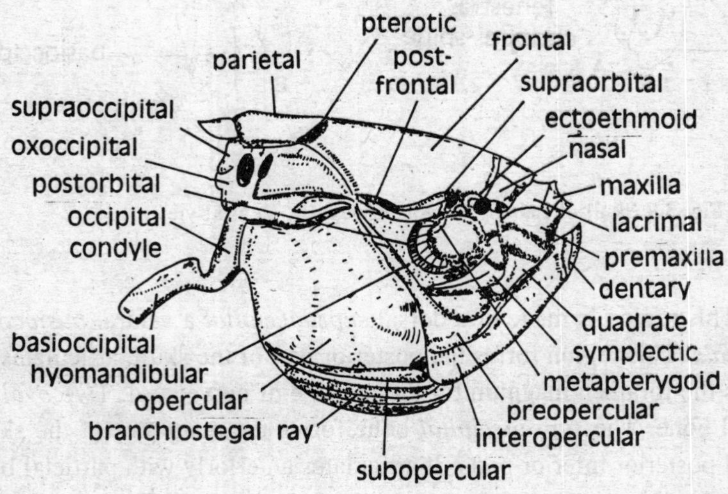

Fig. 3.10 Side view of skull of Labeo.

Ethmoid region: The ethmoid or nasal region includes paired *nasals, ectoethmoids* and *lacrimals* (Fig. 3.9.B), unpaired *mesethmoid, vomer* and *rostral* bones. These bones develop in relation to nostrils and snout. Of these bones, nasal, lacrimal and vomer are dermal bones and mesethmoid and ectoethmoids are compound bones and rostral is a replacing bone. Fig. 3.9B

Visceral Skeleton

The visceral skeleton is made of a series of seven arches encircling the pharyngeal wall. All arches are united with one another on the midventral side to form a basket-like *visceral skeleton.* The first arch is *mandibular*, the second is *hyoid arch* and remaining are *branchial arches*. First branchial arches support gills and the fifth is modified into *inferior pharyngeal bones* with large teeth bearing *masticatory plates.*

Mandibular arch (Fig. .3.10): It is divided into a dorsal palatopterygoquadrate bar forming primary upper jaw and a ventral Meckel's cartilage forming primary lower jaw. The primary upper jaw becomes intimately attached with cranium and ossified into *palatine metapterygoid* and *quadrate* bones. Two dermal bones *premaxilla* and *maxilla* form secondary upper jaw. The lower jaw is formed by paired small *articular* and *angular* bones and large *dentaries.* The dentaries unite in the middle line. The dentary is dermal bone, articular is replacing and angular is of compound origin.

Hyoid Arch: It is divided into two parts, an upper *hyomandibular* and a lower *hyoid cornu.* The hyomandibular forms the suspensorium. The support for jaws is done by paired *sympletic bones* (replacing) attached to metapterygoid and quadrate which form a rigid support for jaws together with the opercular and preopercular bones. Many paired dermal bones are associated with hyomandibular are *opercular, preopercular subopercular* and *interopercular* which make the operculum of the fish (Fig. 3.11A) The *hyoid cornu* includes paired *epithyals, ceratohyals, hypohyals* unpaired

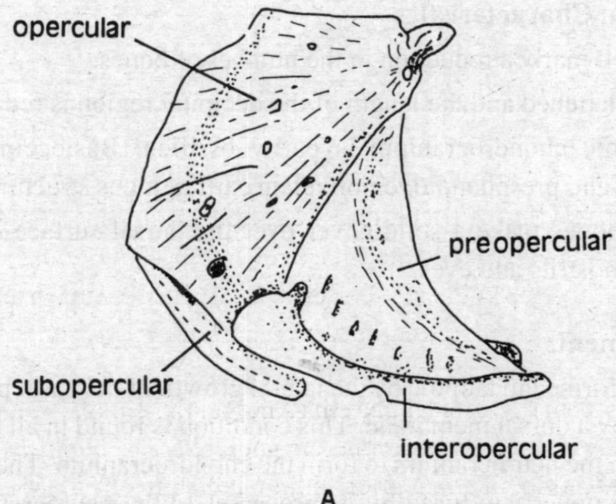

Fig. 3.11 A. An operculum

Fig. 3.11 B. Urohyal

basihyal and *stylohyals* or *interyals.* Three dermal *branchiostegal rays* (dermal bones) are attached to the lower edge of the ceratohyals One large *urohyal* (dermal) bone (Fig. 3.11.B) is present posterior to the basihyal in between epi-cerato and hypohyals of the two sides.

Branchial Arches: Each branchial arch is made of four replacing bones on either side. These are a dorsal *pharyngobranchial,* a lateral *epibranchial,* a ventral large *ceratobranchial* and a small *hypobranchial.* The third branchial arch has an unpaired median piece. The pharygobranchial of fourth arch is unossified and hypobranchial is absent. A single basibranchial piece connect right and left hypobranchials and is also connected to urohyal. The hypobranchials. basibranchials and basihyal constitute together a median ventral plate at the floor of the pharynx.

The skull of Labeo may be summarized by the table 3.6.

Skull in Class Amphibia: Characteristics

1. There is a well marked reduction in the number of bones.
2. The skull is flattened and the length of the occipital region is reduced.
3. The embryonic chondrocranium is partly ossified. Basioccipital, supraoccipital, basisphcnoid and presphenoid region retain cartilaginous structure.
4. The dermal bones make a solid cover over the dorsal surface of the skull leaving passages for nostrils and eyes.

Skull of Rana: Development

The *neurocranium is* formed in tadpole by the lateral growth of the basal plate. The roof of the neurocranium is formed by a dorsal membrane. This condition is found in all tetrapods. Olfactory and auditory capsules join the neurocranium to form the chondrocranium. The jaws are formed by the mandibular arch with upper jaw fused by its processes to the neurocranium and lower jaw supported by quadrate cartilage. This type of jaw suspension in called *autostylic.* The hyomandibular of the hyoid arch is modified into columella of the middle ear. Remaining hyoid arch and first four branchial arches form the *hyoid apparatus.* The last branchial arch forms the *arytenoids* and *cricoids.* Five cartilage bones *sphenethmoid* two *prootics* and two *exoccipitals* ossify in the skull. The Meckel's cartilage of the lower jaw ossifies to form *Mentomeckelian bone*. The sides of the skull, roof and two jaws develop dermal bones.

TABLE 3.6

Skull of Labeo

Part of skull	*Region*	*Replacing bones*	*Dermal bones*	*Mixed bones*
Neurocra- nium	Occipital region	Exoccipitals-2		Supra- occipital-1 Basiocci- pital-1.
	Auditory region (Otic region)	Pterotics-2 epiotics-2 Sphenotics-2 Protics-2		
	Orbito- temporal region	Sphenotic Orbitosphen- oids-2 Alisphenoids-2 (temporal) region	Parietals-2 Frontals-2 Supratempo- rals-2 Parasphenoid-1	
		Orbits	Supraorbitals-2 Postfrontals-2 Preorbitals-2 Infraorbitals-2 Postorbitals-2	
	Ethmoid region (Nasal region)	Rostral-1	Nasals-2 Lacrymals-2 Vomer-1	Mesenthmoid-1 Ectoethmoids-2
Splanch- Nocranium (Visceral 1 skeleton)	Mandibu- lar arch	Upper jaw Palatines-2 Metaptergoids-2 Quadrates-2	Premaxillae-2 Maxillae-2	
		Lower Articulars-2 jaw	Dentaries-2	Angulars-2
	Hyoid arch- II	Hyoman- dibula	Hyomandibulars-2 Symplectics-2	Operculars-2 Preoperculars-2 Suboperculars-2 Interoperculars-2
		Hyoid cornu	Epithals-2 Ceratohyals-2 Hypohyals-2 Basihyal-1	Urohyal-1 3 pairs of branchiostegal rays.
	Branchial arches. III to VII	Pharyngobranchials-2 Epibranchials-2 Ceratobranchials-2 Hypobranchials-2 Basibranchial-1		

Structure of the Skull: (Fig. 3.12)

The *exoccipital* bones are two in number surrounding the foramen magnum and bear two *occipital condyles* for the articulation with atlas vertebra. The *sphenethmoid* bone encircles the anterior end of the cranial cavity and extends forwards to separate the olfactory sacs. This bone supposed to be formed by the fusion of *mesethmoid, ectocthmoids, presphenoids* and orbitospenoids. It is a replacing bone. *Paraspenoid* bone is unpaired dermal bone at the floor of the cranium. The *frontoparietal* bones are two in number, fused with one another in the mid line at the roof of the cranium and with *paraspenoid* at the floor. The *olfactory capsules* are cartilaginous with paired dermal bones called *nasals* and *vomers* situated on the dorsal and ventral sides. The *nasal* makes the dorsal triangular roof of the olfactory capsule and articulates with the sphenethmoid behind and premaxilla infront. The *vomer* has pointed teeth. The *auditory capsules* are also cartilaginous, each having single *prootic* bone which articulates with the frontoparietal. The *upper jaw* consists of paired dermal bones. *Premaxillae* are present at the tip and articulate with maxillae laterally and nasal dorsally. The *maxillae* are situated at the sides and articulate with premaxilla antriorly, quadratojudgal posteriorly and palatine on the inner side. The *quadratojugal* is present posterior to maxilla and articulates with squamosal. The *palatine* connects the sphenethmoid with the maxilla, palatine, and quadrate. The *squamosal* is situated above the pterygoid and articulates with the quadratojugal. The *suspensorium* is formed by quadrate cartilage lying at the angle of the jaw. The *lower jaw* has two dermal bones dentary and *angulosplenial* (Fig. 3.13) and one cartilage bone mentomeckelian. The quadrate of the upper jaw articulates with the angulosplenial of the lower jaw. The *hyoid apparatus* (Fig. 3.14) is situated at the floor of the buccal cavity and provides support to it. It consists of cartilaginous *body* of hyoid with long recurved cartilaginous *anterior cornua* and short *posterior cornua.* The posterior cornua is a dermal bone attached to the glottis.

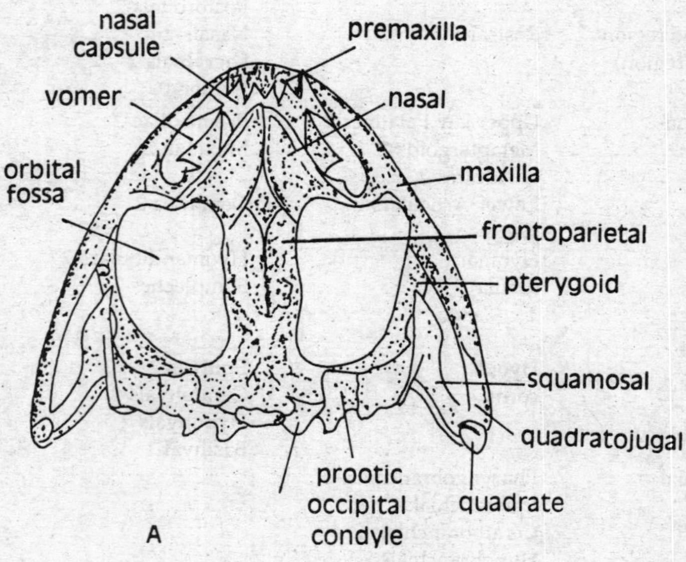

Fig. 3.12 Skull of frog **A.** Dorsal View

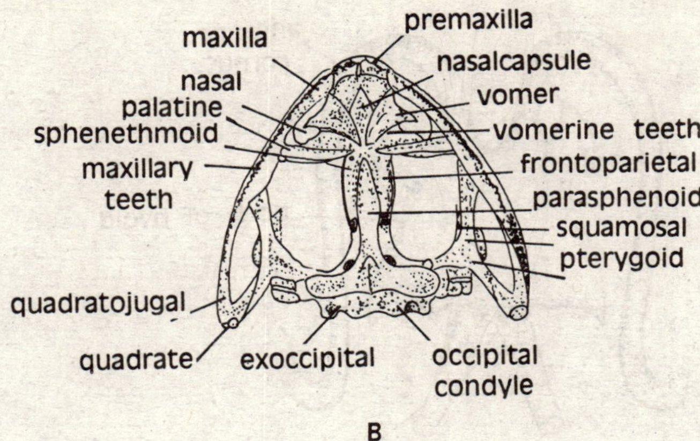

Fig. 3.12 Skull of frog, **B.** Ventral view

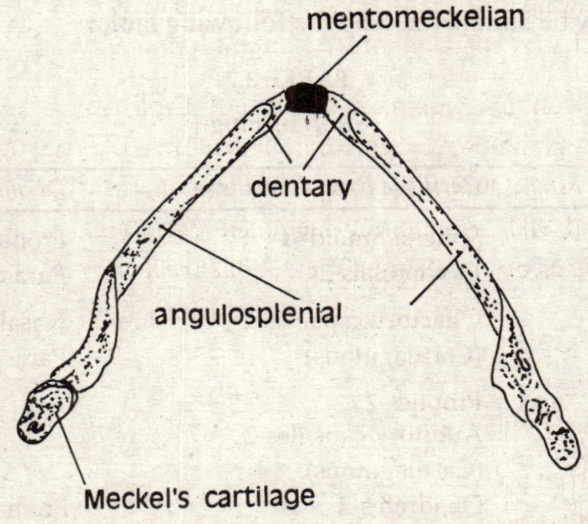

Fig. 3.13 Lower Jaw of frog

Skull in Class Reptilia: Characteristics

1. The larval chondrocranium is fully ossified excepting the *naso-ethmoid region.* The number of dermal bones is larger than in Amphibia.
2. The skull has incorporated two more segments than in Amphibia.
3. The *parietals* are usually fused together leaving an interparietal foramen.
4. An *interorbital septum* is present between two orbits. Usually two *temporal fossae* are present behind orbits providing attachment for the muscles of jaws.

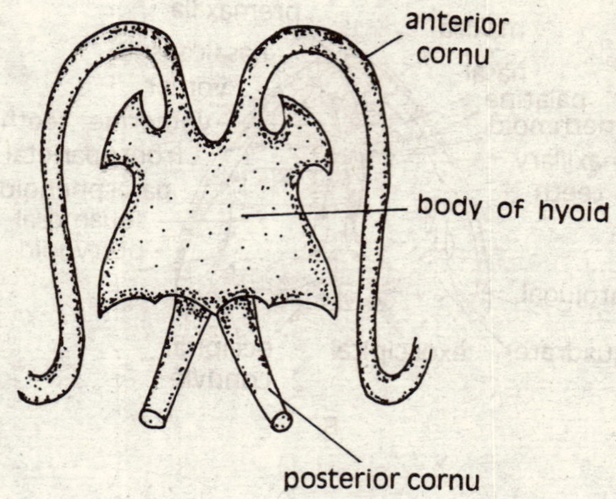

Fig. 3.14 Hyoid apparatus

The skull of frog may be summarized by the following table:

TABLE 3.7
Skull of Frog

Region	*Cartilage bone or cartilage*	*Dermal bone*
Cranium	Sphenethmoid- 1 Exoccipitals-2	Frontoparietuls-2 Parasphenoid-1
Olfactory region	Olfactory capsules-2 (Cartilaginous)	Nasals-2 Parasphenoid- 1
Auditory region	Prootics-2 Auditory capsules-2 (Cartilaginous)	
Upper jaw	Quadrates-2 (Cartilaginous)	Premaxmillae, maxillae quadratojugals, palatines, pterygoids, and squamosals, (all paired)
Lower jaw	Mentomeckelians-2 Meckel's cartilage-2 (cartilaginous)	Dentaries-2 Angulosplemals-2
Hyoid apparatus	Body of hyoid-1 (cartilaginous) (Anterior cornus-2 (cartilaginous)	Posterior cornua-2

5. *Paraspenoid* is fused with the basisphenoid.
6. Single *occipital condyle* is present in the occipital region. All the four bones of the occipital region surround the foraman magnum.
7. The *auditory region* has separate *prootic, epiotic*, and *opisthotic* bones.
8. *Ectopterygoid or transverse* and *epipterygoid* bones are present.
9. The *Meckel's cartilage* is reduced and the lower jaw is having one cartilage bone and five dermal bones.
10. The articulation of jaw is *streptostylic* with quadrate forming a movable union with squamosal on one side and lower jaw on other side.
11. The *quadratogugal* is absent.
12. The *columela* of the middle ear is formed by upper part of the hyoid arch. The lower part of the hyoid arch and the third, fourth arches form the hyoid.

Skull of Varanus

The skull of *Varanus* is more developed than that of frog and contains more bones. It is well ossified and some vacuities or fossae are present in the temporal region behind the orbit. The skull consists of the *cranium* proper, *olfactory* and *auditory* capsules, and visceral skeleton including *jaws hyoid* and *auditory ossicles* (Fig. 15, 16, 17).

The *occipital region* includes one *supraoccipital* two *exoccipitals* and one *basioccipital* which surround the *foramen magnum.* Only one *occipital condyle* is present in the basioccipital bone. The *basisphenoid* or *rostrum* is reduced and fused with the basisphenoid.

The *auditory region* includes three bones on each side, namely, *prootic, epiotic* and *opisthotic.* The prootic remains separate and epiotic and opisthotic are fused with the occipital region.

The *orbitotemporal region* contain several bones. The parietals and frontals are present on the dorsal side of the cranium. The parietals are fused together and enclose an interparietal foramen. The frontals are separate. The orbit has *prefrontal* in front, *postorbital* (post frontal) behind, lacrimal at the inner border and supraorbital above. A V-shaped *squamosal* is present behind the postorbital. Two fossae, *post tempeoral* and *supratemporal*, are present near the squamosal which represent the diapsid condition.

The *olfactory capsule* contain two closely situated *nasals*, above and two *lateral ethmoids* (septomaxillary) surround the nostril partly. The *vomers* are present on the ventral side of the olfactory capsule.

The *upper jaw,* includes *premaxillae, maxillae* and *jugal* bones on its outer side. Teeth are present on premaxillae and maxillae. The inner side of upper jaw has *vomer* articulated to premaxilla and maxilla, a *palatine* posterior to vomer, a *pterygoid* behind the palatine articulating with the basisphenoid and a *quadrate* at the posterior side providing articulation to lower jaw. *Epipterygoid* or *columella cranii* is present between pterygoid and *prootic* bones. *Transverse* or *ectopterygoid* connects maxilla with pterygoid.

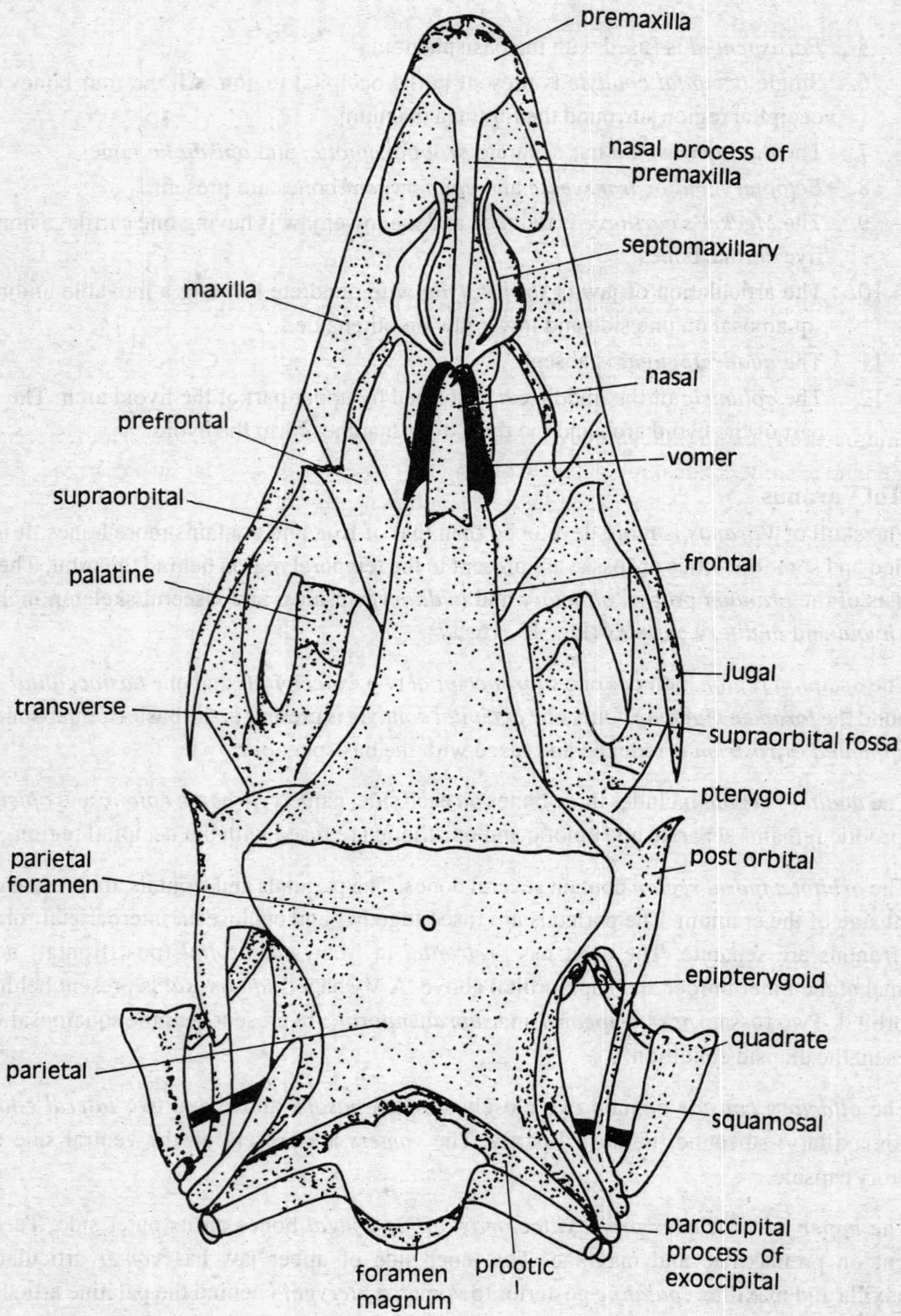

Fig. 3.15 Skull of Varanus in Dorsal View

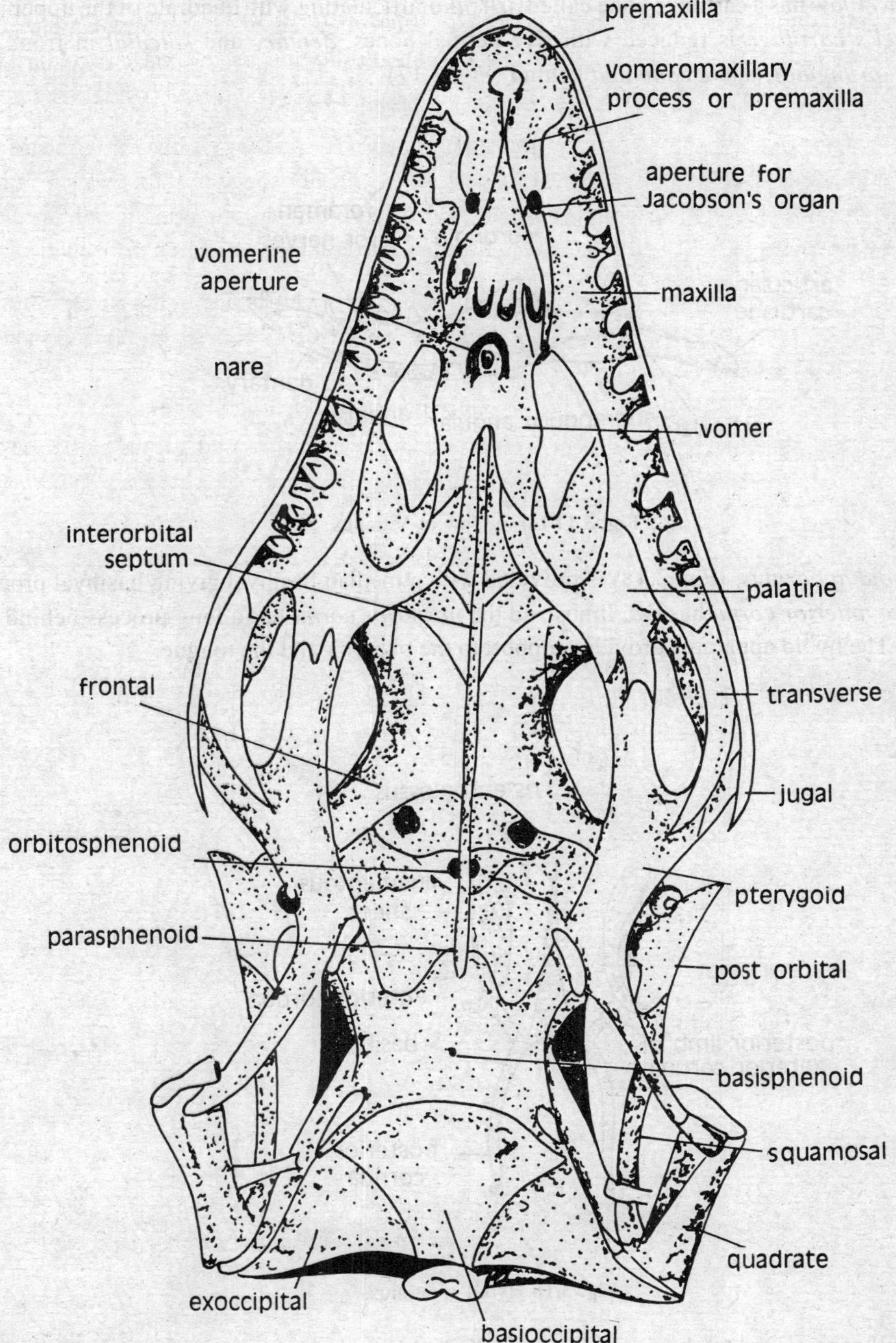

Fig 3.16 Skull of Varanus in Ventral View.

The *lower jaw* has a cartilage bone called *articular* articulating with quadrate of the upper jaw. The *Meckel's cartilage* is reduced with five dermal bones, *dentary* and *splenial* in front and *angular, suprangular,* and a *cononoid* behind (Fig. 3.17)

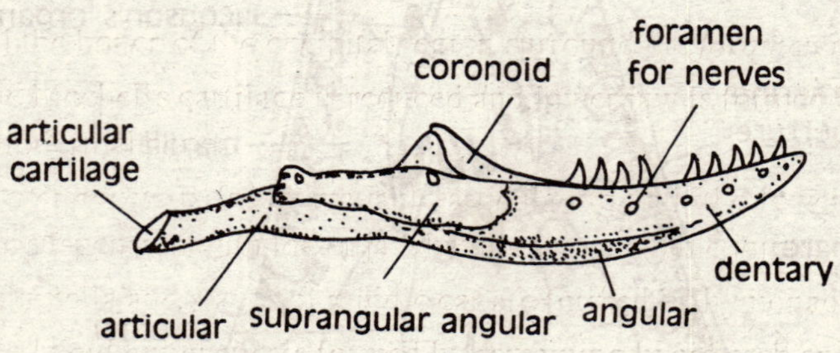

Fig. 3.17 Lower Jaw of Varanus

The *hyoid apparatus,* (Fig. 3.18) is provided with a median basihyal giving basihyal process infront. The *anterior cornu* has two limbs and the *posterior cornu* has a long process behind the basihyoid. The hyoid apparatus provides support to the pharynx and the tongue.

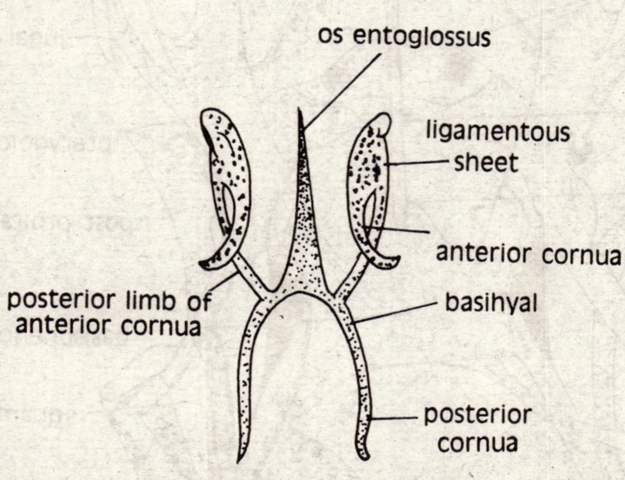

Fig. 3.18 Hyoid apparatus.

The skull of varanus may be summarized by the following table:

TABLE 3.8
Skull of Varanus

Region	*Cartilage bones*	*Dermal bones*
Occipital Region	Supraoccipital (Single) Exoccipitals (paired) Basioccipital (Single) Basisphenoid (Single)	
Auditory region	Prootic, Epiotic Opisthotic (all paired)	
Olfactory region		Nasal, lateral ethomoids (septomaxillary) vomer - all paired
Orbitotemporal region		Parietal, frontal, preferontal, postorbital, lacrimal, supraorbital, squamosal, posttemporal and supratemporal (all paired).
Upper jaw	Epipterygoid quadrate (both paired)	Premaxiallae, maxillae, jugal, vomer, palatine, pterygoid and ectopterygoid (all paired)
Lower jaw	Articular, Mekel's cartilage cartilage	Dentary, splenial, suprangular and coronoid (all paired)
Hyoid apparatus	Basihyal (medium) Posterior cornua (paired) Anterior cornua (paired)	

Skull in Class Aves: Characteristics

1. The skull resembles reptilian plan and is larger with light pneumatic bones without sutures.
2. The teeth are absent in modern birds.
3. The cranium is larger due to greater development of the brain which lies at its posterior part.
4. Large orbits lie in front of the cranium separated by interorbital septum which is more developed than is reptiles. Each orbit has thin dermal bones called *sclerotics* in the sclerotic coat of the eye.
5. Large pointed beak is formed by premaxillae and dentaries covered by horny epidermal sheath called *rhamphotheca.*
6. A single *occipital condyle* is formed by basioccipitals and exoccipitals shifted to ventral side.

7. *Infratemporal fossa* is complete with elongated jugal and quadratojugal surrounding it.
8. *Parietals* and *frontal* are large, covering the roof and lateral walls of the cranium.
9. *Parasphenoid* is replaed by basitemporal plate.
10. The bones of the auditory regtion are fused with auditory capsule, which is sunk inwards.
11. All the bones of the occipital region surround the *foramen magnum.*
12. The Meckel's cartilage is reduced and lower jaw is formed by one cartilage bone and four dermal bones.
13. The *quadrate* is freely mavable articulating with the lower jaw making *streptosylic jaw* suspension with lower jaw having two movable joints.
14. Columella and stapes of middle car are formed by hymonadibular.
15. The *hyoid apparatus* is formed by remaining hyoid arch and the first branchial arch.

Skull of Gallus

The skull of *Gallus* (fowl) has spongy bones and is light with a few sutures. It presents characteristics similar to reptilian skull.

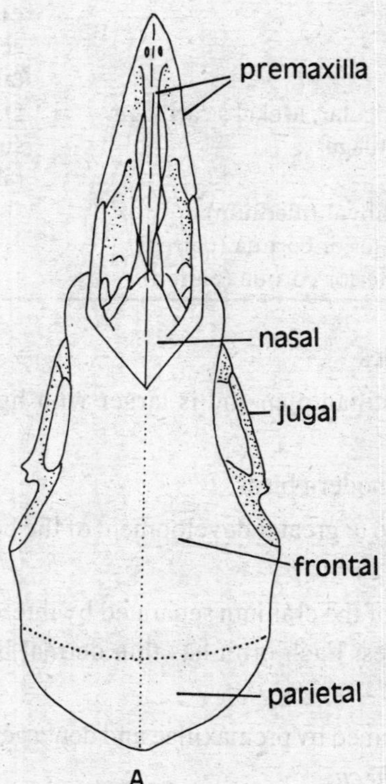

Fig. 3.19 A. Skull of Gallus (Dorsal view)

Cranium

The *occipital region* has downward directed *foramen magnum* surrounded by a dorsal *supraoccipital*, two *lateral exoccipitals* and a *ventral basioccipital* with a single *occipital condyle.*

The *orbitotemporal region* contains a large *basitemporal* plate with *basisphenoid bone* below the basioccipital and a long *rostrum.* The *rostrum* and *basitemporal* have been regarded equivalent of parasphenoid of lower forms. The *orbitosphenoid* and *alishphenoid* bones are present on sides. The skull roof has large paired *parietals* and *frontals.* The articulation of parictals with supraoccipital ' is conspicuous due to the presence of *lamboidal ridge.* The *orbits* are large separated by an *interorbital septum* which is part of mesethmoid. *Lacrimal* is present in fot of orbit and *opisthotic* bones are intimately fused with occipital bones.

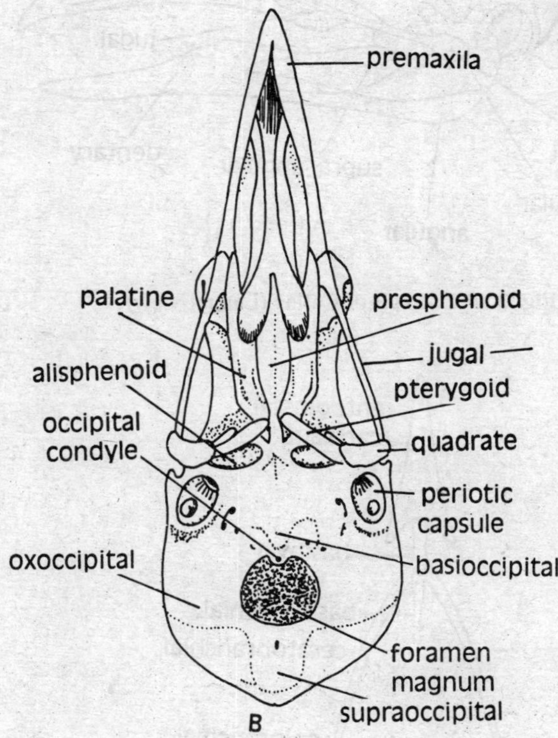

Fig. 3.19 B. Skull of Callus (Ventral view)

Visceral Skeleton

The *upper jaw* of each side has a *premaxilla* and *maxilla.* The premaxillae of two sides unite in middle forming a beak without teeth. The jugal and *quadratojugal* bones are slender and surround and *infratemporal fossa.* A *palatine* and a *pterygoid* are present on the inner side of the upper jaw. The *pterygoid* articulates with vomer which is present in front of palatine.

The *lower jaws* of each side are fused in the middle. *Articular* bone articulates with *quadrate* and four dermal bones *dentary, splenial, supra-angular,* and *angular* are intimately fused together. Teeth are absent in both jaws.

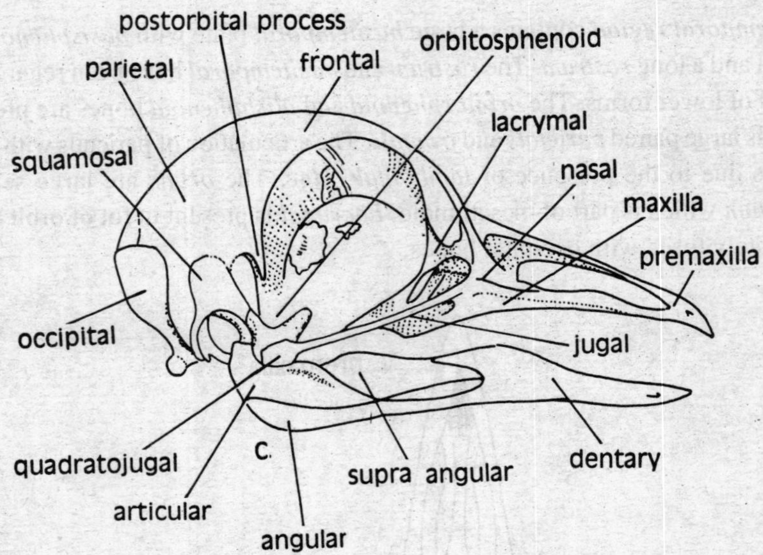

Fig. 3.19 C. Skull of Callus (Lateral view)

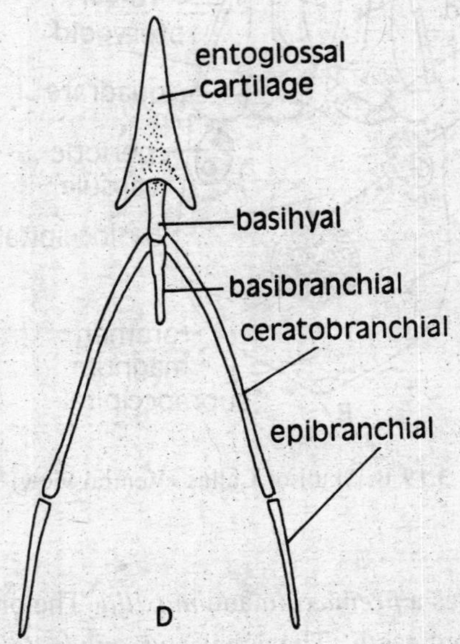

Fig. 3.19 D. Hyoid Apparatus

The *Hyoid apparatus* supports the tongue and contains three median bones, *entoglossal, basihyal* and *urohyal (basi-branchial). Anterior cornuae* are short paired processes given from *entoglossal. Posterior cornuae* are long, paired, posterior processes arising from the junction of basihyal and urohyal.

The skull of Gallus may be summarized by the following table:

TABLE 3.9

Skull of Gallus

Region	*Cartilage bones*	*Dermal bones*
Occipital region	Supraoccipital (Single) Basioccipital (Single) Exooccipital (Paired)	
Orbitote-mporal region	Basisphenoid (single) Mesethmoid (single) Orbitosphenoid (paired) Alisphenoid (paired)	Rostral (Single) Lacrimal, Nasal, Parietal, Frontal (all paired)
Auditory region	Prootic, Epiotic and Opisthotic (all paired)	
Upper jaw	Quadrate (paired)	Premaxillae, Maxillae, jugal Vomer, Quadrato- Jugal, Paltine, Pterygoid (all paired).
Lower jaw	Articular	Dentary, Splenial, Supraangular, Angular (all paired)
Hyoid Apparatus	Anterior and posterior Cornuae (paired) Basihyal (single) Entoglossal (single) Urohyal (single)	

Skull in class Mammalia: Characteristics

1. Considerable reduction in the number of bones by fusion or loss is found in mammals. The *parasphenoid, quadrato-jugal, prefrontal, postfrontal, postorbital quadrate* and lower jaw bones excepting *dentary* are lost. In some mammals all the bones of occipital region are fused into one bone. Bones of temporal region are fused into one bone. Bones of temporal region, *basisphenoid* and *alisiphenoid* may fuse.
2. The dermal bones have become closely joined to cartilage bones forming sutures at the edges.

3. The *cranium* is large to accommodate large brain and its cavity is closed anterior by cribriform plate.
4. Two *occipital condyles* are present and skull remains at right angle to the vertebral column.
5. All the bones of auditory region fuse into *petrosal* which encloses the inner car. The tympanic bone forms an *external auditory meatus* with rounded tympanic bulla enclosing ear ossicles.
6. The *quadrate* forms *incus, articular* forms *malleus* and the *hyomandibular* becomes *stapes.*
7. The jaw sespension is between *squamosal* and dentary (craniostylic).
8. Lacrimal, maxilla, orbitsphenoid, palatine, and ethmoid in some cases form the orbit.
9. Prémaxillae, maxillae, palatines and pterygoids in some cases form the bony palate. The *palate* pushes the internal nares backwards and separates the respiratory channels from the buccal cavity so that mammals are able to chew and breathe simultaneously.
10. An *internasal septum* formed by mesethmoid separates the two respiratory channels. Ectoethmoids, nasals, and maxillae extend into nasal chambers by *turbinals* to provide warming system for the inhaled air.
11. A *zygomatic arch* enclosing a single *temporal fossa* is formed by jugal and a process of squamosal.
12. A single large *dentary bone* makes half of the lower jaw and articulates with glenoid fossa of the squamosal.
13. The hyoid apparatus is formed by the hyoid arches.
14. The cartilages of larnynx and rings of trachea are formed by remaining branchial arches.

Skull of Oryctolagus

The skull of *Oryctolagus* (Fig. 3.20, 3.21) is small with small orbit. Lower jaw articulates with *squamosal*. The brain case is large with double occipital condyle. The dentition is *heterodont* and the teeth are present on premaxilae, maxillae and dentaries.

Cranium

Occipital region: The basioccipital bone forms base of the skull and lower parts of the occipital condyle.

Exoccipitals are present on either side of the foramen magnum forming upper part of the occipital condyle. The *supraoccipital* surrounds the foramen magnum dorsally.

Orbitotemporal region: A *basisphenoid* is anterior to basioccipital. *Parietals* are paired dermal bones forming roof the cranium on the mid dorsal line. Paired *alishphenoids* form side of the cranium ventrally. Between parietals and supraoccipital a small median dermal bone *interparietal* is situated. A median dermal bone *parasphenoid* is situated infront of basisphenoid. Paired *orbitosphenoid,* bones form the sides of the cranium near orbits. Paired *forntals* form the roof and sides of the cranium on anterior side, each giving supraorbital process projecting over the orbit. The *squamosal* of upper jaw fills in the space between exoccipital and alisphenoid bones. The *ethmoid* bone is present on the anterior side of the skull of the skull abd has cribrifrom plate with perforation for the exit of the olfactory nerves. A zygomatic arch protects the orbit from out side and is formed of processes from maxillae, *jugal* and *squamosal.* A small bone *lacrimal* perforated by a foramen is situated between the frontal and maxilla.

Auditory region: The auditory capsule is small in size and encloses internal ear. It consists of *pterotic* bone having apertures. On the outer surface of pterotic bone a *tympanic bone* is closely applied between basisphenoid and squamosal. The tympanic bone is flask shaped with rounded tympanic bulla having tympanic cavity and auditory ossicles. The *auditory meatus* is enclosed in the tubular part of tympanic bone. *Malleus, incus,* and *stapes* are three auditory ossicles.

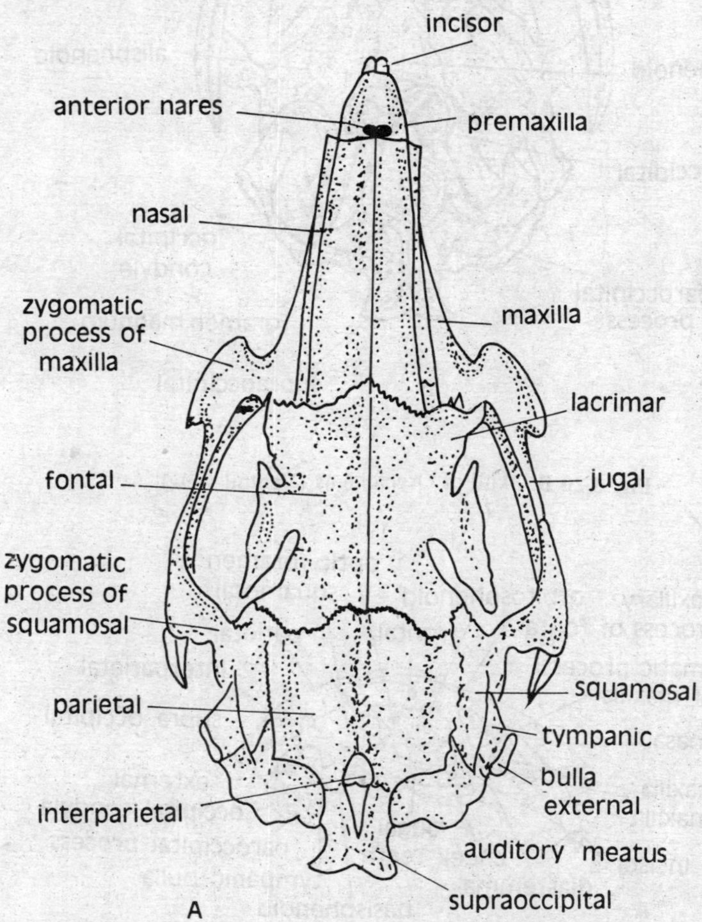

Fig. 3.20 A. Skull of Oryctolagus (Dorsal view)

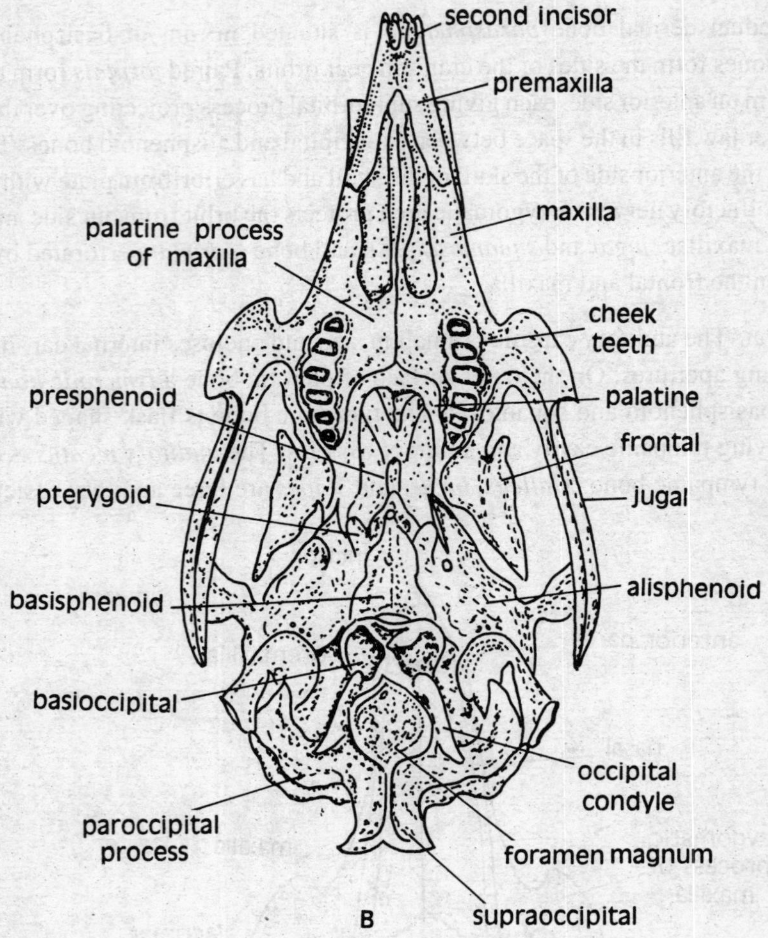

Fig. 3.20 B. Skull of Oryctolagus (Ventral view)

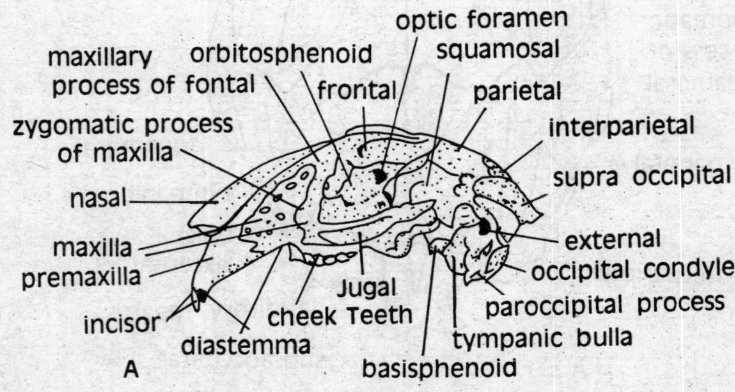

Fig. 3.21 A. Skull of Oryctolagus (Side view)

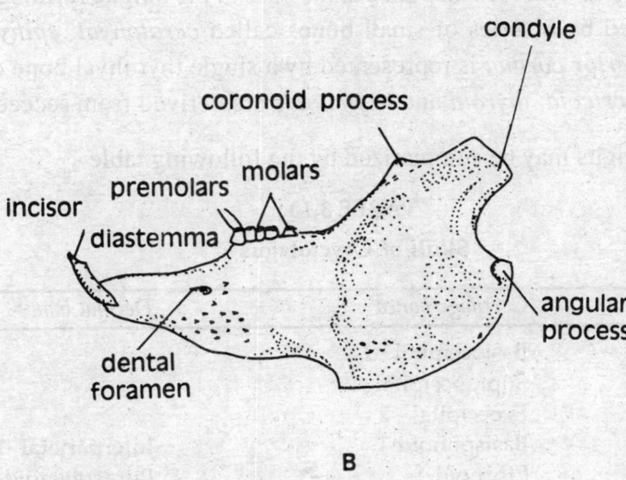

Fig. 3.21 B. Skull of Oryctolagus (Lower jaw)

Olfactory region: The nasal capsule is covered over by the *nasal* and below by the *vomer.* A vertical *mesethmoid* bone lies infront of *Cribiform plate* separating two nasal cavities. Three thin scrolled bones called *turbinals,* extend as outgrowth of maxillae, nasals and ethmoid inside each nasal chamber.

Upper jaw: The jaw articulation is autostylic and craniostylic with ossification of palatopterygoquadrate bar having different structure and functions. The *palatine* is present at the posterior part of palate below basisphenoid and articulates with maxilla and pterygoid. The *quadrate* bone is absent and forms the incus of the auditory capsule. The *articular* is modified into the malleus of the auditory ossicles. The premaxillae are paired with sockets for incisors and two processes; nasal process extends backwards to meet frontal and palatine process extends backward meeting its fellow to form the floor of the nasal cavity. The *maxillae* lie posterior to premaxillae and have sockets for premolars and molars. The maxillae join the palatines on the posterior side. The hard *plate* is formed by the extension of the shelves of premaxillae and maxillae. Between premaxillae and maxillae a pair of *nasopalatine foramina* are present. The *pterygoid* bones are attached to the cranium base near the junction of alisphenoid with basisphenoid. The *jugals* form a part of zygomatic arch.

Lower jaw: A single membrane bone dentary constitutes half of the lower jaw. The two halves are joined in the middle line by a symphysis on the anterior side. The *dentary* has a posterior ascending process with condyle for the articulation with squamosal of the upper jaw. The dentary has sockets for incisors, premolar and molars.

Hyoid arch: The body of hyoid is embedded at the base of the tongue forming bony palate. The anterior cornua is formed by a series of small bones called *ceratohyal. epihyal, stylohyal* and *tympanohyals.* The *posterior cornua* is represented by a single thyrohyal bone on each side. The cartilages of larynx e.g. *cricoid, thyroid* and *arytenoids* are derived from secceeding arches.

The skull of Oryctolagus may be summarized by the following table:

TABLE 3.10

Skull of Oryctolagus

Region	*Cartilage bones*	*Dermal bones*
Occipital region	Basioccipital - 1 Supraoccipital -1 Exoccipital - 2	
Orbitotemporal region	Basisphnoid-1 Ethrnoid-1 Alisphenoids-2 Orbitosphenoids-2	Interparietal -1 Purasphenoid- Parietals-2 Frontals - 2 Lacrimals-2
Auditory region	Pterotics-2 Tympanic-2 Tympanic bulla-2 Malleus -2 (articular) Incus-2 (hyamandibular) Stapes-2	
Olfactory region	Mesethmoid-1	Nasals-2 Vomer-2 Turbinals of Maxillae, nasals and ethmoid
Upper jaw	Incus (Quadrate)-2	Palatines -2 Premaxillae-2 Maxillae-2 Pterygoid-2 Jugals-2
Lower jaw	Dentaries-2	
Hyoid arch	Body of Hyoid-1 Anterior cornus-2 Posterior cornus-2 (Thyrohyal)	
Visceral arches	Cricoid, Thyroid, Carti- Lages of traches-1 each Arytenoid cartilages-2	

Elements of skull in selected vertebrate have been shown in Table 3.11.

VERTEBRAL COLUMN

The vertebral column makes the longitudinal axis of the body posterior to skull. It is made of a series of elements called vertebrae which are metameric. The primary axial skeleton is formed by a flexible rod notochord desribed in section 3.11.

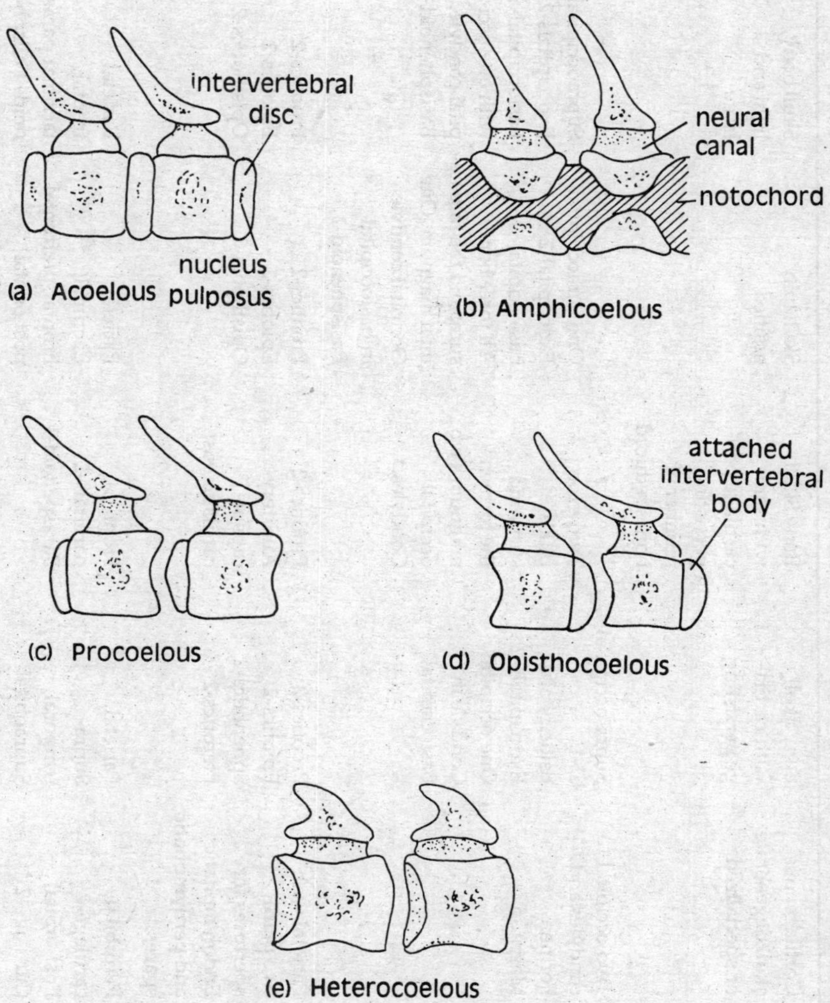

Fig. 3.22 General centra shapes. The shapes of articulating centra ends, as viewed in sagittal sectio, define specific anatomical types: (a) acoelous, both ends are flat; (b) amphicoelous, both ends are concave; (c) procoelous, anterior end is concave; (d) opisthocoelous, posterior end is concave; (e) heterocoelous, saddelike articulating ends. Anterior to the right.

TABLE 3.11

Element of Skull in Selected Vertebrates

Region	*Scoliodon*	*Labeo*	*Rana*	*Varanus*	*Gallus*	*Oryctolagus*
General Characteristics	Cortilagenous skull:degenerate or specialized	Bony Skull with cartilage persisting	Bony Skull with some cartilage per persisting, number of bones reduced	Skull fully ossified	Skull bony, light and large	Skull bony large and heavy
Occipital Region	Two occipital condyles and Formen Magnum	Supra occipit al-1 al-1 Bsiocipital-1 Exoccipitals-2 One occipital Condyle in Basioccipital	Two exocci-pitals surround the formen magnum Two occipital Condyles.	One Suraocoipital Exoccipital-2 basioccipital- 1 All three bones surround the fore-men magnum. One Occipital condyle in basioccipital Basisphenoid-1	Supraoccipital-1 Exoccipitals-2 Basioccipital- 1 with one occi-pital condyle. Basisphenoid- 1	Supraoccipital- 1 Exoccipitals-2 One basioccipital with two occipital Condyles Occipital segement intimately fussed.
Auditory Region	Parietal with two paird apertures for Endolymphatic and perilymphatic spaces.	Prootics-2 Epiotics-2 Sphenotics-2 Pterotics-2	Prootics-2 Auditory-capsule cartilaginous	Prootics-2 Epiotics-2 Opisthotics-2	Prootics-2 Epiotics-2 Opisthotics-2	Pterotics-2 Tympanic bulla-2 Malleus-2 Incus-2 Stapes-2
Orbito Tempornal Region	Perorbital cartilage-2 Post-orbital Cartilage-2 Suborbital Ridge Supraorbital	Parietal. Supra-temporal, Supraobital Post-frontal Preorbitals, Infraorbitals Postorbitals,	Front parietals-2 Paraspenoid-1	Parietal Lacrimal, Frontal, prefrontal post orbital Supraorbital Squamosal, post Temporal, Supra-temporal, -All	Parietal frontal, lacrimal parasphe noid- 1 alisphenoid Orbitosphenoid- All paired. Squamostal, Lacrimal-	Basisphenoid-1 Ethmoid- 1 Interparietal-1 Alisphenoid, Orbitosphenoid, Parietal frontal

	Ridge	Orbito- Sphenoid, Alisphenoid Parasphenoid- 1		paired.		All paired Zyomatic arch made of squamosal and jugal.
Ethmoid Region Olfactory Regional	Two large plfactory capsules Internasal Septum Rostrum	Nasal, lacrimal vomer, Ecto- ethmoid All paired Rostral and mesethmoid- Both unpaired	Nasal,Vomer both paired cartilaginous, olfactory capsules Paired	Nasal, Lateral ethmoid (Septomaxillary) vormer- all paired	Nasal-paired mesethmoid-1 rostral (unpaired)	Nasal-2 Mesethmoid-1 vomer-2 Turbinals of maxilla, nasal and vomer.
Manaibular Arch		Upper jaw:	Upper jaw:	Upper jaw:	Upper jaw:	Malleus (articular)-2 Statpes Hyomandibular)2 Incus(Quadrate)-2
1 pair of visceral arch (Upper and lower jaw)	Palatoptery- goquadrate cartilage (form upper jaw).	Palatines-2 Metaptery- guide-2 Quadrates-2 Premaxillae Maxillae	Palatines-2 Petrygoids-2 Squamostal-2 Quadrates-2 (cartilagnaus) Premaxillae-2 Maxillae-2 Quadratejugal-2	Palatines-2 Pterygoids-2 Epiptery goids-2 Ecotoptery goies-2 Quadrates-2 Premaxillae-2 Maxillae-2 Jugals-2	Palatines-2 Pterygoids-2 vomer -2 Quadrates-2 Premaxillae-2 maxillae-2 . Jugals-2 Quadratojugals-2	Palatines-2 Premaxilae-2 Maxillae-2 Pterygoids-2 Jugals-2
	Meckel's cartilage (forms lower jaw) Hyostylic jaw Suspension.	Lower jaw: Articulars-2 Angulars-2 Dentaries-2 Hyostylic jaw Suspension	Lower jaw: Mentomeckelian-2 (cartilaginous) Angulosplenials-2 Dentaries-2 Meckel's Cartilage-2 Autostylic jaw Suspension	Lower jaw: Articulars-2 Splenials-2 Sppraangulars-2 Dentaries-2 Meckel's cartilage-2 Autostylic jaw suspension	Lower jaw: Articularis-2 Splenials-2 Supraangulars-2 Dentaries-2 Angulars-2 Autostylic jaw Suspension	Dentaries-2 jaw suspension is cranioustylic
Hyoid Arch	Basihyal-1	Basihyal-1 Urohyal-1	Hyoid apparatus	Hyoid apparatus	Hyoid apparatus	Hyoid apparatus

Region	*Scoliodon*	*Labeo*	*Rana*	*Varanus*	*Gallus*	*Oryctolagus*
(II pair of visceral arches)	Ceratohyals-2	Epi-.certo and Hypohyals.	Body of Hyoid-1	Basihyoid-1	Basihyal -1	Body of Hyoid- 1
	Hyomandibular (Dorsal)-2	-2 each, Antirior ocrnus-2 Hyomandibulars.	Antirtor comun-2	Enoogloseal- 1	Anterior corunua-2	
		Symplectics, Operculars, Preoperculars. Suboperculars, Interoperculars -2 each and 3 pairs of Branchiostegal Rays.	Posterior cornua-2	Postenor cornua-2	Urohyal-1 Anterior cornua-2 posterior coraua-2	posterior cornuna-2
Branchial Arches	All paired-pharyngobran-Chials	paired pharygobranchials	Unpaired cricoid and thyroid			cartilages.
III to IV	IV, V fused. Epibranchials, Ceratobranchials Hypobranchial (Vabsent) Basibranchial- 1 (median)	(IV-unossified) Epibranchials Ceratobranchials Hypobranchials Basibranchial- 1 (meian)				paired arytenoids cartilages. tracheal cartilages.

Structure of typical vertebra (Fig. 3.22, 3.23): Considerable diversity of structure is found in the vertebrae of different animals with basic fundamental plan. Each typical vertebra has a *centrum* formed either by enclosing the notochord or by the replacement of the notochord. It may be concave or funnel shaped at each side (*amphicoelous*); concave posteriorly and convex anteriorly (*Opisthococous*); concave anteriorly and convex posteriorly (*precocious*): without cavity (*acoelos* or *amphyplatyan*): or saddle shaped at both sides (*hetrocoelous*) (Fig. 3.23). Above centrum, *neural arch* is situated with a dorsal neural spine through which the nerve cord passes. A *haemal arch* is found in some animals below the centrum, which encloses a *haemal canal* allowing the passage of caudal artery and vein. The haemal arch is produced into a *haemal spine*. The vertebra has different types of articulating surfaces or *apophses* including *zygapophyses, diapophyses, parapophyses, haemapophyses, pleuropophuses* and *hypapophyses*. The zygapophyses are articulating surface arising from the basal region of the neural arch between successive vertebrae. The prezygapophyses are anterior processes pointing upward and postzygapophyses are posterior processes pointing downwards. The diapophyses are called *tansverse processes,* which arise from the neural arch or centrum and provide attachment to the dorsal head of the two-headed ribs. The *basapophyses* arise from the ventrolateral side of the centrum and represent haemal arch or when haemal arch is present, provide attachment to them. The *pleurapophyses* are lateral projection of the centrum providing place for fusion to the short ribs. The *hypapophysis* is mid ventral extension of the centrum for providing attachment to the muscles.

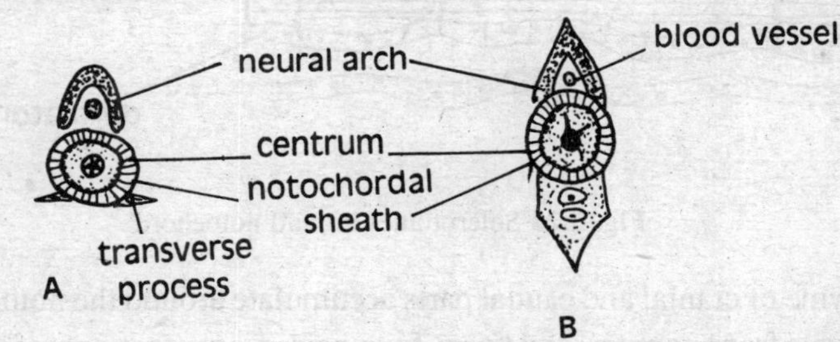

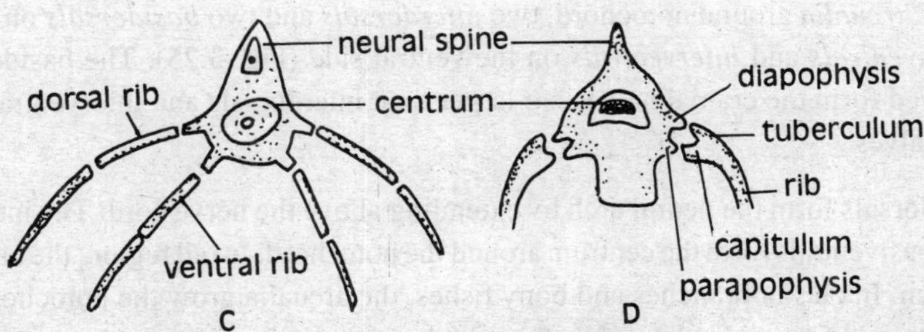

Fig. 3.23 Different parts of vertebral & their articulation facets

Development of the Vertebral Column

In vertebrates, the notochord serves as a foundation for vertebrae formation by its partial or complete replacement by cartilaginous or bony vertebrae. The metameric *sclerotomes* (Fig. 3.24) produce mesenchyme by their proliferation. The mesenchyme migrate into the space between notochord and the mesoderm and eventually surround the neural tube completely. The sclerotome, median to each myotome is differentiated into a posterior dense (caudal half) and an anterior less dense (cranial half) rudiment of the vertebra. Each caudal half fuses with the cranial half of the succeeding vertebra. In this way, the definitive vertebrae come to lie interesegmentally, alternatively with the myotomes, connecting each myotome with two vertebrae and two successive ribs.

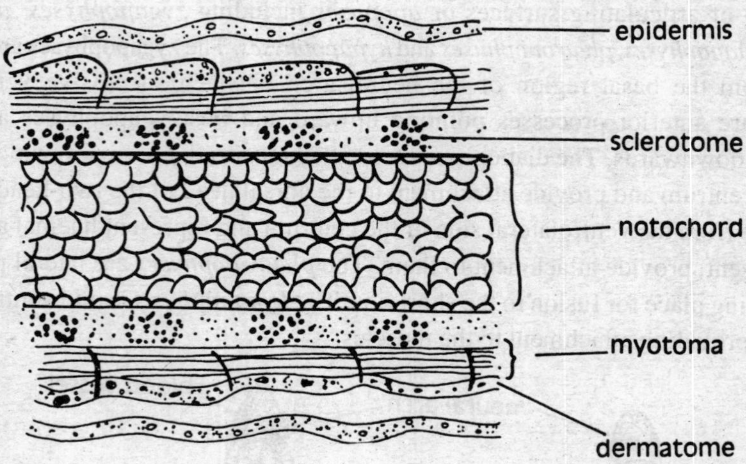

Fig. 3.24 Sclerotomes around notochord

The mesenchyme of cranial and caudal parts accumulate around the notochord and neural tube after breaking away from somites and form four paired aggregations of mesenchyme in each segment on the dorsal and ventral side of the notochord. These aggregations chondrify later and form four pairs of *arcualia* around notochord, two *interdorsals* and two *basidorsals* on the dorsal side and two *basiventrals* and *interventrals* on the ventral side (Fig. 3.25). The basidorsals and basiventrals derived form the cranial halves are larger than interdorsals and interventrals derived from the caudal halves.

The two basidorsals form the neural arch by extending above the nerve cord. The interdorsals. interventrals and basiventrals form the centrum around the notochord. In tail region, the basiventrals form a haemal arch. In clasmobranches and bony fishes, the arcualia grow the notochord outside notochordal sheath and form a *perichordal ring* which grows and ossifies and extends over the notochord making a *perichordal centrum.* In some cases, the mesenchyme lying between sclerotormes and the notochord forms the perichordal centrum and arcualia play a minor role in its formation. Considerable diversity is found in the formation of centra in different groups of vertebrates.

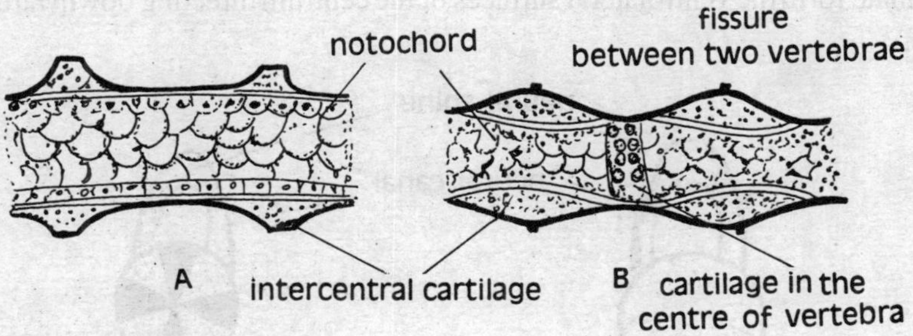

Fig. 3.25 Development of a vertebra

The length of the vertebral column increase by the growth of vertebrae themselves. The notochord becomes beaded in appearance resulting into intervertebral enlargements giving rise to amphioelous centra of fishes. The centra may be calcified fishes in radiating form or in concentric rings. The centra may be completely bony in teleosts and tetrapoda.

Vertebral Column in Vertebrate Series

Class Chondrichthyes

Scoliodon (Fig. 3.26): The vertebral column of *Scoliodon* consists of a chain of cartilaginous vertebrae bound together by fibrous connective tissue. The vertebrae develop around a persistent notochord. A typical vertebra consist of an *amphioelous centrum,* a dorsal *neural arch* around spinal cord and a pair of *transverse processes* projecting ventrolaterally from the centrum. The centra are strengthened by calcified fibrocatilage forming four wedges traversing the body of the centrum from the its periphery upto the centre. Such a centrum is called asterospondylus type. The posterior vertebrae bear *haemal arch* on the ventral side of the centrum enclosing haemal canal and giving haemal spine to support the ventral lobe of the caudal fin. The posterior most end of the vertebral column is bent upwards supporting an asymmetrical heterocercal tail.

Class Osteichthyes

Labeo (Fig. 3.27): The vertebral column of *Labeo rohita* contains 37-38 vertebrae which are fully ossified. It includes an anterior trunk region containing 21 trunk vertebrae bearing movable ribs and a posterior caudal region having 16-17 caudal vertebrae without ribs and haemal arches. The anus marks transition between two types of vertebrae. The first four vertebrae are modified in relation with air bladder and internal ear. The last three caudal vertebrae are modified into an upwardly directed rod-like urostyle to provide support to caudal fin structure.

A typical vertebra is made of deeply biconcave centrum. A pair of backwardly directed processes arise form the anterolateral borders of centrum, enclose the spinal cord and unite above to form the neural arch which is produced into a pointed, backwardly directed neural spine The pre-zygapophyses are present at the bases of the neural arch as small blunt processes. The postzygapophyses arise from the posterolateral edges of the vertebra pointing upwards and backwards. A pair of short parapophyses originate form the ventrolateral surfaces of the centrum directing downwards providing attachment to ribs.

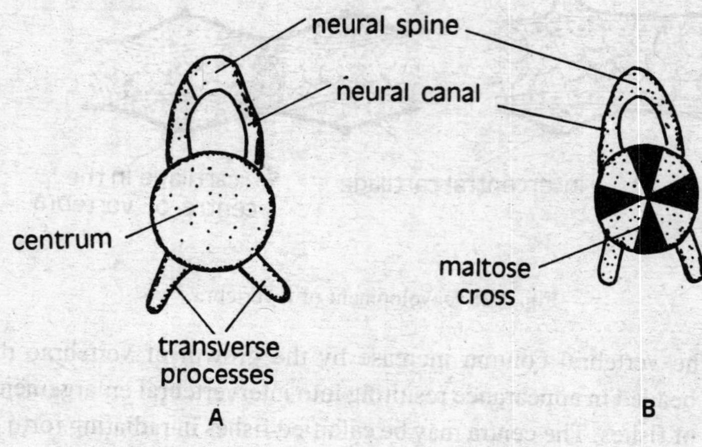

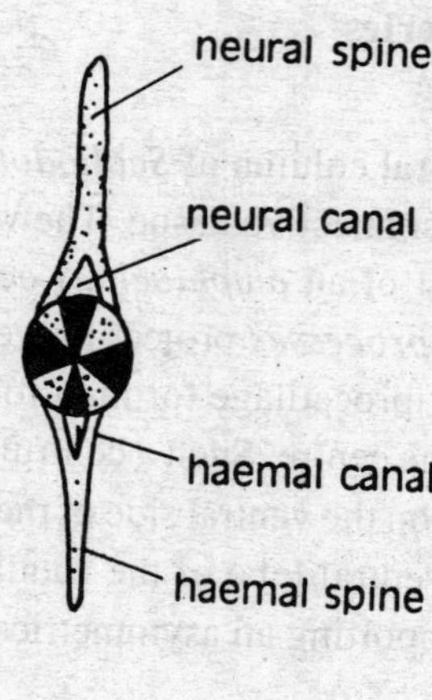

Fig. 3.26 Vertebral of a cartilegenous fish

A typical caudal vertebra has amphicoelous centrum with four depressions, one median dorsal, one median ventral and two laterals. It has a long backwardly directed neural spine and zygapophyses like the trunk vertebra. A pair of backwardly directed processes called haemal arches meet in the midventral line to form a canal for caudal blood vessels, producing a backwardly directed haemal spine. One pair of blunt small processes from the bases of haemal arches and another pair of similar processes from the posterolateral side of the centrum are also present.

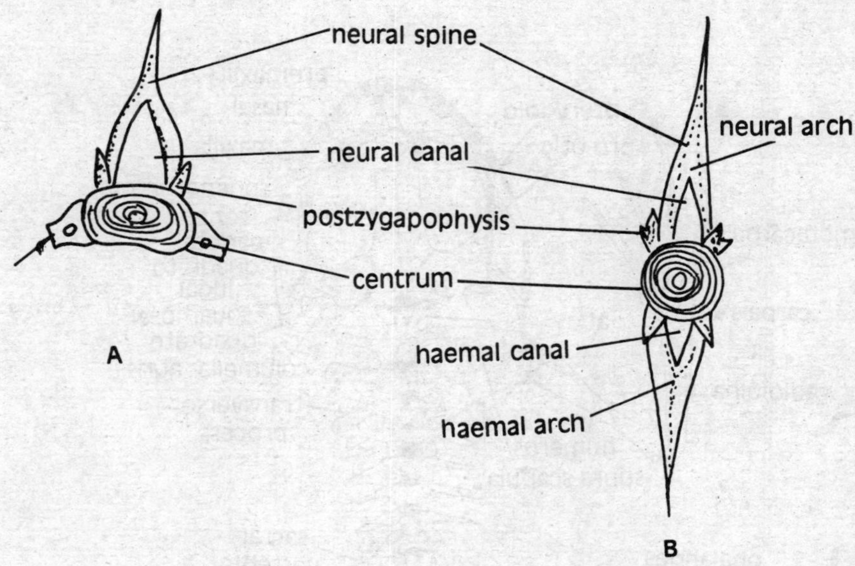

Fig. 3.27 Vertebral of a bony fish.

Class Amphibia

Rana (Fig. 3.28, 3.29): The vertebral column of frog includes 10 vertebrae. A typical vertebra is made of solid procoelous centrum with transverse processes on either side and neural arch on the dorsal side with neural spine. The prezygapophyses and postzygapophyses are present on the neural arches. The first, eighth, and ninth vertebrae have different structure.

The first vertebra is *Atlas vertebra* which lacks transverse processes and prezygapophyses. The centrum and neural spine are reduced.

The *eighth vertebra* has amphicoelous centrum.

The *ninth* or *sacral* vertebra has a centrum with biconvex structure anteriorly and double biconvex structure posteriorly. The two posterior convexities fit into the concavities of the tenth vertebra and its strong transverse processes provide articulation to the ilia of the pelvic girdle. The *tenth* or *urostyle* vertebra is elongated structure tapering gradually on the posterior side. On dorsal side, it is raised, and bears two concavities the anterior side for the attachment of double convexities of the ninth vertebra.

Class Reptilia

Varanus (Fig. 3.30): The vertebral column of *Varanus* is fully ossified and includes a large number of vertebrae. It is divisible into four regions: *Cervical, thoracolumber, sacral,* and *caudal.* All the vertebrae are deeply precoelous.

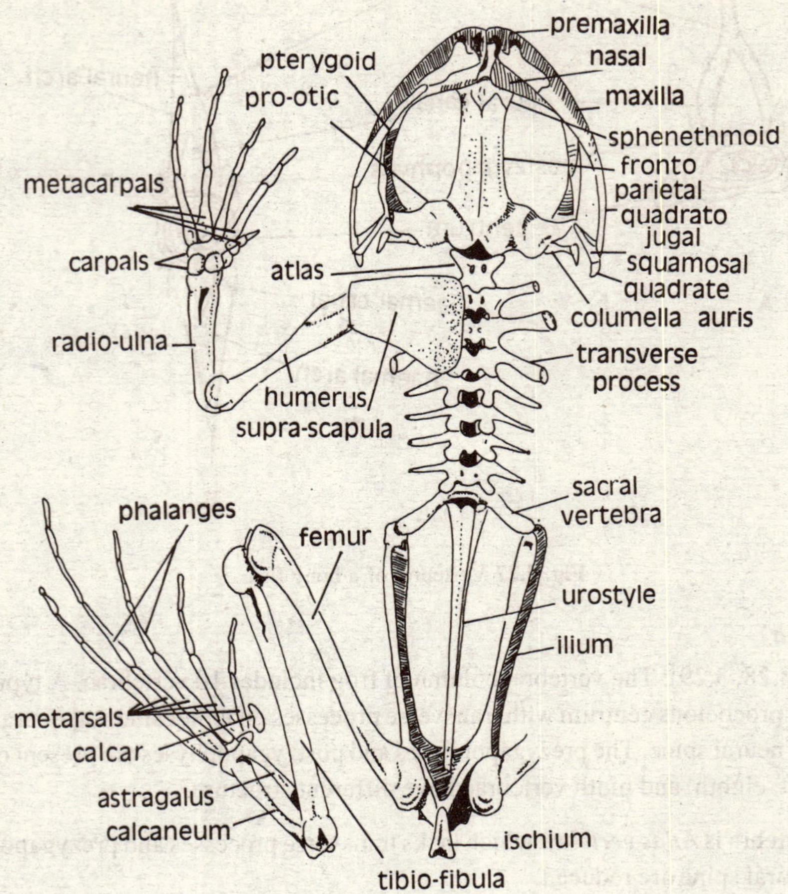

Fig. 3.28 Skeleton of frog (left Side only)

Cervical region: There are eight cervical vertebrae with first two modified and specialised. A typical cervical vertebra has a stout *procoelous centrum,* a pair of neural arches projecting upward and uniting above the centrum to form neural spine. *Pre* and *postzygapophyses* are present. A backwardly directed posterior process, the *hypapophysis* is present on the ventral surface. The first cervical is called *atlas vertebra.* It is ring-like without centrum and articulates with the skull. The ring is made of three pieces; one ventral smaller piece bearing a downward projecting ventral spine, and two dorsolateral larger pieces arching over the ventral piece and meeting on mid-dorsal side. A *transverse ligament* divides the atlas vertebra into two parts. The second cervical is called *axis* vertebra. It has a flat neural spine with small prezygapophyses and well developed postzygapophyses. The centrum bears a borad odontoid process on its anterior face.

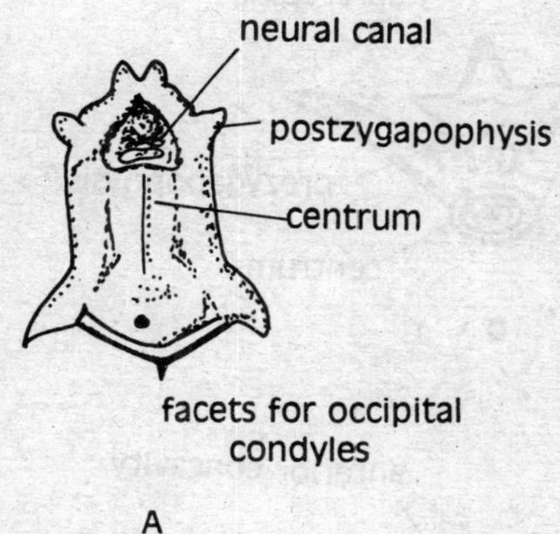

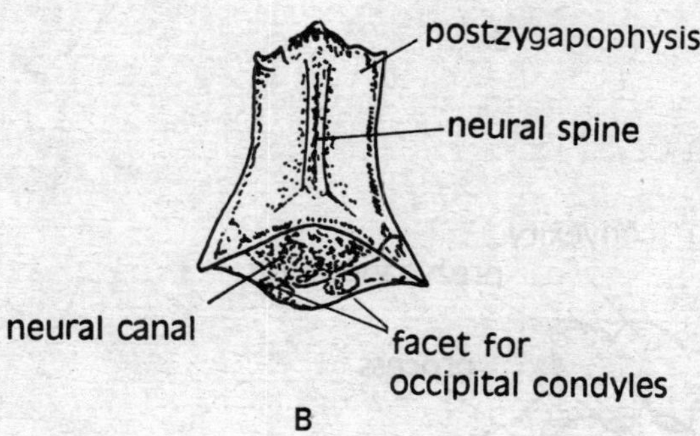

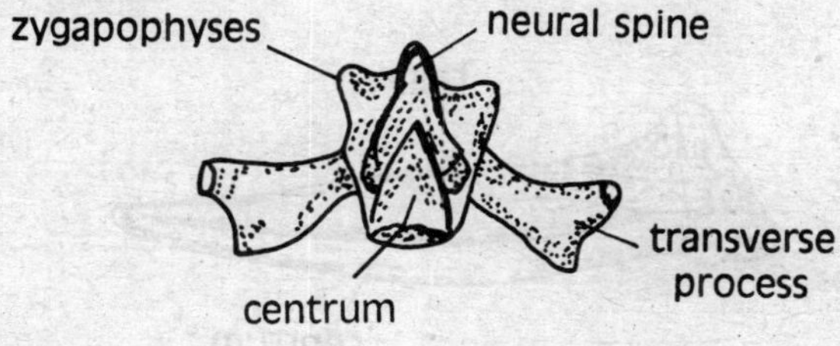

Fig. 3.29 Vertebrae of Rana (*cont...*)

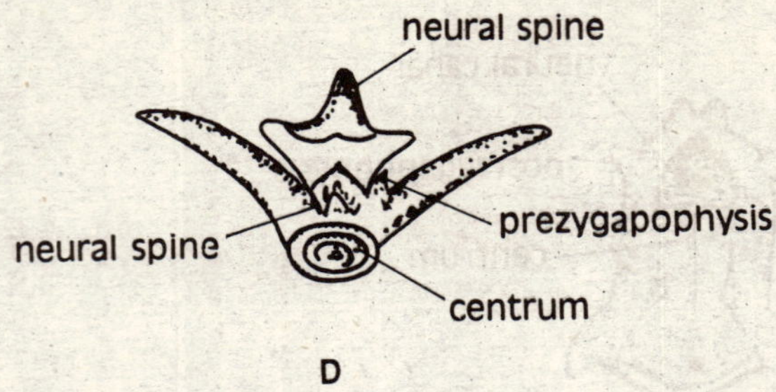

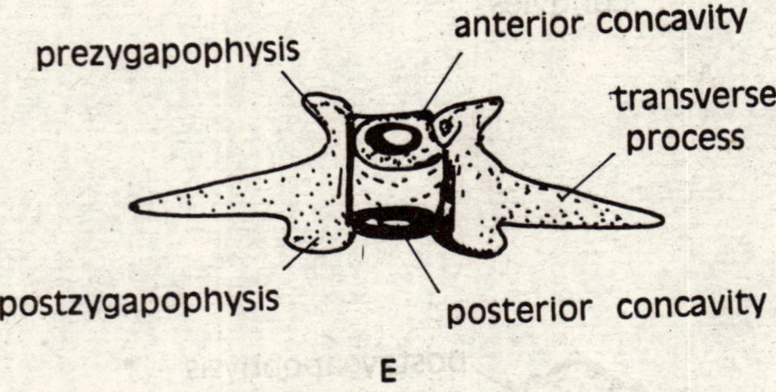

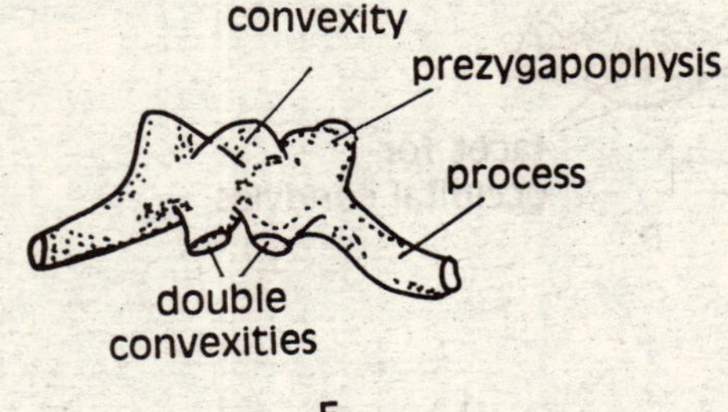

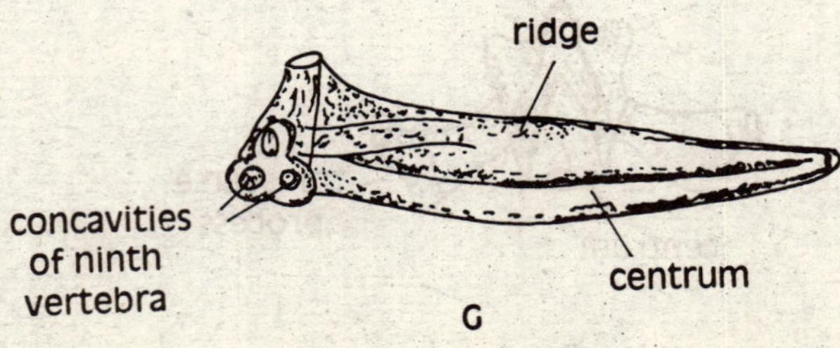

Fig. 3.29 Vertebrae of Rana

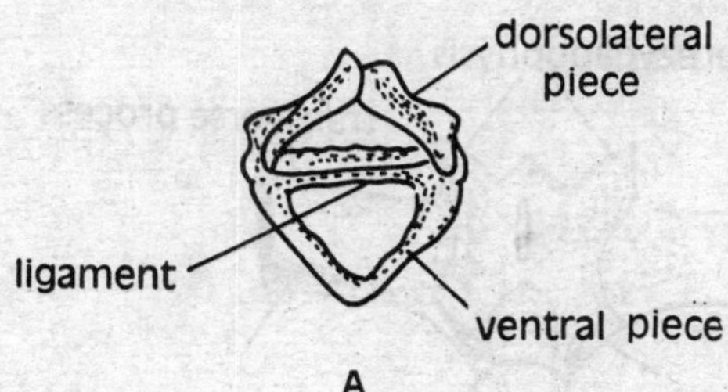

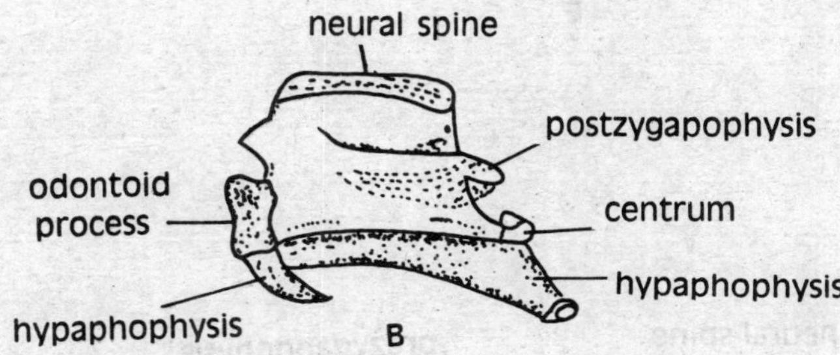

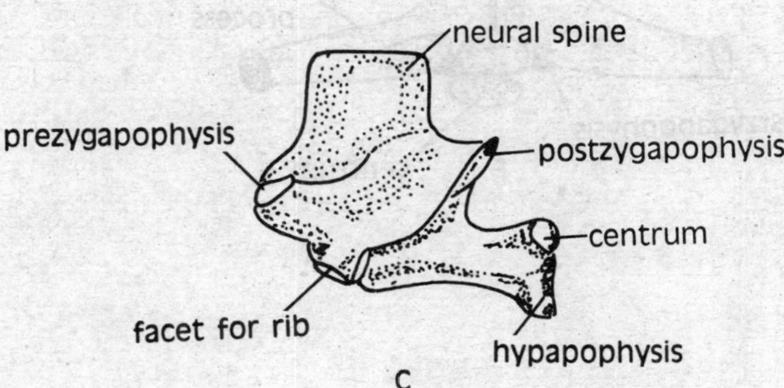

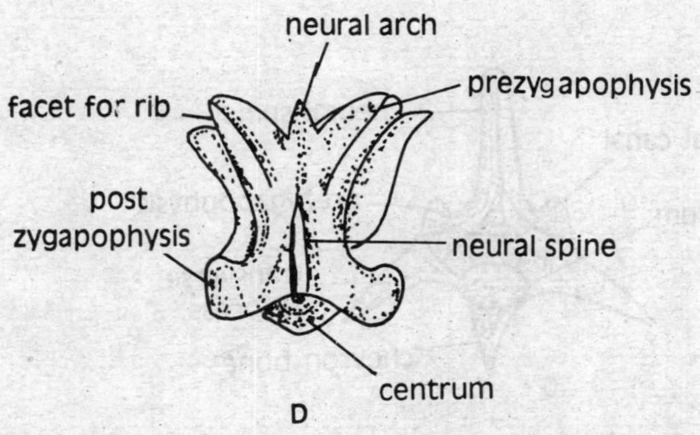

Fig. 3.30 Vertebrae of Varanus (*cont...*)

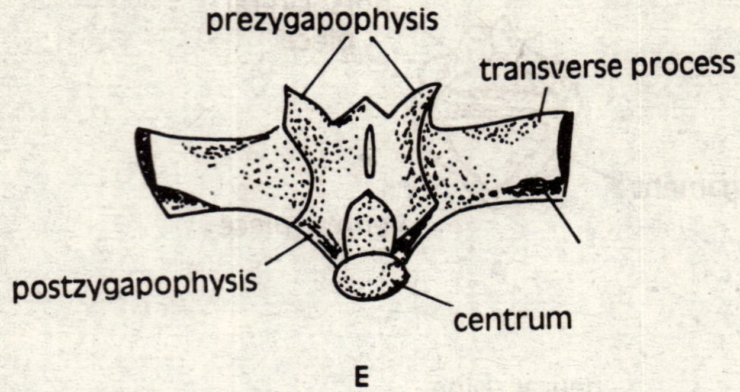

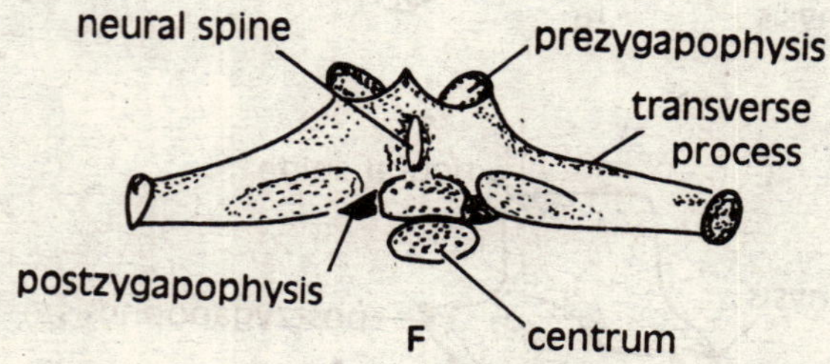

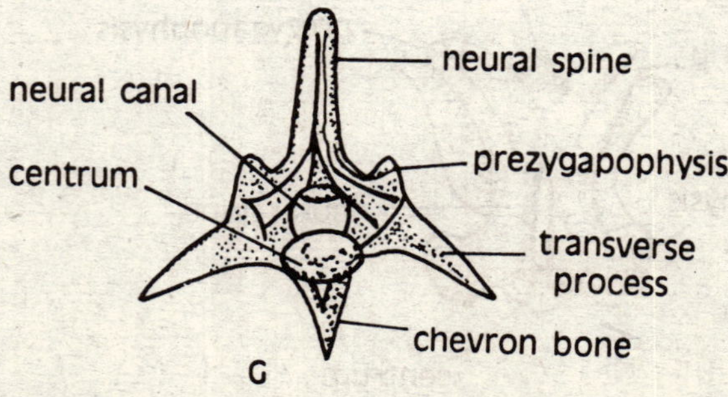

Fig. 3.30 Vertebra of Varanus

Thoracolumber Region

This region includes twenty two vertebrae after cervicals. These thoracolumber vertebrae are similar to cervicals but are larger in size and bear articular facets for the attachment of ribs at the junction of centra and neural arches. First few vertebrae have hypapophyses which disappear gradually in the posterior vertebrae.

Sacral Region

This region contains two *sacral vertebrae,* which are stout and articulate with the ilia of the pelvic girdle. The centrum is short. The transverse processes of the first sacral are strong and notochord distally at their anterior faces to provide articulating surface to the ilia of the pelvic girdle.

Caudal region: The caudal vertebrae are numerous and gradually taper posteriorly into rod-like or nodulated structures. The anterior caudal vertebrae have longer centra, thin transverse processes and long neural spines. They have Y-shaped *cheveron bones* attached to the ventral surface of the centra enclosing the haemal canal.

Class Aves

Gallus (Fig. 3.31): The vertebral column of fowl has cervical region almost equal to the remaining length of the column, rigid trunk region with *thoracic vertebrae fuses and short tail* region having fused caudal vertebrae forming *pygostyle.* The vertebrae are lighter in weight.

The vertebral formula of *Gallus* is as follows:

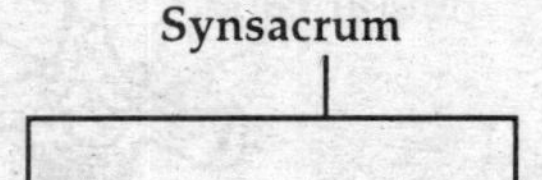

Cervical 16, Thoracic 4+1+2, Lumber 6, Sacral 2, Caudal 5+6+4=46

The *cervical vertebrae* are sixteen in number. Each vertebra is larger than in *Varanus.* The centrum is saddle-shaped or *hetrocoelous.* The neural arch is notched in the middle and is shorter than the centrum. The neural spine is rudimentary. A short backwardly pointed process is formed by the fusion of transverse processes with cervical ribs enclosing a foramen vertebrarterial canal. The prezygapophyses face upwards anteriorly and postzygapophyses face downwards posteriorly.

First two *cervical* vertebrae are specially modified into *atlas* and *axis* vertebrae. The *atlas vertebra* is a light bony ring with ventral thickened centrum notched above to receive odontoid process of the axis vertebra. Cup-like processes are present on its anterior side to provide articulation to the occipital condyle of the skull. The postzygapophyses are present on the posterior border of the neural arch for articulation with the second vertebra.

The *axis vertebra* is larger than atlas, without transverse processes and ribs. Its centrum bears a slender *odontoid process.*

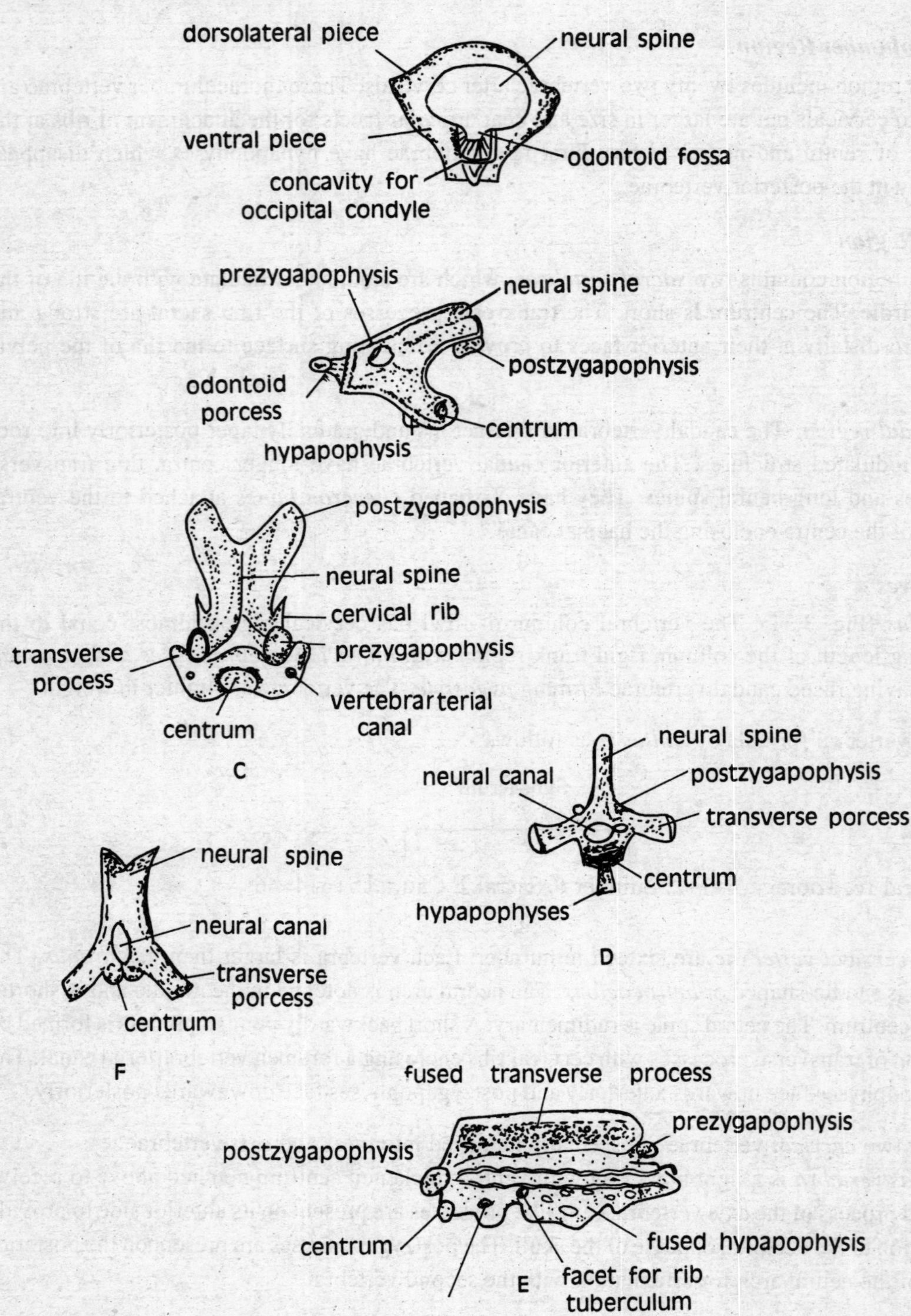

Fig. 3.31 A, B, C, D, E, F. Vertebrae of Gallus

The *thoracic vertebrae,* are 5 in number. Anterior three thoracic vertebrae are fused with the last cervical with their neural arches and processes confluent. The centra of the anterior thoracic and posterior thoracic are produced into a plate with fenestrae for the attachment of the neck muscles. The fourth thoracic is free and fifth is fused with the first sacral vertebra.

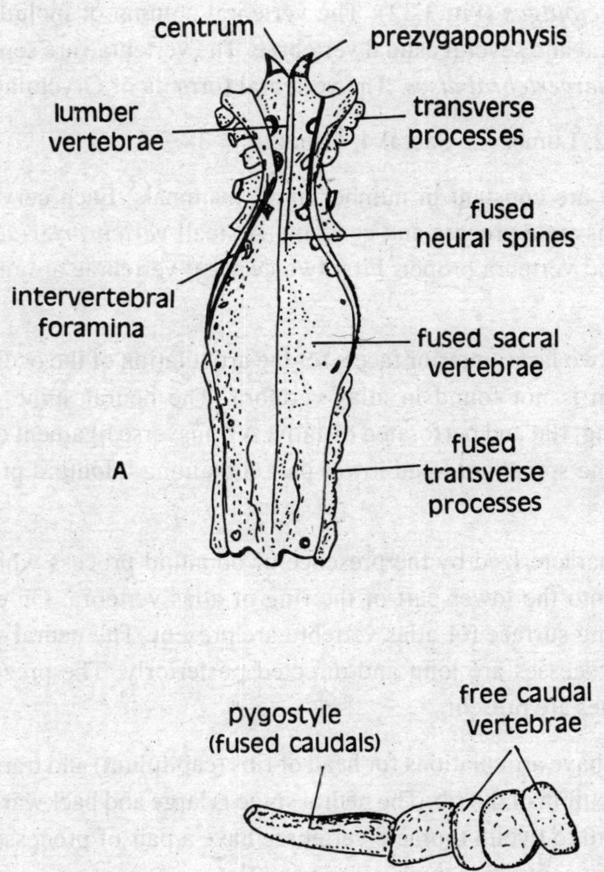

Fig. 3.31 G. Synsacrum & Pygostyle of Callus

A single mass of vertebrae called *synsacrum* is present after thoracic vertebrae which provides attachment to the huge pelvic girdle laterally (Fig. 3.31 A.). The first vertebra of the synsacrum is last thoracic with a pair of free ribs: next five or six are without ribs and may be called lumber vertebrae. All the lumber vertebrae are having a dorsal continuous plate made of transverse process of the vertebrae, haemal arches and ligaments joining them; all being ossified together. Next two vertebrae are sacral having transverse processes coming out of the neural arch. The sacrals bear ventral outgrowths called sacral ribs from each side of the centrum providing articulation to the ilium of the pelvic girdle just inside the acetabulum. Remaining vertebrae of synsacrum numbering five are caudal.

Six free caudal vertebrae are present behind the synsacrum. Last four or more hindermost bones are fused into an upturned compressed, plough -shaped bone called *pygostyle* (Fig. 3.31. B).

Class Mammalia: *Oryctolagus* (Fig 3.32): The vertebral column of includes 7 cervical, 12 thoracic, 7 lumber, 3-4 sacral and several caudal vertebrae. The vertebrae are separated by pads of fibrous cartilages called *intervertebral discs.* The vertebral formula of Oryctolagus is as follows:

Cervical 7. Thoracic 12, Lumber 7, Sacral 4, Caudal 8 = 38.

The *cervical vertebrae* are constant in number in all mammals. Each cervical vertebra has cervical rib fused with transverse process and centrum. A small *vertebrarterial canal* is present between the cervical rib and vertebra proper. First two cervical vertebrae are midified into *atlas* and *axis* vertebrae.

The *atlas vertebra* has two large anterior facets for the articulation of the occipital condyles of the skull. Distinct centrum is not found in atlas vertebra. The neural spine is small and the transverse processes are long, flat and perforated distally. A transverse ligament divides the neural canal, upper part lodging the spinal cord and lower part containing odontoid process of the axis vertebra.

The *axis vertebra* is characterized by the presence of odontoid process which is an anterior extension of the centrum into the lower part of the ring of atlas vertebra. On either side of the odontoid process articulating surface for atlas vertebra are present. The neural spine is long and compressed. Transverse processes are long and directed posteriorly. The prezygapophyses are absent and postzygapophyses are present.

The *thoracic vertebrae* have articulations for head of ribs (capitulum) and transverse processes for the articulation of tuberculum of the rib. The neural spine is large and backwardly directed. The transverse processes in ninth to twelth thoracic vertebrae have a pair of processes metapophyses above prezygapophyses. The centrum is thick, strong and flat.

The *lumber vertebrae* have a flat centrum, neural arches enclosing a neural canal, anterior and posterior zygapophyses, neural spine and transverse processes. Paired processes metapophyses directed upwards from the prezygapophyses and anapophyses directed backwards from the postzygapophyses are present.

The *sacral vertebrae* are fused into a composite structure called *sacrum.* It is present between two pelvic arches and narrows posteriorly. The first vertebra articulates with ilia of the pelvic girdle and has long neural spine.

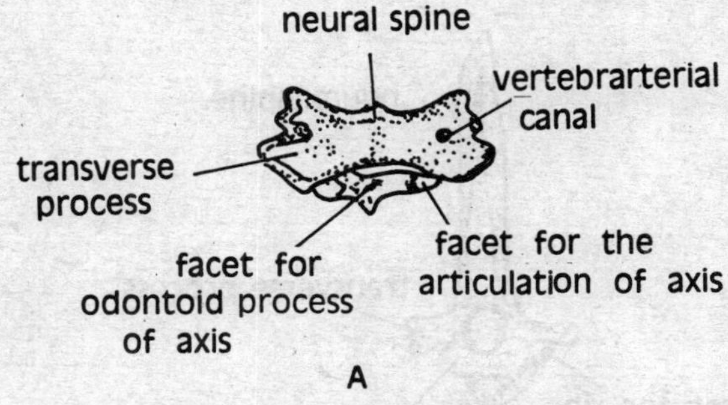

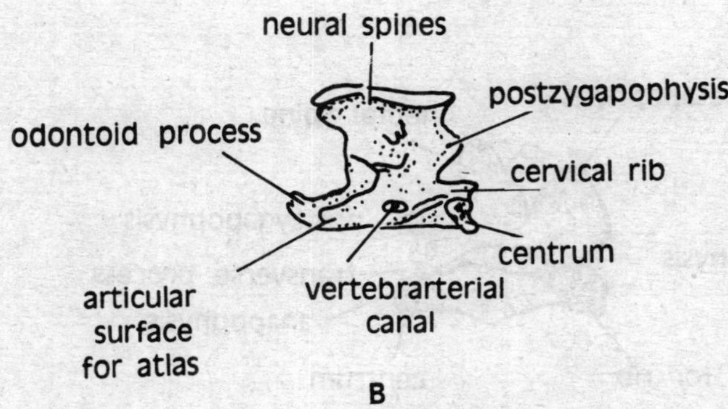

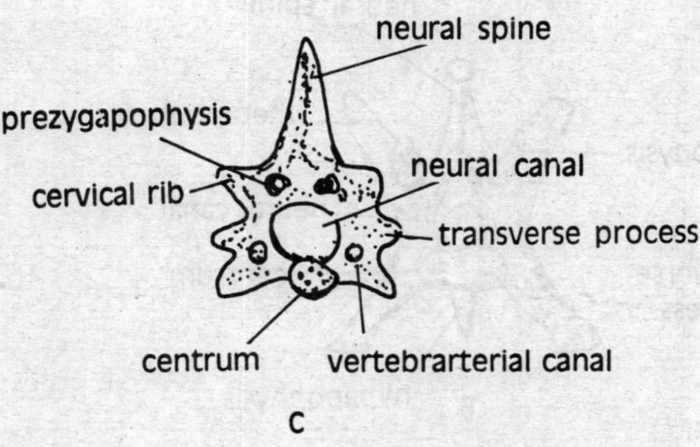

Fig. 3.32 A, B, C Vertebrae of Rabbit

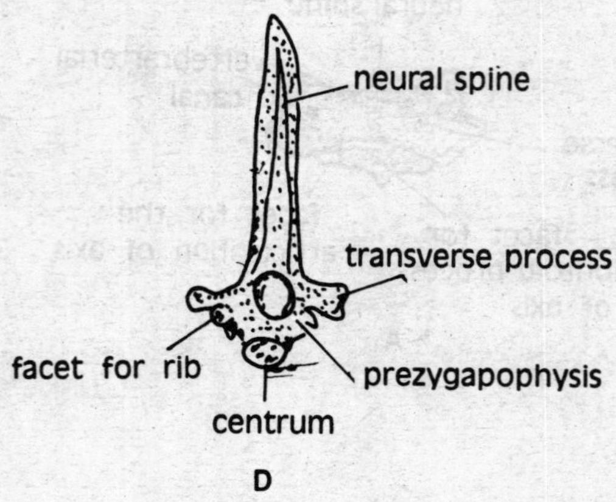

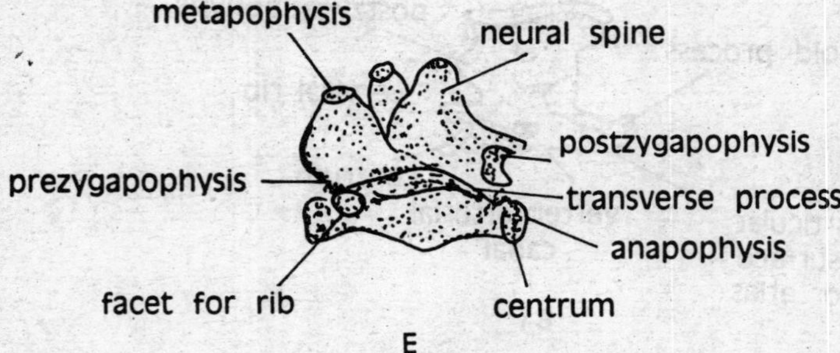

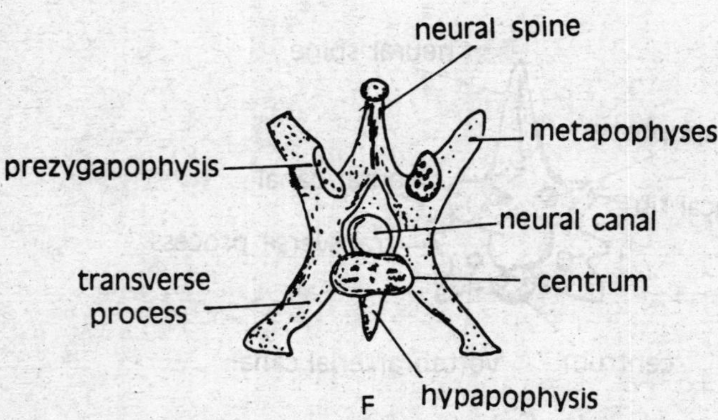

Fig. 3.32 D, E, F Vertebrae of Rabbit

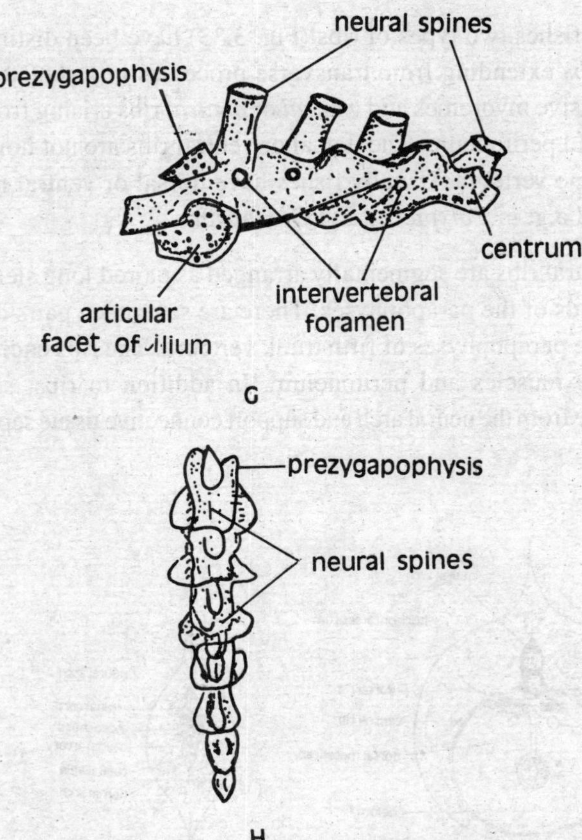

Fig. 3.32 G, H. Sacral and Caudal vertebrae of Rabbit

The *anterior* caudal vertebrae have distinct centra, transverse processes, and neural arches. The posterior caudal vertebrae reduce gradually and the posterior most have only centrum left.

RIBS

Ribs are cartilaginous or bony structures attached to vertebrae in vertebrates. It is not definite whether ribs of tetrapoda are homologous to the dorsal or ventral ribs of fishes. It is generally believed that they are homologous to ventral ribs. In amniotes, double-headed ribs occur as a rule. It is believed that the ancestral Tetrapoda had double headed (bicipital) ribs and the single headed elements must have developed either by the fusion of tuberculum and capitulum or by the loss of tuberculum.

Ribs in Vertebrate Series

Class: Chondrichthyes Scoliodon: The description on the ribs of Scoliodon is not available. Those of sharks and rays called dorsal ribs are short rods of cartilage attached to the centra of corresponding vertebrae in trunk. They extend laterally between muscles and develop in continuity with basiventral arcualia.

Class Osteichthyes: In fishes two types of ribs (Fig. 3.33) have been distinguished. They are *dorsal* or *intramuscular ribs* extending from transverse processes into skeletogenous horizontal septum between two successive myotomes and *ventral* or *pleural* ribs arising from the centrum and lying between body wall and peritoneum. The dorsal or ventral ribs are not homologous and may occur separately on the same vertebra. Usually fishes have dorsal or ventral ribs but sometimes both types of ribs are found e.g. in *Polypterus* and *Gambusia.*

In *Labeo rohita* the pleural ribs are segmentally arranged as paired long slender rods. The ribs are attached to the distal ends of the parapophyses. There are seventeen pairs of ribs and the first pair of ribs is attached to the parapophyses of fifth trunk vertebra. The ribs encircle the abdominal cavity and lie between the muscles and peritoneium. In addition to ribs, series of Y-shaped intermuscular bones originate from the neural arch and support connective tissue septa or myocommata.

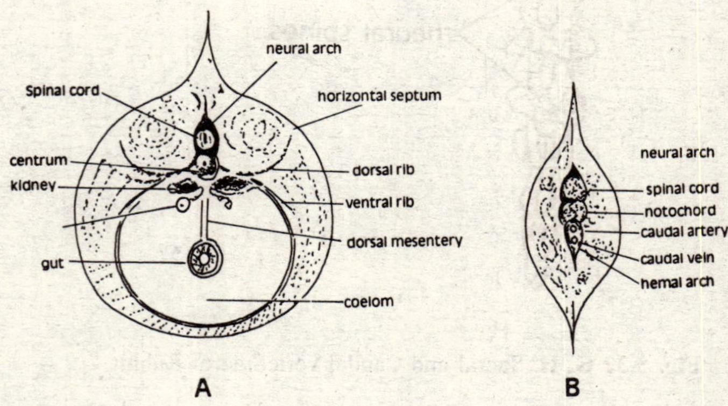

Fig. 3.33 **A.** Diagrammatic cross Actions of typical vertebrate, showing position of dorsal and ventral ribs **B.** Section through tail of fish showing normal arch.

Class Amphibia: In frog, the ribs are totally absent as they are fused with the transverse processes. In Amphibia, the ribs correspond to the dorsal ribs of fishes and develop from the basiventrals. In some Anura, they are reduced, single headed and articulate with transverse processes of the vertebrae. They are free distally and do not reach the sternum. In Urodela and Gymnophiona, they are double headed at vertebral ends and do not reach the sternum.

Class Reptilia: In Lizards, the cervical ribs are small and few anterior thoracic ribs join the sternum and other thoracic ribs remain free. Such ribs have ventral and sternal parts. In snakes, the ribs are similar throughout the thoracic region as there is no sternum. In Chelonia, cervical ribs are absent and ribs of trunk vertebrae are ankylosed to the costal plates of the carapace. The cervical ribs of crocodilia are double headed and posterior thoracic ribs are single headed, lumber vertebrae have no ribs caudal vertebrae have two heads. The gastralia or abdominal ribs occur as dermal bones on the ventral part of the abdomen in *Sphenodon* and crocodilia. (They are parts of exoskeleton).

In Varanus all the precaudal vertebrae excepting first three have ribs which are single headed and curved. The anterior thoracic ribs consist of bony vertebral and cartilaginous sternal portion.

Class Aves: Gallus: In *Callus,* the cervical vertebrae excepting the first two have double-headed cervical ribs with *capitulum* fused with centrum and *tuberculum* with the transerse process. The space thus enclosed is vertebrarterial canal. The ribs of the last two vertebrae are not ankylosed with them. The anterior thoracic ribs consist of vertebral double-headed part attached distally to the sternum. Both sternal and vertebral parts are well ossified and the vertebral part a backwardly directed, bony, uncinate process. In *Archaeopteryx* the abdominal ribs were present.

Class Mammalia Oryctolagus: The ribs of *Orctolagus* and most of the mammals are double headed with well developed capitulum and tuberculum. In monotremes, they are single headed. The anterior thoracic ribs consist of vertebral and sternal parts like in reptiles and birds and articulate with vertebra and sternum respectively. The sternal ribs are cartilaginous. The posterior thoracic ribs remain free in the abdomen and are called *floating false ribs.*

STERNUM

The sternum is a cartilaginous or bony plate like structure in the thoracic region of the body cavity. It is a skeleton part closely connected with the pectoral girdle and ribs excepting in amphibia. The sternum is found in vertebrates with legs and is supposed to be an adaptation for terrestrial locomotion. It provides protection to thoracic region, attachment to anterior limb muscles and flight muscles in birds. It takes part in respiratory movements in amniota. The sternum is absent in cyclostomes and fishes.

Class Amphibia: Rana (Fig. 3.34): In *Rana* the sternum is represented by (our plates present on the mid ventral axis of the body in relation to the pectoral girdle. The plates ate an *omosternum* capped with cartilaginous *episternum* anteriorly' and *mesosternum* tipped with cartilaginous *Xiphisternum* posteriorly.

Class Reptilia: Varanus (Fig. 3.35): The sternum of *Varanus* is a rhomboïdal cartilaginous plate between the coracoids with sternal parts of three pairs of thoracic ribs attached to it. A T-shaped *episternum* is present on its ventral side which is a dermal bone and not a part of sternum.

Class Axes: Gallus (Fig. 3.36): The sternum of *Gallus* is a huge ossified structure with a broad triangular bone extending below the abdomen posteriorly. The mid-ventral portion is produced into a keel or carina for the attachment of large flight muscles forming a T-shaped girder to tolerate the weight of heavy abdominal viscera. A pair of coastal processes are present on the anterior side of the sternum where ribs are attached. Two lateral paired *Xiphoid processes* extend from its posterior side. The keel is absent in running birds.

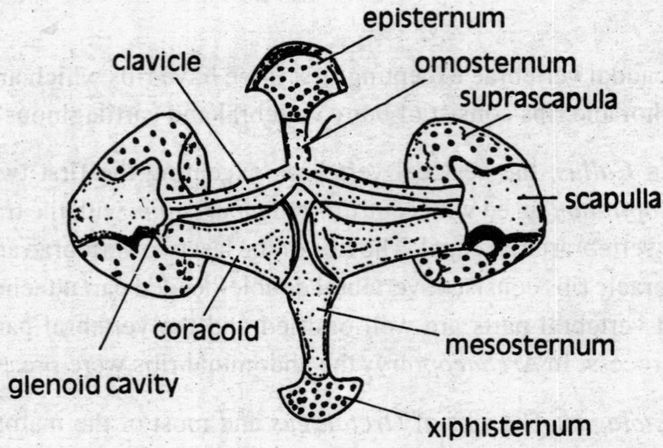

Fig. 3.34 Sternum with pectoral girdle in frog.

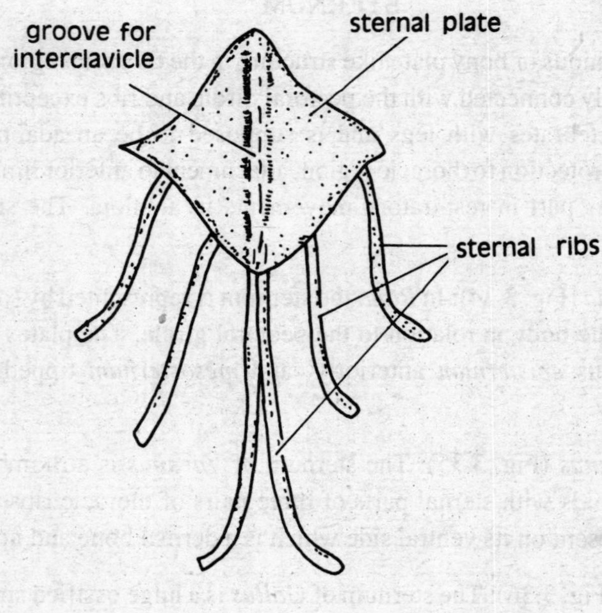

Fig. 3.35 A. Sternum & Ribs in reptiles.

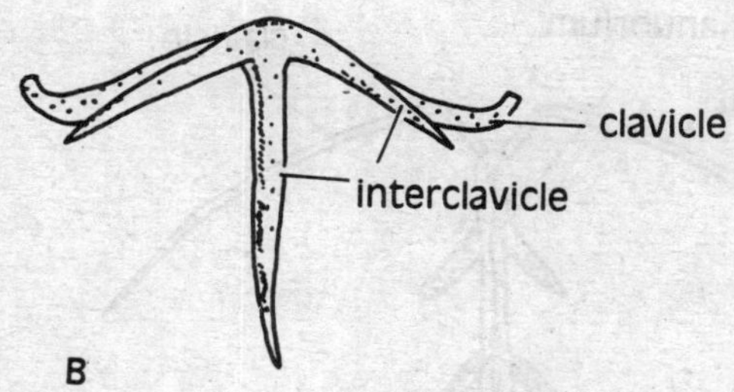

Fig. 3.35 B. Reptilian clavicles & interclavicle.

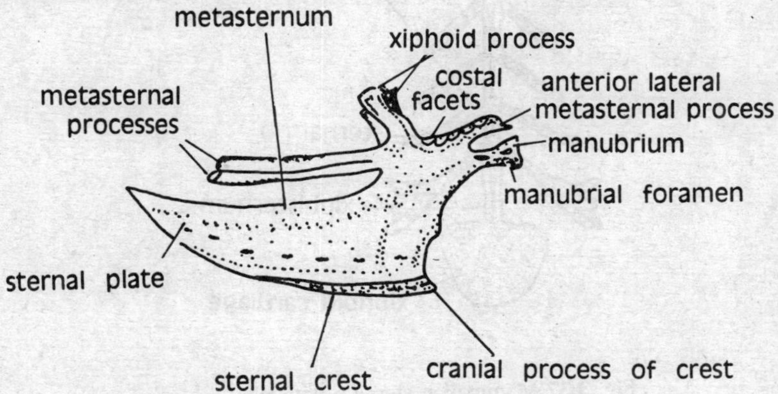

Fig. 3.36 Sternum with keel in Callus

Class Mammalia: Oryctolagus (Fig. 3.37): In rabbit the sternum is made of a mid-ventral series of bones arranged in a row. It is divided into three regions. The anterior region is called presternum or manubrium. The middle region contain five pieces called *sternebrae.* The fifth being very small. The posterior region is called *metasternum* or *xiphisternum* which is a long part containing a terminal expanded xiphoid cartilage. Seven pairs of sternal ribs are attached to the sternum.

3.3.2. Appendicular Skeleton

The appendages may be of two types: *median* and *paired*. The median are azygos type include median fins found in aquatic vertebrates only and the paired appendages are found in all classes excepting cyclostomata and some exceptions. Both kinds of appendages have internal skeleton.

The Meaian Appendages: The median appendages are fins situated on dorsal, terminal, or annal position of the body. They occur in all fishes, in larval and tailed amphibians and aquatic mammals like whales. In Amphibia and higher groups there is no skeleton in median fins. In fish, cartilage, bone, or horny suostance form the skeleton of the median fins.

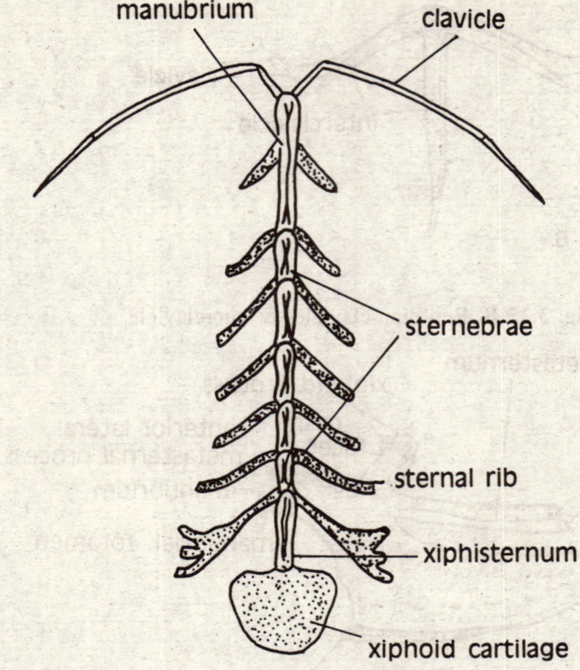

Fig. 3.37 Mammalian sternum with ribs

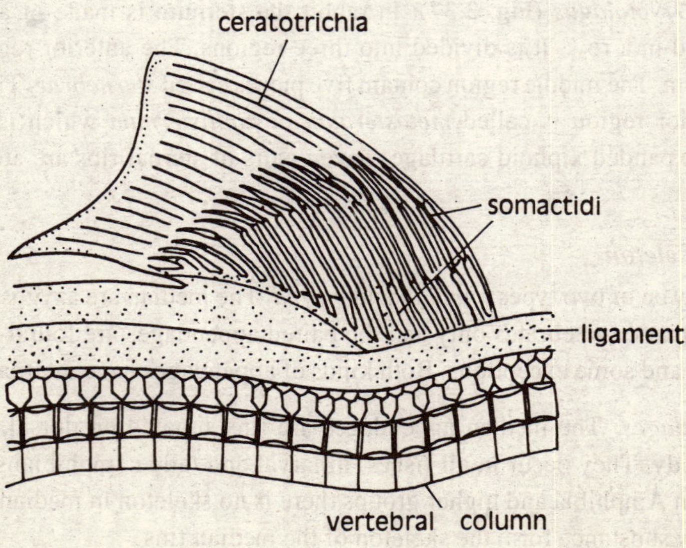

Fig. 3.38 Median fin of Scoliodon

In *Scoliodon* (Fig. 3.38) the median fins have endoskeleton *pterygiophores* forming a series of radial or *somactidia* bearing a double series of distal *ceratotrichia.* In caudal fin, the *somactidia* are absent and their place is taken by neural and haemal spines of caudal vertebra.

In *Labeo* (Fig. 3.39) the skeleton supports of the median fin include *somactidia* or *radials* and *dermotrichia* or *dermal fin rays.* The somactidia are parallel bony rods lying embedded in body muscles and divided into three segments; a proximal, a median and a distal segment. The dermostrichia of *Labeo* are branched and called *lepidotrichia.* In addition to these fin rays, delicate horny *actintrichia* are present at the free edges of the fins

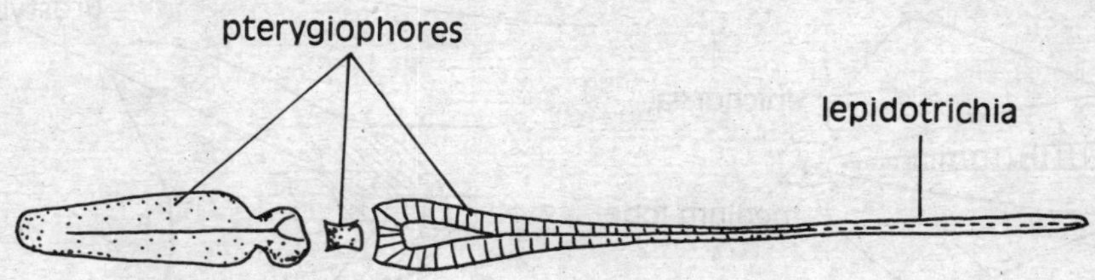

Fig. 3.39 Dorsal fin of Labeo

The dorsal fin of *Labeo* is supported by fifteen or sixteen lepidotrichia seated on fourteen radials. Each radial has enlarged and dragger-shaped proximal segment, called interspinous bone or axonost. The median segment is short and distal segment is reduced.

The anal fin of *Labeo* is supported by a series of eight fin rays supported by seven radials.

The *caudal fin* of Labeo is homocercal and supported by a numbers of flattened bony rods including two epiurals and a radial on the dorsal side of urostyle and nine hypurals on the ventral side. The fin-rays are attached on hypural and epiural elements.

Several types of caudal fins are found in fishes (Fig. 3.40):

1. *Protocercal* caudal fin is ancestral type of fin with notochord or vertebral column ending straight and fin extended on the dorsal and ventral sides equally. It is found in many larval fishes but not in adults.
2. *Heterocercal* fin has vertebral column ending in elongated upper lobe. Such a fin was there in Ostracoderms and old Gnathostome groups. It is found typically in Elasmobranches (Scoliodon).
3. *Hypocercal* fin has vertebral column ending in elongated lower lobe of the caudal fin. Such fins are found in Anapsida, some Ostracoderns and Ichthyosaurs.

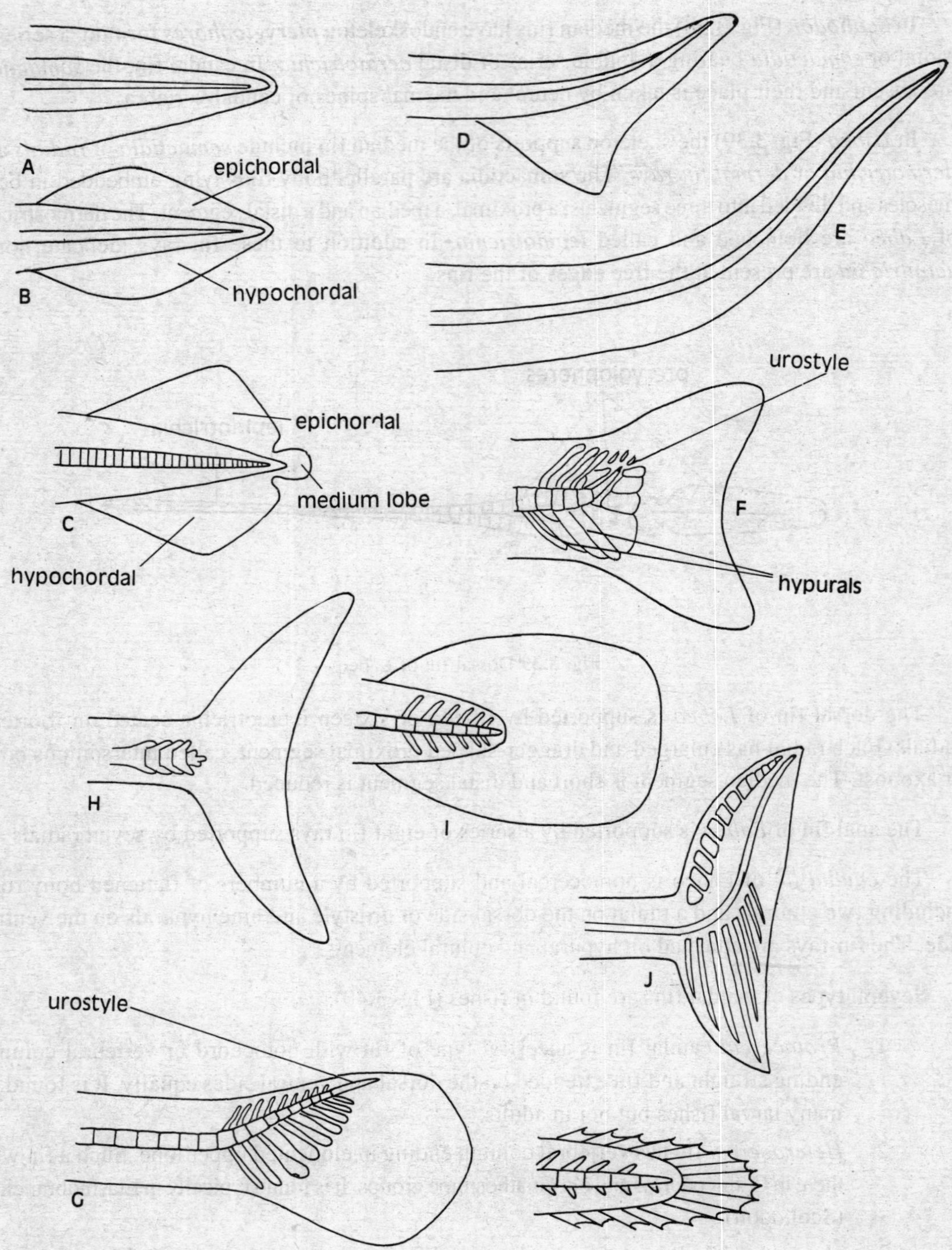

Fig. 3.40 Different types of caudal fins in fishes

4. *Diphycercal* fin has vertebral column going upto the end of the fin with dorsal and ventral parts equally developed. Such fins are found in modern cyclostomes. Holocephali, living lungfishes and later Crossopterygians. When dorsal and anal fins unite with caudal fin to make a continuous median fin. the tail is called gephyrocercal (e.g. *Fieraspis*).
5. *Homocercal* fin (Fig. 3.41) is found in teleosts where the heterocercal condition is found in early development stages and a fully symmetrical caudal fin is secondarily developed beyond the vertebral column. This fin is symmetrical externally but asymmetrical internally.

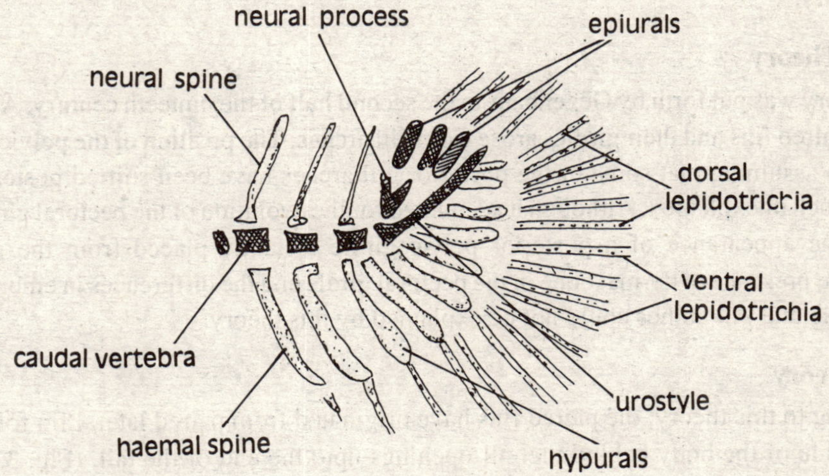

Fig. 3.41 A. Homocercal fin

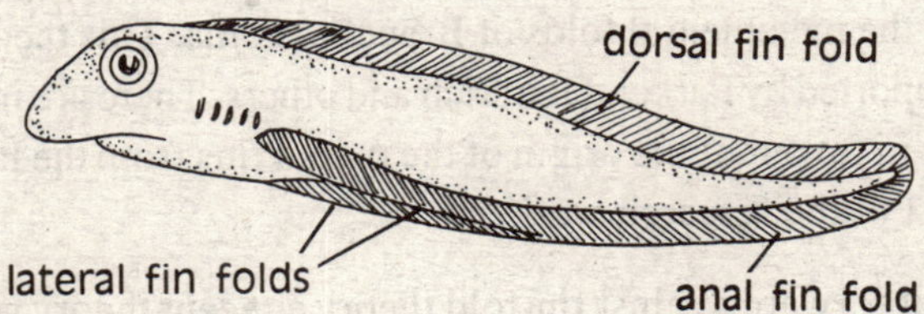

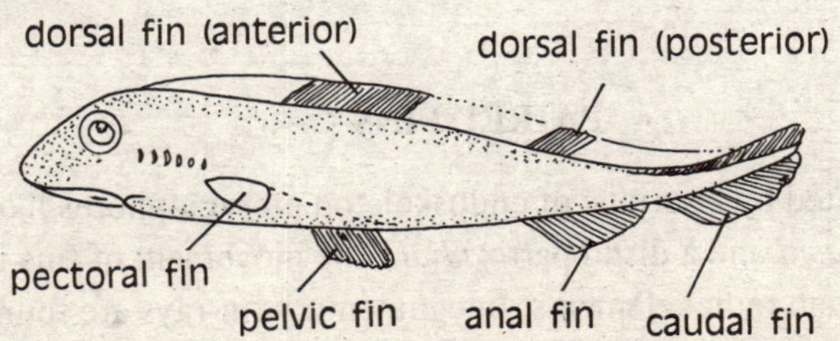

Fig. 3.41 B. Paired and unpaired fins in fishes

Origin of fins and appendages (Fig. 3.41): The median fins have originated form Cartilaginous fold of tissue extending form the posterior region of the head and continuous posteriorly around the tail and forward upto anus. This fold is supported by parallel cartilaginous rods. Such hypothetical condition is seen in the median fins and metapleural folds of *Amphioxus.* Peristence and disappearance of certain parts of fin folds and median fins result in the formation of paired and median fins. In this way, in the course of the evolution of appendages, we find one transition from fin-folds to fins and second from fins to limbs.

Phylogenetic origin of the paired appendages has received considerable attention. Out of several theories advanced two are considered important. These are (A) Gill-arch theory and (B) Fin-fold theory.

Gill -Arch Theory

This theory was put forth by Gegenbaur in the second half of the ninteeth century. According to this theory paired fins and their girdles arose from gill arches. The position of the pelvic fins can be explained on assuming that some of the posterior gill arches have been shifted posteriorly. This theory left much unexplained. Although it accounted for the evolution of the pectoral girdle it failed to explain the appearance of a posterior pelvic girdle distantly placed from the gill arches. Moreover, the presence of dermal bone in the pectoral girdle and the differences in embryologies of pectoral girdle and gill arches could not be explained by this theory.

Fin-Fold Theory

According to this theory, the paired fins have originated from paired lateral fin folds running down each side of the body behind the gill-openings upto the end of the tail. (Fig. 3.41). These lateral fin folds are separated at the anterior part, but become fused posteriorly as a median ventral fin fold. There is another continuous median dorsal fin-fold. The condition of the lateral fin-folds in the ancestral forms was like the metapleural folds of Branchiostoma. This theory was proposed by Balfour and Thacher and supported by Parker, Goodrich and others. There are main palaconological and embryological evidence in favour of the origin of the paired fins from the lateral fin-folds. This theory was accepted by most of the research workers.

After some evidences put forward against fin-fold theory a recent theory suggests that the fins appeared probably in connection with some paired and unpaired median spines, as found in many ostracoderms. In course of evolution membranous structure appeared between these spines and the adjacent body-wall.

PAIRED FINS

Paired fins are also supported by a serious of endoskeleton pterygiophores like median fins, each divided into a basal part *basal* and a distal part *radial.* The movement of fins is performed by the muscles attached to basal and radial elements. Integumentary fin-rays are found attached to these elements supporting blade or propellers of the fin. The fin-rays may be unjointed any horny *ceratotrachia* in elasmobranches, bony and jointed *lepidotrichia* in bony fishes, and fibrous and jointed *campotrichia* in Dipnoi. The paired fins of clasmobranches and Dipnoi have basal lobes and vane-like part. The basal lobe is missing in bony fishes and paired fins have vane-like part only. Such fins are called rayed-fins.

Girdles and Paired Fins of Scoliodon (Fig. 3.41)

The pectoral girdle or scapulocoracoid contains two semicircular cartilages untied together alongwith midventral line. It situated on the posterior side of the last branchial arch. Each half of the pectoral girdle is made of a thick rod-like scapula on the dorsal side and thin flattended coracoid on the juntion of corcoid and scapula for the articulation of three basal cartilages of the pectoral fin.

The pectoral fin has three basal cartilages, proptery-gium, mesoptergium, and metapterygium. Main radial cartilages are present on the basal cartilage to provide support for the pectoral fin.

The pelvic girdle (Fig. 3.41.B) is situated near the cloaca. It is made of flattened cartilaginous rod situated in the transverse plan infront of cloacal aperture.

The pelvic fin has radials supported by a curved basal cartilage basipterygium, which is attached on the anterior side of the pelvic girdle. The distal ends of the radials have small cartilages with ceratotrichia

In males, the claspers are present. Each clasper contains a tublar cartilage grooved on dorsal side and terminates into a sharp style enclosed by two sheathing plates. A small accessory cartilage is present at the upper end of the style.

Girdles and Paired Fins of Labeo

The pectoral girdle (Fig. 3.42) contains replacing and dermal bone elements. Three replacing bones are *scapula, coracoid*, and *mesocoracoid.* Each half of pectoral girdle has scapula on the dorsal side, a coracoid on the ventral side and a mesocoracoid in front of the scapula. Glenoid facets are present between scapula and coracoid. The dermal bones are cleithrum or clavicle, supracleithrum or supraclavicle. posttemporal and postcleithrum. Two halves of the pectoral girdle do not meet at the mid ventral line.

The pectoral fin has four proximal *pterygiophores* articulating with the glenoid facets of the pectoral girdle. Bony jointed fin-rays or *lepidotrichia* are situated distal to the radials for providing support to pectoral fin. The fin does not have endoskeletal support as it is small. The bony radials lie inside the body wall.

The *pelvic girdle* (Fig 3.43) has two halves or pelvic bones lying in the ventral body wall. Each half has an anterior and posterior part. The anterior part is formed and posterior has a rod-like process produced into a narrow cartilage. The two pelvic bones meet mid-ventrally.

Girdles in Tetrapoda

Pectoral Girdle: The tetrapoda pectoral girdle has two halves, each having a precoracoid infront and a coracoid behind the coracoid fenestra limited by an epicoracoid medilly. The ossification of cartilage may result in the formation of scapula, coracoid, precoracoid and epicoracoid bones. The upper part of scapular region persists as cartilaginous suprascapula or gets united with the scapula. A dermal bone clavicle often replaces precoracoid. A cleithrum (dermal bone) is present in lower forms. The mammals have a reduced coracoid. The bones of pectoral girdle meet in the glenoid region to accommodate glenoid fossa.

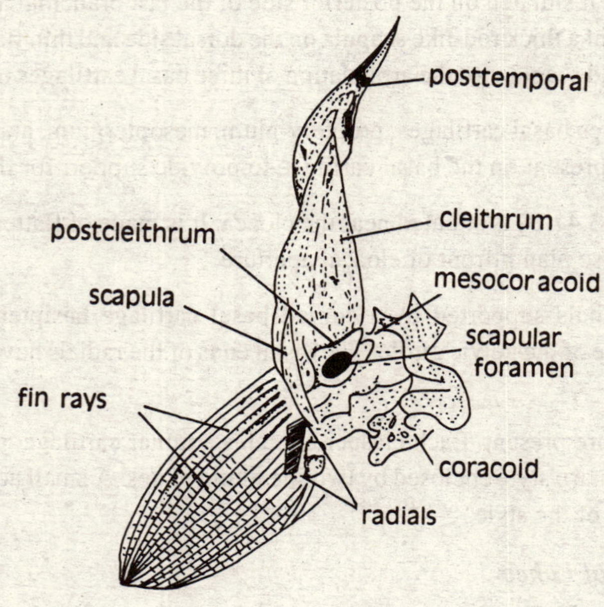

Fig. 3.42 Girdle & pectoral tin of Labeo

Class Amphibia: Rana Fig. 3.44): The pectoral girdle of frog is arch-shaped structure enclosing the chest. The two halves of the pectoral girdle unite in the midventral line and are closely related to the sternum. Coracoid, precoracoid, clavicle, scapula, suprascapula, and paraglenoid cartilage are present in each half. The clavicle (dermal bone) is fused to cartilaginous precoracoid The glenoid cavity is formed by coracoid, paraglenoid, and scapula. A epicoracoid cartilage is present on the inner margin of the coracoid.

Class Reptilia: *Varanus* (Fig. 3.45): Each half of the pectoral girdle has bony scapula and coracoid participating in the glenoid cavity. Suprascapula is an expanded calcified cartilage. The two halves of the girdle are jointed to each other and to the sternum. An expanded epicoracoid cartilage is present between the coracoids of two sides. Three fenestrae are situated between epicoracoid and coracoid. On the mid-ventral side a T-shaped dermal bone interclavicle or episternum is present. The clavicle, another dermal bone, is present on the arm of the interclavicle providing support to the girdle.

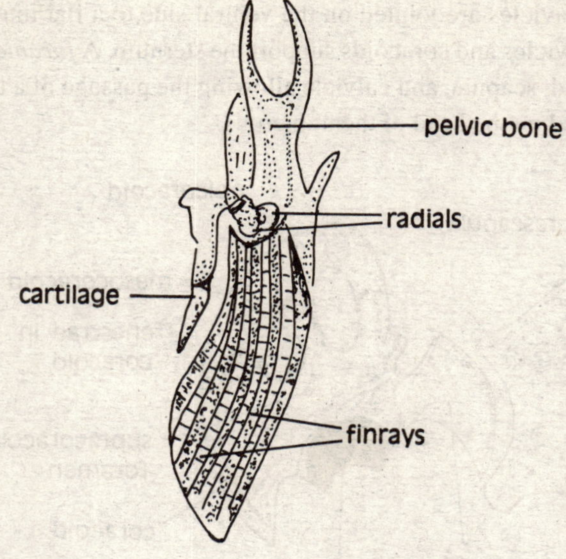

Fig. 3.43 Girdle & pelvic fin of Labeo

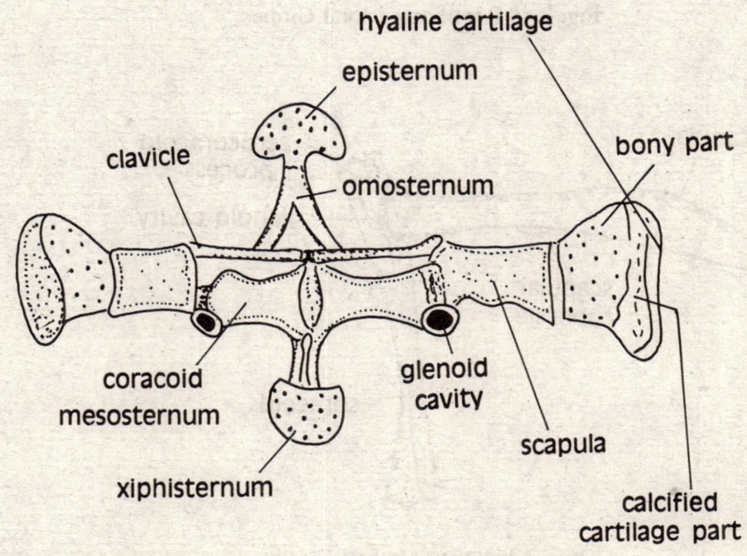

Fig. 3.44 Pectoral girdle of Rana

Class Aves: *Gallus* (Fig. 3.46): The pectoral girdle or scapulocoracoid of fowl lies near the centre of the body. It consists of stoutcoracoid joined at right angle to a sword-shaped scapula. The coracoid is attached to the sternum at the coracoid groove and scapula lies over ribs providing support to them. The glenoid cavity is formed by scapula and coracoid. A clavicle is situated infront of the coracoid. Two clavicles are jointed on the ventral side to a flat interclavicle forming wish bone or furcula. The clavicles and coracoids support the sternum. A *foramen triossenum* is present at the funtion of coracoid, scapula, and calvicle allowing the passage of a tendon of flight muscle, pectoralis minor attached on the head of the numerous.

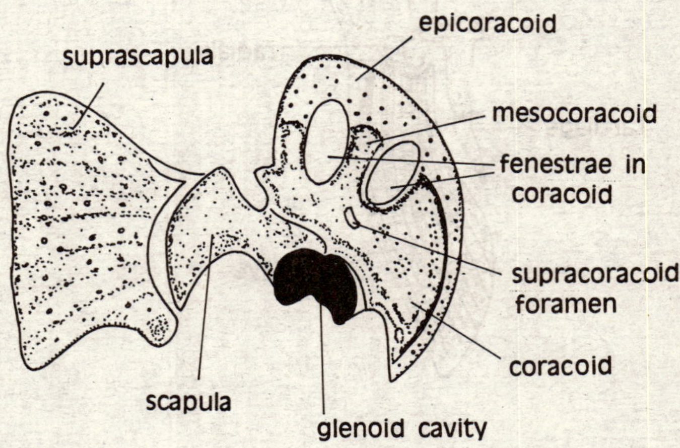

Fig. 3.45 Reptilian pectoral Girdle

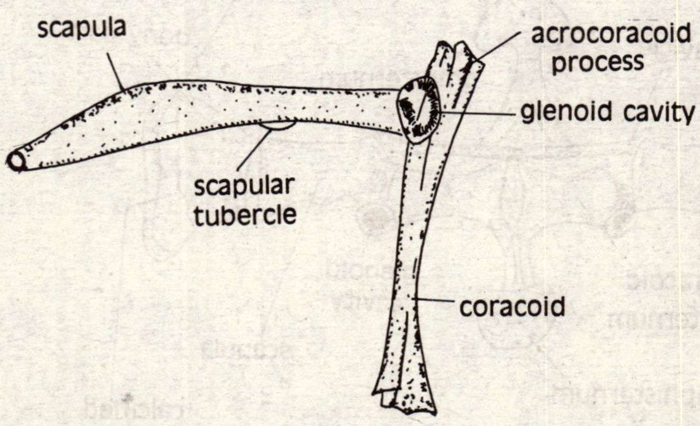

Fig. 3.46 Avian pectoral Girdle

Class Mammalia: Oryctologus (Fig. 3.47): Each half of the pectoral girdle in rabbit consists of a triangular scapula having glenoid cavity at the end. A coracoid process represents the reduced coracoid near the glenoid cavity. The outer surface of the scapula has two processes, an acromian process and a metacromian process for the attachment of muscles. Clavicle is a slender dermal bone attached by ligaments to the acromian process of the pectoral girdle and *manubrium* of the sternum.

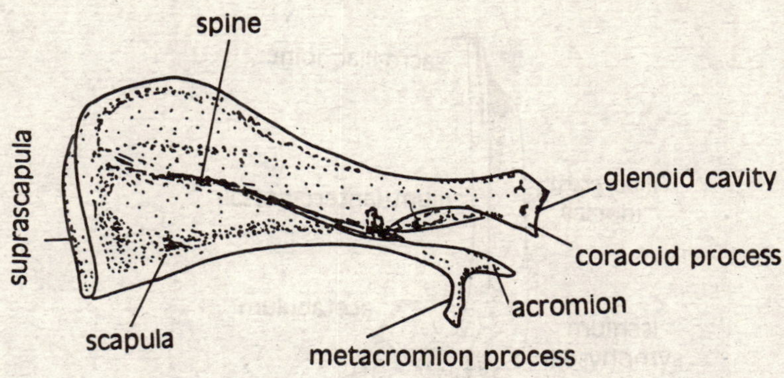

Fig. 3.47 Mammalian pectoral Girdle

Pelvic Girdle: The tetrapod pelvic girdle has three cartilage bones in each half. These are ilium, ischium and pubis. The dermal bones are absent and all the bones participate in the formation of acetabulum cavity. In mammals, acetabular or *cotyloid bone* is present. The two halves of pelvic girdle meet at pubic or *ischiac symphsis* along the midventral line. Between ischium and publis, an obturator formen is usually present. The pelvic girdle is attached to the vertebral column by the ilia articulating with the sacral vertebrae. The ilium, ischium and pubis bones of the pelvic girdle of tetrapoda are homologous to scapula, coracoid, and precoracoid bones of the pectoral girdle respectively. The clavicle which is a dermal bones is not homologous to the *pubis* which is a cartilage bone.

Class Amphibia: Rana (Fig. 3.48): The pelvic girdle is V-shaped with long ilium on each side articulating with the transverse processed of the sacral vertebra. *Pubis* and *ischium* are smaller than *ilium.* All the bones share in the formation of acetabulum cavity meant for the articulation of head of the femur bone. Pubis is partly ossified and partly cartilaginous. The pelvic girdle of frog is specialised for jumping action and its long ilia form long lever for the transmission of jumping force from hindlimbs to the vertebral column.

Class Reptilia: *Varanus:* The *ilium, ischium* and *pubis* are fused intimately to form innominate bone in each half and share in the formation of *acetabulum.* Ilium is articulated with two sacral ribs of the sacral vertewbrae making post-acetabular articulation. The two halves of the girdle meet at public and ischiac symphyses. *Epipubis* and *hyposchium* cartilages are present at these symphyses.

A small foramen for obturator nerve is present in pubis. A large *ischiopubic* or *condiform foramen* is present between two halves of pelvic girdle and is divided by a ligament into two.

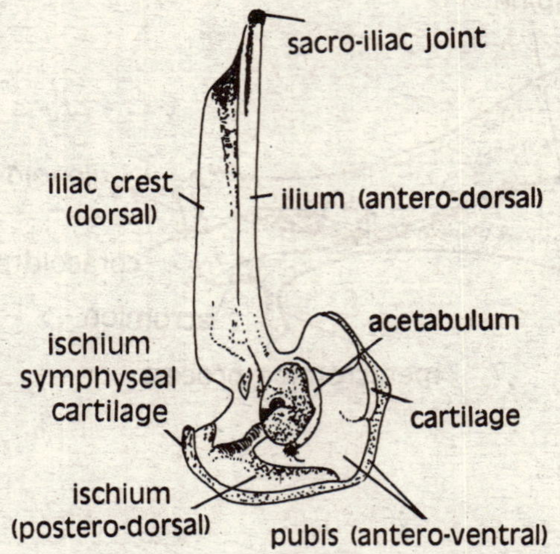

Fig. 3.48 Half of pelvic girdle (os-innominatum)

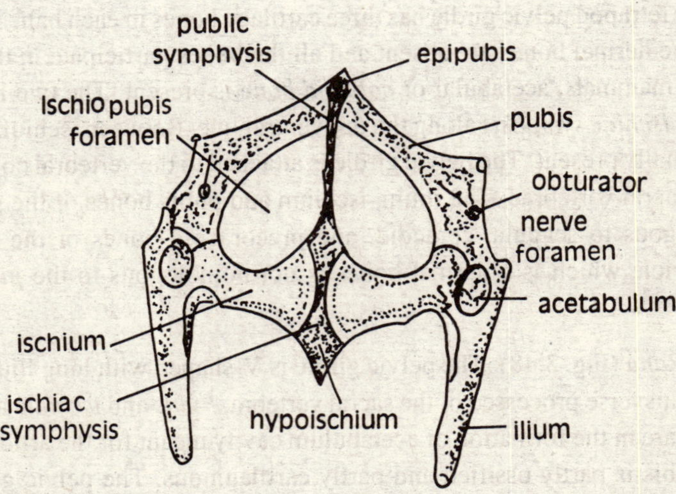

Fig. 3.49 Pelvic girdle of Varanus

Class Aves: Gallus: It has a large expanded *ilium* infront and behind the acetabulum and fused with the entire length of the sacrum vertebra. *Ischium* is a broad bone extending posteriorly and lying parallel to ilium and fused with it. *Ilioischiatic foramen* is present between the two bones. *Publis is* a slender bone extending backwards and ending freely. The *Obturator formen* is situated between ischium and pubis. The acetabulum has a "foramen Bones of the pelvic girdle are fused intimately and have no symphyses.

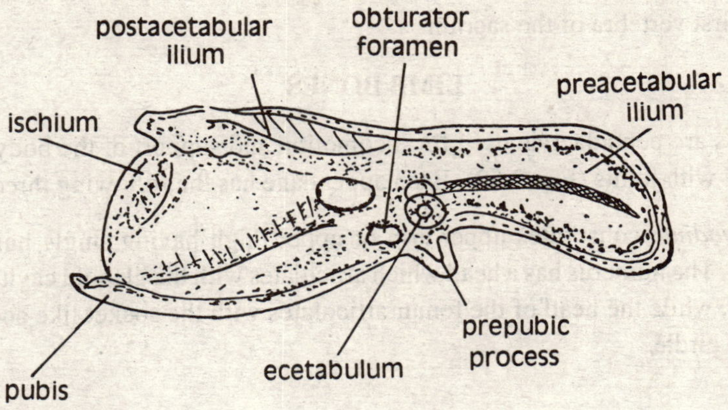

Fig 3.50 Pelvic girdle of Gallus

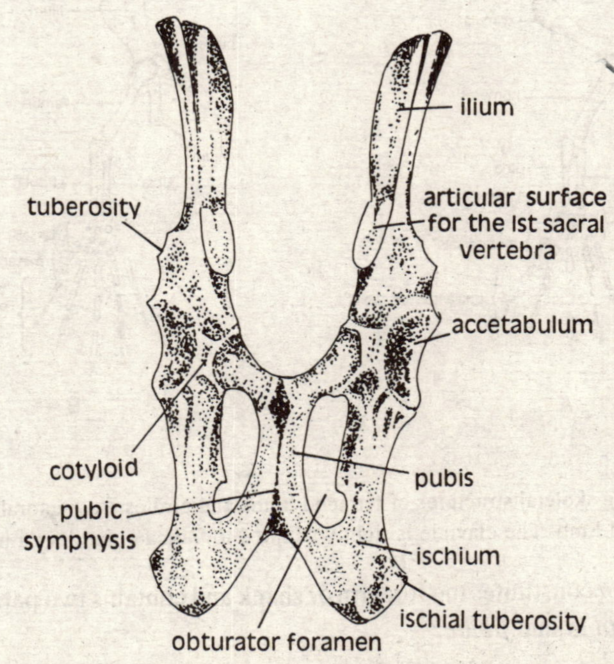

Fig. 3.51 Pelvic girdle of Rabbit

Class Mammalaia: Oryctolagus: The innominate bone formed by ilium, ischium and pubis in each half. The two halves meet midventrally by a *pubic symphsis.* A small *cotyloid* or *acetabular* *bone* is present at the lower margin of the acetabulum. Ilium, ischium, and cotyloid bones meet at acetabulum: pubis does not take part in its formation. A large *obturator, foramen* is present between pubis and ischium on each side. The pelvic girdle articulates by its ilia to the transverse processes of the first vertebra of the sacrum.

LIMB BONES

The tetrapod limbs are pentadactyle used for locomotion and support of the body on the ground. They are provided with joints (Fig. 3.52). Each appendage has the following three segments:

1. *Stylopodium* constitutes upper arm or upper thigh having single humorus or femur bones. The humerus has a head which articulates with the glanoid cavity of the pectoral girdle, while the head of the femur articulates with the socket like acetabulum of the pelvic girdle

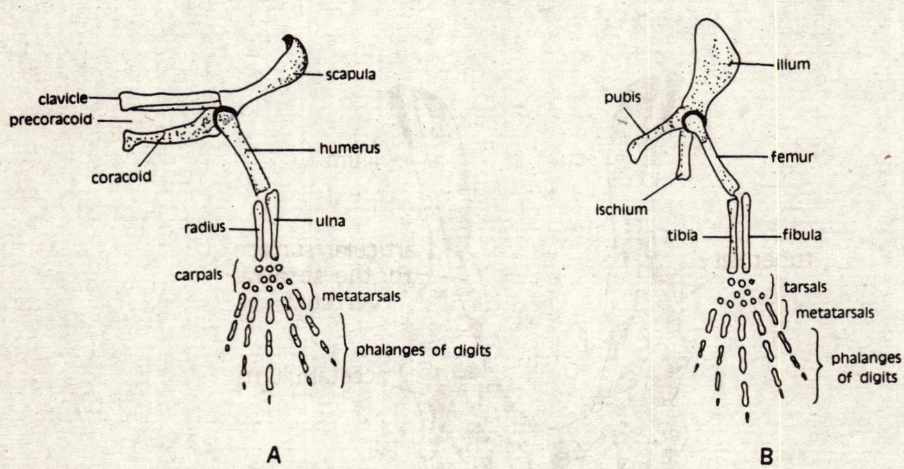

Fig 3.52 Diagram showing skeletal structures of tetrapod limbs and girdles: **A.** pectoral girdle and limbs; **B.** pelvic girdle and limb. The clavicle is lightly stippled to indicate that it is a membrane bone.

2. *Zeugopodium* constitutes the forearm or shank and contains two parallel bones, *radius* and *ulna* or *tibia* and *fibula.*
3. *Autopodium* constitutes three division:
   i) *Carpus* or *tarsus* (wrist or ankle) with 10 bones in three rows.
   ii) *Metacarpus* or *metatarsus* (plain or sole) has five long metacarpals or metatarsals
   iii) *Digits* (fingers or toes) are generally five in number with linear rows of phalanges. Each digit has 2, 3, 4, 5, 6, phalanges beginning from the first to the last.

Fore Limb Bones

In *Rana* (Fig. 3.53) the humerus is provided with prominent deltoid ridge. *Radius* and *Ulna* are fused into a radioulna bone. Five carpals in two rows are present with four figits and a rudimentary pollex encloses the skin. There are five slender metacarpals the first being rudimentary. The second digit bears two phalanges: third and fourth digits contain three phalanges each.

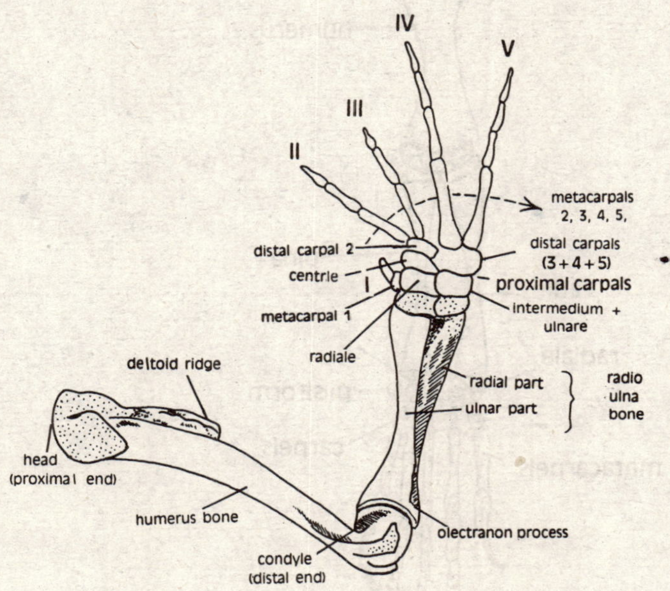

Fig. 3.53 Fore limb bones of Rana

In *varanus* (Fig. 3.54) the radius and ulna are separate. *Humerus* is broad. Carpals are in two rows, four in first row and five in the second row. Each of the five digits ends in a claw. Each digit consists of a metacarpal, and varying number of phalanges.

In *Gallus* (Fig. 3.55) the forelimb is modified for flight and balancing of the body. The humerus is stout with a deltoid ridge and pneumatic foramen for accommodating the air sac. *Radius* and *ulna* are well developed and ulna provides attachment to flight feathers. The first row of carpals is free including radiale and ulnare. The second row of carpals is fused with three *metacarpals* to form the *metacarpus* bone. The first metacarpal is reduced, second and third are fused forming two rods. Three digits are present, first with one phalanx second with two phalanges and third with one phalanx. Digits are one fingered only.

In *Oryctolagus* (Fig. 3.56) the humerus has a stout deltoid ridge and rounded head. At the proximal end of humorous two tuberosities are present for the attachment of muscles. The distal end has pulley-like trochlea for articulation with radius-ulna and a *supratrochlear* or *supracondylar* foramen for the passage of a nerve and artery. The *radius* and *ulna* are separate bones closely attached and capable of movement relative to one another. The ulna is produced into an *Olecranon process* or elbow. A *sigmoid notch* is present infront of the olecranon process for the articulation with trochlea of the humerus. The *carpus* has three bones in front row, a small centrale and four carpals. totaling eight bones. There are five metacarpals with digits ending in claws.

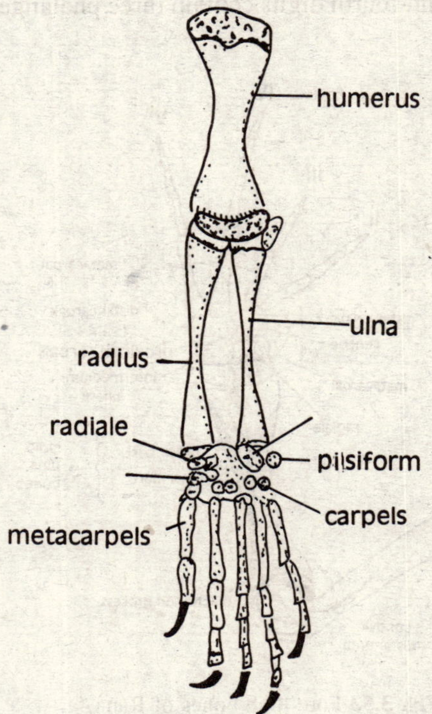

Fig. 3.54 Fore limb bones of Varanus

The digits have 2,3,3,3,3, phalanges willi claws.

Hind-Limb Bones

In *Rana* (Fig. 3.57) the hind limb bones are long and stout, adapted for swimming and jumping. The femur is S-shaped bone, its head fits into acetabulum cavity of the pelvic girdle. The tibio-fibula is a compound bone due to the fusion of tibia and fibula bones. A *nutrient -foramen* passes through this bone The *tibio-fibula* articulates with *astragalus* (tibilale) and *calcaneum* (fibulare) joining one another at the distal ends; these bones make the ankle long for jumping. Five tarsals are present following by five long and slender metatarsals. The first and second digits contain two phalanges each, the third and fifth three each and the fourth four.

In *Varanus* (Fig. 3.58) the femur is a stout bone with its head fitting into acetabalum cavity of the pelvic girdle. Greater and lesser trochanters are present on the ventral side of the head of femur. The *tibia* and *fibula* are separate bones. The *fibula* is slender; tibia is stronger and slightly longer than fibula. The *tarsals* are reduced in number due to fusion. Two tarsal bones in the proximal row, three bones in distal row and intertarsal ankle joint are present in the tarsal region. Five digits with 2, 3, 5, 5, 3 phalanges are present ending in claws.

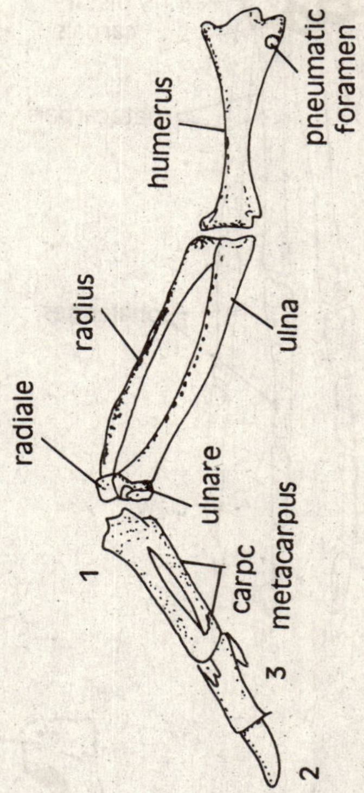

Fig. 3.55 Fore limb bones of Gallus

In *Gallus* (Fig. 3.59) the hind limb bones are specially adapted for bipedal locomotion. The femur is short and strong and articulates with aectabulum by its rounded head. A prominent *trochanter* is present near the head at right angle to the shaft of the bone. The distal end of femur has a pulley-like *condyle* for the articulation of tibiotarsus and fibula. The tibiotarsus is a fused bone in which tibia of other tetrapods is fused with tarsals of the first row at its distal end. The *tibiotarsus* is a long stout bone with *cnemal-crest* at its proximal end for the attachment of thigh muscles. The fibula is a reduced bone tapering to a point at its distal end. A *sesamoid bone* called *patella* is present infront of the juntion of femur and tibiotarsus. The tibiotarsus articulates with tarsometarарsus formed by the complete fusion of the tarsals of distal row and three metatarsals; its distal end has three distinct pulleys for the articulation with toes. The ankle joint is intertarsal between two rows of tarsals. The first metatarsal projects near the lower end of tarsomtatarsus. A bony spur is present projecting backwards from the metatarsus. Only four digits are present, fifth being absent. They have 2, 3, 4, 5 phalanges. The first digit hallux is pointed backwards and others are pointed forwards.

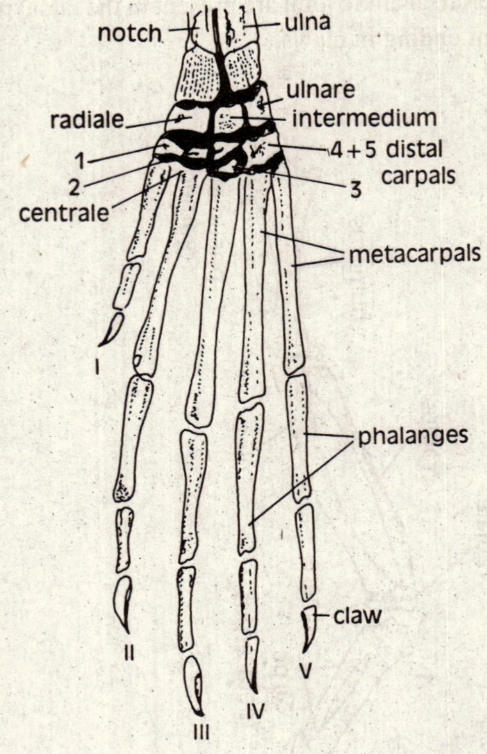

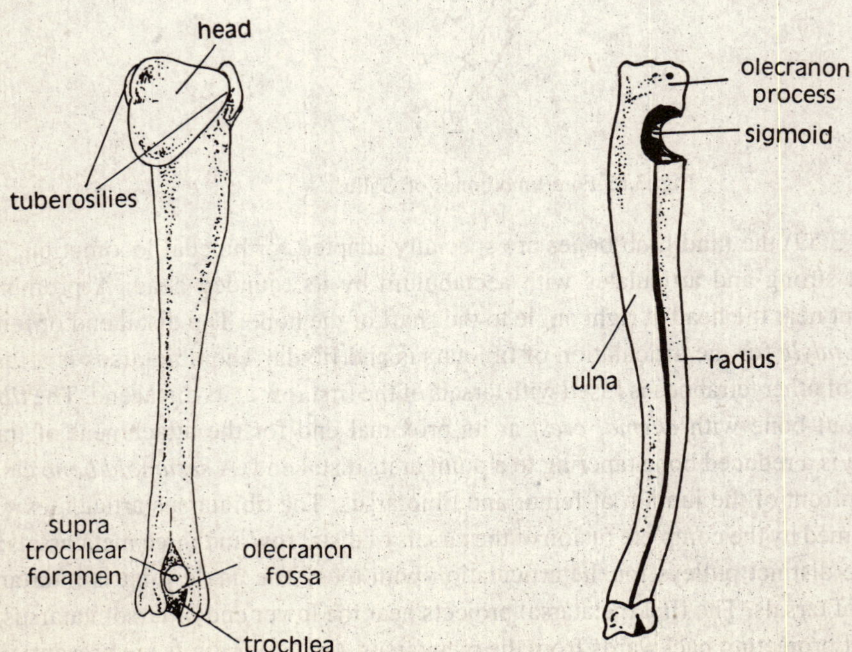

Fig. 3.56 Bones of Mammalian fore limb

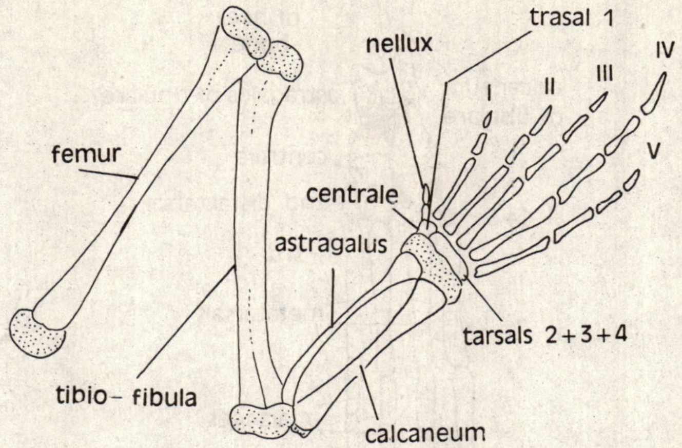

Fig. 3.57 Hind limb bones of Rana

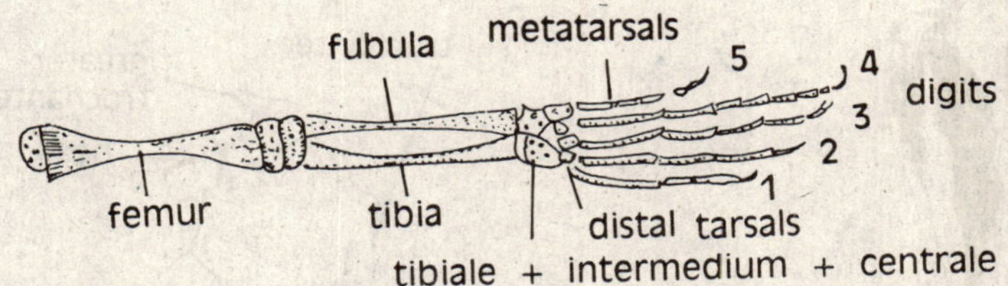

Fig. 3.58 Bones of hind limbs of a Varanus

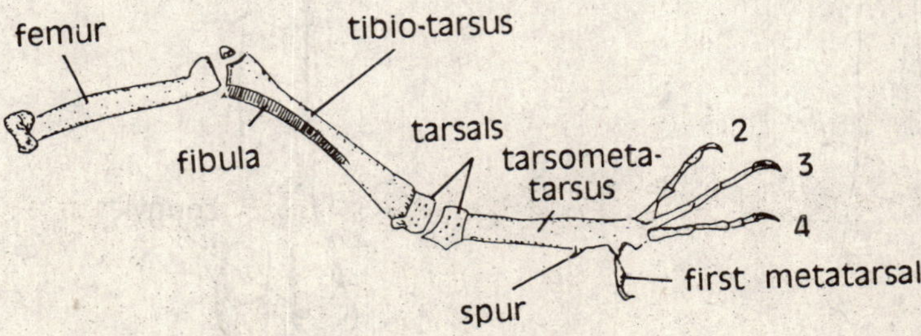

Fig. 3.59 Bones of hind limb of Gallus.

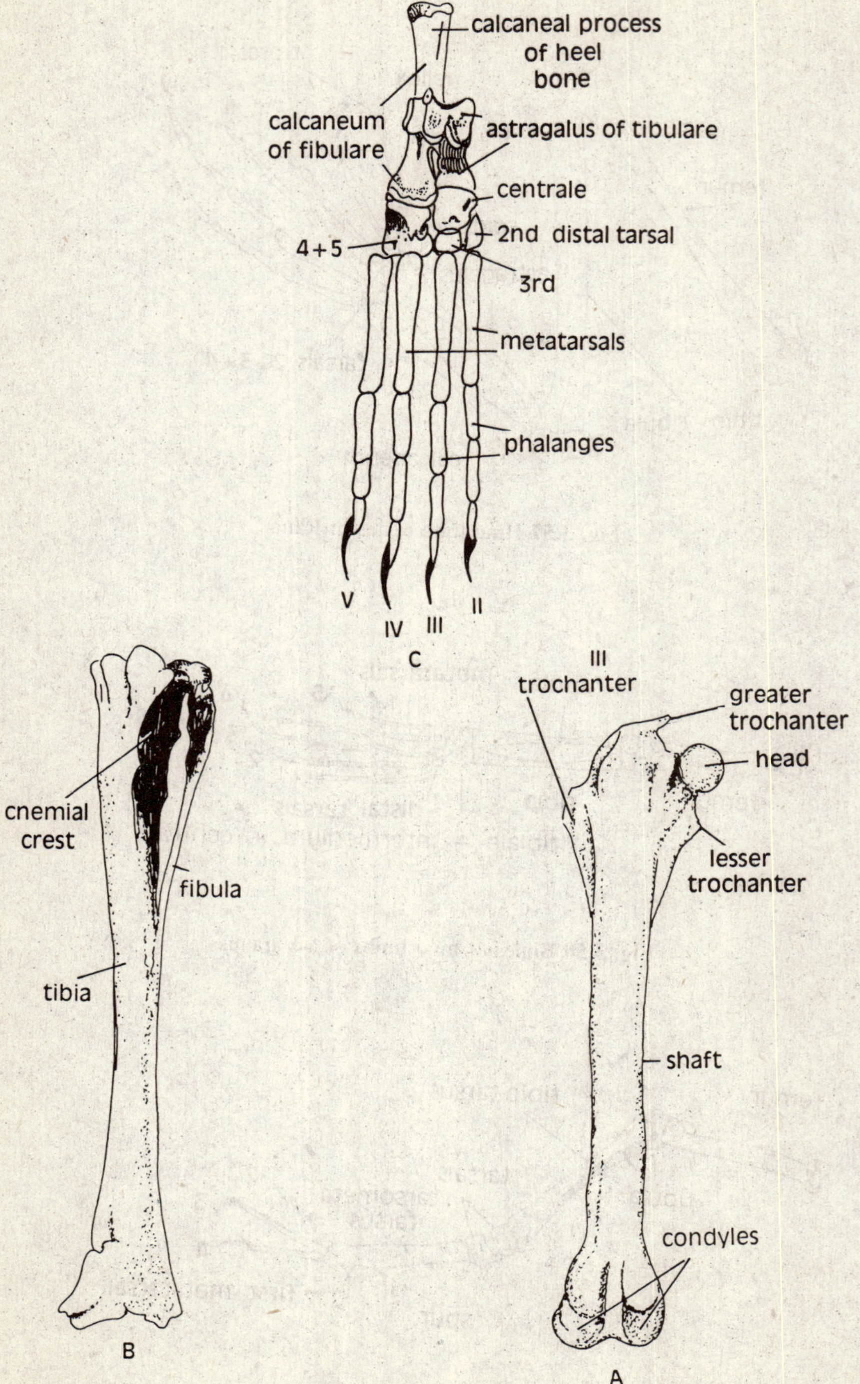

Fig 3.60 Bones of hind limb of Rabbit.

In *Oryctolous* (Fig. 3.60) the femur is strong with a head for articulation with acetabulum and three *trochanters* for the attachment of muscles. The distal end of femur has two smooth pulley-like *condyles* to form a hings-joint with the tibio-fibula. *Tibia* is strong bone with a enemial crest and a *sesamoid* bone or *patella* which forms a knee cap. *Fibula* is reduced and fused with tibia distally. The tarsals or ankle region has six bones arranged in three rows. A proximal row has astragalus calcaneum, and distal row has three tarsals with centrale lying in fornt of astragalus. Four metatarsals with digits are present. Each digit has three phalanges ending in claws (first digit of primates is called hallux and is opposable).

CHAPTER SUMMARY

1. The skeleton present outside the body (*exoskeleton*) and the one inside (*endoskeleton*) form the framework of the body. The skeleton is made of cartilage or bone or both. The endoskeleton of lower forms may consist of notochord only. In Cyclostomes and Chondrichthyes the cartilage forms major skeleton structure. In higher chordates the skeleton consists of dermal and cartilage bones.
2. The dermal skeleton is derived from the dermal parts of the skin. The scales and fin rays of fishes, bony plates under the skin of crocodiles, turtles, snakes and lizards; rib-like gastaralia found in some reptiles; bony shell of armadillo and dermal bones of skull and pectoral girdle are dermal bones.
3. The endoskeleton consists of axial skeleton (skull, vertebral column, ribs sternum) and appendicular skeleton (girdles, limbs).
4. The skull is derived from a cartilaginous *chondrocranium*, a dermal bony *dermatocranium* and the visceral skeleton or *splanchnocranium* supporting gills in lower forms.
5. The attachment of splanchnocranium to the chondrocranium is called splanchnocranium or *jaw suspension.* The jaw suspension is distinguished into autodiastlic (in early acanthodians), amphistylic (in primitive shark *Hepatanchus*), hyostylic (in most of clasmobranches and bony fishes) and autostylic types. The autostylic type includes holostylic in holocephali. monimostylic is tetrapoda excepting mammals, streptostylic in lizards. snakes and birds and craniostylic in mammals.
6. There is well marked uniformity in the basic plan of development and arrangement of various elements of the vertebrate skull. The visceral skeleton is progressively reduced from fishes to mammals. The skull of clasmobranches is cartilaginous called chondrocranium and consists of intimately fused cranium, four sense capsules and visceral skeleton forming jaws and supporting pharynx with gills. Bony fishes have the largest number of bones in the skull of all vertebrates. The cranium of bony fishes is covered over by numerous thin dermal bones, some of them with lateral line canals. In amphibia there is well marked reduction in the number of bones. In reptilian skull increased ossification with several new bones is found. The skull of birds is light in weight and fusion of various bones has taken place to a high degree. The mammalian

skull shows considerable reduction in the number of bones by fusion or loss; the cranium is large to accommodate large brain.

7. The notochord persists throughout life in *Branchiostoma*. cyclostomes and a few fishes but is replaced by the centra of vertebrae in most vertebrates
8. The vertebral column makes the longitudinal axis of the body posterior to the skull. It is made of a series of segmental structures called vertebrae. The vertebrae enclose the spinal cord and provide rigidity to the body and place for the attachment to girdles, ribs and muscles. In lower forms the vertebrae are similar throughout but in higher forms they are grouped in several regions such as cervical, thoracic, lumber, sacral and caudal. In many forms fusion of several vertebrae occurs which is more conspicuous in birds.
9. The ribs are cartilaginous or bony structure attached to vertebrae. In most amniotes the ribs in thoracic region form strong arches which unite with the sternum to protect the thoracic cavity. The ribs of tetrapoda are generally believed to be homolgous to the ventral ribs of fishes.
10. The sternum is cartilaginous or bony plate-like structure in the thoracic region of tetrapoda It is found in vertebrates with legs and provides protection to thoracic region, and attachment to anterior limb muscles and flight muscles in birds. It takes part in respiratory movements in amniots.
11. The appendicular skeleton includes that of the median fins of aquatic vertebrates and paired appendages found in all classes of vertebrates excepting cyclostomata.
12. The caudal fin may be protocercal in larval fishes, heterocercal in sharks, hypoceral in anapsida, some ostraoderms and Ichthyossurs, diphyceral in Fieraspis or homoceral in telcosts.
13. The pelvic and pectoral girdles of fishes are not connected with the vertebral column. Cartilage and cartilage bone of girdles in fishes are homologous to those of tetrapoda. Several dermal bones have become associated with the pectoral girdle of bony fishes. In tetrapoda the girdles and limbs show homologies in basic structure.
14. The tetrapod limbs are pentadactyle having stylopodium or upper portion, zeugopodium or fore portion and autopodium or lower portion. The upper portion of each limb containing single bone articulates with the girdle. The fore portion contains two parallel bones. The third lower portion contains wrist or ankle with several carpal or tarsal bones. The lower portion, hand or foot consists of five long, narrow bones and five digits. Each digit contains two to five small phalanges. Reduction and fusion of limb bones may lake place in some animals or groups.

4

Muscular System

4.1. GENERAL ACCOUNT

The muscles form nearly half of the entire weight of the body. They may vary in shape, size and attachments. Each muscle is provided with capillaries and nerves. The muscles may be divided into a series of similar units in fishes or may spread out in ribbonlike, triangular, pinnate or fan-like sheets or may be massive and spindle shaped in appendages. The main functions of the muscles are:

1. Movement of trunk, limbs, jaws, organs or organ parts.
2. Maintenance of the stability of body parts.
3. Production of body heat.
4. Assistance in major functions of the body such as circulation of blood, digestion, respiration, locomotion, reproduction and behaviour.

The muscular tissue has more power of contraction as compared to other organs. The duration of contraction in muscles corresponds to long axes of muscle cells. The muscle tissue is derived from mesoderm excepting those of mammalian ectodermal glands such as lacrimal, salivary, sweat, mammary glands and iris of the eyes.

4.1.1. Types of Muscular Tissue

The muscular tissue has been distinguishes into three types including smooth, cardiac, and striated on the basis of structural and functional differences.

Smooth muscle tissue *(Fig. 4.1):* These muscles are found in the lining of the digestive tract, ducts of digestive glands, urinary bladder, trachea and bronchi, circulatory vessles, ducts of genital organs and connective tissue of skin and other areas. These muscles consist of long, narrow, spindle shaped cells. The cytoplasm of the muscle cell is called sarcoplasm. The nucleus is elongated and situated in the middle. The sarcoplasm contains longitudinally arranged thread like myofibrils which are actual contractile elements. The smooth muscles fibres or cells have an average length of 0.2 mm The smooth muscle fibres may form bundles or sheets or may occur individually. The smooth muscletissue is mostly involuntary, slow in contraction and activated by sympathetic nervous system.

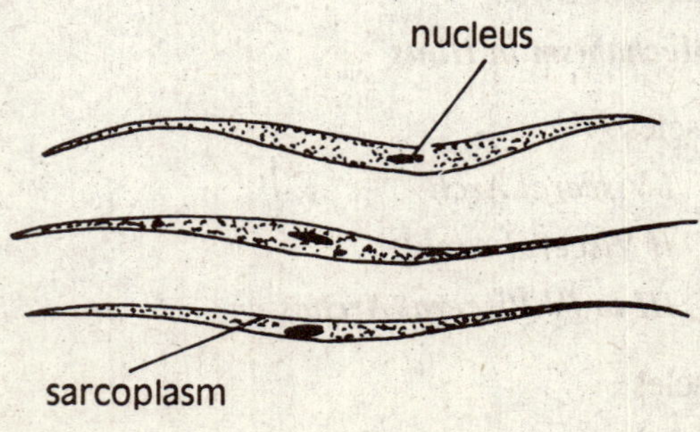

Fig. 4.1 Smooth muscle fibres

***Cardiac muscle tissue** (Fig. 4.2):* This (issue forms the bulk of muscles wall of the heart and roots of the large blood vessels joining the heart. The cardiac muscles show structure of both smooth and straiated muscles. The muscle fibres are usually branched, interlacing, multinucleate and partly striated. These muscles are involuntary with automatic regulation where contraction is initiated within the muscle itself and not by a nervous impulse.

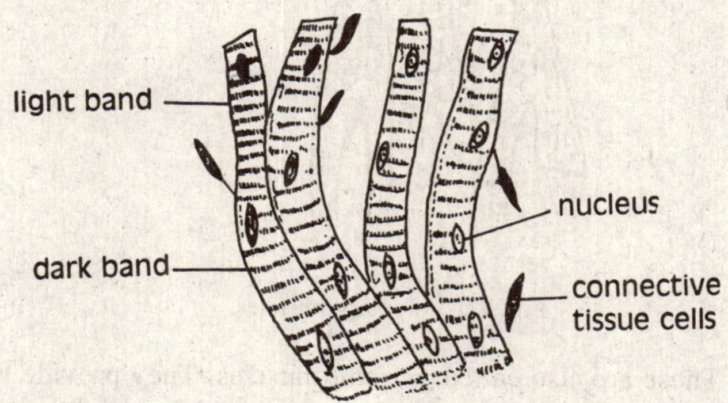

Fig, 4.2 Cardiac muscle

***Striated muscle tissue** (Fig. 4.3):* These are called voluntary or skeletae muscles. They constitute the flesh of the body. The striated muscle fibres are large, multinucleate. elongated or cylindrical with length varying from about a mm to a number of centemetres. These fibres are arranged in parallel to form muscles. Connective tissue are present between the fibres and bind them forming sheaths for fibre bundles called *perimysium.* A common sheath surrounding many bundles is called *epimysium.* The fibre surface is covered by a thin *sarcolemma membrane.* Each fibre contains a large number of closely packed longitudinal *myofibrils,* each containing numerous tiny filaments. Alternative dark and fight regions of the myofibrils give striated or crossbanded appearance to the muscle fibre. A muscle is formed by bundles of striated muscle fibres found together in a sheath of connective tissue. The muscle is usually fastened to some part of the skeleton by means of connective tissue of muscle sheaths. A tendon is a thick cord continuous with epimysium and serves for the attachment of the muscle The striated muscles are controlled by the voluntary nervous system which sends branches to all muscle fibres. The impulse form the central nervous system normally result in the contraction of a muscle. The striated muscle may be distinguished into red and white types. The red and white muscles may be present in different vertebrates or even in different regions of an individual. Red muscles are vascular with abundance of sarcoplasm. The pectoral and leg muscles of pigeon are red muscles and play a significant role in flight. In chickens the pectoral muscles are white and leg muscles red.

4.1.2. Classification and Terminology

The muscles have variously been classified according to structure, embroyonic origin, location, functions or shape. Two main groups of muscles which show adaptive modifications in successive higher vertebrates are somatic and visceral.

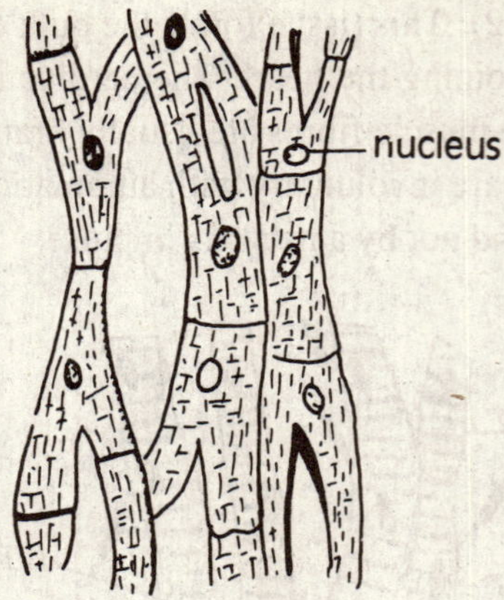

Fig. 4.3 Voluntary muscles.

Somatic muscles: These are also called *parietal* muscles. They provide for locomotion and orientation of the animal in the environment. They are found inserted on the skull, vertebral column and limbs. They are also called *myotomal muscles* as they are derived from myotomes of the mesodermal somites These muscles are skeletal, voluntary and striated and innervated by motor neurons. The somatic muscles attached to skull and vertebral column are called *axial* and those on appendicular skeleton as *appendicular.* Muscles originating outside of the part upon which they act are called *extrinsic* and those entirely belonging to a part are called *intrinsic* muscles. In addition, there are *integumentary muscles* which move the skin.

Visceral muscles: These muscles are found in the alimentary canal, blood vessels, tubes, ducts, in the skin for the elevation of hairs and on visceral skeleton. They regulate the internal environment and assist in homeostasis. They also regulate feeding and respiratory movements. They are mostly smooth, nonskeletal and involuntary excepting branchiomeric muscles which are striated and skeletal. These muscles are deriverd from the mesenchyme of the splanchnopleure and never from the myotomes. They are innervated by visceral motor fibres.

Several teems connected with muscles are given below:

Name of the muscles	*Function*
Extensor	Opening the joints
Flexor	Closing the joint
Adductor	Drawing a segement inward
Adductor	Drawing a segement upward
Levator	Raising of an structure
Depressor	Lowering of an structure
Rotator	Twisting of a limb segment
Supinator	Turning of an appendage upside down or supine

Pronator	Return of an appendage to normal lower surface down or prone
Constrictor (Sphinctor)	Surround apertures like anus, gills and close them on constriction

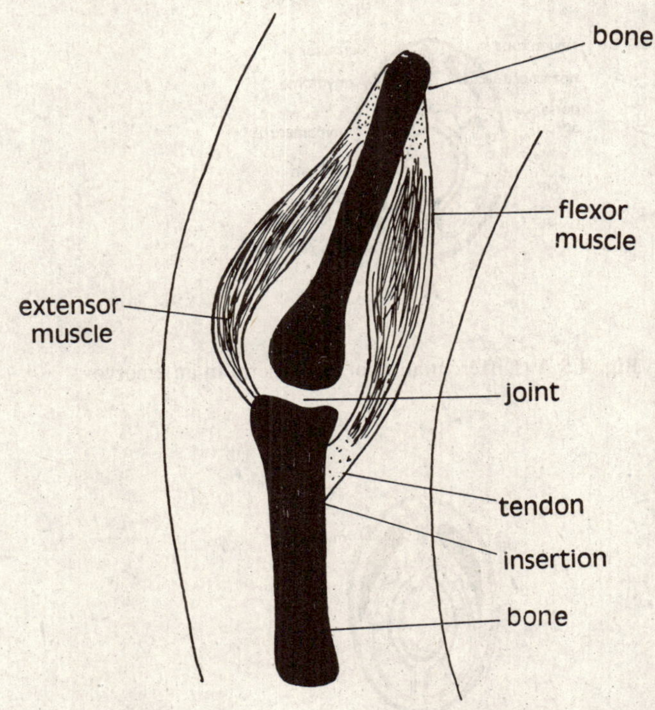

Fig. 4.4 Somatic muscles

4.1.3. Development of Muscular Tissue (Fig. 4.5)

During early development the mesoderm is subdivided into a segmented *epimere* or *dorsal mesoderm,* partially segmented *mesomere* or *intermediate mesoderm* and unsegmented (excepting in *Branchiostoma) hypomere* or *lateral mesoderm.* The dorsal mesoderm or the mesodermal somites subdivide into *myotome, sclerotme* and *dermatome*. The mytome gives rise to somatic muscles ontogenerically or phylogenerically The scleretome gives rise to parts of skeleton system and dermatome to skin. The myotomes or muscle segments are somatic muscles arranged at regular intervals along the longitudinal axis of the body in primitive vertebrates. They extend from the middorsal to midventral line. A connective tissue called *linea alba* occupies the space between the myotomes. A connective tissue partition called *myosepta* or *myocommanta* bind the myotomes. Segmental arrangement of the myotomes is disturbed by girdle and paired appendages when muscles buds are sent out from myotomes to produce muscles. The cells of the original myotome giving rise to muscle tissue are called *myoblasts.* In higher vertebrates the segmental muscles are hardly recognizable due to increase in size of muscles of girdles and appendages. A tendency towards loss of fundamental metamerism of muscles due to splitting and fusion of segmental muscles is thus seen in vertebrates.

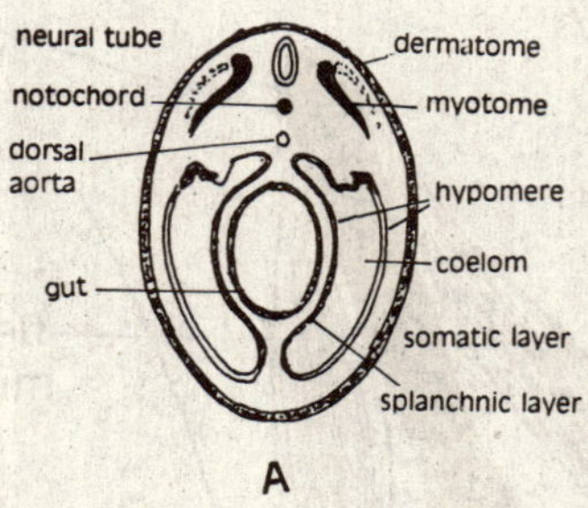

Fig. 4.5 A. Differentiation of mesoderm in an embryo

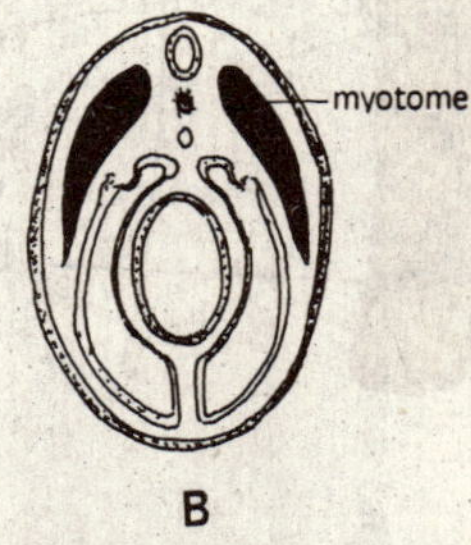

Fig. 4.5 B. Differentiation of mesoderm in an embryo

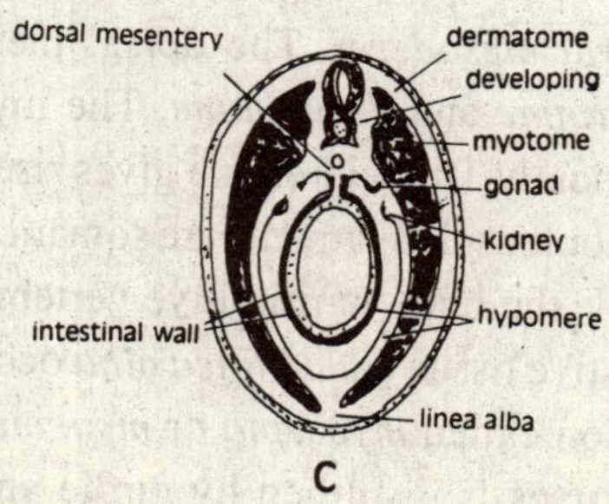

Fig. 4.5 C. Differentiation of mesoderm in an embryo

The splanchinic layer of the hypomere which surrounds the endoderm of the gut gives rise to *smooth muscles* of the gut and cardiac muscles. The branchiomeric muscles which are voluntary and striated also originate from the splanchnic layer of the hypomere and not from the myotomes.

4.2. SOMATIC MUSCLES IN VERTEBRATES

As already stated the musculature consists of a series of muscles segments or *myomeres* by *myosepta* or *myocommata* The myosepta extended from the skin and muscle fibres. Each myomere arises on one myoseptum and is inserted on the next. A *horizontal septum* divideds the musculature into a dorsal and a ventral mass and extends outward from the transverse process of the vertebrate to the skin to form a *fibrous membrane.* Dorsal side of this membrane contains a *epxaxial muscles* and ventral side *hypaxial muscles.* The muscles of two sides are separated by vertical midorsal and midventral septa. A longitudinal ligament called *linea alba* is present at the base of the midventral septum. The epaxial muscle of a given myomere is inverted by dorsal ramus branch of the spinal nerve supplying to the segment and hypaxial muscle is supplied by the ventral ramus branch of the same spinal nerve.

4.2.1. Branchiostoma (Fig. 4.6)

There are sixty or more pairs of V-shaped myotomes, myopmeres or muscle segments on the sides of the body. The myomeres point forward and those of right and left sides are placed alternatively unlike other chordates where they are placed opposite. This arrangement helps in the movement of the body sideways with considerable rapidity during locomotion. A transverse sheet of muscle is present in the wall of the atrium. The myotomes are the blocks of striated muscle fibres extending throughout the length of the body They are separated from one another by V-shaped dense connective tissue *myocommata* pointing forwards. The muscle fibres are arranged longitudinally and run straight. One muscle fibre is attached to two successive myocommatea. The contraction of the fibres of myotomes results into pulling off of the myocommata. thus flexing the body.

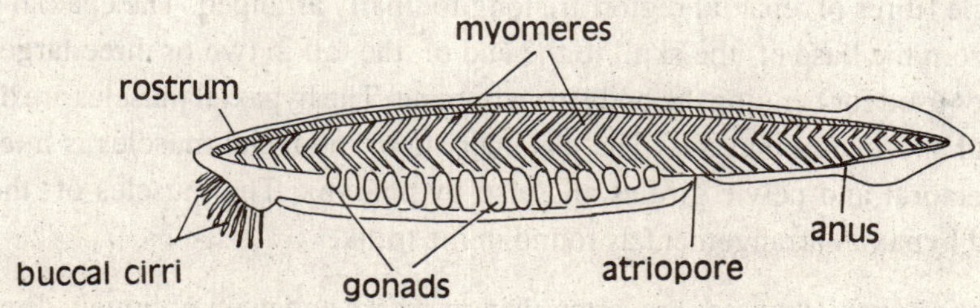

Fig. 4.6 Myotomes in Amphioxus

4.2.2. Cyclostomes (Fig. 4.7)

Due to the absence of paired appendages the muscular arrangement of Cyclostomes has undergoes little change from that of *Branchiostoma.* The myotomes are nearly vertical and slightly bent forwards at their dorsal and ventral ends (Fig. 4.7). There is no interruption of metamerical arrangement anywhere along the length of the body behind the gills. The myostomes are not divided into epaxial and hypaxial masses

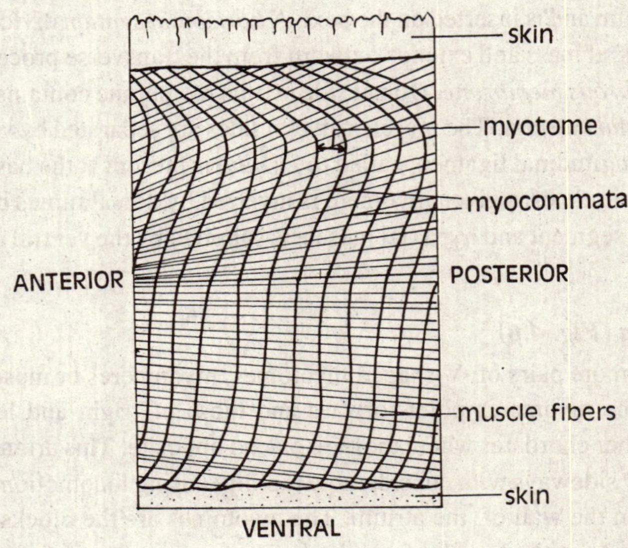

Fig. 4.7 Lateral view of several parietal myotomes of lamprey in region just anterior to the first dorsal fin. The skin has been removed to show the muscle fibres beneath.

4.2.3. Fishes

In sharks the myotomes are separated into epaxial and hypaxial masses separated by a longitudinal *lateral septum* of connective tissue (Fig. 4.8). The myocommata have taken a zigzag shape but the muscle fibres of epaxial region are longitudinally arranged. The epaxial muscles of each side extend from the base of the skull to the end of the tail in two or three large masses of longitudinal bundles used for bending the body in swimming. The hypaxial muscles are divided into ventral longitudinal and lateral muscles. The matamerism of hypaxial muscles is interrupted by musculature of pectoral and pelvic girdles and also by pharynx. The muscles of the tail have similar epaxial and hypaxial arrangement as found in the trunk.

In bony fishes the hypaxial masses are larger than in sharks and main obliquely directed fibres are present. The ventral most fibres form a definite bundle called the *rectus abdominis.* The innermost fibres of the hypaxial muscles are attached to the ventral ribs or myocommata. Certain of the epaxial and hypaxial muscles have become specialized and send portions to median fins. The body wall muscles of fishes serve for locomotion and fins serve mainly as stablizers. The waves of contractions from the anterior end of the trunk to the tail result in swimming movement.

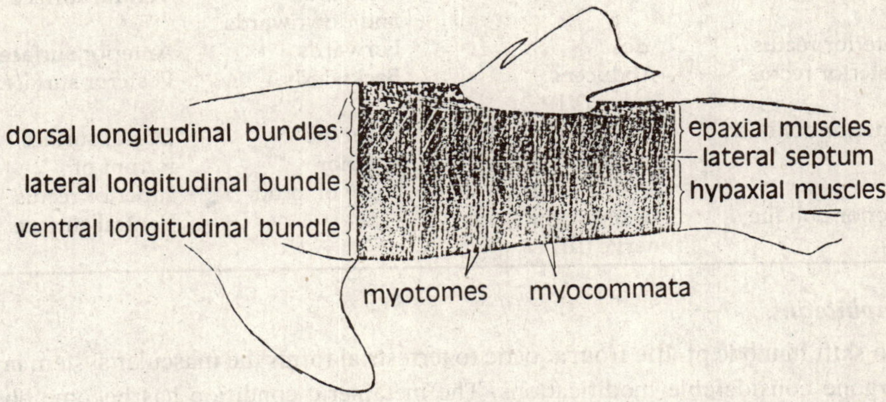

Fig. 4.8 Lateral view of protion of body wall of dogfish *Squalus acanthias*, showing zigzag arrangement of the myotomes. The skin has been removed to show the underlying musclature.

Extrinsic Eyeball Muscles: The muscles that operate the eye ball are striated skeleton and voluntary. They arise from three preoptic somites of the head during embryonic development of the clasmobranches. The eye muscles of dogfish are isolated in two groups including *recti* and oblique muscles and keep the eye in orbit (Fig. 4.9). These muscles arise from the wall of the skull and gel inserted on the eyeball. The recti muscles are four in number and include *superior, inferior* and *posterior*. The oblique muscles are superior and inferior. The details of these muscles are outlined below.

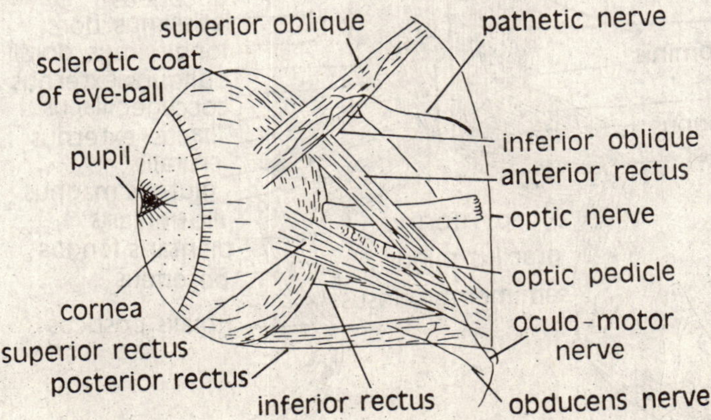

Fig. 4.9. Showing the disposition of different extrinsic eye muscles of *Scolidon* (after Thiallayampalam).

Name	*Nerve supply*	*Course*	*Insertion on eyeball*
Superior rectus	Branches of III cranial	Outwards and upwards	Dorsal surface
Inferior rectus	-do-	Outwards and downwards	Ventral surface
Anterior rectus	-do-	Forwards	Anterior surface
Posterior rectus	Abducens nerve (VI)	Backwards	Postcror surface
Superior oblique	Trochlear nerve (IV)	From the anterior angle of orbit	Dorsal surface infront of superior rectus
Inferior oblique	Oculomotor nerve (III)	-do-	Ventral surface

4.2.4. Amphibians

Due to shift in mode of life from aquatic to terrestrial forms the muscular system in tetrapods has undergone considerable modifications. The metameric condition has become obscure and extensive development of appendicular skeleton accompanied by reduction of axial musculature has taken place (Fig. 4.10).

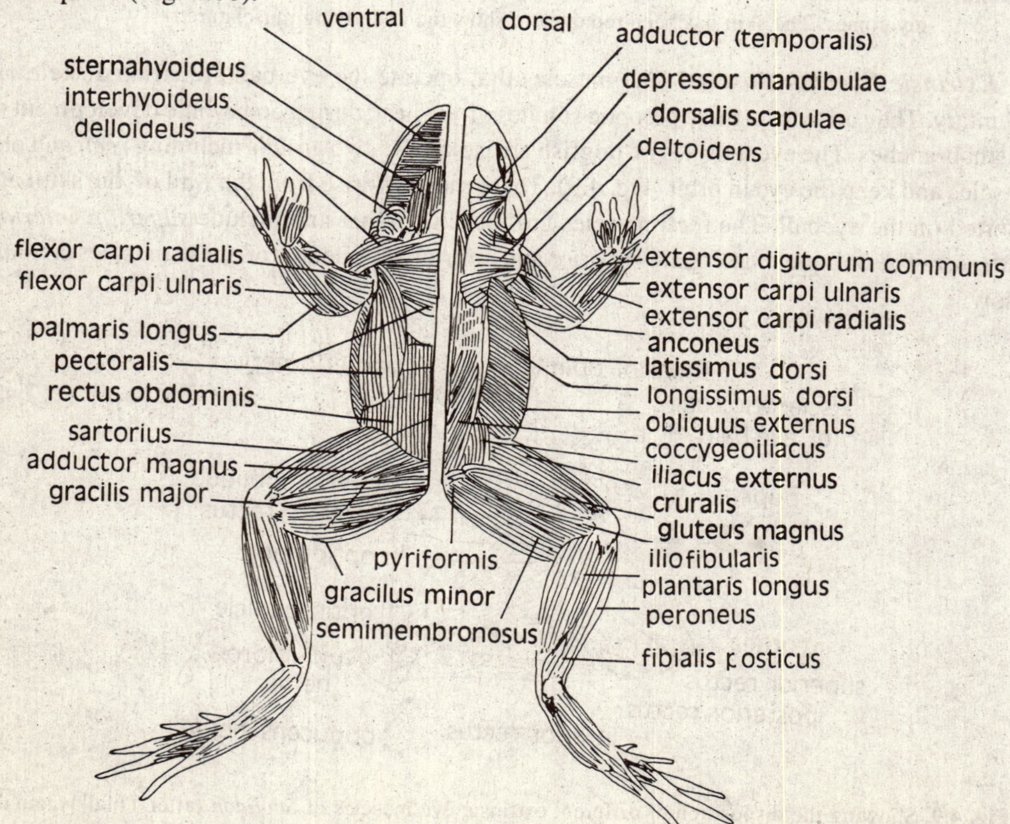

Fig. 4.10 Muscles of green frog, *Rana clamitans;* ventral aspects on left, dorsal on right.

In frogs the epaxial muscles are reduced and used in bending the vertebral column from dorsal side. The epaxial muscles have given rise to *dorsalis trunci* bundles which further differentiated into *interneural muscles* between the neural arches and *intertransvesorial muscles* between the transverse processes. A long muscle, the *longissimus dorsii* passes from head to near the end of the urostyle. Several muscles are attached to the skull for turning the head. The hypaxial muscles are further reduced in anurans. The *rectus abdominis* extends from sternum to the pubis.

4.2.5. Reptiles

The arrangement of body muscles in reptiles is complex and comparable to that of mammals. Lateral septum is absent and muscles have developed in relation to ribs also. The epaxial muscles have differentiated into several groups including *spinalis* and *semispinalis muscles.* The longissimus dorsii of amphibians is differentiated into a *longissimus dorsal proper* in lumber region, a *longissimus capitis* is neck and skull region and *lateral dorsi* at the proximal ends of ribs.

The hypaxial muscles of lacertilians and crocodilies have external obique and transverse layers and additional layers in the thoracic region. The obique muscles have given rise to intercostals layer subdivided into *external* and *internal intercostals* used in respiratory movements. The *rectus abdominalis* is well developed.

In snakes small muscles derived from the oblique muscles and rectus abdomonis pass from the ribs to the skin underlying ventral scales and help in locomotion with the help of certain dermal muscles. The hypaxial muscles are poorly developed in turtle due to development of a rigid shell.

4.2.6. Birds

Due to specialized mode of life in birds the epaxial muscles are poorly developed. The oblique muscle are reduced and transverses

4.2.7. Mammals

The trunk musculature of mammals is similar to that of reptiles with reduction of hypaxial musculature. The musculature of limbs and girdles is well developed. The metameric arrangement of trunk mytomes is slightly visible in *intercostals serratus, scalene, intervertebral* and *rectus abdominis trunk muscles.*

The epaxial muscles have reptilian arrangement. A rector spinae muscle of sacral region is divided into a lateral *iliocostal,* an *intermediate longissimus dorsi* and a *medial spinalis dorsi.* The transverse spinalis system reptilian is represented here as *multidus muscle* filling the groove on either side of the vertebrae. The *interspinal* and *intertransversorial* muscles present between adjacent vertebrate are formed by deeper epaxial muscles.

The hypaxial museles include an *inner transverses abdominis* a middle *internal oblique* and an *outer external oblique*. The rectus abdominis lies on either side of the *linea alba* enclosed in a sheath formed by the *aponeuroses* of the *oblique* and *transverse* muscles *Sternohyoid, sternothyroid, geniothyroid, amothyroid* and *thyrohyoid* muscles derived from rectus abdominis are present in neck region. The thoracic muscles of mammals are reduced and almost entirely covered by large appendicular muscles. *External intercostals and internal intercostals muscles* are attached on ribs and help in respiration. A *transverses thoracic* is confined to thoracic wall, *scalene* from anterior rib to transverse processes of cervical vertebrae and *serratus* to thoracic and anterior lumber vertebrae.

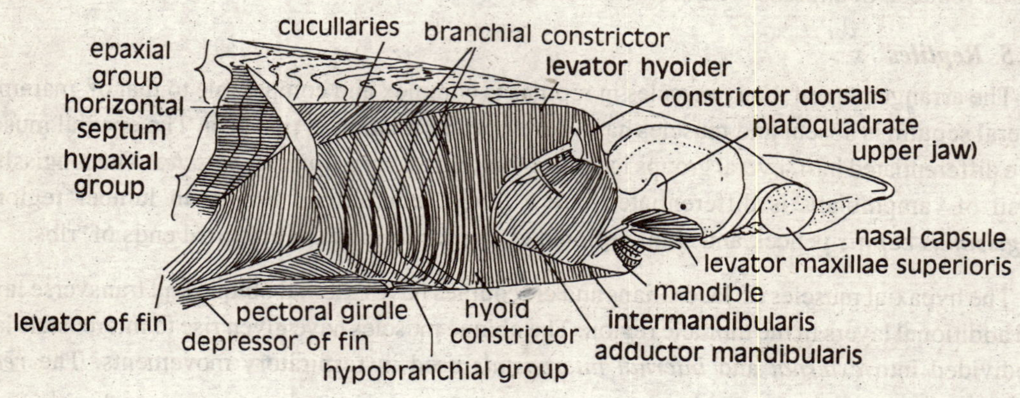

Fig. 4.11 Head muscles of shark, Squalus acanthias

Diaphragm in Mammals: The muscular diaphragm is a transverse voluntary muscle found in mammals which divides the body cavity behind the pericardium and lungs. It is a dome-shaped structure attached to the vertebral column and to the ribs. It is derived embryologically from the septum transversum and contains other memebranes. The portion of the diaphragm adjacent to sternum, ribs and vertebral column is invaded with muscle tissue contributed by myostomes in the cervical region. Oesophgus and large blood vessels pass through the diaphragm. The contraction of the muscles of the diaphragm enlarges the pleural cavities and thus aids in respiration. The diaphragm is innervated by branches of cervical spinal nerves which throws light on the embryonic origin of its muscle constituents.

4.3. APPENDICULAR MUSCLES

The appendicular muscles are attached to girdles and limbs for locomotion. They are divided into *extrinsic* and *intrinsic* types. The extrinsic muscles are attached to axial skeleton and inserted on girdles and limbs. The intrinsic muscles arise on girdles or on proximal skeleton elements of the appendage and insert on more distal elements. The appendicular muscles are striated, skeletal voluntary and somatic. They play insignificant part in the locomotion in fishes, which is mainly done by the body wall muscles. In amphibian and reptiles the appendicular muscles overgrow other muscles and play a major role in locomotion. In birds and mammals these muscles are highly developed for flight and support on legs.

4.3.1. Fishes

Extrinsic muscles at the base of the fin form two opposing muscle masses, *dorsal extensors* and *ventral* flexors. The intrinsic muscles are non existent. Movement of fins is simple

4.3.2. Tetrapods

Considerable evolution of extrinsic and intrinsic muscles has taken place in tetrapods as they are required for propulsion on land and for the movement of independent segments of the appendages. The tetrapod limb muscles do not show direct relationship between them and body myotomes. They arise from undifferentiated mesenchymatous cells in the limb itself.

Muscles arising in the dorsal part of the appendage become *extensors* in most cases and ventral derivatives become *flexors.* Outgrowth form these muscles to trunk form abductors and adductors. The dorsal and ventral muscles found in tetrapods have been shown in table 4.1. The homologies of some of the muscles are obscure in different tetrapods.

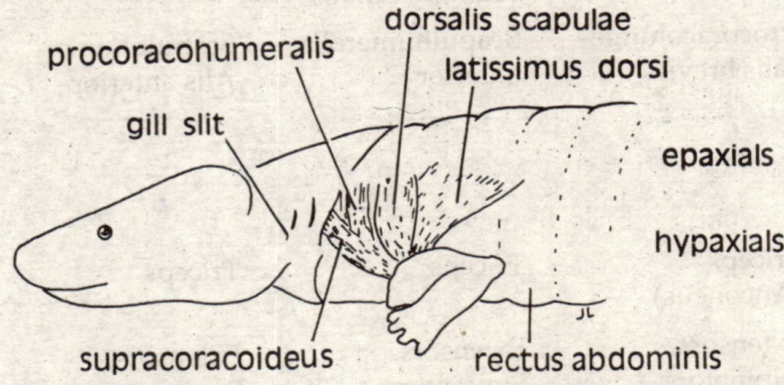

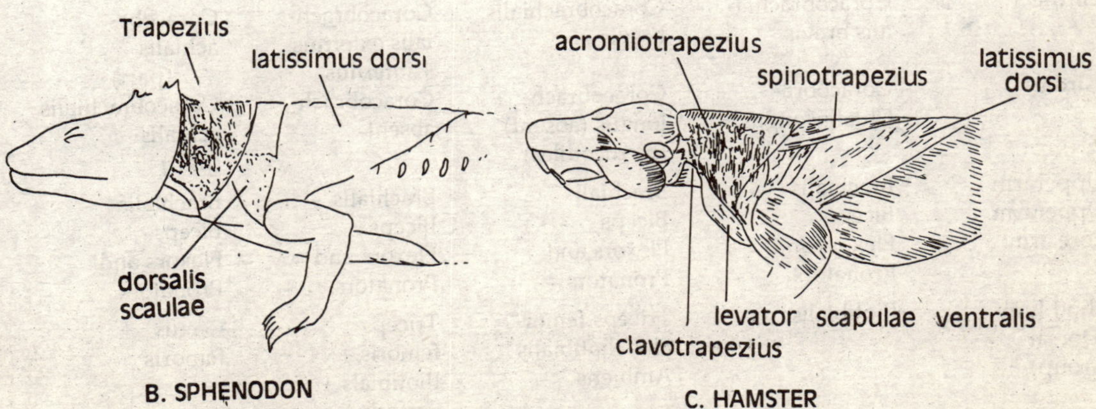

Fig. 4.12 Superfical shoulder muscles of an amphbian, reptile, and mammal, Illustrating increased extent of should muscles in successively higher tetrapods. Of the four shoulder muscles shown in the necturus, only the latissimus dorsi is extrinsic. The other three are intrinsic appendicular muscles.

TABLE 4.1

Appendicular Muscles in Tetrapods

Region	*Amphibians*	*Reptiles*	*Birds*	*Mammals*
Fore arm Dorsal Group	Latissimus dorsi	Latissimus dorsi + teres major	Latissimus-dorsi	Latissimus-dorsi
	Subcoraco-scapularies	Subcoracoscapu-laris+scapulo-humeralis posterior	Subcoraco-scapulars+ Scapulohu-maralis posterior	Subscapu-laris
	Deltoides-Scapularis Procoracohu-Meralis longus	Deltoides-scapularis Deltoides-clavicularis + Humeroradialis	Deltoid-scapularis Deltoides-propatagialia	Deltoid scapularis Deltoid-clavicularis
Girdle	Procoracohume-ralis brevis	Scapulhumeralis anterior	Scapulohumer-Alis anterior	Teres-minor Infrasp-inatus, Supraspinatus
Upperarm	Triceps (Anconcus)	Triceps	Triceps	Triceps
Fore arm	Extensors, Supinators	Extensors, Supinators	Extensors, Supinators	Extensors, Supinators
Forearm (Ventral Group) Girdle	Pectoralis Supracora-ciodcus	Pectoralis Supracora-ciodcus	Pectoralis Supracora-ciodeus	Pectoralis Supracora-ciodcus
Girdle	Coracobrachia-alis brevis	Coracobrachialis brevis	Coracobrach-ialis externus + internus	Coracobr-achialis (part)
Girdle	Coracobra-Chialis longus	Coracobrach-longus (absent) In crocodiles)	Coracobrach-absent	Coracobrachialis achialis (part)
Upperarm Upperarm Fore arm	Brachialis Biceps Flexors and Pronators	Brachialis Biceps Flexors and Pronators	Brachialis Biceps Flexors and Pronators	Brachialis Biceps Flexors and Pronators
Hind limb (Dorsal Group)	Iliotibialis	Triceps femor-alis, Iliotibialis Ambiens	Triceps-femoris + Iliotibials + Sartorius	Rectus femoris
			Ambiens	Ambiens
	Femoratibialis Iliofibularis	Femoratibialis Iliofibularis	Femoratibialis Iliofibularis	Vastus gr. Tenuissius

	Ilioextensorius	Iliofemoralis	Iliofemoralis-eternus + Iliotrochantericus	Glute+ Tensor fasciae
	Pubioschiofemoralis	Puboischiofemoralis internus	Iliofemoralis-internus	Iliacus, Psoas, Pectineus
Lower leg and foot	Extensors	Extensors	Extensors	Extensors
Hind limb (Ventral Group)	Puboischiotibialis	Flexor cruris Puboischiotibialis	Ilioflexorius Caudiloflexorius	Gracilis Semi membranosus
		Flexor tibialis-internus Flexor tibialis-externus		Semitendinosus Bicepsfemoralis
	Pubotibialis	Pubotibialis	Puboischio femoralis	
	Adductor Femoris	Adductor femoris		Adductor femoris
	Puboischiofemoralis externus	Puboischiofemoralis externus Ischiotrochantericus	Obturator Ischiofemoralis	Obturator externus Quadratus femoris Obturator internus
Caudal vertebrae	Caudofemoralis	Caudofemoralis Longus +	Coccygeofemoralis	Caudofemoralis
Lower leg and foot.	Flexors	Flexors	Flexors	Flexors

4.3.3. Flight Muscles in Birds

The following muscles are used for flight in flying birds (Fig. 4.13).

Pectoralis Major: It is a very large muscle attached to the edge of the sternum and carina, to the clavicle inform and to the head of the humerus ventrally by a tendon. It depresses the humerus resulting into powerful downstroke of the entire wing.

Pectoralis Minor: It arises from the sternum on the dorsal side of the pectoralis major, its tendon passes through the foramen triosseum between junction of clavicle, scapula and coracoid and is inserted on the dorsal side of the humorous. It is responsible for the elevation of the wing, which is brought about by the backward pull of its tendon through the foramen triosseum.

Coracobranchialis longus, and Coracobranchialis brevis: These are two small muscles arising from coracoid and scapula and joined to the head of humerus. They are responsible for the rotation of the wing in the glenoid cavity.

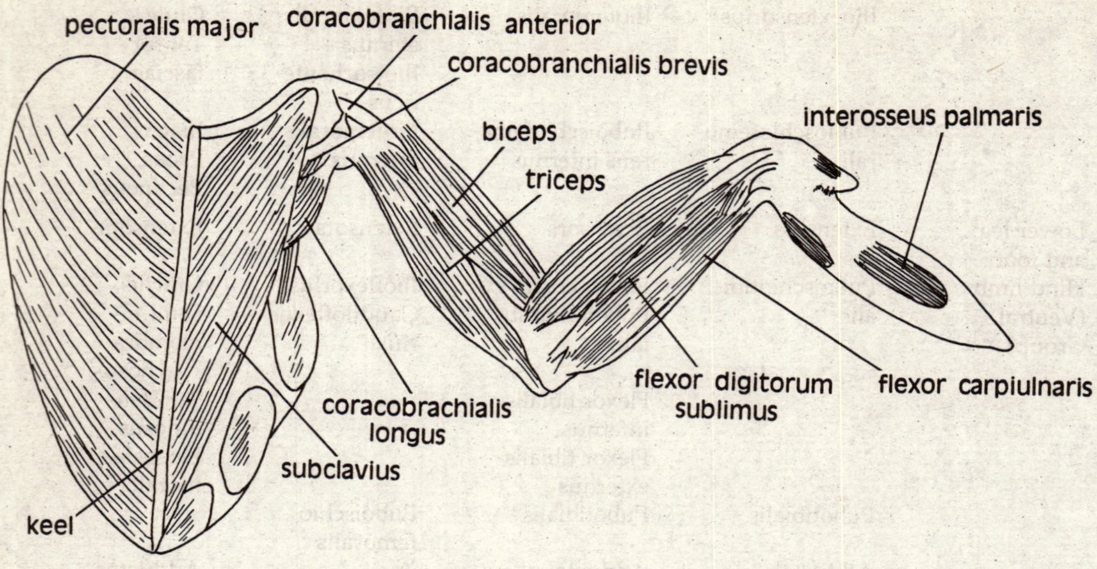

Fig. 4.13 Ventral view of breast and wing musculture of *Columba.*

Tensor longus, Tensor brevis and Tensor accessories: These are three muscles responsible for keeping the patagium tensely stretched during flight.

Tensor posterioris alae keep the postpatagium tensely stretched during flight.

The adjustments during flight are performed by muscles present inside the upper arm called biceps and triceps which operate the elbow.

4.3.4. Parching Mechanism in Birds

Some birds including pigeon have ability to perch on the branches of the tree. Some leg muscles are specially adapted in such a way that toes can close round the twig when the bird sits on the tree. These muscles are given below (Fig. 4.14).

Peroneus longus: It is single muscle of the shank with a tendon which divides into three tendons reaching three front digits.

Gastrocnemius or Calf muscle: It is also single and its tendon passes behind the ankle, joins the tendons of peroneus longus and goes to three front digits. These tendons are flexors of the digits.

Flexor Perforans or Shank muscle: It has a tendon which goes to the hallux and is joined by a slip from the peroneus longus. A pull upon the tendon flexes all the toes. When a bird sits on its perch the legs are bent at the knee and ankle resulting into flexing of the tendons and closing of the digits around the perch. When the bird sleeps over the twig the ankle is bent further due to the weight of the body and a firm grip is maintained around the perch. When the bird raises its body the tendons are relaxed and unlocking of the digits takes place. In some birds one more muscle called ambiens arises from the ilium and its tendon joins the flexor tendons of some front digits. The action of this muscle in perching is negligible.

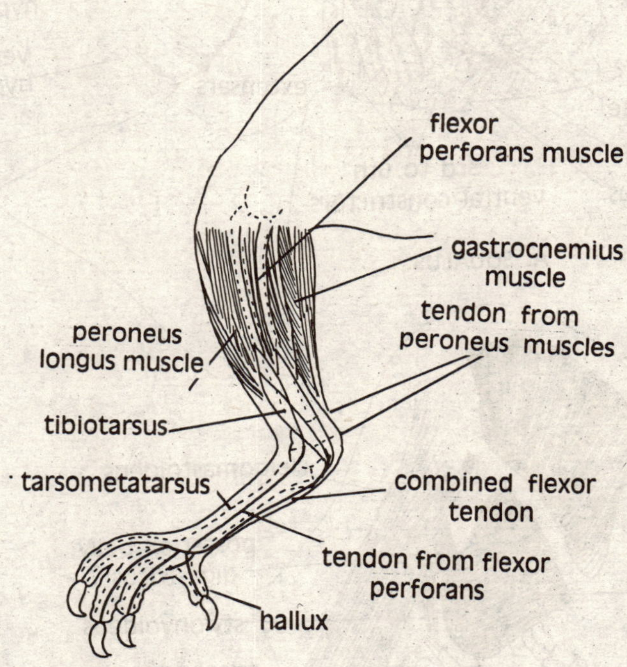

Fig. 4.14 Leg muscles with tendons of *Columba*. Note that the muscles and the tendons help in perching.

Locking of the digits furnishes second perching device. At the junction of the phalanges with the metatarsus the lower surface of the tendon sheath is ridged. When the bird sits two sets of ridges get locked firmly and digits are closed.

4.4. BRANCHIOMERIC MUSCLES

The branchiomeric muscles are associated with visceral arches of vertebrates. They help in the operation of the jaws and successive gill cartilages in dogfish. In tetrapods, due to loss of gills the posterior branchiomeric muscles have assumed new functions. These muscles are visceral but striated and arise from splanchnic mesoderm. The nerve supply is by visceral efferent fibres.

4.4.1. Muscles of I Visceral Arch

In dogfish (Fig. 4.15) the branchiomeric muscles of the first visceral arch operate the jaws. The

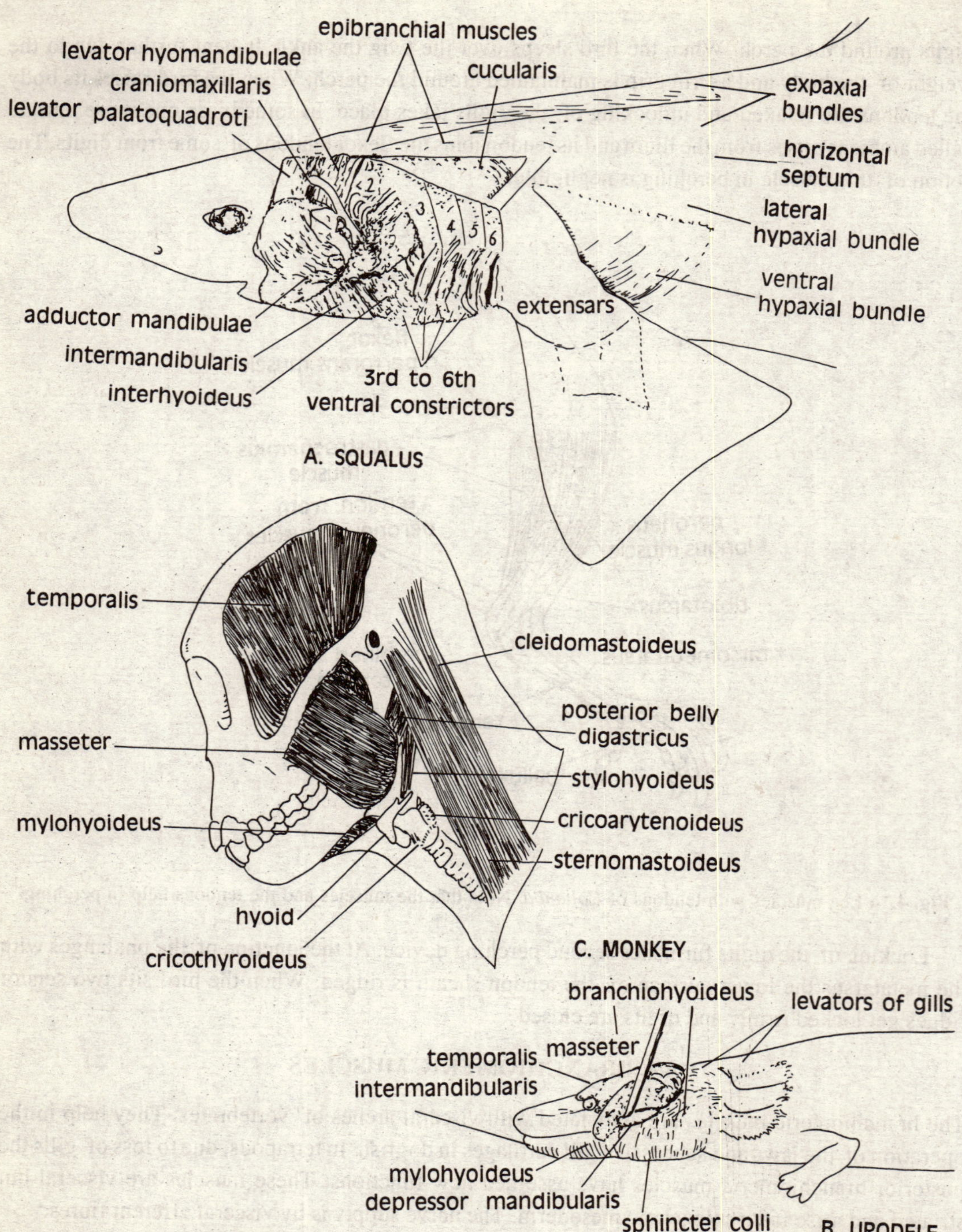

Fig. 4.15 A, B, C. Selected branchiomeric muscles of three vertebrates. Muscles of the first, second, and remaining pharyngeal arches are indicated by three shades of red. 2 to 6, Dorsal constrictors of the hyoid and successive arches.

raising of the jaw is done by a large *adductor mandibulae* and *intermandibularis* or *mylohyoideus*. The raising of the upper jaw is done by a levator maxillae superioris. The mandibular branch of fifth cranial nerve innervates these muscles. In tetrapods the operation of the jaws is done by muscles of the first arch including *internal* and *external pterygoideus muscles,* a large *adductor mandibulae* and *intermandibularis* or *mylohyoideus* and *elevators* of eye balls and depressors of the lower lid. The muscles of the first arch are associated with the operation of the jaws along with the modification of gill arches. In mammals these muscles operate the malleus.

4.4. 2 Muscles of II Visceral Arch

These are responsible for controlling the movement of the hyoid cartilage and its derivatives. The muscles of this arch include constrictors such as *epihyoideus* extending from otic capsule to hyomandibular and ceratohyal cartilages and *interhyoideus* extending from mid ventral side of the pharynx to the hyoid cartilage. In bony fishes the *constrictors* of the hyoid arch operate the operculum. In tetrapoda some of the muscles of second arch are intimately associated with those of the first arch. A large *depressor mandibular* muscle operates the lower jaw in most vertebrates excepting mammals. In mammals, the lower jaw is operated by the posterior belly of the digestric muscle. The poster most fibres of interhyoideus give *rise to sphincter celli* in amphibians and reptiles. The muscles of facial expression called *miemetic muscles* are derived from sphincter celli. The *stylohyoideus* and *stapedius muscles* of mammals are also of second arch origin. All the muscles of the second arch are innervated by motor branches of seventh cranial nerve.

4.4.3. Muscle of III to VI visceral Arches

These are represented by a series of constrictors above and below the gills. In 'fishes and amphibia with gills they are responsible for the movement of gill cartilages. In bony fishes the operculum mostly controls respiratory movements and muscles of third to sixth arches are reduced. In amniotes *stylopharyngeus muscle* of the third arch runs from the styloid process of the skull to the pharyngeal wall. *Intrinsic muscles* of the larynx are derived from muscles of fourth and fifth arches. These are *thyroarytenoideus* (on thyoid cartilage), *cricoarytenoideus* (on cricoid cartilage); they receive nerve supply from the tenth cranial nerve. Large *cucullaris muscle,* represented by *trapezius* and *sternocleidomastoid* in mammals is partly derived form the posterior branchiomeric muscles.

4.5.INTEGUMENTARY MUSCLES

The skin of vertebrates is not loose but firmly attached to the trunk musculature. The muscle slips attached to skin are called *integumentary* or *dermal muscles.* Most integumentary muscles are derivatives of branchiomeric and myotomal muscles. These are innervated by visceral nerve fibres.

In fishes the *myocommata* between myotomes attached to dermis and outermost muscle fibres of the myotomes are closely applied to the dermis. Integumentary muscles, as such, are not present in fishes.

In amphibians a few integumentary muscles occur. In anura *gracilis minor, gracilis major, cutaneous pectoralis* and muscles in the region of external nares are integumentary muscles.

In reptiles prominent integumentary muscles capable of twitching the skin have emerged. Locomotion in snakes is accomplished by alternate elevation and lowering of the *ventral* and *ventrolateral scales* operated by *costocutaneous muscles* extending from ribs to scales.

In birds the movement of individual feathers and movement of patagium is accomplished by *dermal muscles. Patagial muscles* are inserted on the skin of the wing membrane.

In lower mammals the integumentary muscles have reached their peak of evolution. In monotremes marsupials and edentates the trunk region is almost completely wrapped in a muscle sheath *panniculus carnosus* derived from latissamus and pectoral muscles. The rolling of hedgehogs and armadillos is accomplished by this muscle. The *cloacal sphincter* of monotremes, *sphincter* of massupial pouch in marsuplials, some muscles for tail movement, axillary and inguinal muscles slips of primates, and *sternalis muscle* overlyiong. pectoralis major in man are examples of integumentary muscles. The erector muscles of hairs called *arrectores pilorum* are intrinsic muscles confined to the dermis.

4.6. ELECTRIC ORGANS OF FISHES

The electric organs of fishes are highly specialised muscle masses which produce, store and discharge electricity. These organs are found in a few elasmobrachs and a wide variety of marine and freshwater teleosts. (The electric organs of an African teleost are derivatives of a skin gland embedded in the dermis)

In *Torpedo* (electric ray) the electric organs are probably of branchiomerous origin and lie in each pectoral fin near the gills. They are innervated by branches of the seventh and ninth cranial nerves.

In *Raia* (ray) and *Electrophorous* (electric eel) the electric organ is modified hypaxial musculature and lies in the tail region.

The electric organs are not distributed in systematic manner in fishes and probably represent convergent evolution. Each electric organ consists of a large number of electric discs called *electroplax* (Fig. 4.16) arranged in vertical up to 20,000 in the tail of one ray. In eels each electroplax is a modified multinucleate muscle cell embedded in a gelatinous material and surrounded by connective tissue. The discharge is caused by nerve endings terminating on each disc. The vascular capillaries develop in the jelly layer. The charge produced by this organ in eel may vary from 50 to 600 volts at several amperes for brief periods and gives a severse stunning shock. These organs are used for killing or stunning the prey and self defence. The high voltage is a result of the polarization of one side of the electroplax and not the other in series.

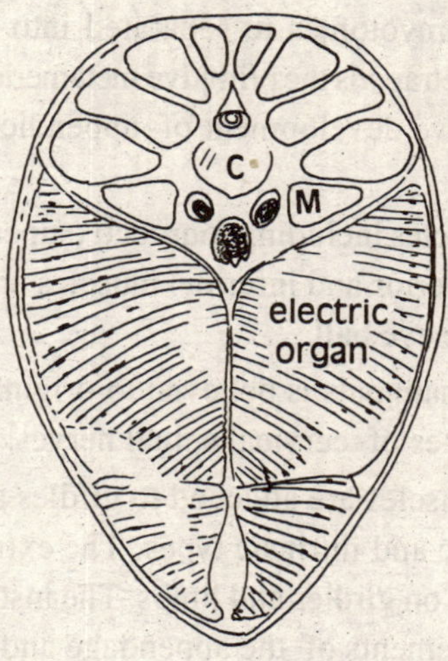

Fig. 4.16 Electric organs in tail of electric eel. Each nucleated horizontal disc (Electroplax) is a modified hypaxial muscle cell. C, Centrum; M, epaxial myomere.

CHAPTER SUMMARY

1. The muscles form nearly half of the entire body weight of the vertebrate body They vary in shape, size and attachments. They may be divided into a series of units in fishes or may spread out in ribbon- like, triangular, pinnate or fan-like sheets or may be massive and spindle shaped in sppendages. The muscles function for movement of trunk, limbs, jaws, organs or organ parts; maintainance of the stability of body parts; production of body heat assistance in major functions of the body such as circulation of blood, digeston respiration, locomotion, reproduction and behaviour.
2. The muscle tissue is derived from the mesoderm. The epimere, marked off as mesoblastic somites form segmentally, arranged myotomes which eventually give rise to voluntary, striated and parietal musculature. The hypomere forms somatic and splanchnic layers of coelom. The splanchnic layer gives rise to smooth, involuntary muscle of the gut and cardiac muscles. The voluntary, striated muscles of gill region are derived from the hypomere and not from the myotomes.
3. The muscular tissue has been distinguished into smooth, cardiac and striated types on the basis of structural and functional differences.
4. The muscles have been classified variously according to structure, embryonic origin, function or shape. Two main groups of muscles are somatic and viscrel. Somatic or parietal or myotomal muscles are skeletal, voluntary and striated; they may be axil or appendicular. Visceral muscles are found on the alimentary canal, blood vessels ducts, skin and viscerl skeleton.

5. Somatic musculature in Branchiostoma and cyclostomes consists of V-shaped myotomes. In higher forms the myotomes re separated into a dorsal or epaxial and ventral or hypaxial portions. In tetrapods the primitive metameric arrangement becomes progressively obscure with extensive development of appendicular musculature, and reduction of axial musculature.
6. Six extrinsic eye muscles including four recti (superior, inferior, anterior and posterior) and two oblique (superior and inferior) muscles are found in vertebrates which effect the movement of the eye ball.
7. The diaphragm of mammals is believed to originate from cervical myotomes as it is innervated by branches of cervical spinal nerves.
8. The appendicular muscles are attached to girdles and limbs for locomotion. They are dīvided into extrinsic and intrinsic types. The extrinsic muscles are attached to axial skeleton and inserted on girdles and limbs. The instrinsic muscles arise on girdles or on proximal skeleton elements of the appendage and insert on more distal elements.
9. The branchiomeric muscles are associated with visceral arches of vertebrates. In tetrapods where visceral arches greatly modified the branchiomeric muscles modify into certain muscles of the neck and shoulder regions.
10. The integumentary or dermal muscles are muscle-slips attached to the skin. Most of such muscles are derived from branchiomeric or myotomal muscles. These muscles are important in locomotion. The movement of feathers in birds, erection of ventral scales in snakes, twitching movements of skin and facial expressions in mammals are brought about by dermal muscles.
11. The electric organs found in certain fishes are highly specialized muscle masses which produce, store and discharge electricity for self defence or stunning or killing of the prey.

5

Body Cavity and Digestive System

5.1. THE BODY CAVITY

5.1.1. General Account

The *coelom* or body cavity is single cavity or more than one cavities which separate the viscera from the body wall. The body cavity is formed by the *hypomere* or lateral plate of the mesoderm. The visceral organs lie in this cavity, their surfaces wet with a serous fluid and occupy almost the entire space. The friction of visceral organs is minimized and the body wall may also change its shape in relation to viscera. A smooth transparent *peritoneal membrane* called *parietal layer* lines the coelomic cavity and another similar layer called *visceral layer* covers the visceral organ. The visceral organs are usually suspended from the dorsal side by double layers of peritoneum called *mesenteries* through which nerves and blood vessels connect the organs. In all vertebrates with lungs, excepting mammals, the lungs are suspended in the general body cavity. In mammals, the body cavity is divided by a diaphragm into paired pleural cavities and median peritoneal or abdominal cavity (Fig. 5.2).

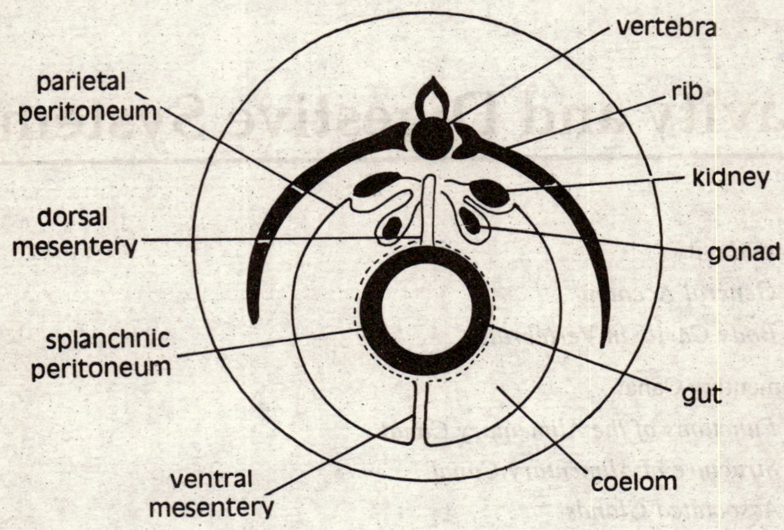

Fig. 5.1 Diagramatic representation of body cavity

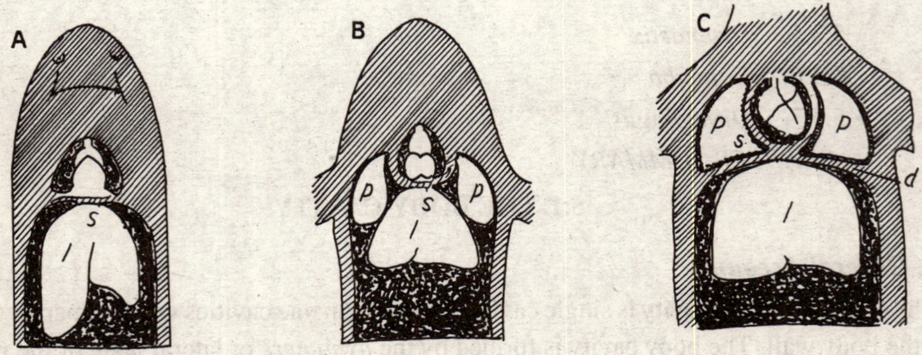

Fig. 5.2 Coalom **A**, fish; **B**, amphibians, reptilies, and birds; **C**, mammals.

5.1.2. Body Cavity in Vertebrates

***Branchiostoma** (Fig. 5.3):* In *Branchistoma* the body cavity is true coelom lined by mesoderm. In the posterior part of the body it lodges midgut and hindgut which are suspended by a mesentery.

In the pharyngeal region, the coelom is confined to a pair of *dorsal longitudinal canals,* a mid-ventral *subendostylar coelom,* and vertically running spaces in the primary gill bars which connect the dorsal and subendostylar coelomic spaces. In early development, the pharyngeal region is uniform and pharynx is surrounded by the coelom. In later stages the coelom becomes restricted due to the development of atrium and formation of gill slits.

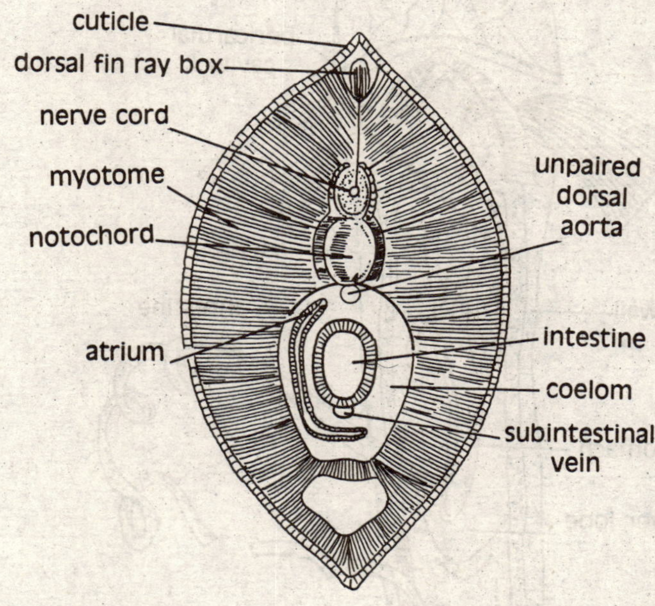

Fig. 5.3 Body cavity in Branchiostoma

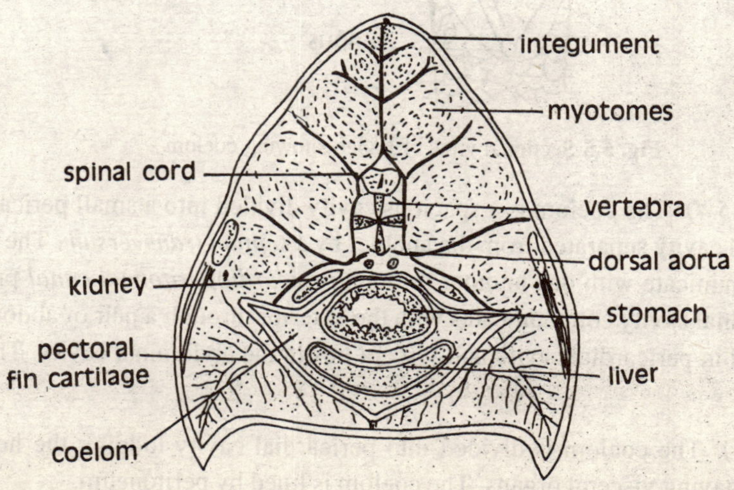

Fig 5.4 Cross section of Scoliodon showing coelom

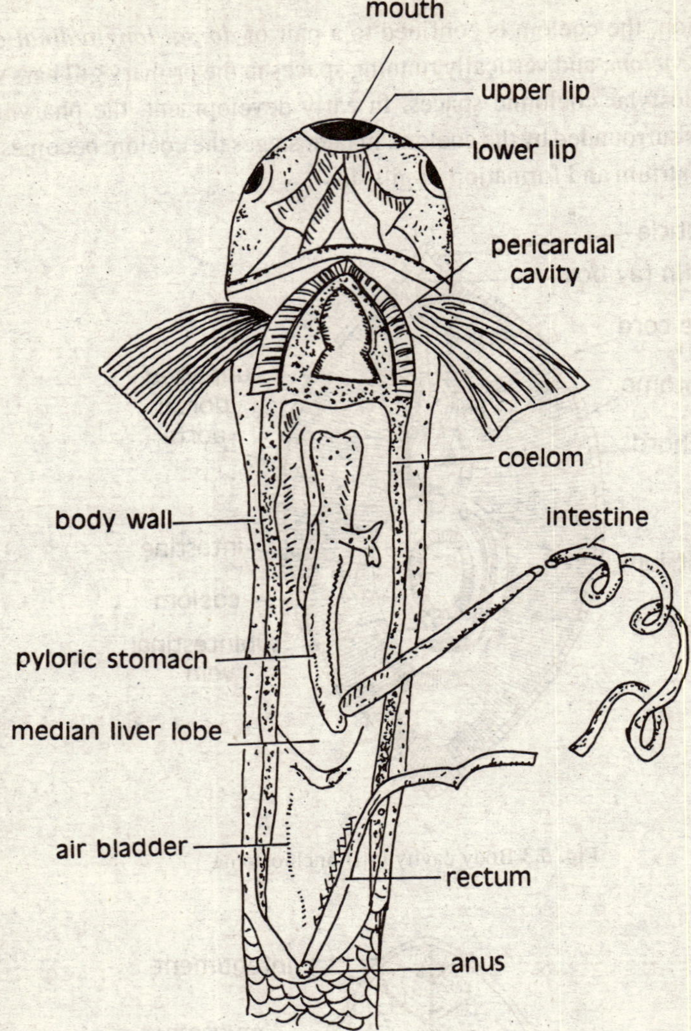

Fig. 5.5 Sectional view of Labeo showing coelom

***Scoliodon** (Fig. 5.4):* The coelom is a spacious cavity divided into a small pericardial and an extensive abdominal cavity separated from the another by a *septum transversum.* The two cavities of the coelom communicate with one another through *pericardioperitoneal canal* present in the septum. The abdominal cavity communicates with the exterior through a pair of abdominal pores. The heart is situated in pericardial cavity and viscera lies in the abdominal cavity. The coelom is lined by peritoneum.

***Labeo** (Fig. 5.5):* The coelom is divided into pericardial cavity lodging the heart and two abdominal cavities having visceral organs. The coelom is lined by peritoneum.

***Rana** (Fig* 5.6): The coelom of *Rana* is a large cavity without partition as in higher vertebrates. The visceral organs are suspended in the body cavity from its dorsal wall by mesenteries and are covered by the peritoneum. A small *pericardial cavity* enclosing the heart is cut off by extensions of the peritomeum, called *pericardium,* on the anterior side of the coelom. The coelom is filled with celomic fluid secreted by the epithelium which acts as lubricant and enables the viscera to avoid friction.

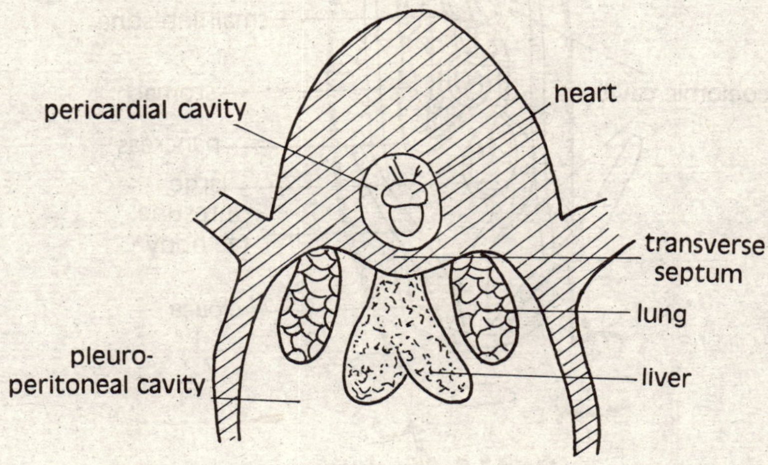

Fig. 5.6 Coelom of Rana

***Uromastix** (Fig. 5.7):* The coelom of *Uromastix*, is a cylindrical cavity without partition and lined with *peritoneum*. The heart is enclosed in pericardium. The digestive tract is attached to the dorsal median line of the coelom by *dorsal mesentery.* The part of the mesentery which supports the stomach is called *mesogaster* and attaches the lung also by a short mesentery. A ventral mesentery is also present and a part from it forms a long *falciform, ligament* of the liver. Mesentery of the urinary bladder is called *median ligament* of the bladder. The gonads have *mesorchium* in female and *mesorchium* in the male. A transverse septum separates the anterior end of the coelom enclosing the pericardial cavity.

***Columba** (Fig. 5.8):* The coelom of *Columba* is lined by visceral and parietal peritoneium as in other vertebrates. It lodges visceral organs which are suspended by mesenteries. The mesentery of the gizzard is called *ventral ligament* of the gizzard. The liver is suspended by *falciform ligament.* The pericardial sac lies between two lobes of the liver attached to the pericardial sac by coronary ligament. A *gastrohepatic ligament* passes between gizzard and the liver. *Mesogaster* connects the gizzard with the dorsal body wall. A *hepatoduodenal ligament* provides attachment to the beginning of the duodenum to the right lobe of the liver. A mesoduodenum stretches between the two sides of the duodenal loop where the pancreas lies.

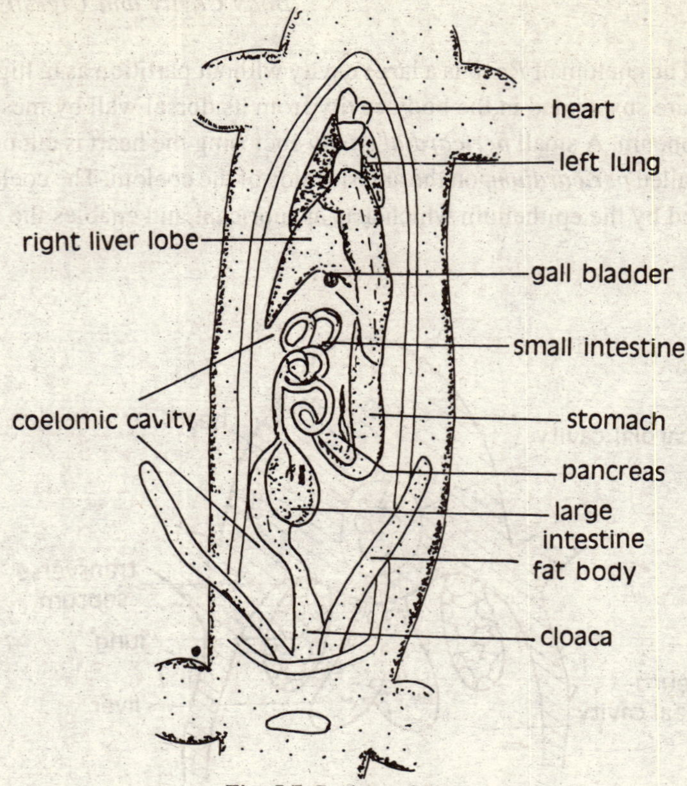

Fig. 5.7 Coelom of Uromastix

The coelom of pigeon is having four chambers, the *pericardial cavity,* two *pleural cavities* and *peritonial cavity.* The heart is enclosed in a delicate pericardial sac with pericardial cavity, a part of coelom. The pesicardial sae is formed at the anterior face of the transverse septum and is in contact with the inner surface of the body wall, making its posterior part free. An *oblique septum* is present at the point where pericardial sac meets latereal body wall and extends obliquely on the posterior side; it contains a large air sac and divides the coelom into anterior and posterior portions. The anterior has two *pleural cavities,* on each side of the pericardial cavity. The posterior cavity is *peritoneal cavity.*

***Oryctolagus** (Fig. 5.9):* The coelom has anterior thoracic and posterior abdominal parts. The thoracic coelom has two lateral pleural cavities and median pericardial cavity. The abdominal or peritoneal cavity contains visceral organs. The pleural cavities accommodate the lungs and pericardial cavity contains the heart. The outer parietal membrane of the pericardial cavity as called *pericardium* and the inner *epicardium.* The pleural cavities are separated from each other by a *mediastinal cavity* containing trachea and oesophagus. A median *mediastinal septum* is formed by the visceral pleurae of the two sides which are united closely on the ventral part of the mediastinal cavity. The pleural cavity is walled posteriorly by a dome-shaped muscular partition called *diaphragm.* The peritoneal cavity has parietal and visceral peritoneal membranes. The mesenteries are formed as double layered structure and support the alimentary canal.

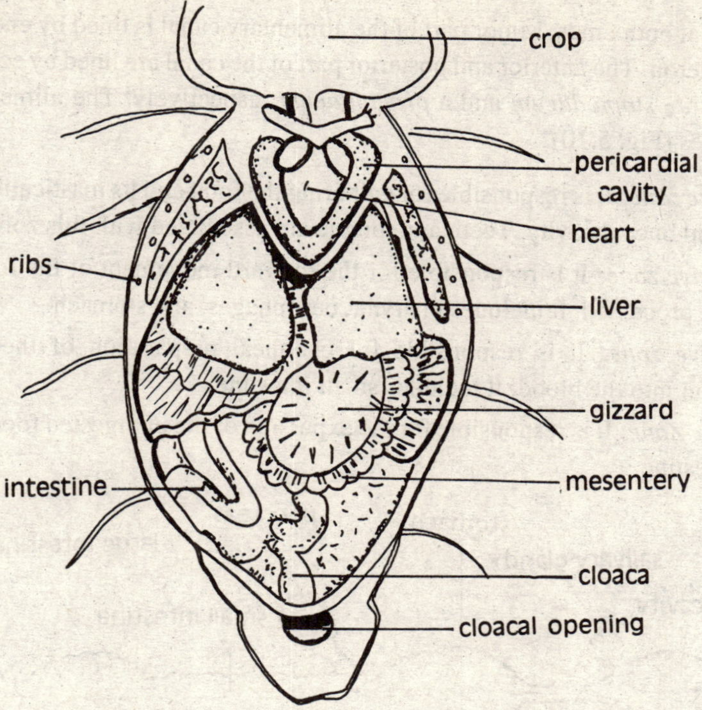

Fig. 5.8 Coelom of Columba

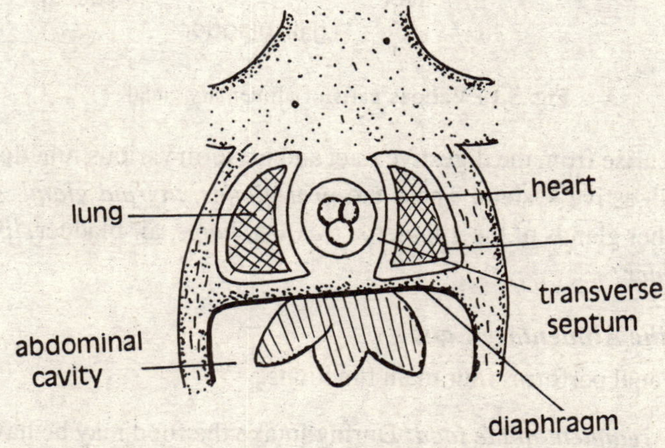

Fig. 5.9 Mammalian Coelom

5.2. THE ALIMENTARY CANAL

The Alimentary canal is a tube beginning at mouth, passing through the body, and terminating at the anus. It is open at both emds. Major part of the alimentary canal is lined by endoderm and is formed by the archenteron. The anterior and posterior part of the canal are lined by ectoderm which is invaginated to form a *stomodaeum* and a *proctodaeum* respectively. The alimentary canal is divided into four zones (Fig. 5.10).

i. *Ingressive zone:* It is responsible for gathering the food and its mastication. If includes mouth and buccal cavity. Teeth and tongue are associated with this zone.

ii. *Progressive zone:* It is responsible for the forward movement of the food and initial digestive processed. It includes pharynx, oesophages, and stomach.

iii. *Degressive zone:* It is responsible for chemical preparation of the food and its absorption into the blood. It includes small intestine.

iv. *Egressive Zone:* It is responsible for the expulsion of the undigested food and includes large intestine.

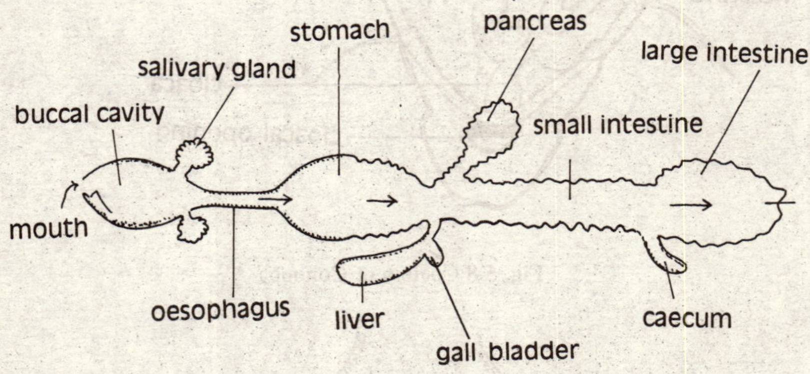

Fig. 5.10 Various parts of alimentary canal

Some outgrowths arise from the digestive tract and perform various function connected or not connected with the digestive system. These are *oral glands, thyroid gland, gill slits, tympanic cavity, thymus* and other glands of the gill clefts, *trachea, lungs,* air bladder, *liver, pancreas, yolk sac* and *urinary bladder.*

5.2.1. Functions of the Alimentary Canal

The Alimentary canal performs four main functions.

i. *Physical treatment of the food:* During intake the food may be having large particles which are chewed by the teeth in the buccal cavity, mixed with secretions to make it pasty and passed on to the pharynx, oesophagus and stomach for further action.

ii. *Chemical treatment:* The chemical treatment is the digestion proper where breakdown of food material into simpler substances takes place. The simple substances are utilized by the body cells.

iii. *Transportation:* It includes gathering of the food materials their transportation to different part of the canal and elimination of waste materials through anus.

iv. *Absorption:* After chemicai treatment food is absorbed by the intestinal wall for circulation to body cells and storage.

5.2.2. Structure of Alimentary Canal

Buccal Cavity: The buccal cavity is a space between mouth and pharynx. The mouth is a space between the jaws and opens in the buccal cavity. In cartilaginous fishes and tetrapoda, nasal cavities open into the buccal cavity by *internal nares.* There are two types of multicellular glands, *mucous gland* and *enzymatic glands* opening into it. In fishes and aquatic amphibia, only mucous glands are present. In reptiles, palatine, lingual, sublingual and labial glands are present in groups. In birds, sublingual glands and a gland at the angle of the mouth are present. Mammals usually have many small *mucous glands,* and *salivary glands* which secrete an enzyme called *ptyalin.* The buccal cavity has a tongue which is not homologous structure in vertebrates. In mammals, the tongue has four types of *papillae* on the upper surface, including *filiform, fungiform, foliate* and *circumvallate* types. The filiform papillae are present on the greater part of the tongue as small conical projections. The fungiform papillae are on the margins, foliate papillae near the base of the tongue as broad leaf-like structure and circumvallate papillae are few in number but large structures.

Dentition: The teeth are meant primarily for grasping, cutting, and grinding of the food in the mouth cavity. The teeth are two types; epidermal and true teeth. The *epidermal teeth* are hard conical structures derived from the epidermis. Such teeth are found in cyclostomes and some mammals such as duck billed platyus. The true teeth are derived from placoid scales or bony dermal plates of ostracoderms and placoderms according to latest theory. True teeth are homologous structure in vertebrates. Following terminology is used for describing teeth:

Homodont	:	All teeth similar in structure.
Heterodont	:	A set of teeth having different structures
Polyphyodont	:	Teeth are replaceable continually on indefinite number of times e.g. lower vertebrates.
Diphyodont	:	Two sets of teeth, milk teeth or deciduous set of teeth, and a permanent set e.g. Man.
Monophyodont	:	When only one set of teeth is present, e.g. aquirrel, moles etc.
Acrodont	:	Teeth attached on the creast of the jaw bone e.g. most vertebrates.
Pleurodont	:	Teeth fixed on the inner margin of the jaws.
Thecodont	:	Teeth with one or more roots embedded in socket of jaw bone with crown projecting above the socket e.g. some fishes crocodiles, and mammals.

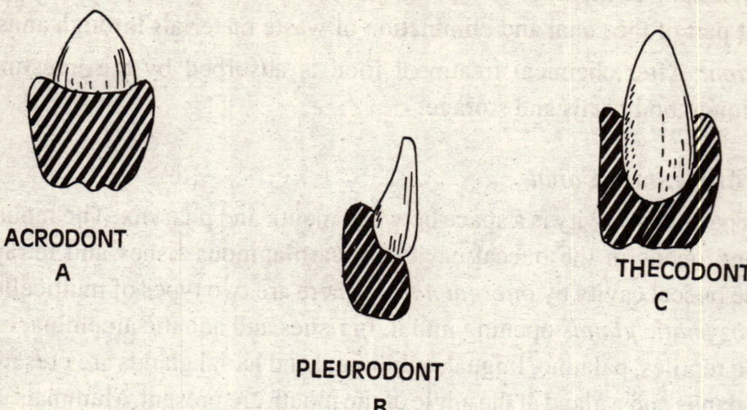

Fig. 5.11 Diagram showing three methods of teeth attachment

A typical true vertebrate tooth has the following structure (Fig. 5.12).

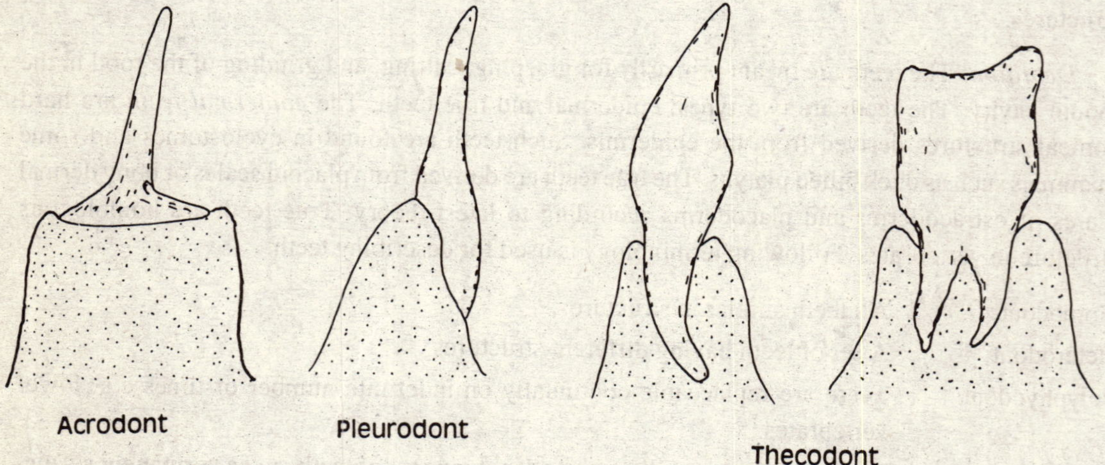

Fig. 5.12 Variations in the relationship of teeth to jaws. Acrodont teeth are attached either at the outer surface of the jaw or, as shown, at the summit. Pleurodont teeth are attached to the inner surface of the jaw. Thecodont teeth occupy alvecli. The bone of the jaw is seen in cross section.

It is embedded in the germ with its root embedded in the *socket* or *alveolus* of the jaw bone. The crown is projecting outside the germ. The neck of the tooth is enclosed by the germ. The outermost layer of teeth is a hard enamel present over the crown. Next layer is of bone like substance, *dentine* which is produced by a layer of odontoblasts laying outside the pulp cavity. The dentine has numerous thin *canaliculi* with protoplasmic processes of odontoblasts. The dentine encloses a pulp cavity having connective tissue, blood vessels and nerve fibres. The neck and root of the tooth are covered by bone like substance called cement.

In some mammals, the teeth are *heterodont* and differentiated into *incisors* canines, *premolars,* and *molars.* The incisors are flat teeth with a single root and work as chisels. The canines are sharp pointed teeth with one root which are used for tearing and cutting Premolars and molars have two or more roots and are used for mastication. The number and arrangement of teeth in mammals is expressed by dental formula in each half of the jaws. The dental formula contains the number of teeth present on the upper jaw and lower jaw as numerators and denominators respectively, with their total number given at the end. The dental formula of some mammals is given below:

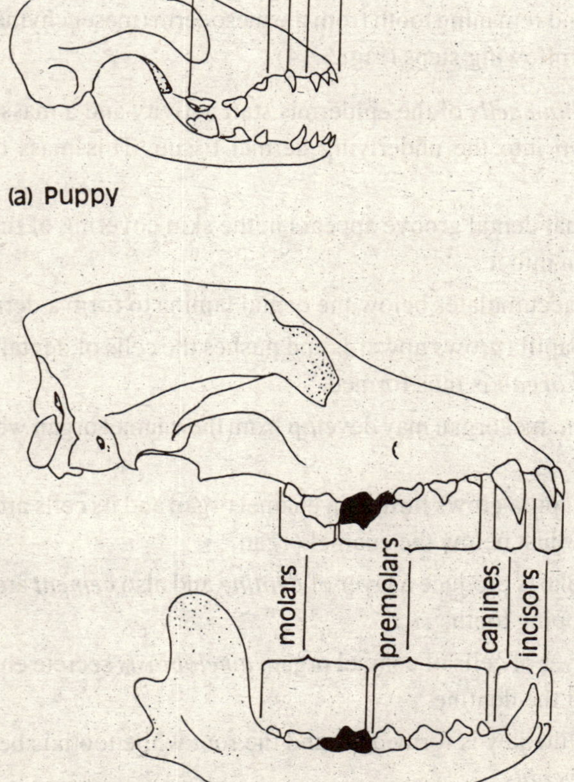

Fig. 5.13 Deciduous (a) and permanent teeth (b) in a dog. The carnassials (shaded teeth) are specialized carnivou teeth derived from last premolar (upper) and first premolar (lower)

Pig and horse $\dfrac{3,1,4,3}{3,1,4,3} = 44$

Dog $\dfrac{3,1,4,2}{3,1,4,3} = 42$

Rabbit $\dfrac{2,0,3,3}{1,0,2,3} = 28$

Cat $\dfrac{3,1,3,1}{3,1,2,1} = 30$

Rat $\dfrac{1,0,0,3}{1,0,0,3} = 16$

Man $\dfrac{2,1,2,3}{2,1,2,3} = 32$

The early development of the tooth takes place below the germ. The enamel of the tooth is derived from ectoderm and remaning tooth from the mesoderm (mesenchyme). The development of tooth takes place in the following steps (Fig. 5.14).

(i) The *malpighian cells* of the epidermis start activity and a mass of cells is formed and pushed down into the underlying dermal tissue. This mass of tissue forms *dental lamina.*

(ii) A longitudinal dental groove appears in the skin covering of the jaw and epidermis is pushed down into it.

(iii) The dermis accumulates below the dental lamina to form a dermal dental *papilla.*

(iv) The dental papilla grows upwards and pushes the cells of dental lamina. A two layered cup, *enamel organ* is thus formed.

(v) Secondary enamel organ may develop from the enamel organ which may produce teeth later in life.

(vi) The dental papilla grows further in enamel organ and its cells are differentiated to form *odontoblasts* just below the enamel organ.

(vii) The odontoblasts produce a layer of *dentine* and also *cement* around the root and neck of the developing tooth.

(viii) The inner layer of cells of enamel organ, *ameloblasts* secrete enamel which covers the upper part of the dentine.

(ix) The bone of the jaw is formed around the root while tooth is being formed, and make *socket* or *alveolus.*

(x) The tooth comes out of the germ after the accumulation of more dentine.

(xi) The dental papilla becomes the pulp in the *pulp cavity* of the tooth.

(xii) The odontoblasts may continue the growth of the teeth e.g. incisors of rodents and tusks of elephants, or may stop functioning after the tooth is fully formed.

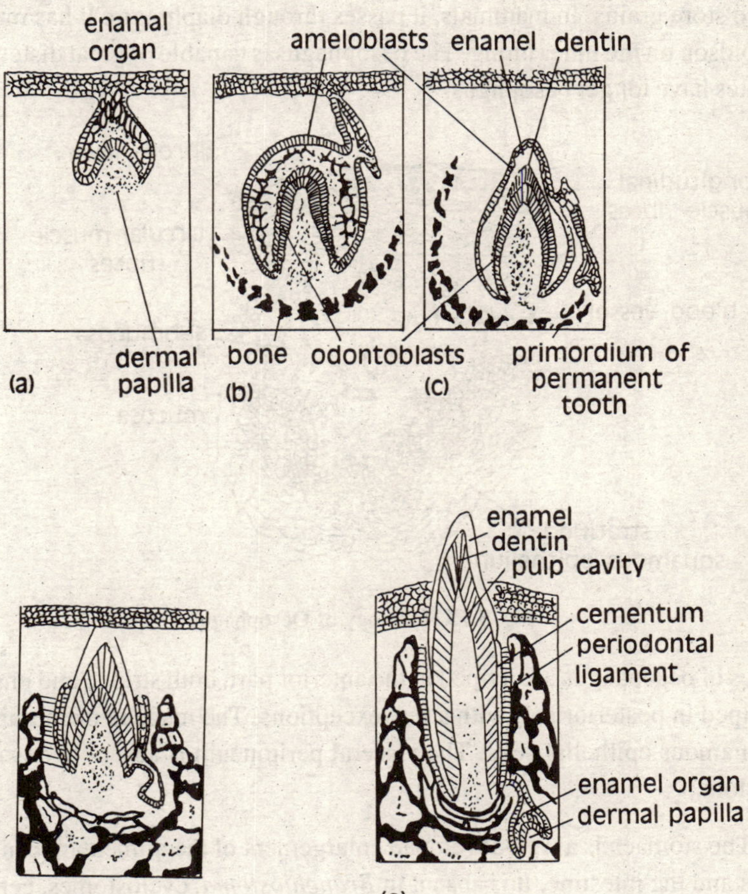

Fig. 5.14 Mammalian tooth development. (a) Enamel organ (from the epidermis) and dermal papilla (from the dermis) appear. (b) Ameloblasts are the source of tooth enamel and form from the enamel organ. Odontoblasts are the source of dentin and form from the dermal papilla. Bone appears and begins to delineate the socket in which the tooth will reside. (c) The primordium of the permanent tooth appears. (d) Tooth growth continues. (e) The deciduous tooth erupts and is anchored in the socket by a well-established periodontal ligament. The enamel organ and dermal papilla of the permanent tooth primordium will not begin to form the tooth until shortly before the deciduous tooth is lost.

Pharynx: The pharynx lies between the mouth cavity and oesophagus. It is lined by the endoderm and serves as a common passage for digestive and respiratory tracts. In digestive system, it is used as all passage for food from mouth cavity to oesophagus and its muscles help in swallowing . Numerous pharyngeal glands including thyroids, parathyroids, thymus etc. take their origin from the pharynx.

Oesophagus: The oesophagus is a simple, tube like structure after pharynx lined by endoderm. It is short in most fishes and amphibians. The oesophagus of many birds has a sac-like pouch crop which serves to store grains. In mammals, it passes through diaphragm. It has mucous glands and longitudinal foldson on the inner lining. The oesophagus is capable of great distensions. The neck-bearing amniotes have longer oesophgus.

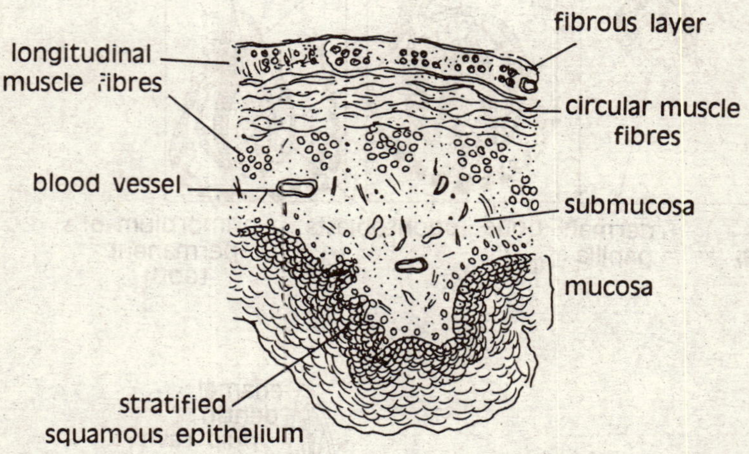

Fig. 5.15 Histology of Oesophagus

The muscles of oesophagus are striped in the anterior part, both striped and unstriped in middle part, and unstriped in posterior part with a few exceptions. The mucous membrane lining is made of stratified squamous epithelial cells. The visceral peritoneum is absent in oesophagus as it lies outside the coelom.

Stomach: The stomach is a well developed enlargement of the alimentary canal lying between the oesophagus and the intestine. It is absent in *Branchiostoma*, cyclostomes, certain teleosts and Dipnoi. A true stomach is that structure in which gastric glands are present. Such stomach is found in fishes, amphibian, and higher animals. The stomach of fishes and amphibians may be curved on itself to form U-shaped loop with two limbs called cardiac and pyloric parts running parallel to one another. In reptiles the stomach is bent upon itself (chelonians), spindle-shaped (snakes) or specialised (crocodile). In birds, it is usually divided into an anterior glandular *proventricular* and posterior muscular stomach or *gizzard.* In mammals. It is sac-like transverse structure having cardiac, fundle and pyloric parts (Fig. 5.16).

Evolution of Stomach in Vertebrates: Stomach is mainly concerned with storage, physical treatment and initial chemical treatment of food. It is absent in *Amphioxus,* cyclostomes, chimaeras, lungfishes and some telcosts. It seems probable that in ancestral vertebrates, as in fishes lacking in stomach, the physical and chemical treatment of food were dealt by the intestine alone. When developed in phylogeny the stomach may have been concerned with the storage of food materials, which could not be accommodated in the intestine. A pouch for food storage and physical treatment would have become necessary at the time of the appearance of predaceous jaw bearing forms. The introduction of digestive enzymes in stomach might have been a later phylogenetic development. The shape of the stomach is dependent upon the shape of the body upto some extent. In lower vertebrates with elongated body it lies in the antero-posterior axis of the trunk. With the increase in the width of the body the stomach becomes more transverse. The lengthening of the tube to the right has resulted into a smaller anterior sac called cardiac part and a larger posterior pyloric part.

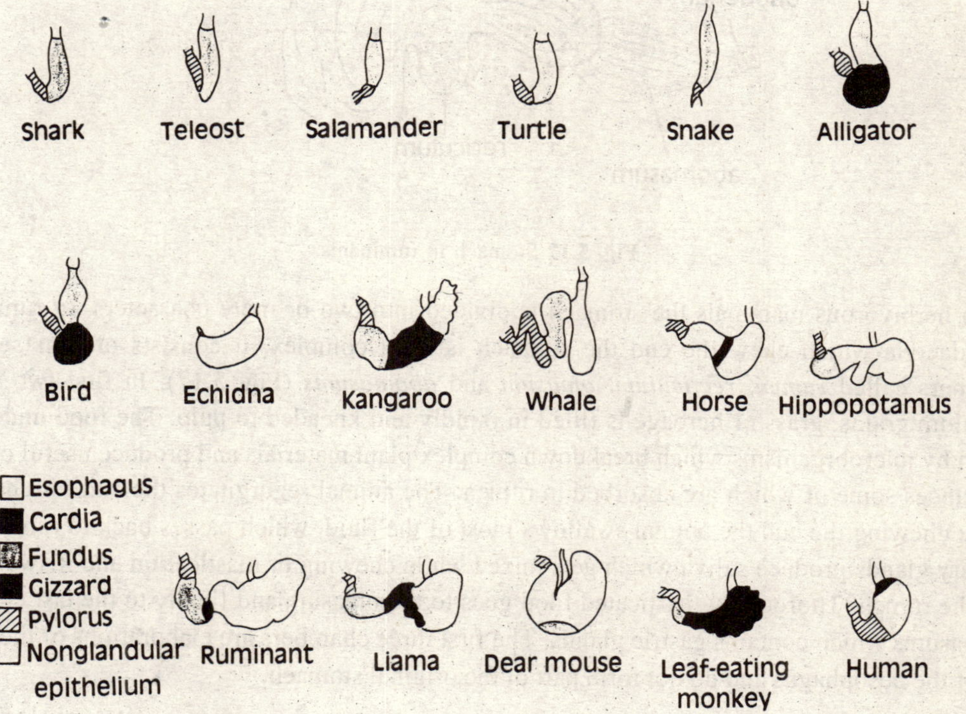

Fig. 5.16 Stomach in various vertebrates. Two regions of the stomach may be recognized graatular and nonglandular regions. The glandular region of the stomach includes gartic glands and often exhibits three divisions: cardia, fundus, pylorus. The nonglandular region of the stomach is lined with an epithelium devoid of gastric fianat that in some species also may be kerainzed. The walls of the stomach are composed of layers of smooth muscle, but in some species, these muscular coats are enlarged into a specialized gizard.

In fishes the stomach may be straight, saccular or with blind sac (Fig. 5.16). In teleosts a great variety of stomach is found in relation to their feeding habits. In amphibians and reptiles the distinction between oesophagus and stomach is well marked and only conspicuous in crocodiles. In birds due to the absence of the teeth and type of food eaten the stomach is modified and differentiated into two regions, the first continuous with the oesophagus proventriculus and the second posterior muscular gizzard. The gizzard is well developed in grain eating birds and less developed in insectivorous birds and birds of prey. The mammalian stomach are typically transversely arranged with considerable variation in different groups. In monotremes the stomach is simple pouch like structure lacking in epithelial glands and serves for the storage of food. It is not regarded as true stomach as it is entirely oesophageal in character.

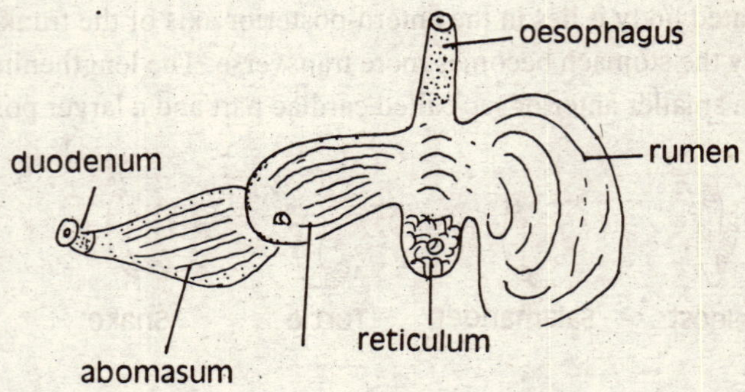

Fig. 5.17 Stomach in ruminants

In herbivorous mammals the stomach is divided into two or more characters. In ruminants (artiodactyla which chew the cud the stomach is most complex. It consists of four separate chambers called *rumen, recticulum, omasum* and *abomasums* (Fig. 5.17). In first two rumen reticulum grains, grass or herbage is filled in rapidly and kneaded to pulp. The food undergoes action by microorganisms which breakdown complex plant materials and produce useful organic substances some of which are absorbed in rumen. The animal regurgitates the cud for chewing. While chewing the cud the animal swallows most of the fluid, which passes back to rumen. The salivery glands produce saliva which gets mixed while chewing or mastication and it swallowed into the rumen. Thoroughly masticated food goes to the omasum and finally to the last chamber abomasums which contains gastric glands. The first three chambers are elaborations of the lower and of the oesophagus and do not form part of the original stomach.

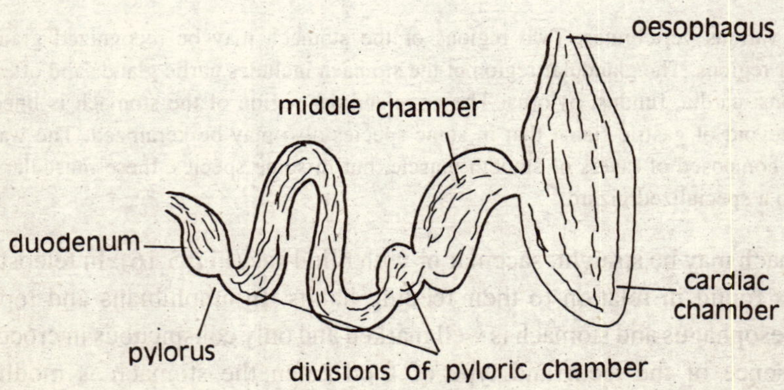

Fig. 5.18 Stomach of porpoise

In camels the oamsum is absent and pouch like diverticulca arise from the rumen and the reticulum. These are called *water cells* which have their openings guarded *by* sphincter muscles, metabolic water drawn from other parts of the body is contained in water cells. Most of the water comes from the breakdown of the glycogen stored in the muscles and fat stored in the hump which may be upto 40 litres in volume.

In cetacea (e.g. propoise) the stomach contains several chambers (Fig. 5.18). The stomach of hippopotamus also contains several chambers. In vampire bats a caecum like structure is found in pyloric stomach for the storage of blood sucked from the victims. The stomach of kangaroos contains peculiar sacculated folds in its walls.

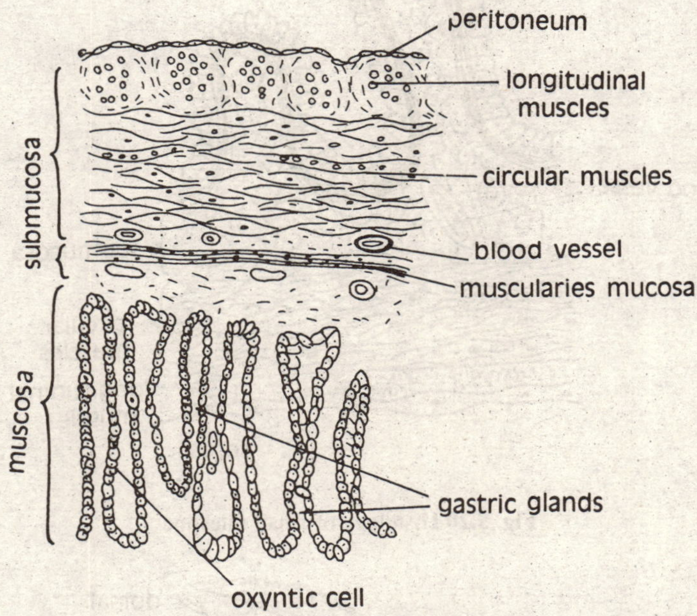

Fig. 5.19 Histology of stomach

The stomach is made of outermost layer of visceral *peritoneum; muscular layer* composed of longitudinal and circular fibres; *submucosa* made of connective tissue, elastic fibres, fat, blood vessels, nerve cells etc., *muscularis mucosa* made of outer layer of longitudinal muscles and inner layer of circular muscles; *mucosa* and *basement membrane* of epithelial cells of several types which secrete gastric juice.

***Small Intestine*:** The small intestine lies next to stomach and serves for the conversion of food mass into soluble form and its absorption by blood stream. It is a large, narrow and coiled tube divided into *duodenum* and *ileum* in lower vertebrates and duodenum, and ileum in mammals. In duodenum, it has many folded *villi* with branching *Brunner's glands* which secrete mucus and enzymes. The duct from liver and pancreas also open into duodenum. The ileum contains a large number of small digestive glands. The lining of the small intestine is produced into small villi, which increase the surface area of the intestine. In cyclostomes the epithelial lining forms a longitudinal typhlosole. In lower fishes, a *spiral valve* is present in the intestine. Some bony fishes have one or more pyloric caeca arising from the small intestine. The small intestine of herbivorous animals is longer than that of carninorous animals.

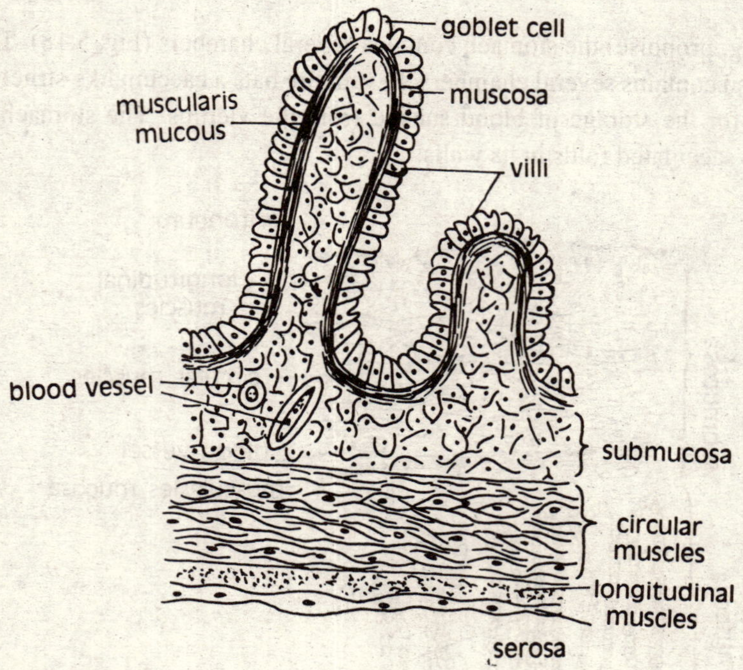

Fig. 5.20 Histology of small intestine

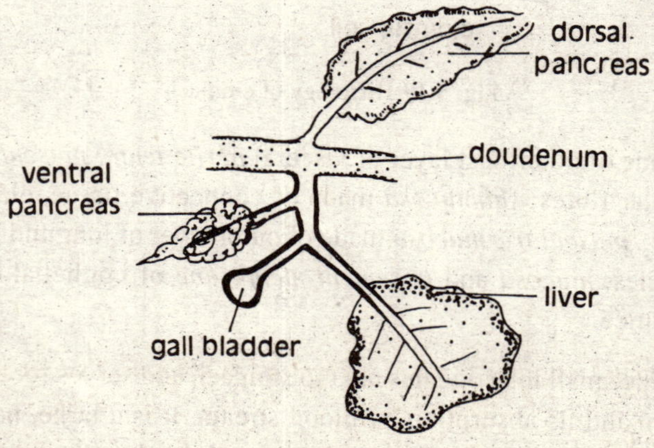

Fig. 5.21 Showing digestive glands

Large intestine: The large intestine constitute egressive part of the digestive tract and is lined by ectoderm. Its diameter is more than that of small intestine. The villi or other such processes to increase its surface are absent or reduced in large intestine. It serves for the bacteriological putrifaction of undigested food and absorption of water in terrestrial animals. In lower animals, the large intestine forms a rectum but in tetrapoda, it is divided into a *colon,* and a *rectum.* In most vertebrates, excepting mammals, the rectum, excretory, and genital ducts open into a terminal part of the rectum called *cloaca* opening to the exterior by a *cloacal aperture.* In bony fishes and mammals, excepting monotremes, the alimentary canal and urogenital ducts open separately to the exterior by anus and urogenital openings. In Amphibia, the large intestine opens into a cloaca and urinary bladder is developed as an outgrowth of its ventral wall. In reptiles, it is usually straight and opens into cloaca with a blind caecum present at the anterior side opening into cloaca. In birds, a pair of blind caecae open at the beginning of the cloaca which is divided into three sections *coprodaeum, urodaeum* and *proteodaeum.* An unpaired growth *Bursa-Fabricii* is present in the birds. In mammals, the large intestine is very long, coiled structure divided into *colon* and *rectum.* A *colic caecum* is present in amniotes at the junction of small and large intestines on *iliocolic valve.* The *vermiform appendix* of man is degenerated remnant of the distal end of the colic caecum

5.2.3. Associated Glands

Liver: The liver is the largest gland of the body present on the ventral side of the digestive tract. It is generally present in all vertebrates. In cyclostomes, a gall bladder and bile duct are present in larval condition but absent in adults. The gall bladder is present in fishes, amphibia and reptiles; in birds, it is absent, and most mammals excepting Cetacae and Ungulata have gall bladder.

The liver arises as tubular outgrowth from the endodermic epithelium of intestine at the junction of stomach and intestine. The outgrowth divides again and again and differentiates into two parts. The anterior part eventually becomes liver and its bile duct and posterior part gives rise to gall bladder and cystic duct (Fig. 5.21). The endodermal epithelial cells are soon joined by mesodermal cells to form the adult liver.

The liver (Fig. 5.22) has branched columns of polygonal liver *trabeculae* or *hepatic cords* separated from one another by blood spaces or sinusoids. The sinusoids have blood capillaries of endothelial cells and receive blood from branched of hepatic artery and hepatic portal vein, the blood pases into branches of the hepatic vein. The liver has larger canals called portal canals at intervals. Each portal canal has a *Glisson's capsule* of connective tissue with interlobular branched of blood and lymph vessels and bile ducts. Each liver cell is surrounded with minute bile capillaries, anastomosing with each other, and convey bile into larger ducts which unite to form a hepatic duct. A *gall bladder* is present in the liver for the storage of bile. A duct from gall bladder called cystic duct joins the hepatic duct from liver and the two ducts form common bile duct which opens into the duodenum.

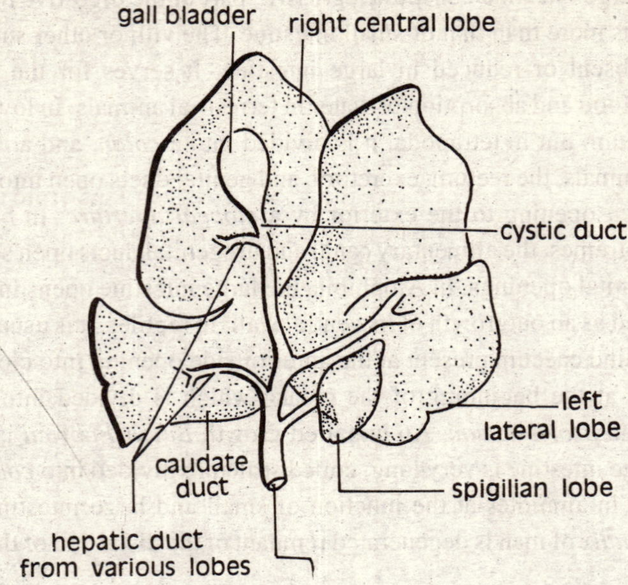

Fig. 5.22 A. Mammalian liver and Gall bladder

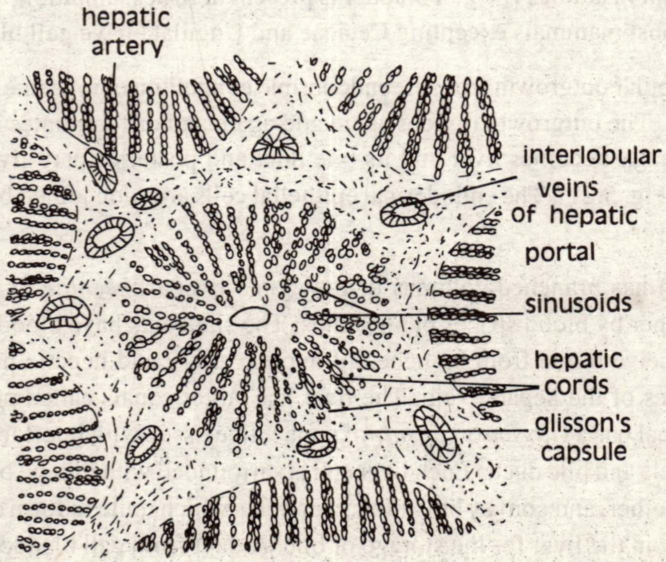

Fig. 5.22 B. Histology of Mammalian liver

Functions of Liver: The liver performs the following functions:

i. *Secretion of bile:* It secretes bile juice which is alkaline, watery substance containing bile pigments, cholesterol, and lecithin. Bile helps in digestion by making the emulsification of fats.

ii. *Conversion of insoluble fatty acids into soluble:* The fatty acids produced after the digestion of fats are insoluble in water. Bile salts combine with fatty acids to make them soluble and help in their absorption.

iii. *Storage of glycogen:* The liver stores excess of carbohydrates in the form of glycogen (Glycogenesis) which may be changed into glucose when required (glycogenolysis).

iv. *Elimination of urea:* Ammonia and excess of proteins are converted into urea by the liver (deamination): they are transported to the kidneys and eliminated as urine.

v. *Removal of some substances:* When the haemoglobin of red blood corpuscles is broken down, certain excretory products are formed. Such products become green or red bile-pigments and enter into the duodenum with the bile and are passed out with the faeces.

vi. *Storage of inorganic salts:* The inorganic salts if iron and calcium are stored in liver.

vii. The liver forms and stores fibrinogen and prothrombin essential for the clotting of the blood. It also forms heparin which prevents clotting of the blood.

viii. In embryo the blood corpuscles are manufactured by the liver.

ix. The liver kills bacteria and eliminates foreign substances from the blood.

Pancreas: The pancreas is the second, largest gland in vertebrates formed from the endoderm of the embryonic archenteron. The pancreas arises from the embryonic mid-gut near liver in the form of several sea—like diverticulae. Three main diverteculae, one dorsal and two ventral are formed which develop glandular part at their distal ends, leaving proximal parts as ducts. The ventral diverticula unite to form a single pancreatic duct. The mesodermal derivatives join the budding pancreatic cells and form a single gland of diffuse or compact nature.

The pancreas has cluster of glandular cells or *acini* bound together by delicate strand of connective tissue. These cells secrete a watery, alkaline, *pancreatic juice* containing enzymes (trypsin, steapsin, amylopsin) which digest carbohydrates and proteins. Compact groups of cell called *islets of Langerhans* are scattered in the pancreas and form an endocrine gland secreting insulin useful for carbohydrate metabolism in the body.

The pancreas is found in all vertebrates. In *Petromyzon* it is diffuse structure embedded in liver. In hagfishes, it lies around the bile duct as a small structure. In elasmobranches, it is well developed bilobed structure. In bony fishes, it is a widely diffused structure. In tetrapoda, it is a well developed compact structure opening in the duodenum by one or two pancreatic ducts either after joining the bile duct or independently.

5.3. ALIMENTARY CANAL IN VERTEBRATES

5.3.1. Branchiostoma (Fig. 5.23)

The Alimentary canal of *Branchiostoma* is simple and straight tube consisting of oral hood, pharynx, oesophagus, midgut, and intestine. It is lined by ciliated epithelium throughout its length.

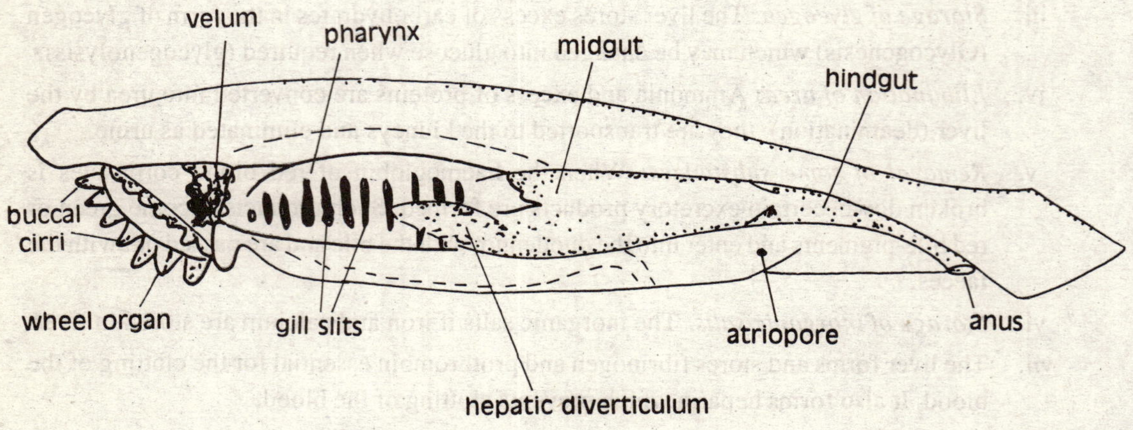

Fig. 5.23 Alimentary canal of Branchiostoma

The *oral hood* (also called buccal cavity or vestibula) lies on the anteroventral side of the body. It is lined by ciliated ectodermal tracts having large cilia. These cilia constitute a *wheel oragan* which produces a current of water in a circular manner towards enterostome. The oral hood has 22-sensory *oral cirri* or tentacles at its free edge and is closed posteriorly by a thin membranous *velum.* The oral cirri block the mouth at the time of diving and serve as sieve to avoid sand particles from food currents during feeding. An aperture called *enterostome* (Sometimes called mouth by analogy) is situated at the centre of velum fringed with 12 velar tentacles. The velum serves as partition between stomadaeum and pharynx. The roof of oral hood has s Hatscheck's pit for mucous secretion. Anterior opening of the oral hood is regarded mouth and a large space below oral hood lined by the ectoderm is stomodaeum.

The *pharynx* (Fig. 5.24) is larger part of the gut lined by endoderm unlike pharynx of higher vertebrates, which is lined by ectoderm. It is laterally compressed sac occupying about 2/3rd of the body, it is divided into a small *peribranchial region* and a long *branchial region.* The wall of the branchial region has more than 150 pairs of oblique and vertical *gill slits* or *gill clefts,* their number increase with the age of the animal. The skeleton of the pharynx consists of gill-bars. The gill slits are separated by *primary gill-bars* which are forked and alternate regularly with secondary gill bars or tongue bars which are straight. The two types of gill bars are connected together by transverse connections or *synapticulae.* The gill-slits open into the atrial cavity. The cilia are associated with gill bars as lateral cilia on the anterior and posterior faces of the gill bars, frontal cilia on inner faces, and atrial cilia on outer faces. The central core of each bar has blood vessels, fibrous connective tissue and coelomic channels. In secondary gill bars, the coelomic channels are absent. The pharynx has long ciliated tracts. A shallow groove called *endostyle* (Fig. 5.25) lies along the mid ventral wall containing a tract of four groups of glandular cells alternating with groups of ciliated cells with cilia of median tract very long. The glandular cells are connected with mucus secretion. A ciliated *epipharyngeal groove* is situated on the mid dorsal side which leads to the opening of the oesophagus at the posterior end of the pharynx. There are two gelatinous skeletal

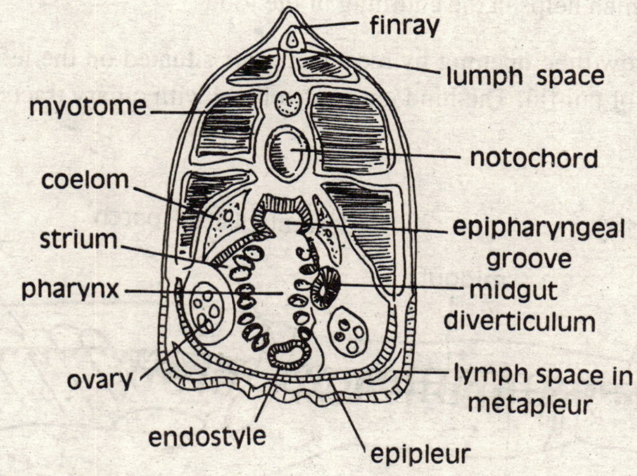

Fig. 5.24 T.S. of Amphioxus through pharynx

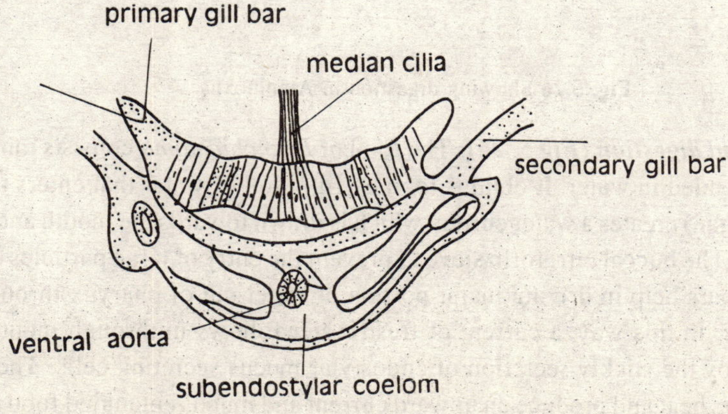

Fig. 5.25 Endostyle of Amphioxus

plates below the endostyle. The two ciliated pharyngeal bands which run on the side wall of the pharynx behind velum connect the epipharyngeal groove and endostyle.

The oesophagus is a short tube which connects the pharynx with the mid gut.

The midgut is a wide tube from where the intestine starts tapering gradually terminating at anus. The midgut gives off a large anteriorly directed midgut diverticulum or hepatic diverticulum, which is a digestive gland. This gland is comparable to the pancreas of vertebrates and is lined with glandular epithelium. The midgut diverticulum secrets enzymes (a lipase and a protease), it is generally called the liver. The midgut is lined with ciliated tracts. An *ilicolic ring* is present at the hind end of the midgut, which helps in the churning of the food.

The hind gut is a narrow tube opening by means of *anus* situated on the left side of the mid ventral line near the base of tail fin. The hind gut is also lined with ciliary tracts.

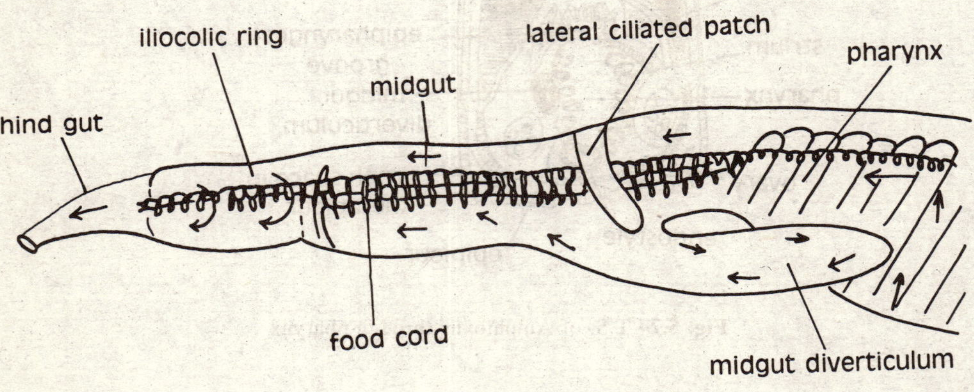

Fig. 5.26 Showing digestion in Amphioxus

Food, feading and digestion (Fig 5. 26): The food of *Branchiostoma* consists minute organic particles found suspended in water. It obtains food by filtration of water that enters the pharynx. The velum (wheel organ) creates a water current which is drawn towards the mouth and controls its entry in the pharynx. The buccal cirri form sieve to prevent the entry of large particles of food. The cilia present on gill bars help in driving major portion of water out of pharynx through gill slits, strium, and atriopore. In this way, a current of fresh water follows in through mouth. The food particles are caught by the stickly secretion of endostylar mucus secreting cells. The cilia of the endostyle and gill bars beat and produce an upward current and mucus entangled food particles are pushed towards the epipharyngeal groove. The food along the peripharyngeal ciliated tracts is pushed to the epipharyngeal groove by the beating of the cilia of the endostyle. The cilia of the epipharyngeal groove beat backwards and push the food backward. A cord-like structure *food-cord* is formed by the secretion of the glandular cells of the endostyle on mucus entangled food particles. The food cord present in the pharynx passes into the midgul diverticula through the oesophagus. The digestive enzymes of the diverticulum act upon this cord which is pushed backward by its cilia. In the ilio-colic ring the food cord is rotated by the ciliary action.

The digestion in *Branchiostoma* is of *intracellular* and *extracellular* types. In intracellular digestion *(phagocytosis)* the secretory cells of the midgue diverticulum engulf the food particles and digest them like in *Amoeba.* The extracellular digestion takes place in the midgut. The absorption takes place in the hindgut and undigested food is expelled out of the body through the anus.

5.3.2. Scoliodon

The digestive system of scoliodon includes the alimentary canal and digestive glands. The alimentary canal consists of mouth, buccal cavity, pharynx, oesophagus, stomach, intestine, and rectum (Fig. 5.27). The digestive glands are liver, pancreas and rectal gland. Its food consists of fishes, crabs and lobsters.

Mouth: The mouth is situated on the anteroventral side of the body as a crescentric opeing surrounded by upper and lower lips.

Buccal Cavity: The buccal cavity is a spacious dorsoventrally compressed cavity lined with thick mucous membrane. A nonmuscular nongranular "tongue" is produced by the lining of the floor of buccal cavity. Dermal denticles or teeth are present in the mucous membrane making it tough.In the jaws, the dermal denticles are modified into oblique. sharp compressed teeth with smooth nonserated edges. Teeth are similar (homodont) situated on the upper margins of the upper and lower jaws arranged in parallel rows. The condition of teeth in Scoliodon is lyodont in which several sets of teeth function in succession. They are used for catching the prey and preventing its escape. The teeth do not function for mastication. The buccal cavity has no glands.

Pharynx: The buccal cavity leads to pharynx. The internal cavity of the pharynx is lined by mucous membrane having a large number of dermal denticles. The internal openings of the spiracles and 5 pairs of gill pouches are present on either side of the pharynx. (The spiracle is vestigial and not visible externally).

Oesophagus: The oesophagus is present after the pharynx. It is a narrow short structure with thick muscular wall. The internal mucous lining membrane us raised into longitudinal fold which closes the cavity if the oesophagus to prevent the entry of water from the pharynx.

Stomach: The oesophagus leads into a dilated large stomach which is highly muscular. It is a J-shaped structure with two limbs. The proximal is cardiac stomach and the distal is pyloric.

Stomach: A blind sac is present at the junction of cardiac and pyloric stomachs. Between cardiac and pyloric parts of the stomach a valve-like structure is formed by the longitudinal folds against a circular fold. A crescentric fold is present at the entrace of the oesophagus into the

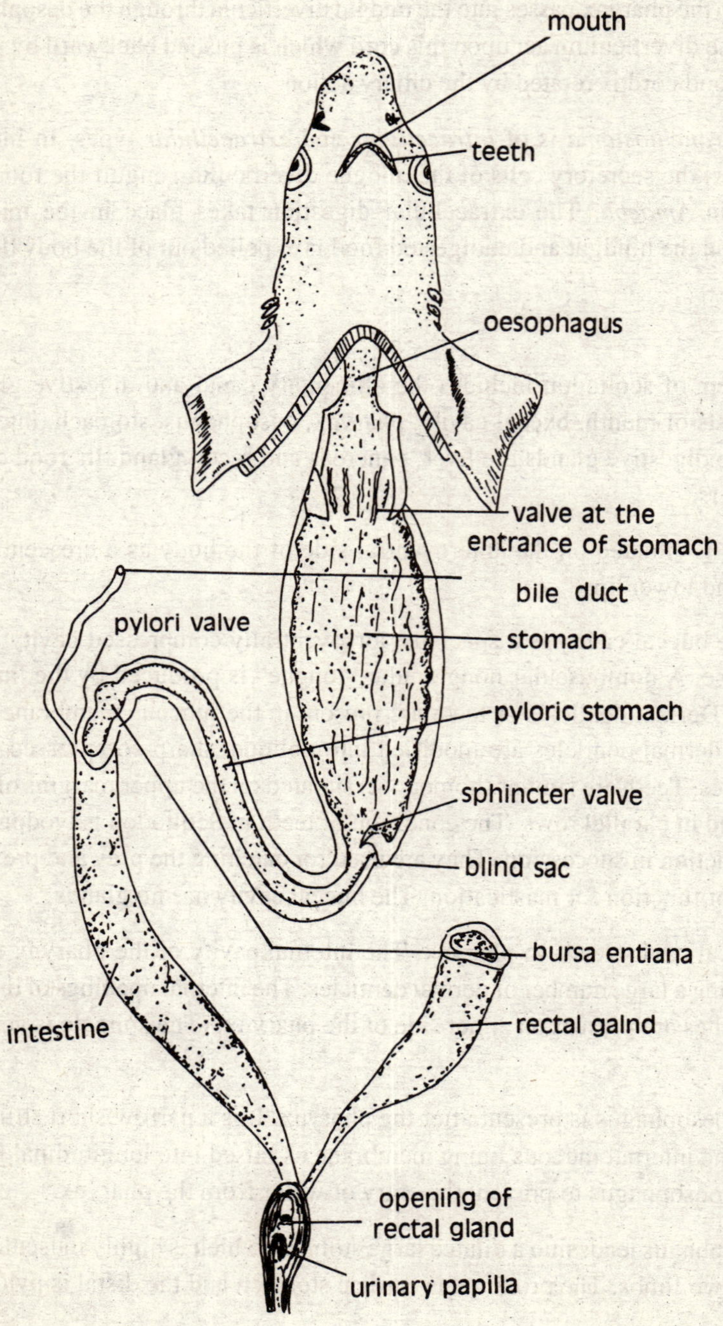

Fig. 5.27 Alimentary canal of Scoliodon

stomach which serves as a valve. The inner lining of the cardiac stomach is folded and that of pyloric stomach is mostly smooth excepting at the distal end. At the end of pyloric stomach a valve is present guarding its entrance into a thick walled small sac called *bursa entiana.* The bursa entiana continues into the intestine.

Intestine: The intestine is a wide tube with its diameter in the middle equivalent to that of the cardiac stomach. It runs straight backwards in the abdominal cavity and joins the rectum behind. A characteristics *scroll valve* is formed by the fold of mucous membrane of the intestine (Fig. 5.28). One end of the scroll valve is joined with the inner wall of the intestine and the other rolled up longitudinally on itself to form a scroll with anticlockwise spiral of about 2½ truns. The scroll valve increase the absorptive capacity of the intestine ans checks the rapid flow of digested food through the intestine.

Rectum: The intestine continues into the rectum which opens behind into the cloaca. The rectal gland opens into the rectum from the dorsal side.

The liver is a massive yellowish gland made of two lobes united anteriorly. A *gall bladder* is present on the anterior side of the right lobe. The bile-duct receives few smaller ducts from the liver lobes and opens into the intestine at the anterior and near the starting point of the scroll valve.

The *pancreas* is a compact, irregular, pale structure made of two lobes. A dorsal lobe is situated parallel to the posterior part of the cardiac stomach and a ventral lobe is attached to the pyloric stomach. The pancreatic duct opens into the intestine opposite the bile duct.

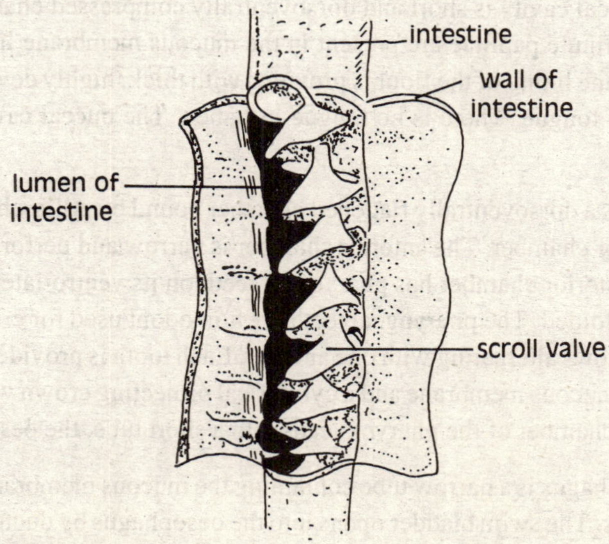

Fig. 5.28 Scroll valve in intestine

The *rectal gland* (caecal or digitiform gland) discharges a fluid into the lumen of the intstine. It helps in the excretion of sodium. It has a central cavity lined with cuboidal cells. It is highly vascular structure made of lymphoid tissue.

Digestion: The *Scoliodon* is carnivorous and its teeth are not meant for mastication but help in swallowing the food. In buccal cavity *mucin* in secreted by the mucous membrane which lubricates the food. The actual digestion begins in the cardiac stomach where its mucous membrane secretes the gastric juice containing *pepsin* and hydrochloric acid which converts proteins into peptones, synotin and proteoses. The pyloric stomach activites the pancreas and does not have any digestive function. The bile and pancreatic juices act upon the food when it enters the intstine in semidigested condition. The action of bile makes the food alkaline for the action of pancreatic juice. The pancreatic juice contains trypsinogen secreting trypsin, amylopsin and lipase. The absorption takes place in the intestine which has great absorptive surfaces. Undigested or waste food materials is passed out through the cloacal aperture.

5.3.3. *Labeo*

The digestive system of *Labeo* includes the alimentary canal and digestive glands. The alimentary canal is extremely long as the fish is herbivorous in habit and includes mouth, buccal cavity, pharynx, oesophagus, stomach, intestine, and rectum (Fig. 5.29). The digestive glands are liver and pancreas.

Mouth: The mouth is terminal and surrounded by soft upper and lower lips with four or five pairs of pigmented conical papillae. The mouth leads into a buccal cavity.

Buccal Cavity: The buccal cavity is short and dorsoventrally compressed chamber with a flat floor and an arched roof. Minute papillae are present in the mucous membrane lining the buccal cavity. The mucous membrane lining of the floor is provided with thick, highly developed muscles which serve the function of tongue. There is no tongue in Labeo. The buccal cavity leads into a pharynx.

Pharynx: The pharynx is a dorsoventrally flattened chamber bound by gill-arches and divisible into an anterior and posterior chamber. The anterior chamber is narrow and perforated by gill slits on the lateral sides. The posterior chamber has pharyngeal teeth on its ventro-lateral walls and its ventral wall is transversely folded. The pharyngeal teeth are homodont used for crushing the solid food, and arranged in three rows alternating with one another. Each tooth is provided with a narrow basal root embedded in the mucous membrane and a cylindrical projecting crown which is laterally compressed. The posterior chamber of the pharynx leads into a short tube, the oesophagus.

Oesophagus: The oesophagus is a narrow tube containing the mucous membrane thrown into a number of longitudinal folds. The swim bladder opens into the oesophagus by ductus pneumaticus. The oesophagus leads into the stomach.

Stomach: The stomach is also called the *intestinal bulb* which is enclosed in the coils of the intestine. It is divided into cardiac and pyloric stomachs. The *cardiac stomach* is broader chamber with small folds of mucous membrane making honeycomb appearance. The *pyloric stomach* is narrower posterior chamber with longitudinal folds of the mucous membrane. The pyloric stomach leads into the intestine.

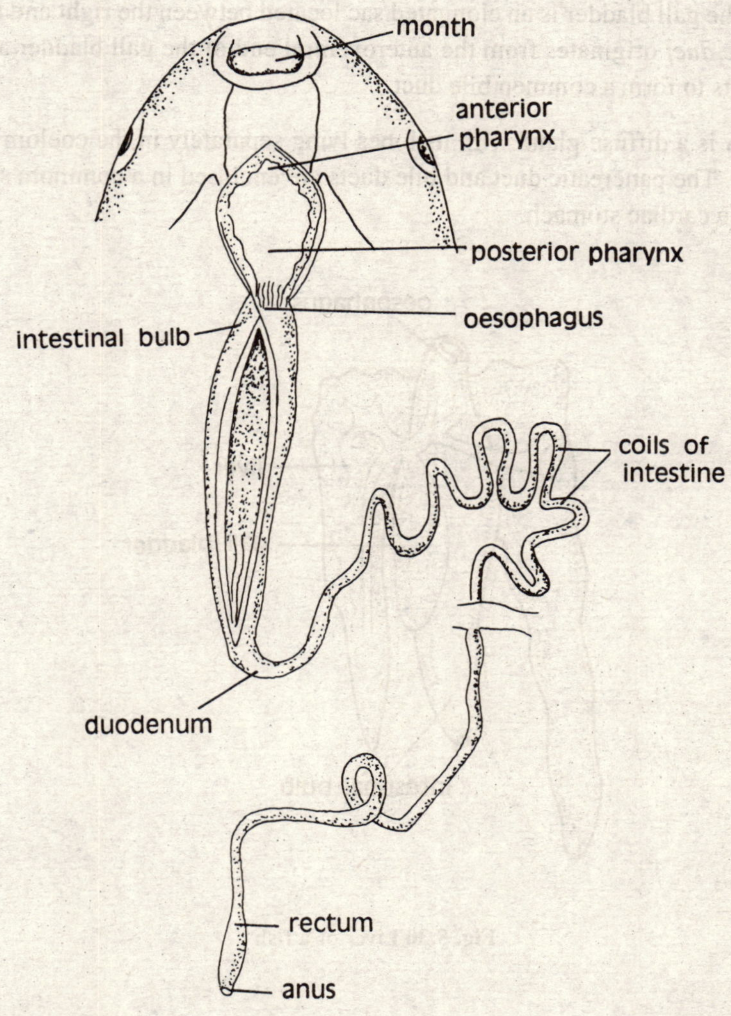

Fig. 5.29 Alimentary canal of Labeo

Intestine: The intestine is very narrow and extremely elongated, measuring about 8 meters in length. It is thin—walled with almost uniform diameter and a number of coils, the mucous membrane of the intestine forms oblique transverse folds in the anterior part and distinct longitudinal folds in the posterior part.

Rectum: The intestine leads into the rectum. The mucous lining of the rectum shows indistinct oblique transverse folds. The rectum communicates with the exterior through the anus situated close to the urinogenital opening infront of the anal fin. The pyloric caeca is absent in Labeo.

The *liver* (Fig. 5.30) is huge, dark brown gland divided into a narrow right lobe and broader left lobe. The two lobes of liver are connected transversely at three place, anterior, middle, and posterior ends. The gall bladder is an elongated sac located between the right and the left lobes of the liver. A *cystic duct* originates from the anteroventral end of the gall bladder and is joined by three hepatic ducts to form a common bile duct.

The *pancreas* is a diffuse gland with its lobes lying separately in the coelom and also in the lobes of the liver. The pancreatic duct and bile ducts are enclosed in a commom sheath and open separately into the cardiac stomach.

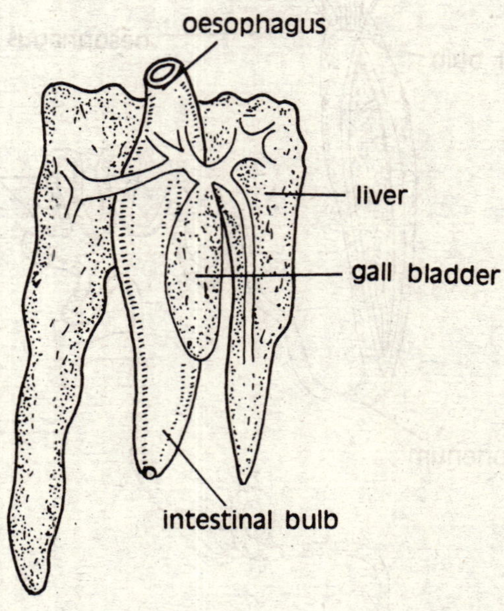

Fig. 5.30 Liver of a fish

5.3.4. Rana

The digestive system of frog includes the alimentary canal and associated digestive glands (Fig. 5.31). The alimentary canal consists of mouth, buccopharyngeal cavity, oesophagus, stomach, duedenum, ileum rectum, and cloaca. The digestive glands are liver and pancreas.

Mouth: The mouth is terminal and opens laterally. It is a large structure capable of ingesting large objects. It is situated between upper and lower jaws and contains a wide gape. The mouth open into the bucco—pharyngeal cavity.

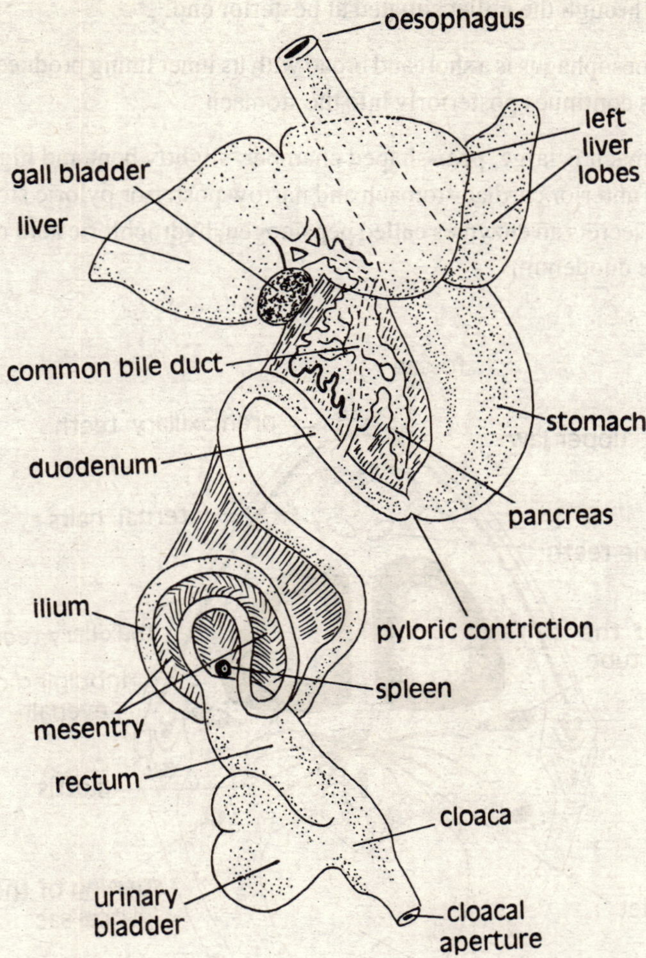

Fig. 5.31 Digestive system of frog

Buccopharyngeal cavity (oral cavity) (Fig. 5.32): There is no distinction between the buccal cavity and pharynx in frog and the oral cavity is called buccopharyngeal cavity. It is a wide cavity with bifurcated tongue covered with mucus. The tongue is attached to the anterior margin of the lower jaw at its anterior end, and its posterior end is free. The tongue is used for the capture of prey. A single row of *maxillary teeth* is present on the upper jaw, and two patches of vomerine teeth are present on the roof of the oral cavity. The teeth are small, sharp, and curved backwards and serve for preventing the escape of the prey and holding it. The teeth are *homodont* (similar) and *polyphyodont* (replaceable). Paired internal nares or nostrils are situated just infront of The vomerine teeth ans communicate with the external nares through the nasal sacs. Two openings, *eustachean tubes* are situated on the roof of the pharyngeal part of the oral cavity, which open into ears. The floor of pharynx has vertical slit called *glottis* through which air enters into the lungs. The male frogs have a pair of openings of vocal sacs at the posterior end of the pharynx. The pharynx joins the oesophagus through the gullet situated at posterior end.

Oesophagus: The oesophagus is a short and broad with its inner lining produced into longitudinal folds. The oesophagus continues posteriorly into the stomach.

Stomach: The stomach is large, pear-shaped chamber, slightly bent and highly muscular. It is divided into a broader anterior cardiac stomach and narrow posterior pyloric stomach. The gastric glands of the stomach secrets an enzymes called pepsinogen, hydrochloric acid mucus. The pyloric stomach opens into the duodenum.

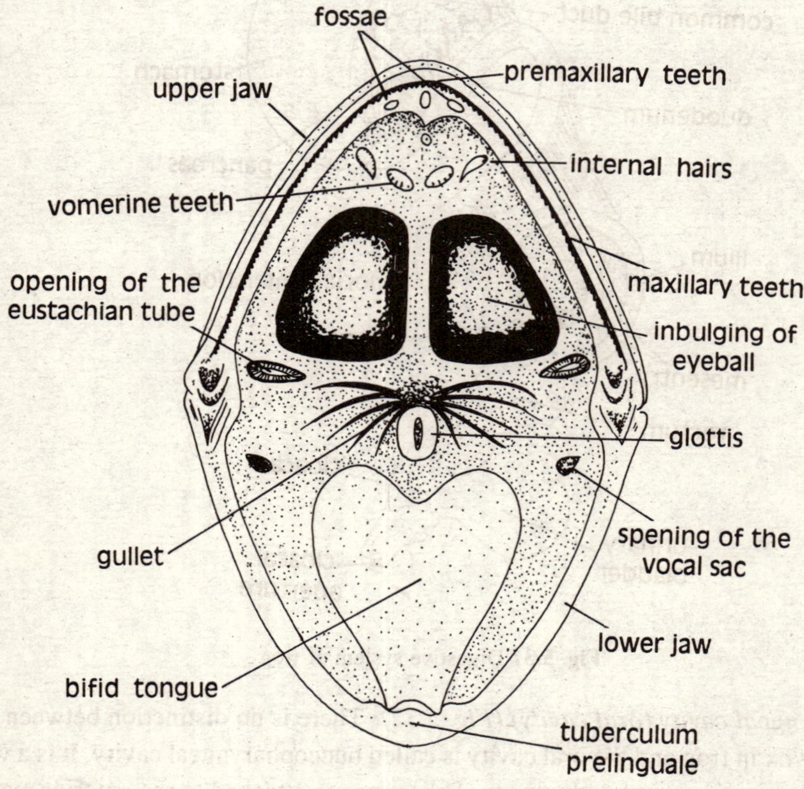

Fig. 5.32 Bucco-pharyngeal cavity of a male frog.

Duodenum: A *pyloric constrictor* is present between the pyloric stomach and the duodenum which regulates the flow of food into the intestine. The duodenum is first part of the intestine. It is U-shaped structure attached to the stomach by membrane called gastroduodenal omentum. The duodenum continues into the ileum. The duodenum and ileum consitute the small intestine.

Illeum: The ileum is a narrow long tube forming several loops. It continues posteriorly into the rectum.

Rectum: The large intestine or rectum has a broad rectum and a cloaca communincating with the exterior by a cloacal aperture.

Digestive glands: The liver is a large gland with two large and a mediua small lobe. The bile duct opens into the duodenum.

The *pancreas* is made of several lobes between the stomach and the duoedenum. The pancreatic duct meets the bile duct to form a common *hepatopancreatic duct* or *ductus chloedochus.*

The spleen is present near the rectum.

5.3.5. Uromastix

The digestive system of *Uromasstix* includes the alimentary canal and associated glands (Fig. 5.33). The alimentary canal consists of mouth, buccal cavity, pharynx, oesophagus, stomach, small intestine and large intestine. The assessory glands are liver and pancreas.

Mouth: The mouth is bound above and below by immovable and muscular upper and lower lips covered by scales. The mouth has a wide gape and teeth on maxillae, premaxillae and lower jaw. The teeth are *homodont, acrodont, polyphyodont* and attached on the outer borders of the jaw bones.

Buccal cavity . The mouth opens into the buccal cavity lined by mucous membrane. A well developed bifid, fleshy tongue is present on the floor of the buccal cavity. The tongue has voluntary muscles, mucous glands, and taste buds. The mucous gland is present in four groups. Near internal nares, a pair of apertures of Jacobson's organs are present.

Pharynx: The pharynx is situated posterior to the tongue. Its linging is produced into longitudinal folds.

Oesophagus: The pharynx leads to the oesophagus which is a narrow, long tube opening into a wide stomach.

Stomach: The stomach is a wide cylindrical structure with well developed *cardiac* and *pyloric* regions. The *pyloric sphincter* valve is well developed. The stomach leads into the intestine.

Small intestine: The small intestine has a duodenum on its anterior side. The ileum is coiled structure and leads into an iliocolic valve which connects the large intestine .

Large Intestine: It is divided into *colic caecum, colon, rectum* and *cloaca.* True colic caecum shows its appearance for the first time in reptiles. It joins a narrower colon which continues into a rectum. The opens into a broader cloaca. There are three chambers in the cloaca called coprodaeum, rodaeum, and proctodacum. The *coprodaeum* is the first chamber where rectum opens. The ureters, urinary bladder and genital ducts open into the urodaeum. The proctodaeum is the last chamber which opens to exterior by a cloacal aperture.

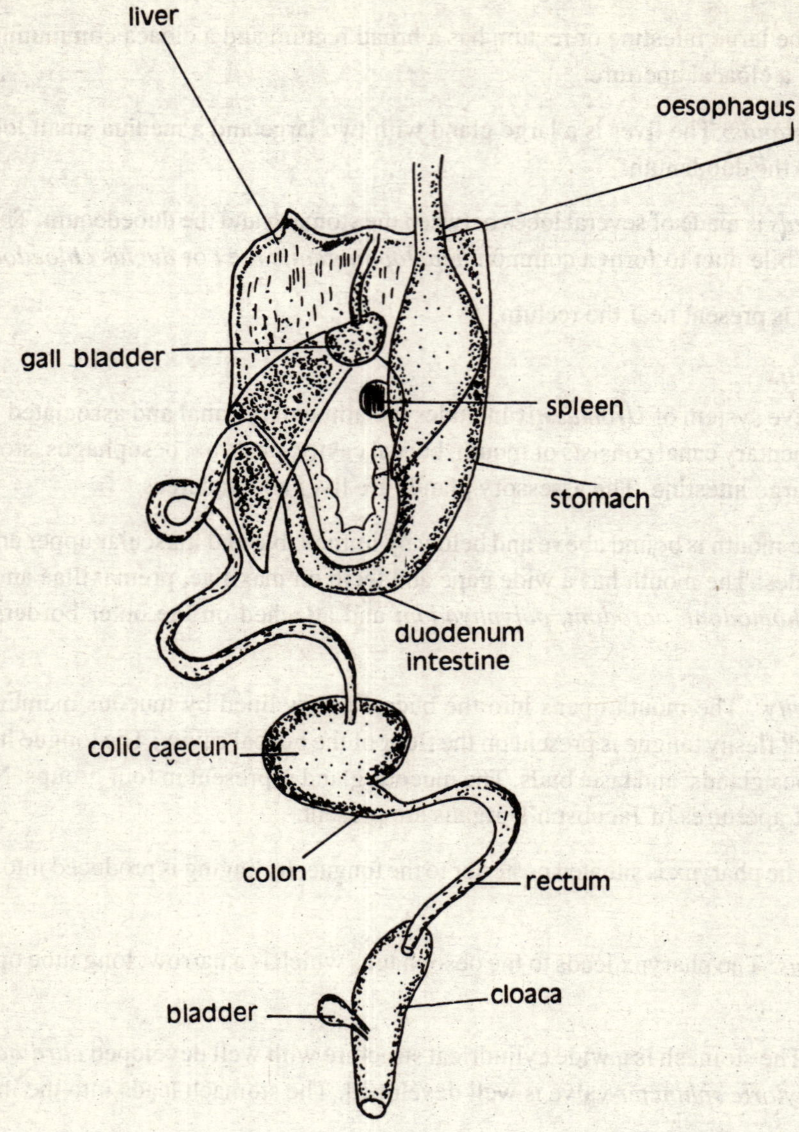

Fig. 5.33 Alimentary canal of Uromastix

Associated glands: The liver is massive bilobed gland with right lobe long and narrow and left lobe shorter and broader. A rounded gall bladder is present and bile duct opens into the duodenum.

The *pancreas,* lies between the stomach and duodenum as narrow, curved. structure. The pancreatic duct joins the bile duct.

The spleen is situated near the cardiac stomach.

5.3.6. Columba

The digestive system consists of the alimentary canal and associated glands (Fig. 5.34). The alimentary canal inclues mouth, oral cavity, oesophagus (including crop), stomach (including proventriculus and gizzard), intestine and cloaca. The digestive glands include salivary glands, liver, and pancreas.

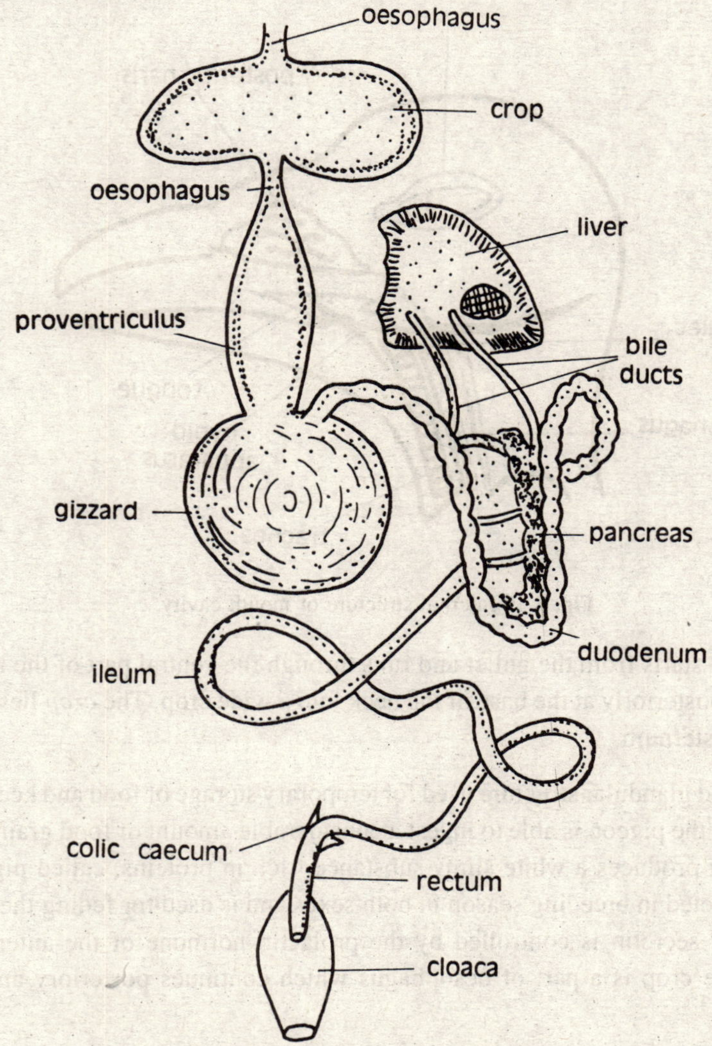

Fig. 5.34 Alimentary canal of columba

Mouth: The mouth is inlet aperture situated at the tip of the head. It is bound by horny upper and lower beak. The mouth leads to the oral cavity.

Oval cavity (Fig. 5.35): The oral cavity is provided with one opening of internal nostrils on the pharynx through which both nostrils open externally. The tongue is prominent structure with anterior end free and pointed. Taste buds and numerous mucous glands are present on the tongue surface. An aperture, called glottis is present neare the base of the tongue, which communicates with the trachea. The gullet is a part of the pharynx, which connects with oesophagus. The teeth are absent in pigeon.

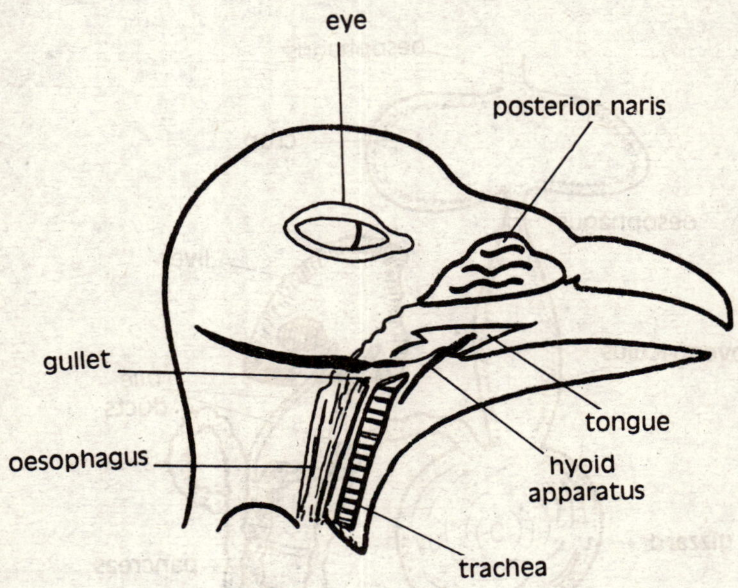

Fig. 5.35 Internal structure of mouth cavity

Oesophagus: It starts from the gullet and runs through the ventral part of the neck as straight tube and enlarges posteriorly at the base of the neck into a wide crop. The *crop* lies below the skin and anterior to the sternum.

It is a soft walled glandular structure used for temporary storage of food and keeping itm moist. Due to its presence the pigeon is able to ingest a considerable amount of food grains very quickly. Its epithelial lining produces a white slimy substance, rich in proteins, called pigeon milk. The pigeon milk is secreted in breeding season in both sexes and is used for feding the young ones by regurgitation. This secretin is controlled by the prolactin hormone of the anterior part of the pituitary body. The crop is a part of oesophagus which continues posteriory and opens into a stomach.

Stomach: The stomach consists of two parts called *proventriculus* and *gizzard.* The proventriculus receives the oesophagus and constitutes the anterior part of the stomach . It is having numerous gastric glands. The gizzard or ventriculus is posterior part having highly muscular walls, small lumen and numerous small tubular glands. A few stones are present in the gizzard which are swallowed by the pigeon for purpose of grinding the food. Due to the presence of gizzard, the role of teeth is compensated. (In flesh eating birds, the gizzard is simple and thin walled.)

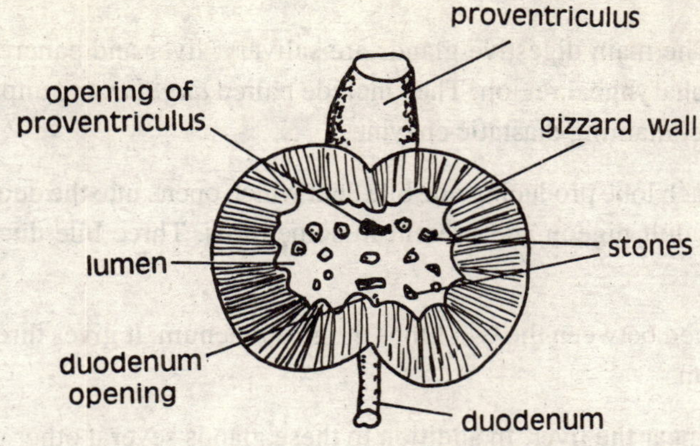

Fig. 5.36 Internal structure of Gizzard

Intestine: The gizzard opens into the intestine which is divisible into duodenum, ilum and rectum. The duodenum is U-shaped first part of the intestine and contains villi, crypts of Leiberkunn and gablet cells. The ileum is next part of the intestine comparatively shorter than in other vertebrates. The anterior and posterior parts of the ileum are loop-like and middle part is coiled. The rectum resembles the ileum in thickness. A pair of blind leaf-like caecae is present at the junction of ileum and rectum. The rectum opens into the cloaca.

Cloaca (Fig. 5.37): The cloaca is a large, spaceous, and muscular sac divided into *coprodaeum urodaeum* and *proctodaeum* like in *Uromastix.* The coprodaeum connects the rectum. urinogenital ducts open into the urodaeum, and the protodaeum opens outside through a cloacak aperture. A small, glandular sac called bursa fabricii opens into the proctodaeum in young stages and disappears in adult stages. The *cloacal aperture* is present on the ventral side near the base of the tail.

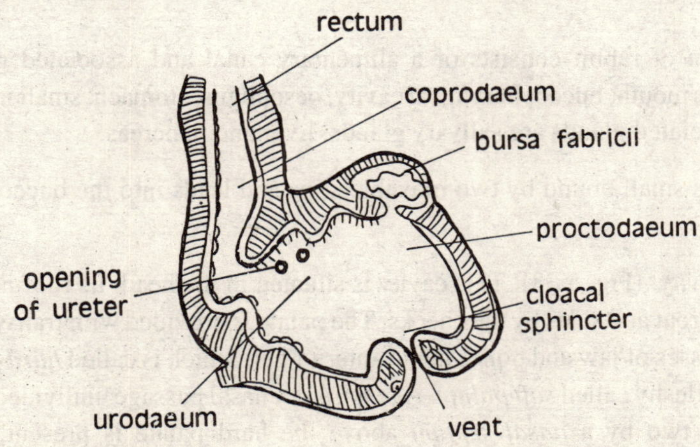

Fig. 5.37 Avian cloaca in section

Associated Glands: The main digestive glands are salivary, liver and pancreas. The salivary glands are located in the pharyngeal region. They include paired *angular* and unpaired *sublingual* glands. It secretes saliva containing diastatic enzyme.

The *liver* is bilobed, each lobe producing one bile duct which opens into the duodenum. The gall bladder is absent in the adult pigeon and present in some birds. Three bile ducts open into the duodenum.

The *pancreas* is situated between the two limbs of the duodenum. It gives three separate ducts opening into the duodenum.

The *spleen* is present near the liver. In addition to these glands several other glands are found in the digestive tract. These are *mucous glands, crop glands, gastric glands, tubular glands, crypts of Lieberkuhn, glands of ileum* and *caecal glands.* These glands have been described alongwith the different parts of the alimentary canal.

Process of Digestion: The ingestion takes place by the beak which is very efficient in picking the grains which are quickly swallowed. The swallowed food is moistened by the secretion of salivary glands and stored in the crop where it is moistened and macerated. The food passage to proventriculus is regulated. In proventriculus the gastric juice containing hydrochloric acid and pepsin act upon the food. The food passes on to the gizzard where it is grinded by the action of muscular walls and stones. In duodenum, the food is acted upon by the bile and pancreatic juices like in other vertebrates. The broken down food is absorbed by the lining of the intestine and residual food passes on to the rectum. In rectum, the water is absorbed and caecal juice breaks down to fibres. Residual food is stored in the cloacam gets mixed with urine, and periodically dropped out through the cloacal aperture.

5.3.7. Oryctolagus

The digestive system of rabbit consists of a alimentary canal and associated glands. The alimentary canal includes mouth, buccopharyngeal cavity, oesophgus, stomach, small intestine and large intestine. The associated glands are salivary glands, liver and pancreas.

Mouth: The mouth is small bound by two movable slips and leads into the buccopharyngeal cavity.

Buccopharyngeal cavity: (Fig. 5.38): This cavity is situated in the head. Its roof is formed by the *palate,* floor by the throat and sides by the checks. The palate is provided with transverse ridges supported by bony processes of jaw and bones in the anterior half which is called *hard palate.* The posterior half is soft and fleshy called *soft palate.* A respiratory nasal passage undivided above soft palate and divided into two by a *nasal septum* above the hard palate is present. A pair of *nasopalatine canals* open into nasal cavities. *Jacobson's organ* is situated in the nasopalatine canal, which is used to detect different types of food. The posterior part of the soft palate has a hanging flap called velum palate. The tonsils are paired structures made of lymphoid and connective

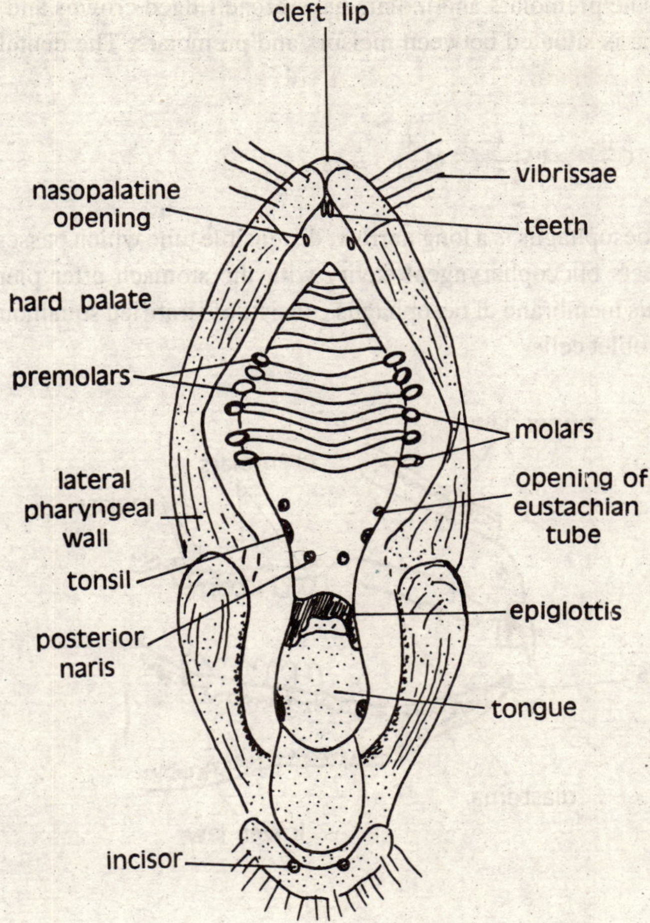

Fig. 5.38 Oral cavity of Rabbit

tissue. They are present in either side of the velum palate. The tongue is highly muscular structure with taste buds. It is attached on the floor of the buccopharyngeal cavity along the greater part of its length and free rounded tip infront. Three different types of papillae are present on the tongue. These are *fungiform* (mushroom-like), *foliate* (broad leaf-like) and a few large *vallate* papillae. The papillae may have taste buds. There is no line of demarcation between the buccal cavity (lined by ectoderm) and the pharynx (lined by endoderm); hence it is called buccopharyngeal cavity. Eustachean tubes open into the pharynx and communicate with the middle ear. The pharynx is divided into upper nasal and lower buccal parts, which merge into a single space behind the soft palate. The pharynx leads into a dorsal gullet (food passage) and a ventral glottis (air passage) posteriorly. An epiglottis cartilage is present above the glottis. The teeth are present in both jaws, which are hetrodont, thecodont (embedded in sockets) and polyphyodont (Fig. 5.39). Three types of teeth, *incisors, premolars* and *molars* are present. The canines are absent and incisors are chisel-shaped cutting teeth. The premolars and molars have broad ridged crowns and grind the food. A wide space or diastema is situated between incisors and premolars. The dental formula is given below:

$$I\frac{2}{1},\quad C\frac{0}{0},\quad PM\frac{3}{2},\quad M\frac{3}{3},\quad = 28$$

Oesophagus: The oesophagus is a long narrow, distensible tube which passes through the neck and thorax and connects buccopharyngeal cavity with the stomach after piercing through the diaphragm. The mucous membrane of oesophagus consists of stratified squamous epithelium cells and mucus-secreting goblet cells.

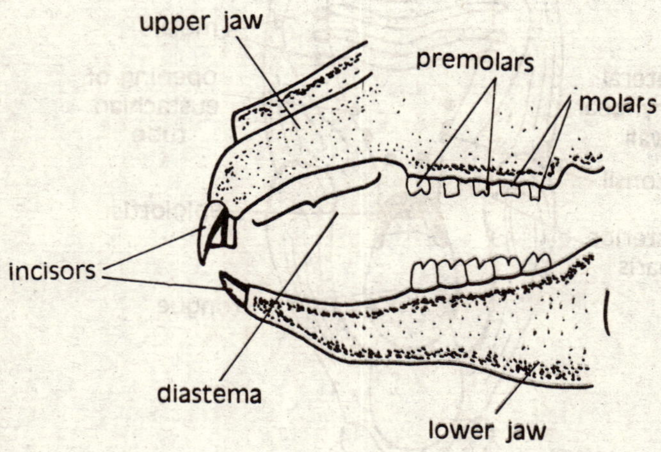

Fig. 5.39 Dentition of Rabbit.

Stomach: The stomach is a large curved sac in the abdominal cavity behind the diaphragm. It is divisible into three parts, an anterior broad cardiac part next to oesophagus, a middle fundic part and a narrow posterior pyloric part which joins the intestine (Fig. 5.40). The openings of the stomach are controlled by cardiac and pyloric valves. The walls of the stomach are highly muscular and its mucous lining is provided with different types of gastric glands described in section 5.2.

Small Intestine: The first part of the small intestine is U-shaped *duodenum* which starts form the pylorus and goes upto the posterior end of the abdomen. It receives bile duct in the beginning and pancreatic duct near the beginning of its forward limb. The *ileum* is the second part of the small intestine about 2.5 metres long. It is narrow and coiled tube with its internal lining produced into villi to increase the surface for absorption. Several miscroscopic intestinal glands are present in the wall of the intestine. The ileum joins a rounded sac called *sacculus rotundus,* which opens into the caecum. An *iliocaecal valve* is present in the sacculus rotundus, which forwards the contents of the ileum into the caecum before passing into the large intestine. The *caecum* is a thin walled diverticulum of about 50cms length. It has pouch-like constrictions with its internal lining forming a spiral valve. The caesum ends distally into a blind, about 10cms long, finger-like process called the *vermiform appendix,* which is made of thick walls.

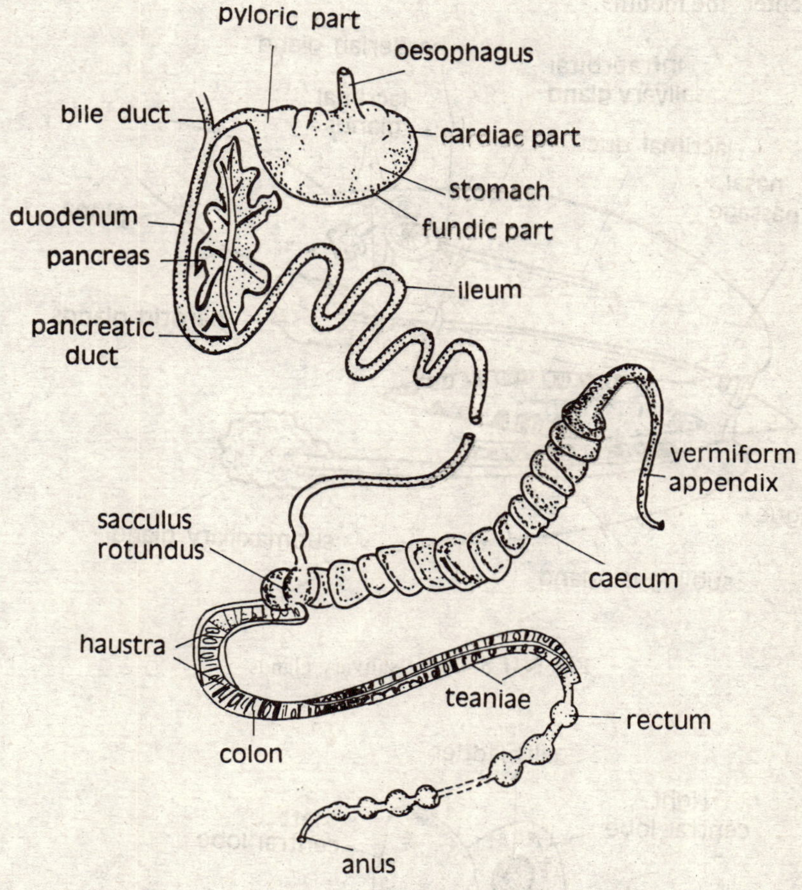

Fig. 5.40 Alimentary canal of Rabbit

Large Intestine: The large intestine starts from the caecum. It is divided into *colon* and *rectum*. The colon is wider tube about 45cms long with sacculated walls. The constrictions produce a series of small pockets called haustra. The *rectum* is distal part about 75cms in length having beaded appearance due to the presence of pallets of faeces. The internal walls of the large intestine do not have *villi.* Small folds containing *glands of Lieberkuhn* and many mucous secreting goblet cells are found in the large intestine. It communicates with the exterior by anus present on the undersides of the tail. The *anus* is provided with a sphincter muscle to control its opening and closing.

Associated Glands: The glands situated outside the alimentary canal are salivary gland, liver, and pancreas.

The *salivary glands* (Fig. 5.41) are mucous glands situated on the palate and the tongue. In addition to these, three pairs of salivary glands are present in the head with their ducts opening into the oral cavity. These are *parotid, sublingual and infraorbital glands.* The salivary glands secrets saliva containing mucilaginous mucin and watery ptyalin. The saliva is produced by reflex action when the food enters the mouth.

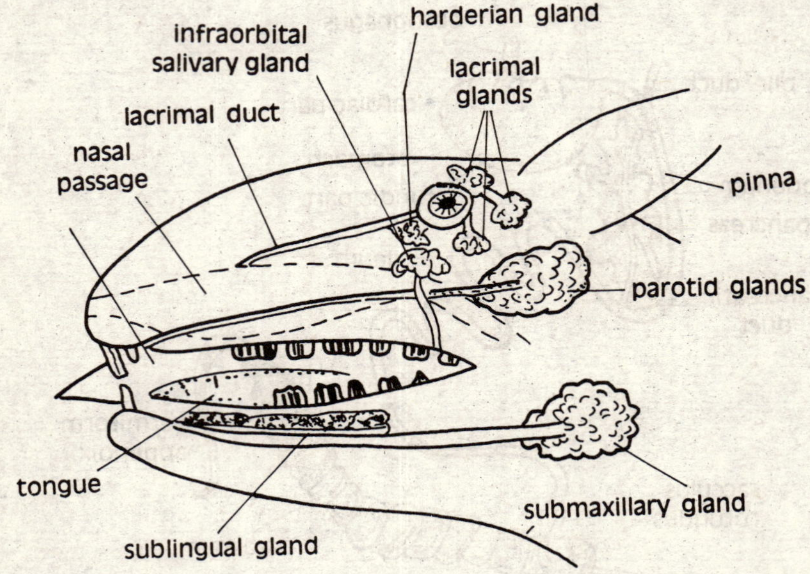

Fig. 5.41 Showing salivary glands

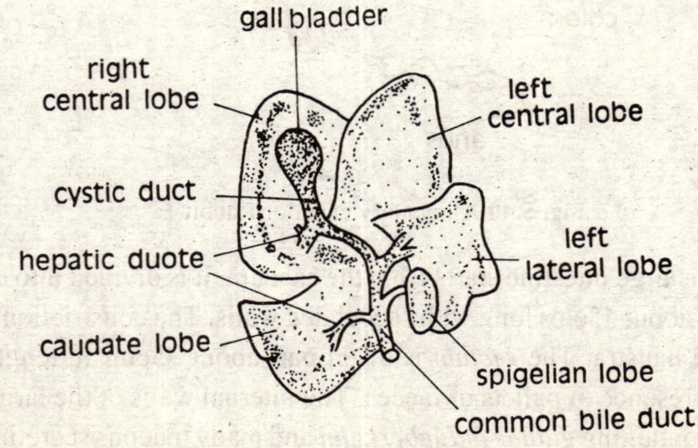

Fig. 5.42 Mammalian liver and gall bladder

The *liver* is the largest gland of the body consisting of five lobes, three on left and two on right side (Fig. 5.42). Left lobes include *left lateral, a left central* and a *spigelian lobe.* The right lobe includes a right central and caudate lobe. The *gall bladder* lies in a groove on the right central lobe. Several small hepatic ducts which bring bile from different liver lobes join a cystic lobe from the gall bladder forming a large common bile duct which opens into the termjnal part of the duodenum.

The *pancreas* is an irregular, diffuse, gland made up of several branching tubes ending in a blind *acini* or *secretory sac.* The pancreatic juice containing enzymes etc. is produced by the pancreas and poured into the distal limb of the duodenum.

CHAPTER SUMMARY

1. The body cavity is formed by the hypomere of the mesoderm. In lower classes the coclom is having pericardial and pleuroperitoneal cavities. In mammals the coelom is partitioned into a pericardial, paired pleural and a median peritoneal or abdominal cavities. Visceral organs are usually suspended by mesenteries through which nerves and blood vessels connect the organs.
2. The digestive system of all vertebrates is built on the same fundamental plan. There are variations in different groups in accordance with particular adaptations. Unlike some systems evolutionary trends are not apparent in digestive system.
3. The alimentary canal is divisible into ingressive, progressive degressive and egressive zones. Mouth and nasal passage are progressively separated from lower to higher groups. The dentition is homodont in lower classes and hetrodont in some reptiles and mammals. Teeth are absent in modern birds.
4. There is clear distinction between the oesophagus and the stomach.
5. True stomach is found in fishes to mammals. It is divided into separate chambers in different groups to suit the food requirement. The stomach is most complicated in ruminants and cetacea.
6. The intestine is differentiated into distinct regions in higher groups. The alimentary canal is progressively lengthened to provide more space for secretion and absorption.
7. The cloaca of mammals is differentiated into colon and rectum. The opening of digestive and urinogenital system are separate in mammals but in lower forms they open through a common cloaca.

6

Respiratory System

6.1. GENERAL ACCOUNT

The respiration is a process by which all animals carry on metabolism in their cells by taking oxygen and releasing carbon dioxide. This exchange of gases is necessary for the oxidation of cellular substances and liberation of energy (ATP), heat, carbon dioxide and water. It involves two stages: *external respiration* which is the exchange of gases between the blood and the environment: and *internal respiration* which is the gaseous exchange between the blood and the tissues or body cells. The inta`ᵏ`e of oxygen in animals takes place by the following ways:

i. Simple diffusion from water or air through a moist surface into the body e.g. Protozoa. Helminthes.
ii. Diffusion from water or air through blood vessels or thin body wall e.g. annelids.
iii. From air through spiracles or tracheal gills to a system of tracheae to the tissue e.g. insects.
iv. From water to blood vessels through external or internal gill surfaces e.g. fishes and amphibians (Branchial system).
v. From air to blood vessel through moist lung surfaces e.g. terrestrial vertebrates (Pulmonary system).

The main respiratory organs of vertebrates are internal gills and lungs. Some additional structure which help in respiration are given below:

1. External gills found in some fishes and amphibia.
2. Air bladder found in some fishes.
3. Mucous membrane of oral cavity and pharynx in amphibia.
4. Skin of amphibia.
5. Intestinal epithelium of certain silurid fishes.
6. Extrabranchial respiratory organs of certain silurid fishes.

6.2. RESPIRATORY ORGANS IN VERTEBRATES.

6.2.1. External Gills

The external gills are usually found in the larvae of most amphibians and rarely in fish larvae. They are lined by ectoderm and are formed by the outer surface of visceral arches. They are retained throughout life in some urodeles and lost in most amphibia during the metamorphosis. The exchange of gases takes place by diffusion through the epithelial lining of these gills as they are in direct contact with water (Fig. 6.1)

6.2.2. Internal Gills and Gill Slits (Fig. 6.2)

The gill slits constitute a fundamental character of chordates and occur in adult or embryonic stages. They are found in adult condition in Protochordata, cyclostomes, fishes and some amphibians whereas in higher vertebrates they are found in embryonic stages. The pharyngeal walls on both lateral sides are invaginated during the development to form a series of blind pouches. These pouches meet similar ectodermal depressions formed by the invagination of ectoderm at corresponding positions. The *gill slits* or perforations appear on the pharyngeal walls when thin partition membrane separating the corresponding ectodermal and endodermal pouches break down. The first gill slit is *spiracle,* smaller than other gill slits and is not found in the adult stages of some fishes and tetrapoda. The gill slits connect the pharynx with the outside and internal lining of this passage develops folds in the form of endodermal filaments and gets richly vascularized. The entire structure containing filaments is called gill. The gill slits are separated from one another by mesodermal *interbranchial septa.* Each interbranchial septum is supported by *visceral arch.* The gills if such structure are called *internal gill* where the respiration takes place.

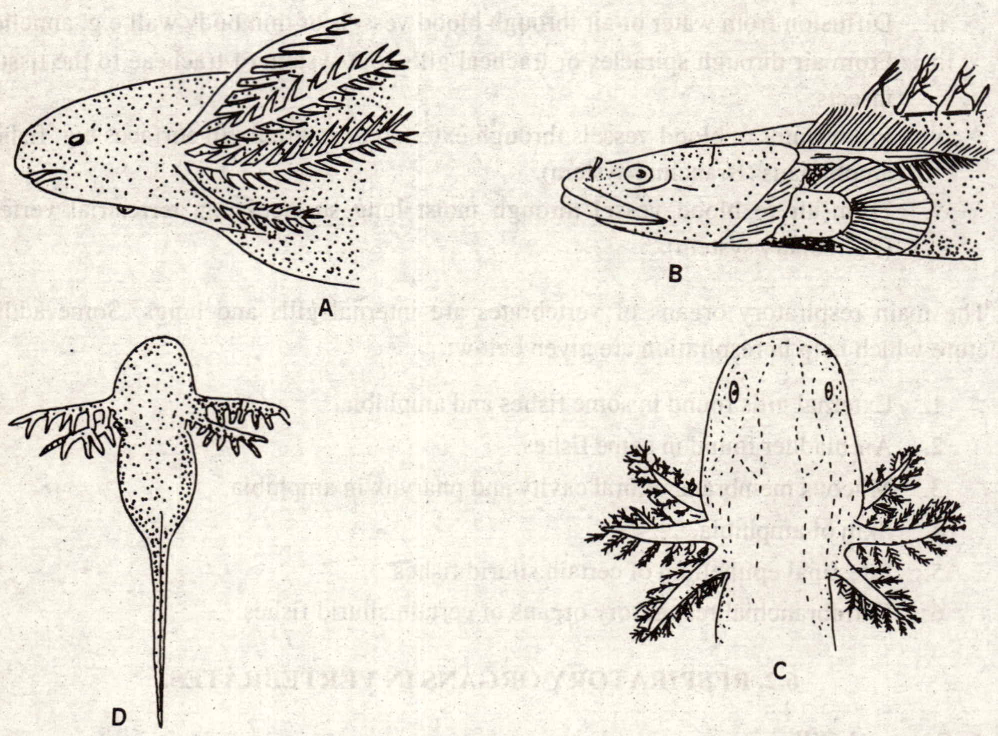

Fig. 6.1 External gills: **A.** larval lungfish, Lepidosiran, **B.** Polypterus, **C.** salamander. Siren, **D.** frog Rana

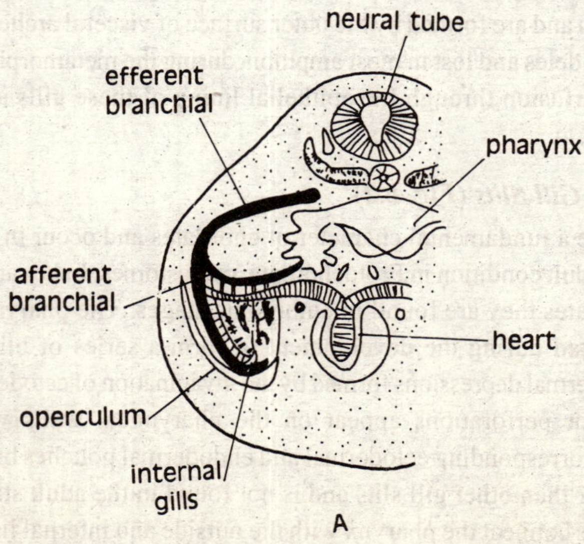

Fig. 6.2 A. Internal gills showing vascular supply

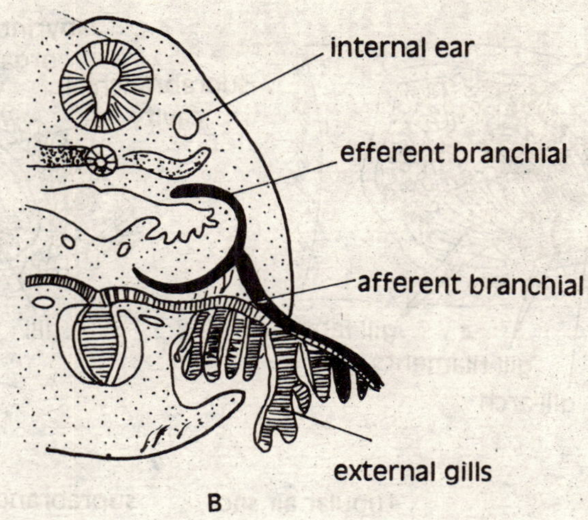

Fig. 6.2 B. External gills showing vascular supply

6.2.3. Extrabranchial Respiratory Organs of Fishes

Sonic fishes are provided with accessory or extrabranchial respiratory organs to meet the extra requirement of oxygen for respiration and survival outside water for short or long duration. The accessory respiratory organs may be for aerial or aquatic respiration. Some accessory respiratory organs are given below: (Fig. 6.3)

Labyrinthiform Organs: The Indian climbing perch. *Anabas scandens*, has three concentrically folded bony lamellae developed from the first epibranchial bone on each side in the air chambers situated above the gills. They are called labyrinthiform organs or branchial outgrowths and have vascular muscous membrane which helps in the aerial respiration. *Anabas,* may survive on land for long periods.

Suprabranchial Organs: The catfish *Clarias* has a pair of suprabranchial organs, one on each side; each divided into two parts. The first part is highly branched *arborescent organ* made by second and fourth branchial arches. The second part is a *vascular sac* enclosing the arborescent organ. The entrance of the suprabranchial organ is closed by several gill fans formed by gill-filaments. The air is taken in by the mouth continuously and passed on to the organ.

***Pharyngeal diverticulae and air sacs*:** The snake headed eels and cuchia eels have air breathing system in which the pharynx gives a pair of sac like diverticula for gaseous exchange. In the Indian catfish *Heteropneustes (Saccobranchs)* a pair of large air sacs which may be filled with air for respiration arise from the branchial chamber and extend laterally backwards into the trunk muscles. In *Channa*, the accessory respiratory organs are a pair of air chambers or *suprabranchial cavities* developed from the pharynx. The air chamber have thick, highly vascularized epithelium. The air chambers are simple sac-like structures and function as lung like reservoirs.

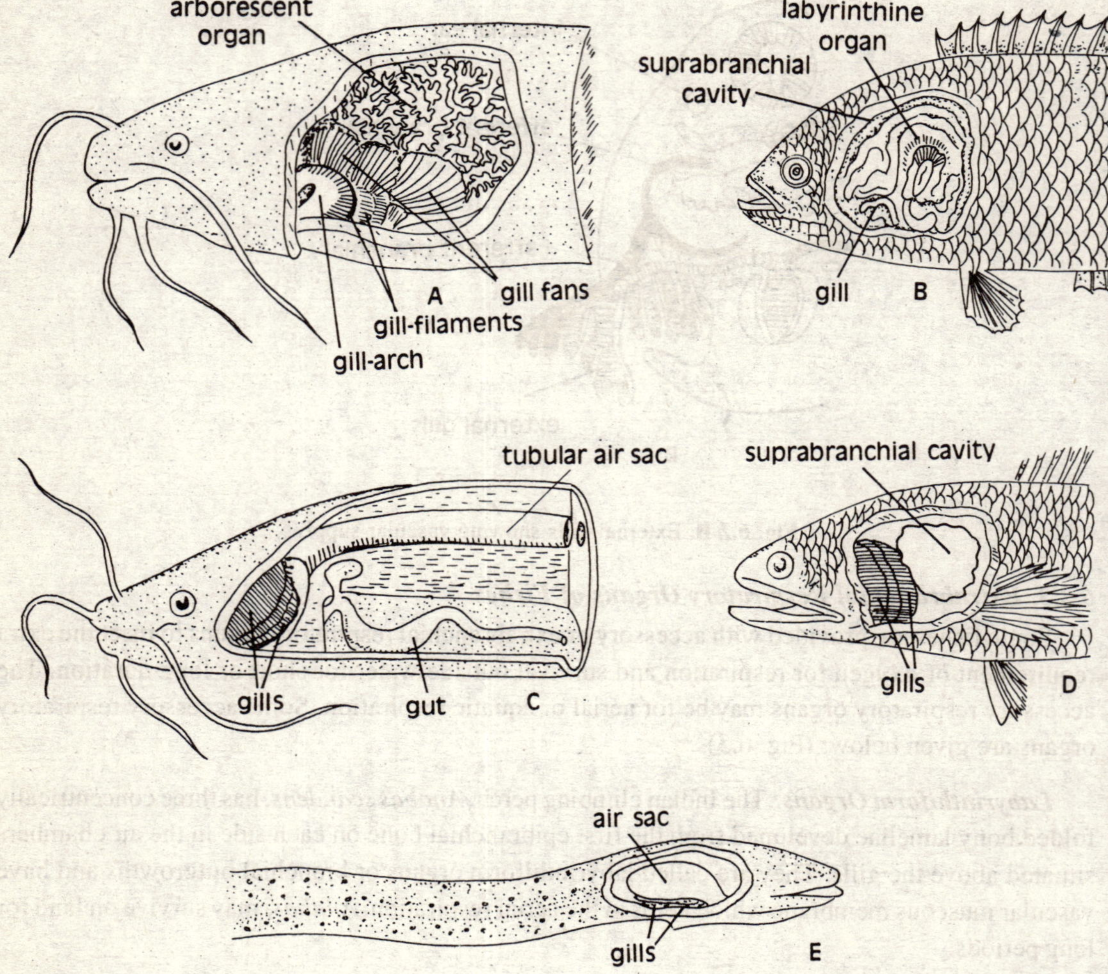

Fig. 6.3 Accessory respiratory organs in air-breathing teleosts. **A.** *Clarias batrachus.* **B.** *Anabas testudinieus.* **C.** *Hetropneustes fossilis.* **D.** *Ghanna punctatus.* **E.** *Amphipnous cuchia.*

Other accessory respiratory organs for respiration are (i) vascular membrane of the buccopharyngeal region in most fishes (ii) integument (iii) gut epithelium and (iv) outgrowths of pelvic fin.

6.2.4. Air bladder (Fig. 6.4)

The *gas bladder* is usually called air bladder or *swim bladder*. It is usually an unpaired structure found in teleosts. It is not a respiratory organ but is considered for its possible relations with lungs. It lies dorsal to the alimentary canal and below the vertebrae and arises as a diverticulum of the alimentary canal. In physostomaous fishes e.g. *Polypterus.* Dipnoi, Amia and lower teleosts it is connected with the alimentary canal by the pneumatic duct throughout life. This connection is lost in most of the higher teleosts called *physoclystous fishes.* In some teleosts parts of some anterior vertebrae are modified into a chain of bones called the Weberian apparatus which conveys differences of bladder pressure to the internal cars. The gas bladder is regarded as hydrostatic organ aiding in the recognition of differences of pressure due to depth changes Possible functions of the air bladder are accessory in respiration as hydrostatic organ, detection of depth changes, and sound production.

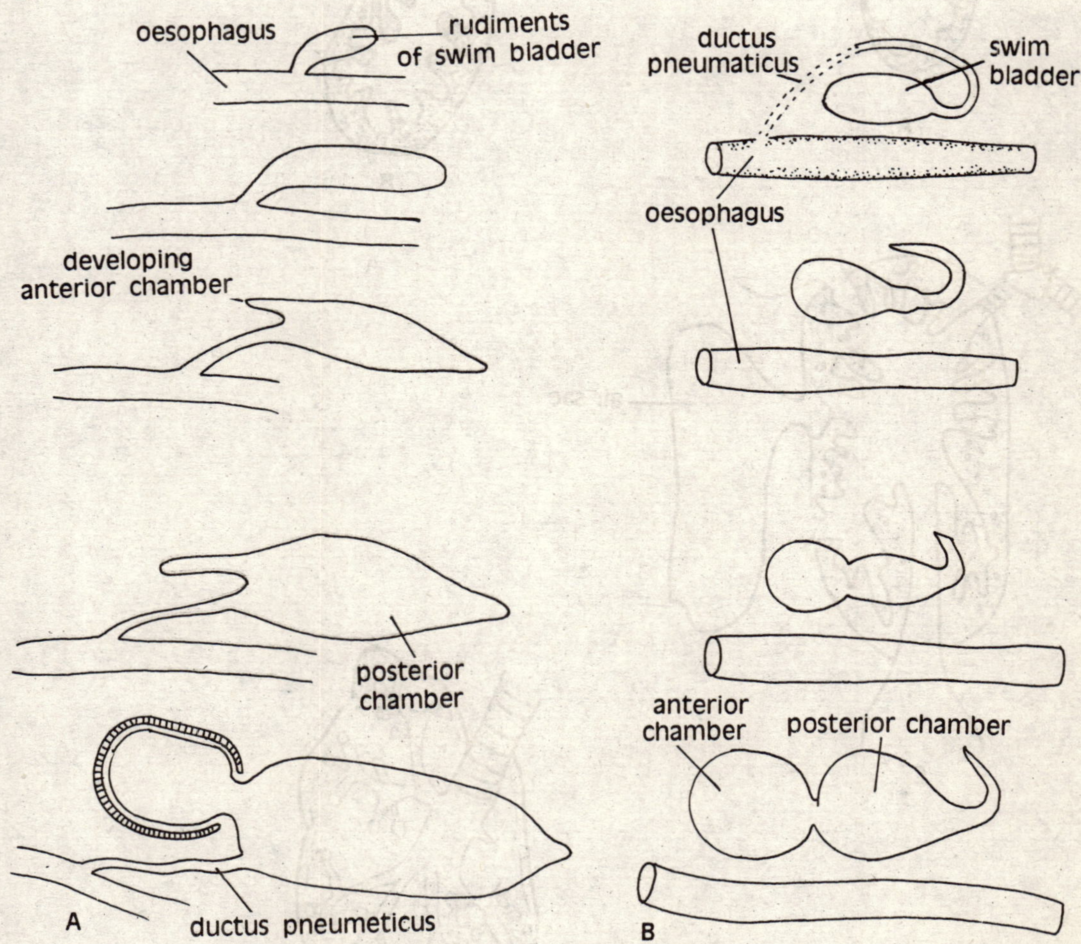

Fig. 6.4 Showing the derivation of the swim-bladder of the fishes from the gut. **A.** Stages of formation of physostomous type of swim-bladder in *Catostomus.* **B.** Stages of formation of the physoclistous type of swim-bladder.

6.2.5. Lungs and Air Ducts

Lungs are found in most adult amphibians, amniotes and some fishes (Fig. 6.5). They arise as a diverticulum from the ventral side of the pharynx immediately behind the last gill pouch. The

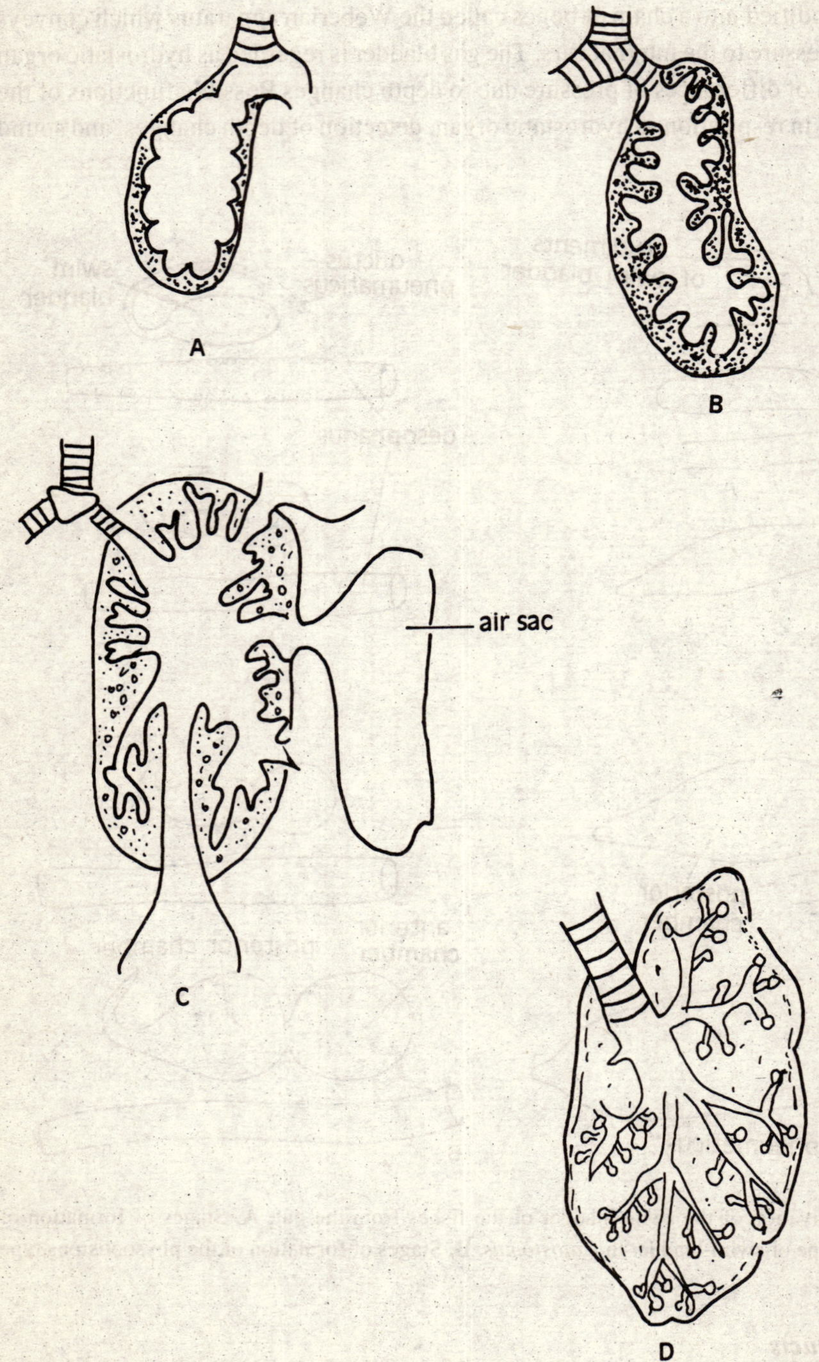

Fig. 6.5 Lungs in various vertebrates

diverticulum grows backwards and divides into two parts which are covered by the mesoderm. The inner epithelial lining of the lungs is raised into a network of ridges to increase vascularized surface for gaseous exchange. In lower animals the lungs are hollow, but in higher forms the ridge increase in number and unite with one another across the lumen of the lung thus making solid spongy structure with countless air-spaces or *alveoli.* The lungs are formed in this way and two parts are thus differentiated first being lungs which are actual seat of gaseous exchange and second air-ducts which lead from pharynx to lungs. The ducts include *trachea* or wind pipe, an unpaired structure divided into two *bronchi* posteriorly leading to two lungs. The bronchi are divided into a large number of *bronchioles* which carry air into the alveoli where the exchange of gases takes place. In most air breathing vertebrates a *larynx* or *sound box* is present which is specialized part of the trachea situated at its anterior end. The respiratory organs include the mechanism by which the air is drawn in or expelled out of the body.

6.3. RESPIRATORY ORGANS IN VERTEBRATES TYPES

6.3.1. Branchiostoma

Special respiratory organs are absent in *Branchiostoma* and the respiration lakes place probably in pharynx (not confirmed), wall of the atrium, and through general body surface. The gill-bars are richly supplied with blood vessels and exchange of gases appears to take place between water currents and blood through gill clefts. It is however, doubtful whether actual respiration takes place in the gill bars as the blood does not contain respiratory pigment. The oxygenation is considered to be chiefly taking place via general body surface (cutaneaes respiration).

6.3.2. Scoliodon

In *Scoliodon* (Fig. 6.6) the respiration takes place by gills situated in gill clefts or branchial clefts. Five pairs of gill clefts are present behind the hyoid arch. The *spiracle* lies in front of the hyoid arch and opens into the pharynx with its external opening closed. Each gill pouch is compressed chamber communicating with the pharynx by a large *internal branchial aperture* and with exterior by the *external branchial aperture* or *gill slit.* The gill clefts have endodermal mucous membrane thrown into a series of parallel folds called *branchial lamellae.* The branchial lamellae are richly supplied with blood vessels and have their covering membrane to faciliatate the exchange of gases in the sea water. Two sets of branchial lamellae, anterior and posterior, are present in each gill cleft, on its anterior and posterior walls respectively. These sets of branchial lamellae are called *hemibranchs* and each gill cleft is thus provided with two hemibranchs. A partion called *interbranchial septum* separates two adjoing gill slits and is made of connective tissue. It extends the branchial lamellae and bends posteriorly to provide protection to the hemibranchs. The interbranchial septum is supported by a visceral arch having cartilaginous gill rays in a single row projecting into it providing further support. The internal branchial apertures are protected by rigid comb-like *gill-rackers* given out by the visceral arches which do not permit the entry of food in gill clefts. Each interbranchial septum which is attached on the visceral arch has two hemibranchs belonging to two adjascent gill clefts; such a gill is called *holobranch* in which the posterior hemibranch is large than the anterior one. The gill with gill lamellae attached along the whole length of the interbranchial septa is called *lamelliform gill* (Fig. 6.7). (In bony fishes the interbranchial septum is reduced and gill lamellae are free on their dista side. Such gills are called filiform or pectinate gills).

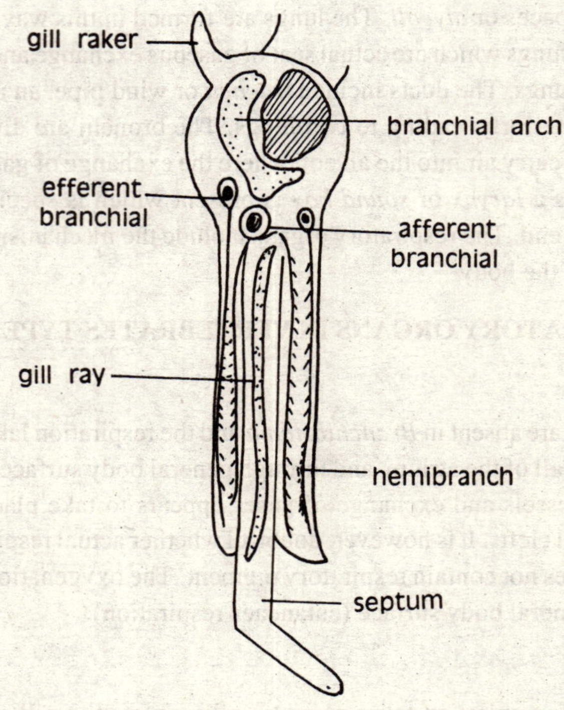

Fig. 6.6 Gill of a cartilageman fish

Mechanism of Respiration: The respiratory movements are conducted by pharyngeal muscles innervated by V, VII, IX, and X cranial nerves and hypoglossal spinal neves. For *inspiration* the floor of the buccopharyngeal cavity is depressed by the hypobranchial or hypoglossal muscles and the mouth is opened. Simultaneously, the visceral arches expand the wall of the pharynx and water containing dissolved oxygen enters through the mouth. Water is not allowed to enter through external branchial apertures by an anterior fold of skin or each gill cleft. The mouth is closed and water is forced into internal branchial apertures by the contractions of the pharyngeal wall The oesophagus remains closed and water washes branchial lamellae in the gill cleft and goes out through the external branchial aperatures. The oxygen dissolved in sea water diffuses into the blood through the thin permeable walls of the blood capillaries and the carbon dioxide of the blood passes out into the water. In this way, the gaseous exchange takes place by the respiratory water currents.

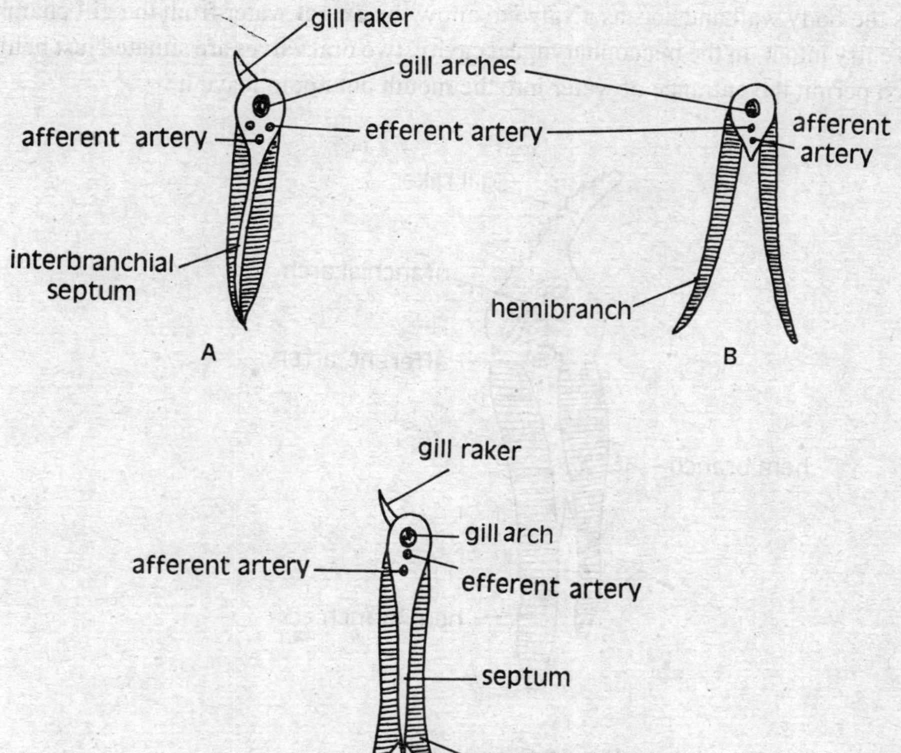

Fig. 6.7 Gills of Teleost fishes

6.3.3. Labeo

The respiratory organs of *Labeo* include five pairs of gill clefts opening into the pharynx by *internal branchial apertures.* The spiracle is absent and the hyoid arch bears no gills. The first four branchial arches have one *holobranch* each and the fifth branchial arch is without gill. The gill clefts do not open to the exterior by separate external branchial apertures. The mucous membrane of the gill clefts is thrown into parallel branchial lamellae. The interbranchial septum present in *Scolidon* in absent in *Labeo* and the branchial lamellae are free at their distal ends (Fig. 6.8). Such gills are called *filiform* or *pectinate* gills. Each branchial arch supports two rows of gill rays, each row supporting one set of gill lamellae. There are four complete holobranchs on either side of the fish. Each holobranch has a supporting visceral arch with two rows of free gill lamellae Comb-like gill raker arise from the visceral arch and project the internal branchial apertures.

The gills lie in the gill chamber situated on the lateral sides of the body. The gill chamber is covered over by an *operculum* formed by the arch which covers the gill clefts. The gill chamber communicates posteriorly by a single crescentic *gill aperture.* The inner surface of the operculum is lined by a branchiostegal membrane supported by *branchiostegal rays.* This membrane extends behind towards the body wall and acts as a valve to allow the exit of water from the gill chamber but prevents its entry into it. In the buccopharyngeal cavity, two oral valves are situated just behind the mouth which permit the entrance of water into the mouth but not to leave it.

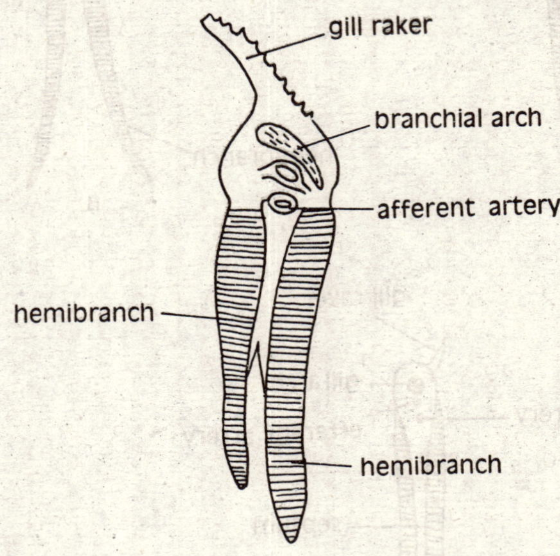

Fig. 6.8 Gill of Labeo

Mechanism of respiration: (Fig. 6.9): The respiration takes place by the lateral movement of the opercula. Reusing of the opercula alongwith closing of gill clefts by branchiostegal membranes results in the creation of vacuum, opening of the oral valves, and entrance of water in the mouth Lowering of the opercula increases pressure on water in the buccopharyngeal cavity, closes oral valves, opens the branchiostegal membranes, and water is forced on the surface of the gills and passes out of the gill chambers through crescentic gill openings. The gaseous exchange takes place by gill lamellae richly supplied with blood capillaries.

6.3.4. Rana

In frog, the respiration takes place by three respiratory organs. These are *skin* for *cutaneous respiration,* buccopharyngeal cavity for *buccopharyngeal respiration* and *lungs* for *pulmonary respiration.* Major part of respiration is done by the skin and the buccopharyngeal cavity, the lungs are used when the oxygen requirement is more.

Cutaneous respiration: The skin functions for the cutaneous respiration in water and on land. It is a major respiratory organ and remains functional during hibernation. The skin is richly supplied with blood vessels and remains moist by mucus or water. The respiratory exchanges take place by atmospheric air in contact with blood caillares of the skin. The oxygen is dissolved in the mucusis diffused into blood capillarises and carbon dioxide is passed out by diffusion. It has been estimated that the intaking of oxygen by skin in frog in more than that of the lungs.

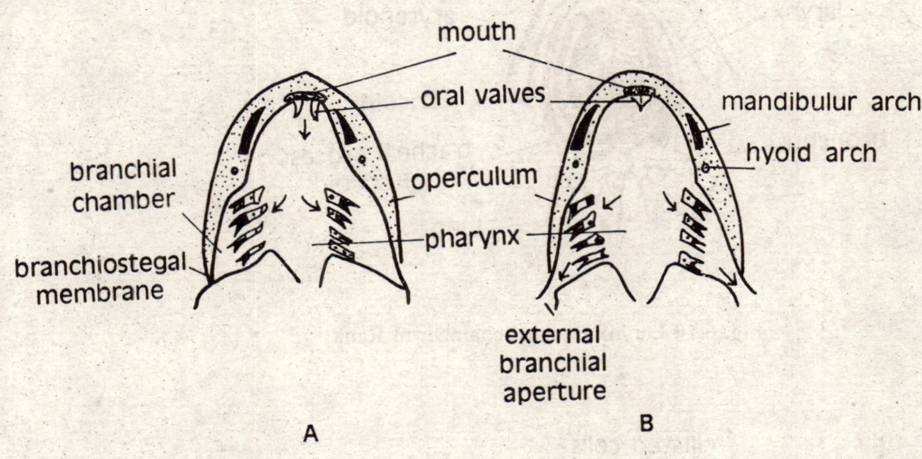

Fig. 6.9 Showing mechanism of respiration in a teleost

Buccopharyngeal respiration: This type of respiration is performed by the rhythmic lowering and raising of the floor of buccopharyngeal cavity. The air containing oxygen enters through the nostrils when the floor is lowered and oxygen is diffused into the blood through the mucous layer.

Pulmonary respiration: The organs connected with pulmonary respiration are internal and external nares, glottis larynx, branchus, and lungs. The glottis leads into larynx which is situated between the posterior cornua of the hyoid. It is supported by a ring-like cricoid and two crescentic arytenoids cartilages surrounding the glottis. The cavity of the larynx has two vocal cords projecting into it which are used for sound production (Fig. 6.10). In male frog, there are two vocal sacs working as resonators and the larynx is larger. For sound production, the air is forced between lungs and air sacs to and fro and croaking sound is thus produced the larynx leads into a short *bronchus* which is connected to two lungs situated in the coelom. The *lungs* (Fig. 6.11) are clastic, hollow sacs having peritoneal epithelial covering, a layer of unstripred muscles and connective tissue; the inner lining is a layer of ciliated epithelial cells with numerous mucous gland cells scattered in it. The wall of lungs has septa which divide its lumen into numerous *alueoli or air spaces.* Blood capillaries travers the septa and the wall of lungs.

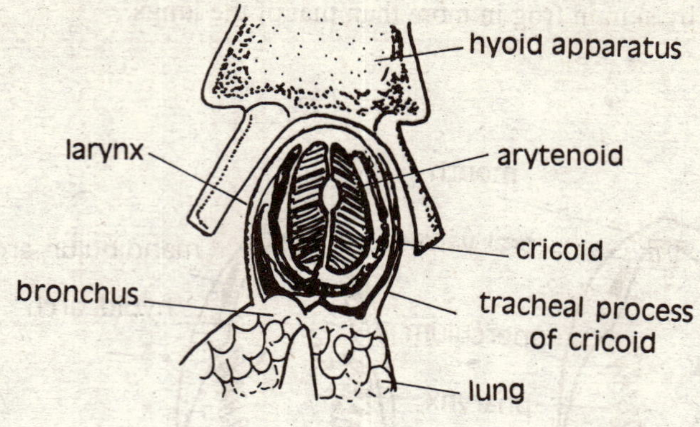

Fig. 6.10 Larynx tracheal chamber of Rana

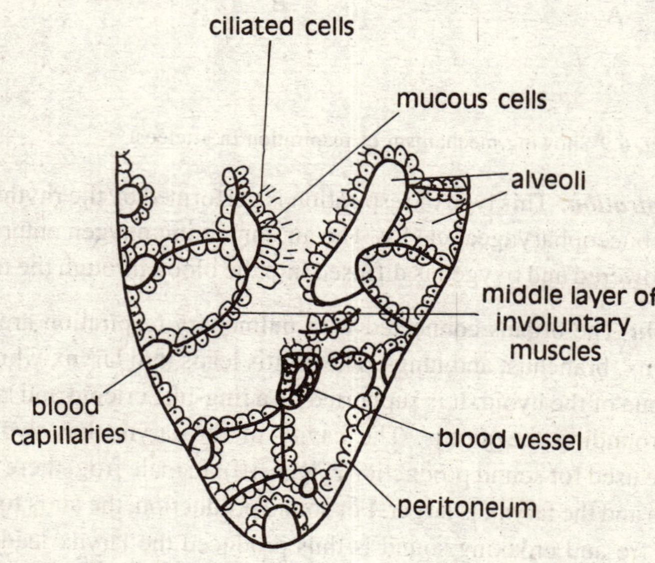

Fig 6.11 Amphibian lung

Mechanism of respiration (Fig. 6.12): The hyoid apparatus present at the floor of the buccal cavity has two pairs of muscle called *sternohyal* and *petrohyal* muscles. The sternohyal muscles contract and lower the hyoid and floor of the buccopharyngeal cavity resulting in the entry of air in the cavity through the *nostrils.* The external nostrils are closed by the tightening of jaws and pushing up of premaxillae. The petrohyal muscles raise the hyoid and the floor of buccopharyngeal cavity. A pressure is thus created forcing the enclosed air into the lungs through the glottis. The oesophagus remains closed during this action preventing the entry of air in the stomach.

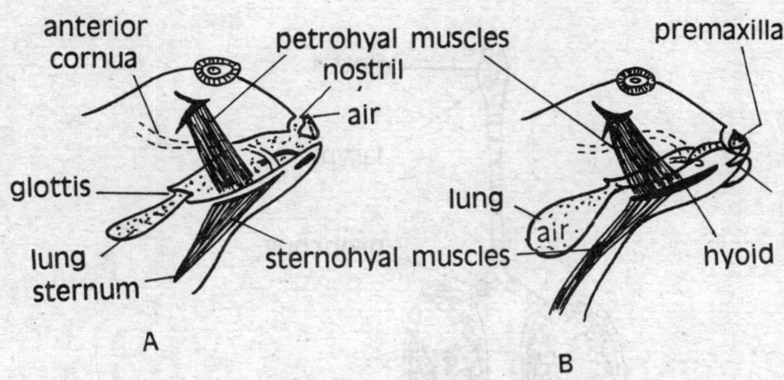

Fig. 6.12 Mechanism of breathing in frog

For expiration, the floor of the buccopharyngeal cavity is lowered by the relexation of sternohyal muscles and air from lungs is partly drawn in the buccopharyngeal cavity and partly pumped by the walls of the lungs. This process is repeated several times and glottis is closed. The raising of the floor of the buccopharyngeal cavity by petrohyal muscles and opening of the external nostrils expels the air.

6.3.5. Uromastix

The skin of *Uromastix* is dry, tough and has horny layers. The respiration is effected by the pulmonary system which includes external and internal nares, glottis, larynx, trachea, branchi and lungs (Fig. 6.13).

The glottis is a slit-like structure opening into a short larynx. The larynx is supported by an incomplete cartilage ring the cricoid and two crescentic arytenoids cartilages. The vocal cords are absent and lizards do not produce sound. Larynx leads into a short trachea. The *trachea* is supported by cartilaginous rings which are complete on the anterior side of the trachea and incomplete dorsally on the posterior part. The trachea is divided on its posterior side into two *branchi.* The bronchi enter into the lungs and do not divide into bronchioles. The lungs are thin walled sac-like structures present in the body cavity. They are unequal in size and have horny comb-like ridges with their cavities divided into chambers by septa. The anterior part of the lung is more sacculated with hollow air sacs or alveoli and the posterior part is smooth without ridges. The bronchi extend to the anterior part of the lungs and posterior part does not participate in respiration.

Mechanism of respiration: The mechanism of respiration in Uromastix differs from that of frog due to the presence of ribs provided with rib muscles. The raising of ribs backward increases the volume of pleuroparietal cavity (coclom) and results in the expansion of the lungs due to reduced pressure. Air enters through nostrils and fills the lungs. Lowering of ribs increases pressure on the lungs and air is expelled out.

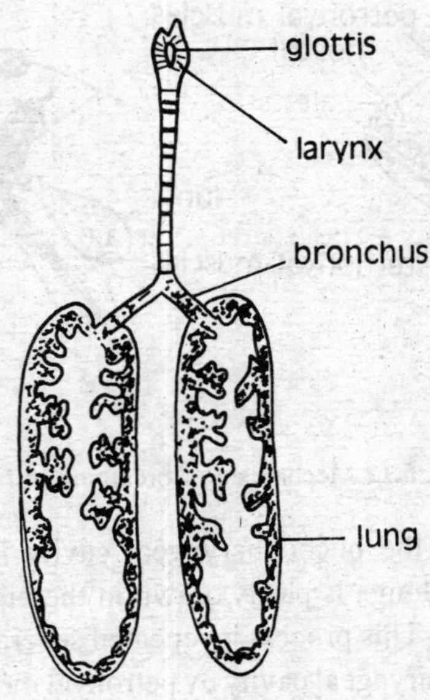

Fig. 6.13 Respiratory organs of Uromastix

6.3.6. Columba

In pigeon the respiration is aerial and the respirator, organs include external and internal nares, glottis, larynx, trachea, syrinx, bronchi, lungs and *air sacs* (Fig. 6.14). The respiratory organs are peculiar in the possession of compact non elastic lungs and air sacs. The external nares are paired openings situated within the cere near the base of the upper beak and communicate with the *single internal nare* which is a single opening at the roof of the pharyngeal cavity. The *glottis* is a slit-like aperture present at the floor of the oral cavity near the base of the tongue and leads to the trachea The trachea is elongated tube starting from the glottis running along the neck on the ventral side of the oesophagus. The trachea is made of long tracheal rings, which are complete. At the anterior side near its beginning, the trachea is produced into the *larynx.* The larynx is a wide chamber supported by *cricoid, cartilage* divided into four pieces and a pair of *arytenoids cartilages.* The larynx does not function for sound production in birds. Larynx opens into a long trachea, which expands into a *syrinx* and then divides into two *bronchi.* The bronchi are supported by rings, which are incomplete on the dorsal side. The first ring of the bronchus is bony. The *syrinx* (Fig. 6.15) is the voice box found in birds only and is formed by the dilation of last three or four tracheal rings and first bony ring of each bronchus, making a resonating chamber called *tympanum* having external and internal *tympaniform membranes.* A cartilaginous ridge, the *pessulus* is present at the junction of bronchi with syrinx, which supports *semilunar membrane.* The syrinx has two intrinsic muscles inserted into it, which lie along the trachea. A pair of sternotracheal muscles is inserted on trachea which arises from the sternum. The sound is produced by the vibration of pessulus, semilunar membrane and probably tympaniform membranes also. The pitch of the sound may be changed by sternotracheal and intrinsic muscles. The bronchi are called *primary bronchi* or *mesobronchi.* Each bronchus enters the lung of its side through a small space *vestibulum,* extends to the posterior extremity of the lung and sends *secondary branchi* giving off a network of *tertiary bronchi* or *parabronchi* which send branches to air sacs (Fig. 6.16). Secondary bronchi of air sacs give small *recurrent bronchi,* which connect the air sacs with parabronchi and the capillaries of the lungs. The tertiary bronchi subdivide again into numerous tubules calloed air-capillaries, which remain in close contact with blood capillaries. The bronchus enters into abdominal air sac by an opening called *ostium.* The lungs are paired, pink coloured structures and small in size in comparison to the body. They are spongy with solid walls with little elasticity and having practically no capacity for expansion. They are situated in the pleural cavity separated from the abdominal coelom by an *oblique septum.* The lungs are attached dorsally to the ribs and thoracic vertebrae. The peritoneal lining is absent on the dorsal surface and the ventral surface is covered over by fibrous peritoneal membrane called *pleura* or *pulmonary aponeurosis.* The pleural wall has special fan-shaped *costopulmonary muscles* originating from the junction of vertebral and sternal ribs.

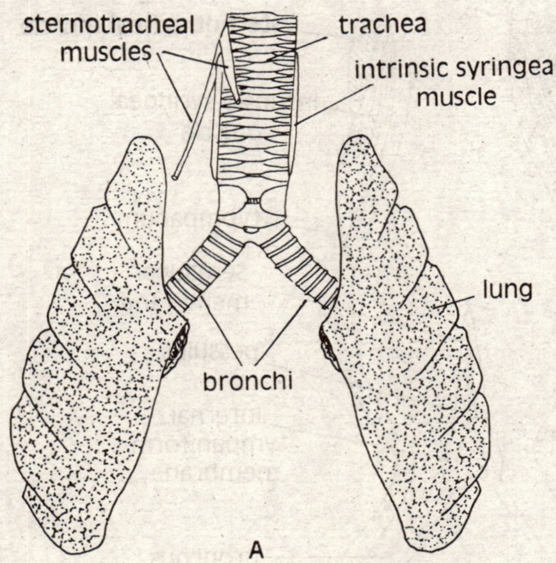

Fig. 6.14 Respiratory organs of *Columba.* **A.** Showing trachea, bronchi and lungs. Longitudinal sectional view of syrinx.

Air sacs: The air sacs are thin walled bladder like sacs formed by the dilation of mucous membrane of the bronchus. They contain poor blood supply and do not receive oxygenated blood.

The air sacs project into the viscera and bones but, do not furnish an increased respiratory surface. There are nine major and four accessory air sacs in pigeon (Fig. 6.17). The major air sacs are four paired and one unpaired. Out of nine air sac of major type; a median unpaired into *clavicular air sac* is connected to the secondary bronchi of both lungs. Two *accessory* or *auxiliary air sacs* originate from each side of the interclavicular air sac, one of which enters humerus bone through *pneumatic foramen*. The cervical air sac arises from the anterior side of each lung, which gives out branches to the neck. *Anterior* and *posterior thoracic air sacs* arise from the sides of each lung, lie below the lungs with their ventral walls covered by the oblique septum. The *abdominal air sacs* arise from the distal end of each lung. The air sacs are not respiratory organs but act as thin reservoirs of air helping in respiration. They act as bellows forcing their air into lungs for ventilation at each expiration to remove unrespired air from the lungs. The interclavicular, cervical and anterior thoracic air sacs are more active during the fight and are considered expiratory in function. The posterior thoracic and abdominal air sacs are more active when the bird is not flying and considered inspiratory . The air sacs provide buoyancy in flight, serve as balloons and reduce the specific gravity of the bird as they contain warm air. They also function for the maintenance of equilibrium of the body, lessening of mechanical friction, regulation and maintenance of the body temperature, containers of reserve air, as resonator and regulation of the water contents of the body.

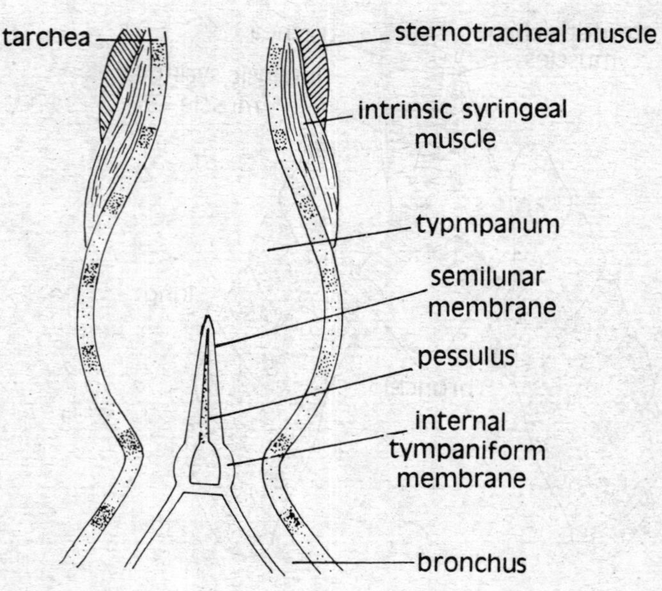

Fig. 6.15 Sound producing organs in pigeon

Mechanism of respiration: In pigeon, improved aeration of lungs is performed by the double supply of the oxygenated air and the respiration is called *double respiration. Inspiration* takes place when fresh air enters into lungs and air sacs through trachea and bronchi and is diffused into the air spaces of the bones. The respiratory exchange takes place only in lungs. In *expiration* the deoxygenated air from lungs goes out and the air from the air sacs enters lungs. In this way, the lungs are filled up again with fresh air from the air sacs. The aeration of blood is complete in pigeon and the respiratory efficiency yields extra energy. The respiration is effected by respiratory movements caused pectoral, intercostals and abdominal muscles. During rest the intercostals muscles induce inspiration and abdominal muscles induce expiration.

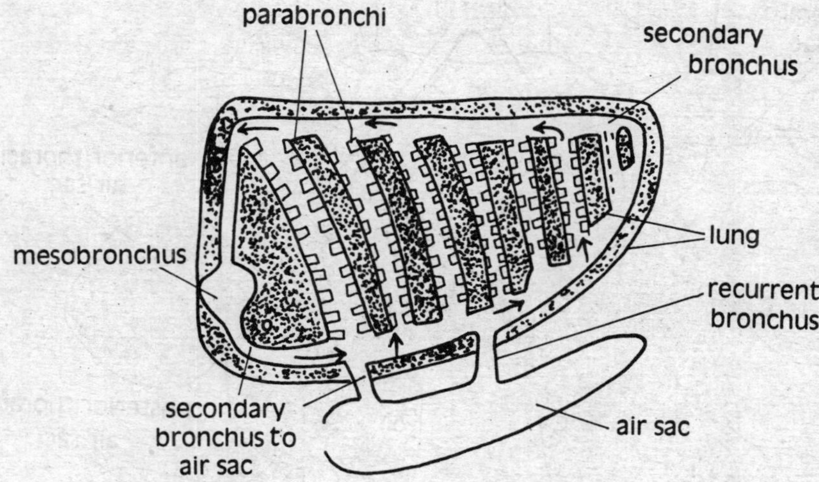

Fig. 6.16 Internal structure of lung and an air sac

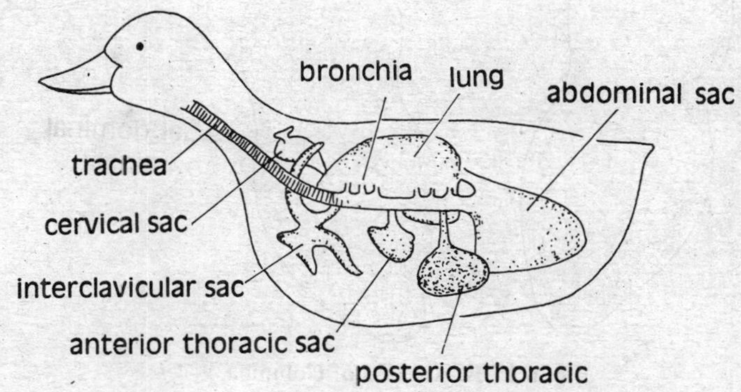

Fig. 6.17 A. Diagrammatic section through bird showing relationship of lung; air sacs and tracheal-bronchial system.

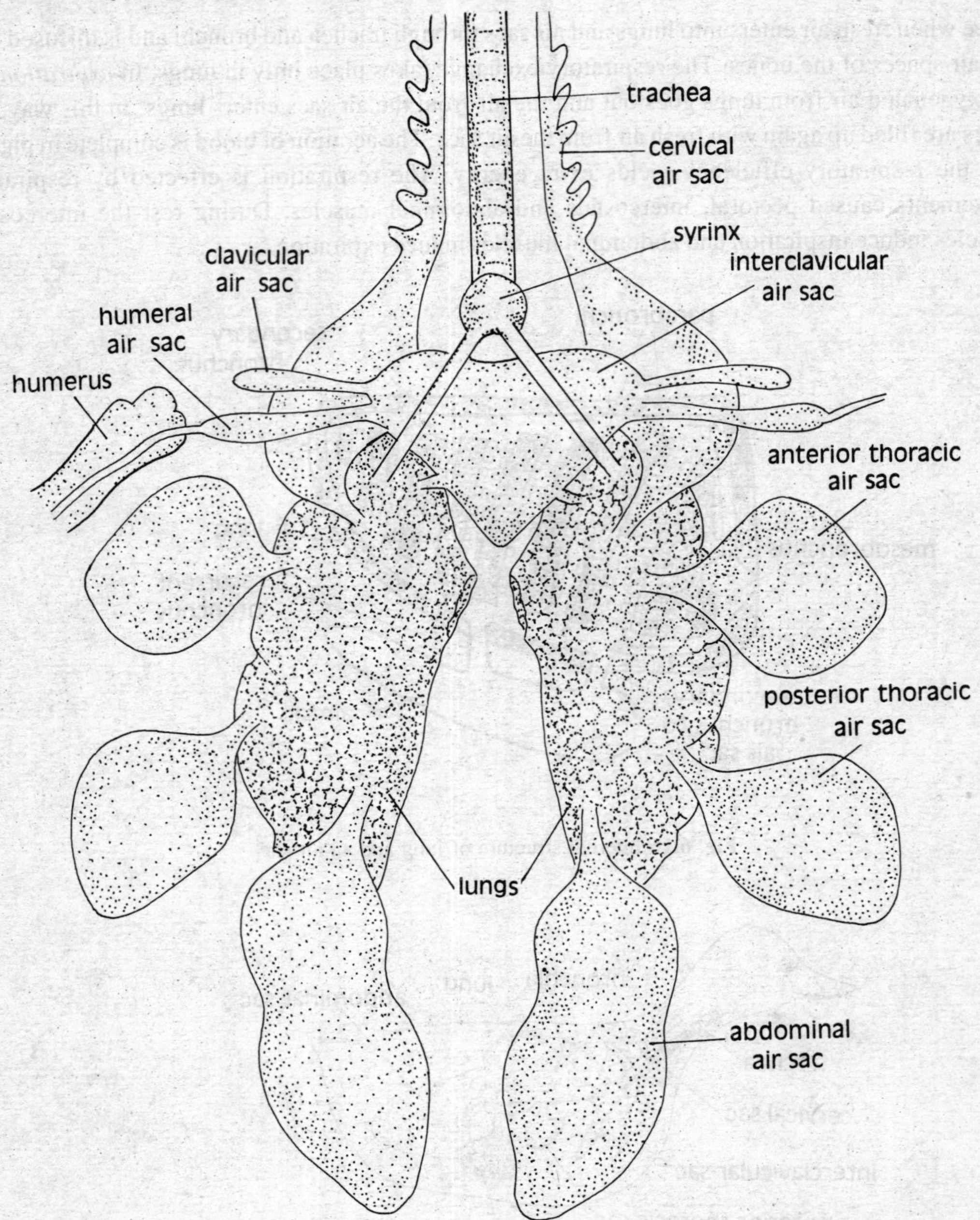

Fig. 6.17 B. Air sacs of Columba.

6.3.7. Oryctolagus

The respiration is aerial in *Oryctologus* and the respiratory organs include external nares, nasal cavities nasal chambers, internal nares, nasopharynx glottis larynx trachea, bronchi, and lungs.

The *external nares* are paired openings present at the lip of the snout. The *nasal cavities* are two in number separated from one another by a *nasal septum.* The *nasal chamber* at each side is situated below the nasal bone and above the hard palate, it is provided with *scroll bones* or *turbinals*, which extends from ethmoid, maxilla and nasal bones. The turbinals are covered over by a ciliated mucous secreting epithelium with rich blood supply. The nasal chamber and respiratory passage warm, the air before it enters the lungs filter the air holding the dust particles in the mucous cavity, and help in smelling due to the presence of olfactory epithelium. The nasal chamber communicate with the internal nares behind the soft palate by a respiratory tube having glandular ciliated epithelial lining. The internal nares are paired internal openings present on the nasopharynx part of the pharynx. The *glottis* is a slit like opening at the base of the buccopharyngeal cavity, which communicates with the larynx.

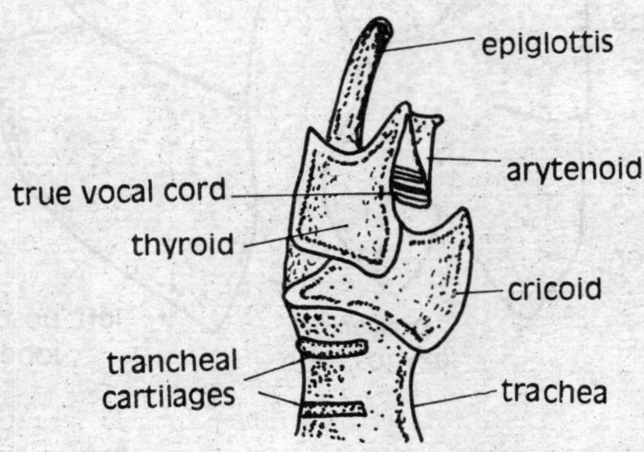

Fig. 6.18 Mammalian larynx

The larynx (Fig 6.18) is supported by four cartilages. The largest cartilage is *thyroid cartilage* situated on the ventral side and is incomplete dorsally. A *cricoid cartilage* makes a ring at the base of the larynx. Two small, *arytenoids cartilages* are present on the dorsal side of the criocoid. An clastic cartilaginous structure *epiglottis* is present in front of the glottis and arises from the ventral edge of the thyroid cartilage; it prevents the entry of food in the glottis. A pair of clastic membranous, *true vocal cords* and another pair of *false vocal cords* are present inside the larynx, which help, in sound production. The larynx leads to the trachea. The *trachea* or *wind pipe* is a large tube, which passes through neck and enters the thoracic cavity where it divides into two bronchi (Fig. 6.19). It is supported by cartilages, which are incomplete dorsally. It is lined with ciliated epithelium. Each *bronchus* enters the root of a lung and divides into numerous *bronchioles* inside the lungs. The brochi are supported by incomplete cartilaginous rings. The bronchioles do not have rings of cartilage. Each bronchiole divides into smaller tubes called *alveolar ducts.* Each alveolar duct ends in a blind *infundibulum.* The walls of the infundibulum form hollow air sacs or alveolar sacs each bearing main minute *alveoli.* Each alveolus has a network of capillaries of the pulmonary artery and vein around it. The alveoli provide immense respiratory surface for the exchange of gases (Fig. 6.20).

The *lungs* are spongy masses of tissue without hollow cavities, and have greatly increased internal surface. The left lung has *two lobes* and right lung has *four lobes* (Fig. 6.19). The left lung lobes are left anterior and left posterior, and right lung lobes are *anterior azygos, right anterior,* and *right posterior* and posterior azyos. The respiratory surface is of ectodermal origin and rest of the lung is mesodermal.

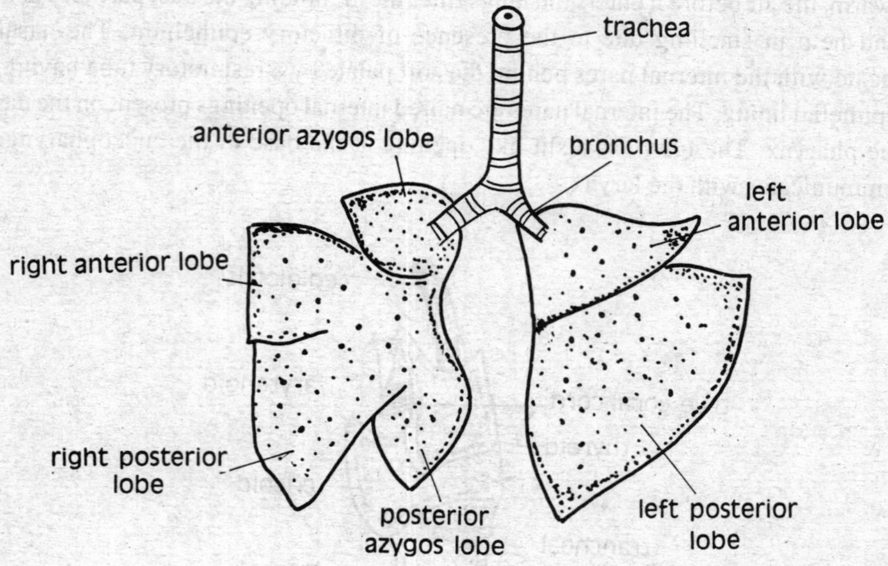

Fig. 6.19 Mammalian lungs

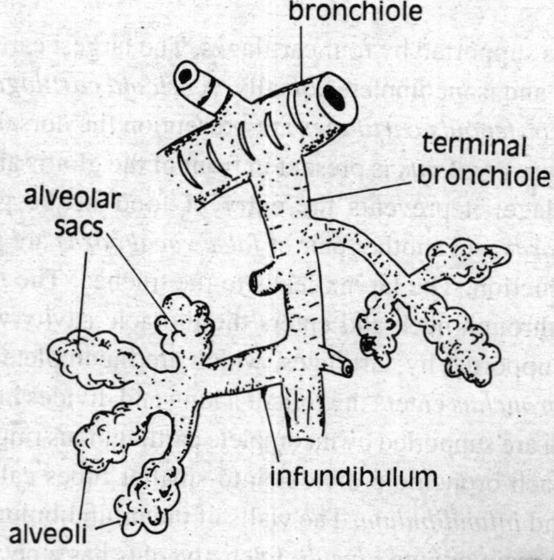

Fig. 6.20 Bronchioles and alveoli

***Mechanism of Respiration** (Fig. 6.21):* The thoracic cavity is separated from the abdominal cavity by a muscular dome shaped diaphragm. The lungs are enclosed in pleural cavities, which are in contact with the diaphragm Any change in the shape of pleural cavities brings similar change in the lungs. *Inspiration:* The central dome of the diaphragm moves backwards and the diaphragm becomes flat. The external intercostal muscles attached to ribs elevate the ribs. In this way. The pleural cavities increase in volume and lungs are expanded resulting into inspiration. *Expiration* The decrease in the pleural cavities is brought about when the diaphragm moves to its original domeshaped condition and the internal intercostals muscles bring the ribs inwards to their original position. In addition to this, the elasticity of the walls of the expanded lungs pushes the air out of the nostrils. The lungs are partially emptied during the respiratory movements and some residual air is left inside the alveoli, which is called "dead space".

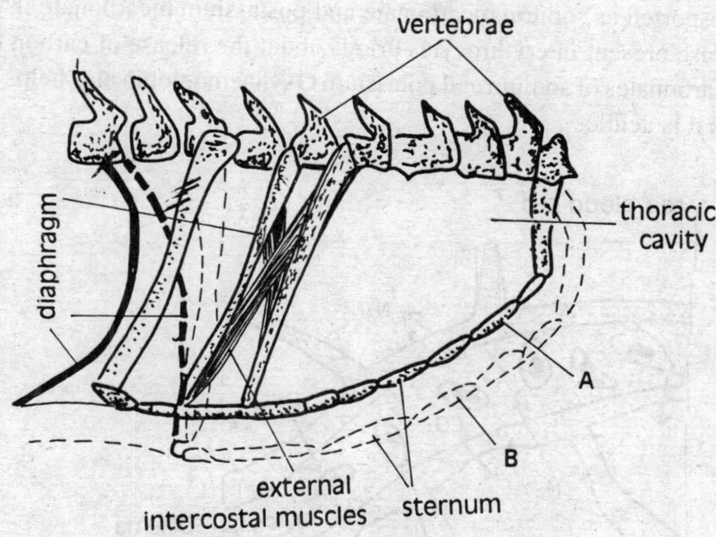

Fig. 6.21 Mechanism of Breathing

6.4. MECHANISM OF RESPIRATION

The main respiratory organs of Vertebrates are internal gills or lungs. Other structures such as external gills in fishes and ampibia, air bladder in some fishes, mucous membrane of oral cavity in amphibia and skin in amphibia etc. also help in respiration. Details of respiratory movements have been given elsewhere in this chapter.

The respiration, which involves taking of oxygen and passing out of the carbon dioxide from lungs or other respiratory organs, is called external respiration. The internal respiration or tissue respiration takes place inside tissue cells.

6.4.1. External Respirations

The respiratory system takes care of the external respiration in which oxygen is taken in by the blood from the environment and carbon dioxide is given out to the environment (Fig. 6.22). This part of respiration takes place in the blood capillaries of the respiratory organs (lungs, gills, or skin). A high perventage of haemoglobin, a respiratory pigment is found in the red blood corpuscles (erythrocytes) of the blood. A high concentration of oxygen is present at the respiratory surfaces and the haemoglobin combines with it to form oxyhaemoglobin. The oxyhaemoglobin gives off most of its oxygen to the tisues during its transportation by the blood. The blood plasma has higher concentration of oxygen than in tissues and continuous diffusion of oxygen from blood plasma to the tissues takes place. The carbon dioxide present in high concentration in tissues after respiration is passed on to the exterior by the blood through respiratory surfaces. About ten percent of carbon dioxide combines with water to form carbonic acid, which is transported by the blood plasma. Another ten per cent is carried as bicarbonate in combination with blood proteins, and remaining eighty percent is transported as sodium bicarbonate and postassium bicarbonate in erythrocytes. The carbonic anhydrase present in erythrocytes brings about the release of carbon dioxide from carbonic acid and bicarbonates of sodium and potassium Oxyhaemoglobin also helps in the release of carbon dioxide, as it is acidic.

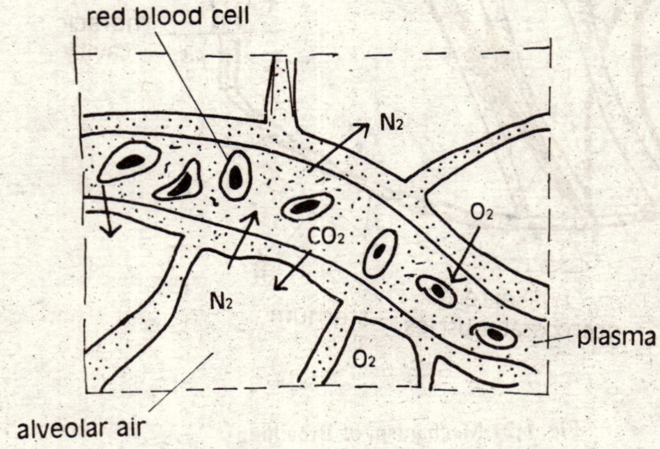

Fig. 6.22 exchange of gases in lungs

6.4.2. Internal Respiration (Tissue Respiration)

It occurs in tissue cells (Fig. 6.23) and results in the liberation of carbon-dioxide water and energy. This process is brought about by complicated chain of chemical reactions by respiratory enzymes present in the protoplasm. These enzymes are cytochrome, dehydrogenase, and oxidase. In presence of oxygen obtained from blood, the cytochrome is oxidized by oxidase. Dehydrogenase produces hydrogen from glucose present in the protoplasm. The hydrogen combines with oxidised cytochrome and forms water, carbon dioxide and energy. The water is set free in the cells, the energy is utilized by tissues for various activities, and carbon dioxide is released out through respiratory surfaces.

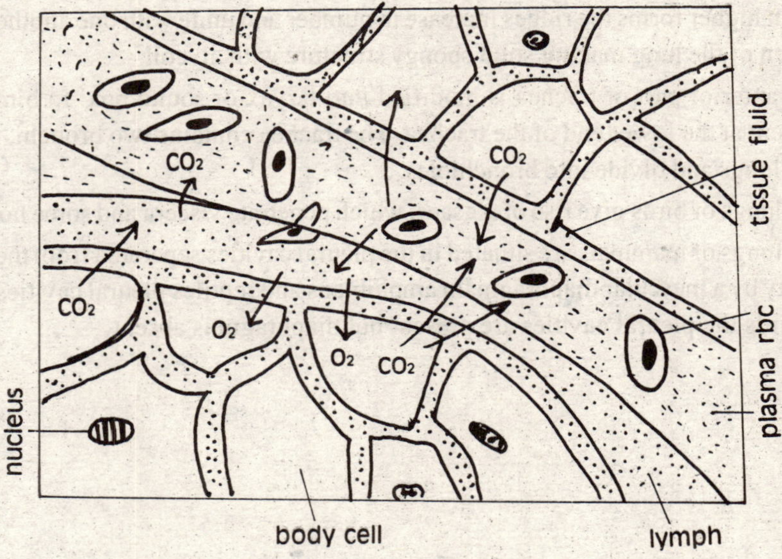

Fig. 6.23 Showing Tissue Respiration

CHAPTER SUMMARY

1. The main respiratory organs of vertebrates are internal gills and lungs. Some other organs such as external gills in some fishes and amphibians, air bladders in some fishes, mucous membrane of oral cavity and pharynx in amphibia, skin in amphibia, intestinal epithelium of certain silurid fishes and extrabranchial respiratory organs of certain fishes also help in respiration.
2. The process of respiration is effected by branchial or pulmonary systems. The respiratory organs are mainly formed in connection with the pharynx; they are lined or covered by epithelia derived from the endoderm.
3. The gill slits are found in adult condition in protochordates, cyclostomes, fishes and some amphibians and in higher vertebrates in embryonic stages. External gills, covered over by ectoderm usually project from the outer surface of visceral arches. Internal gills are supported by visceral arches and form vascular or lamellar extensions of the epithelial surface of the gill pouches.
4. Some fishes are provided with accessory or extrabranchial respiratory organs to meet extra requirement of oxygen and survival outside water for short or long duration. These organs may be labyrinthiform in *Anabas*. suprabranchial in *Clarias*, pharyngeal diverticulae in snake headed els and air chambers in *Channa*.

5. The swim bladder of fishes is homologous to lungs. It is used for respiration in some fishes but mostly serves as hydrostatic organ.
6. The lungs are found in most adult amphibians, amniotes and some fishes. They arise as a diverticulum from the ventral side of the pharynx. In lower animals they are hollow but in higher forms the ridges increase in number and unite with one another across the lumen of the lung making solid spongy structure with alveoli.
7. The anterior part of trachea is modified into larynx or sound box. In birds, syrinx is present at the lower end of the trachea. The trachea contains two bronchi, which enter into lungs and divide into bronchioles.
8. The lungs of birds give rise of air sacs, which penetrate viscera and some hollow bones.
9. The lungs of mammals are situated in the pleural cavities separated from the abdominal cavity by a muscular diaphragm. In amphibians and reptiles pleural cavities are absent. In birds the pleural cavities are present but diaphragm is absent.

7

Circulatory System

7.1. GENERAL ACCOUNT

In complex animals where organs and tissues are separated from the external surface and alimentary canal, a system of transporting oxygen and food materials and elimination of waste materials is essential for metabolic processes. The circulatory system serves such requirements and consists of blood vascular system and lymphatic system. The term circulatory system is used for blood vascular system which consists of:

1. The blood having plasma fluid and blood corpuscles, which serve as the conveying medium for food and oxygen throughout tissue and organs.
2. The heart with muscular parts, which contract periodically and pump the blood to different part of the body.
3. A system of elaborately branched tubes through which the blood passes.

The circulation of blood was demonstrated in 1616 by William Harvey for the first time. Malphighii (1661) discovered the interconnection of arteries and veins by capillaries. The structure and course of circulation show variation in different groups of vertebrates, which are functional and have no connection with the phylogeny of the group.

7.2. THE VERTEBRATE BLOOD

The blood of vertebrates has plasma, white blood carpuscles, red blood corpuscles, and platelets.

7.2.1. Plasma

The plasma is almost colourless and carries dissolved food, waste materials, internals secretions, and some gases. It contains *serum* and fibrinogen. The *fibrinogen* becomes changed into solid fibrin during the clotting of the blood. The plasma contains proteins, enzymes, antibodies, antithrombin, salts, hormones etc. Human blood plasma contains about 92 per cent water, proteins and other organic compounds, and about 0.9 per cent inorganic salts, mainly sodium chloride.

7.2.2. The White Blood Corpuscles Or Leucocytes

These are of several kinds which have main activity in tissues and originate in bone marrow, spleen, or lymphoid structures and pass out to different tissues through blood. They are colourless, amoeboid, single or multinucleate corpuscles and may pass through thin endothial linings of capillaries. The leucocytes have phagocytic properties and may ingest or destroy bacteria and thus protect the body. There are about 8,000 leucocytes per cubic millimeter in the blood of man. The average life of a white blood corpuscle is 12 to 13 days. The leucocytes are of two types, *granulocytes* and agranulocytes. The granulocytes may be (1) polymorphonuclear cells or neutrophils (2) eosinophils or acidophils and (3) basophils. The agranulocytes may be (1) lymphocytes and (2) monocytes of large and small sizes.

Granulocytes: These are granular white blood corpuscles with granular cytoplasm and tabulated nucleus. The acidophils or eosinophils are spherical with bilobed nucleus. The basophils have elongated S-shaped nucleus. The *heterophils* or *neutrophils* are spherical highly polymorphic with cytoplasm differentiated into nongranular peripheral and granular inner lavers.

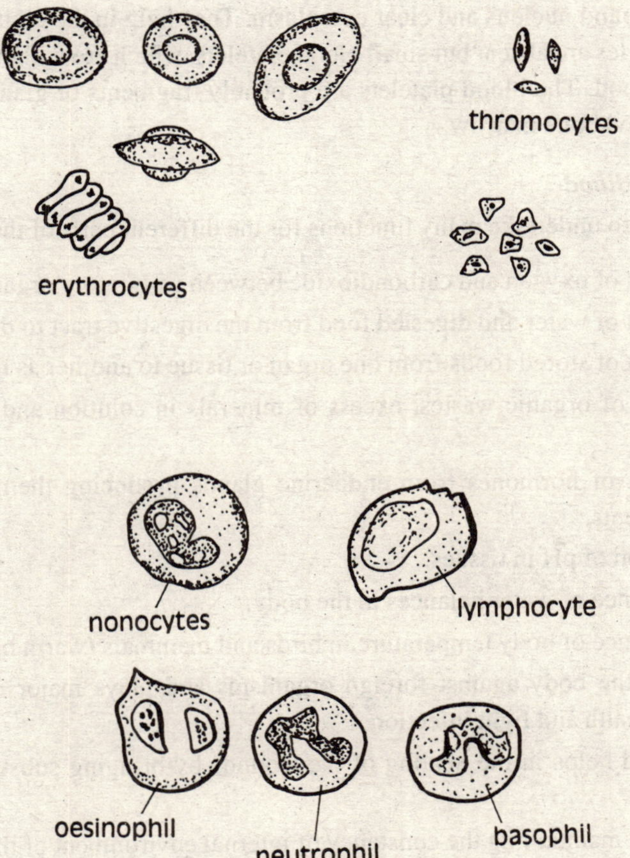

Fig. 7.1 Various blood corpuscles.

Agranulocytes. These are nongranular with homogenous cytoplasm without granules. The lymphocytes are small cells with relatively large spherical nucleus. The monocytes have small oval or kidney-shaped nucleus and large amount of cytoplasm. The large monocytes are called mononuclears or macrophages and medium or small cells are monocytes or transitional leucocytes.

7.2.3. The Erthrocytes or Red Blood Corpuscles

There are nucleated and oval in all vertebrates. In mammals, they are nucleated in developmental stage but non-nucleated, biconcave and circular in adults. In camels, they are oval. There are about

5 million erthrocytes in man and 4.5 million in woman per cubic millimeter. They are produced chiefly in red bone marrow and their excess supply is often stored in spleen. The old cells are also destroyed in spleen.

7.2.4. Thrombocytes and Platelets

There are spindle-shaped cells found in most vertebrates excepting mammals. They are smaller in size with oval or round nucleus and clear cytoplasm. They help in the clotting of the blood. In mammals, thrombocytes are absent but small blood platelets, which are without nucleus, work for the clotting of the blood. The blood platelets are probably fragments of giant cells called megakaryocytes found in red bone marrow.

7.2.5. Functions of Blood

The blood serves to undertake many functions for the different parts of the body. These are:

(i) Transport of oxygen and carbondioxide between respiratory organs and tissues.

(ii) Transport of water and digested food from the digestive tract to different organs.

(iii) Transport of stored foods from one organ or tissue to another as required.

(iv) Removal of organic wastes, excess of minerals in solution and water to excretory organs.

(v) Transport of hormones from endocrine glands producing them to places of their requirements.

(vi) Regulation of pH in tissues.

(vii) Maintenance of water balances in the body.

(viii) Maintenance of body temperature in birds and mammals (warm blooded animals).

(ix) Defends the body against foreign organisms and plays major role in maintaining normal health and fight infection.

(xi) The blood helps in the healing of the wounds by bringing substances to the site of injury.

(xii) It helps in maintaining the constancy of internal environment of the body.

7.2.6. The Clotting of Blood

The blood coining out of a cut is stopped soon by a protective *clot* within seconds. The platelets stick the edges of the cut and form a plug. *Thromboplastin* (thrombokinase) is released by injured tissues and disintegrating blood platelets, which combine with calcium ions present in the plasma, and act upon prothrombin of the blood and produce thrombin. A soluble blood protein fibrinogen is converted into *fibrin* by thrombin. The fibrin is converted into a mass of fine fiberes and entangles corpuscles to form a clot. The blood flowing in the blood vessels is prevented from clotting and formation of therombin by *heparin.* The clotting of blood may be illustrated by the following.

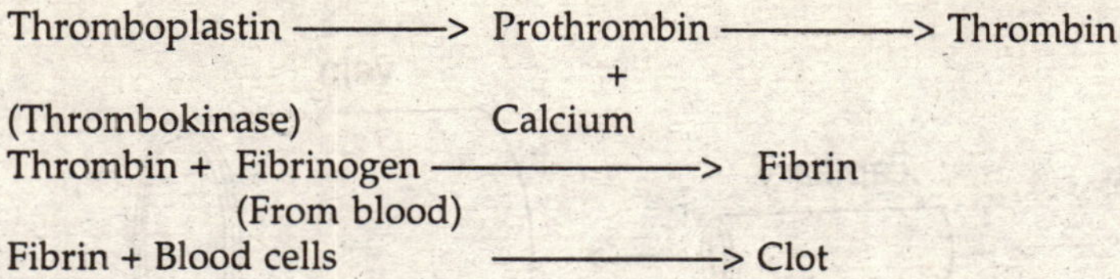

Thromboplastin ———> Prothrombin ———> Thrombin
(Thrombokinase) + Calcium
Thrombin + Fibrinogen (From blood) ———> Fibrin
Fibrin + Blood cells ———> Clot

7.3. BLOOD VESSELS

In vertebrates, the blood vascular system is closed type in which the blood flows from heart to blood vessels and returns to it. The blood vessels are arteries and veins, which are interconnected with one another by thin capillaries. The blood vessels are derived from the embryonic mesenchyme The inner lining of the blood vessels is called *endothelium.* The blood vessels are as follows: (1) heart; (2) arteries: (3) capillaries and (4) veins. The heart will be discussed in section 7.4.

7.3.1. Arteries

The *arteries* are the blood vessels which carry blood from heart to the body tissues. Each artery is made up of three layers, *tunica adventitia, tunica media,* and *tunica intima.* The outermost layer is tunica adventitia or tunica externa made of connective tissue having collagen fibres and longitudinal clastic fibres. The tunica media is thick middle layer made of circular smooth muscles and clastic fibres The tunica interna or intima is innermost layer made of endothelial cells and clastic membrane.

7.3.2. Veins

The veins are mainly responsible for returning blood to the heart. They are also made of three coats like arteries but their walls are thinner. The diameter of a vein is much larger than that of a corresponding artery. The tunica and some clastic fibres form valves in large veins to prevent backward flow of the blood.

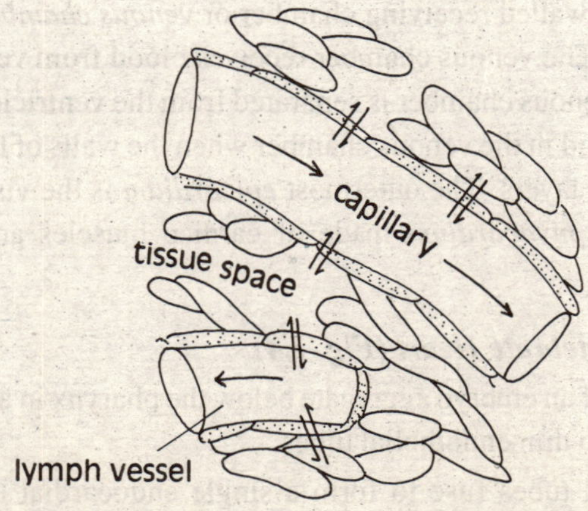

Fig. 7.2 **A.** Diagram of relationship between blood capillary, tissue space and lymph vessels. Arrows show flow of blood and lymph; double arrows show diffusion through membrane.

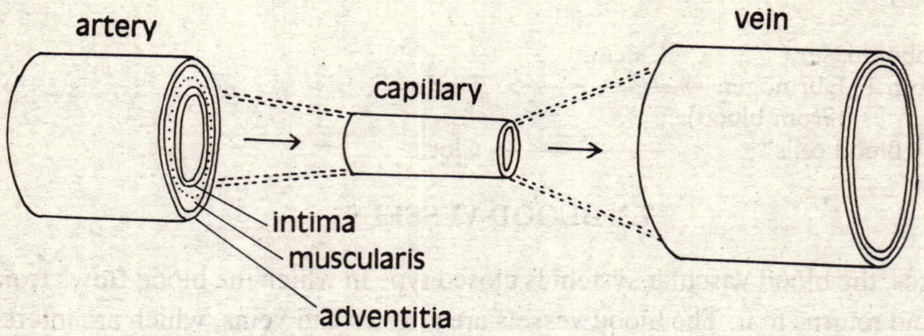

Fig. 7.2 B. Differences between artery, capillary and vein and continuity of lining.

7.3.3. Capillaries

The capillaries are minute microscopic vessels having single layered walls made of endothelium, which allow the exchange of materials with tissues. The lumen of the capillaries is slightly more than the blood cells so that the blood cells may pass through them in single rows. The capillaries are very important part of the circulatory system and make complicated network connected with arteries and vein. There are about 400 capillaries in per square millimeter in frog muscles and 1 cc of blood passes through about 2,700 sq. mm of capillary surface. In man the capillaries of skeletal muscles may have a total length of 60,000 miles and surface area of about 1½ acre.

7.4. THE HEART

The circulation of blood is effected by the heart which is a modified blood vessel meant for the efficient circulation of the blood.

The heart works as a muscular pumping device like a force pump (Fig. 7.3). A generalized vertebrate heart is present in the pericardia cavity whose walls are made of *pericardium.* It is made of two types of chambers, a thinwalled receiving chamber or *venous chamber* and a thick walled muscular chamber or *ventricle.* The venous chamber receives blood from veins and ventricle and forwards it to the arteries. The venous chamber is separated from the ventricle by valves, which do not permit backward flow of blood in the venous chamber when the walls of the ventricle contract. The heart is provided with three layers. The outermost *epicardium* is the visceral pericardium of fibrous nature, the middle one *myocardium* made of cardiac muscles and the inner most is *endocardium* or *endothelium.*

7.4.1. The Development of Vertebrate Heart (Fig. 7.4)

(1) The mesenchyyme of an embryo aggregate below the pharynx in a group of endocardial cells which form two thin endothelial tubes.

(2) The two endothelial tubes fuse to form a single endocardial tube lying below the pharynx longitudinally.

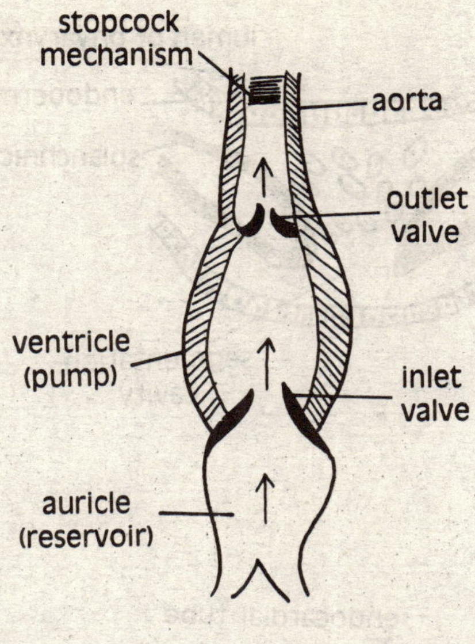

Fig. 7.3 A. Simplest vertebrate heart

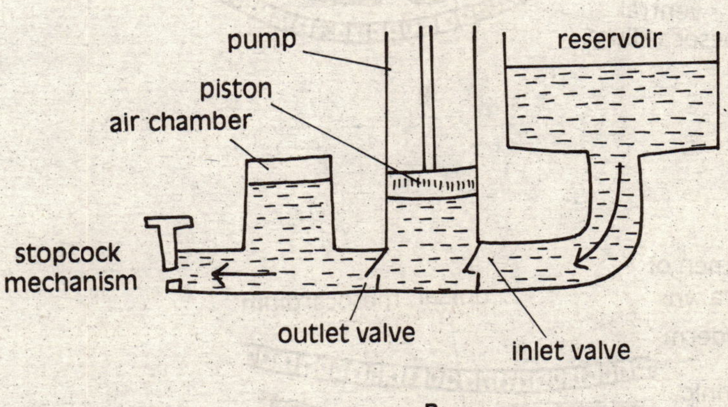

Fig. 7.3 B. Basic principle of circulation

(3) The splanchnic mesoderm present below the endoderm makes longitudinal folds around the endocardial tube.

(4) The two layered tube thus formed eventually gives rise to heart. The splanchnic mesoderm thickens to form myocardium and epicardium, and endocardial tube makes the lining of the heart of endocardium.

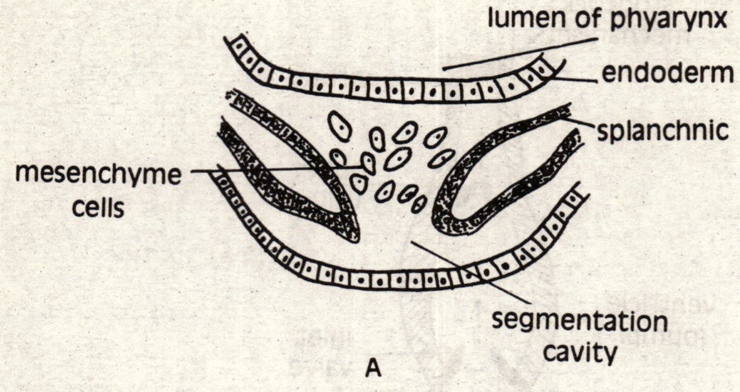

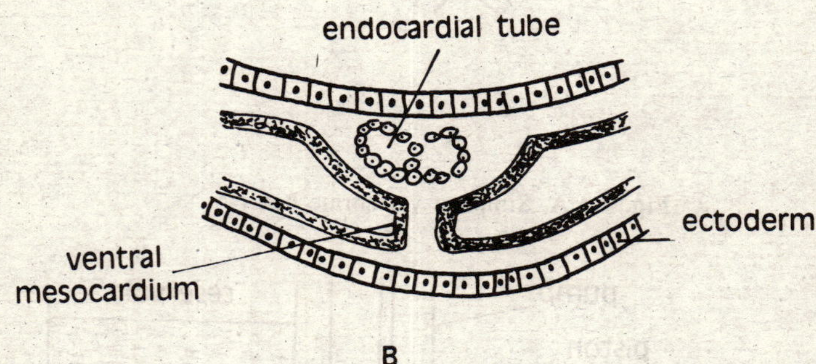

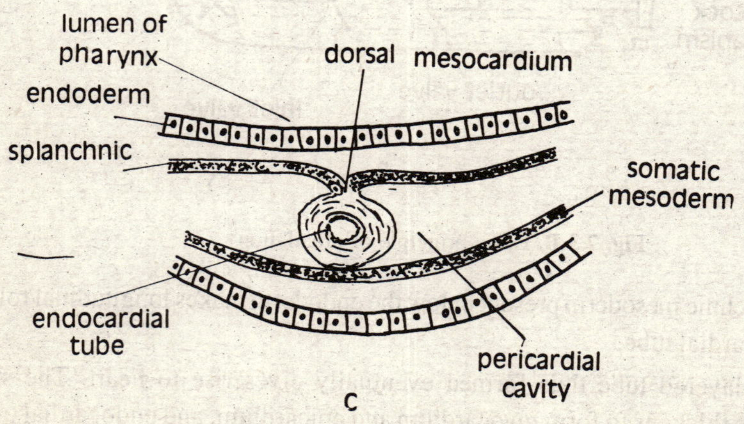

Fig. 7.4 A, B, C Development of Heart

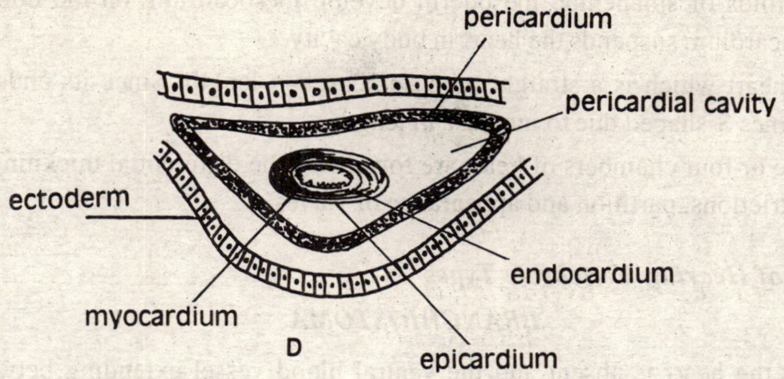

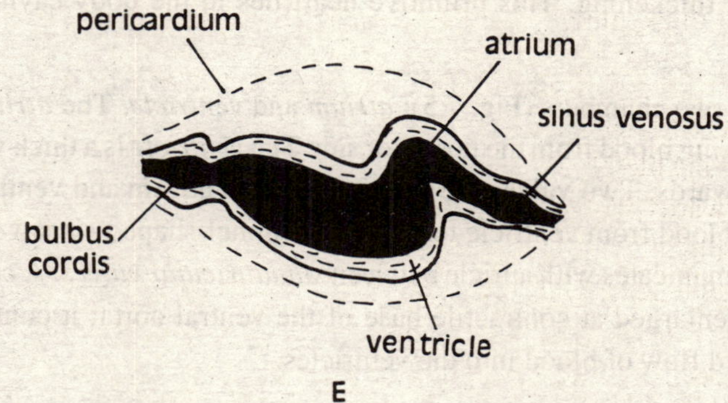

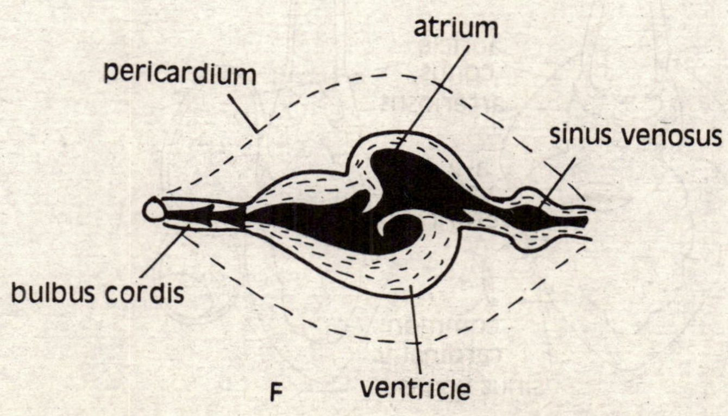

Fig. 7.4 D, E, F Development of Heart

(5) The folds of splanchnic mesoderm develop mesocardium on the dorsal side. The mesocardium suspends the heart in body cavity.

(6) The heart which is a straight tube, increases in length. Since its ends are fixed, it becomes S-shaped due to increase in length.

(7) Three or four chambers of heart are formed by the differential thickning of its walls, constrictions, partition and appearance of valves.

7.4.2. Structure of Heart in Vertebrate Types

BRANCHIOSTOMA

In *Branchiostma* the heart is absent and the ventral blood vessel extending between the liver diverticulum and gills is contractile and forces the blood to the anterior side.

In cyclostomes the heart is represented for the first time in the larval amocoetes stage of lampreys. The ventral aorta lying between the liver and gills is specialized by enlargement, constrictions and unequal thickening. This primitive heart lies in the body cavity with visceral organs.

In fishes the heart has two chambers (Fig. 7.5); *atrium* and *ventricle.* The *atrium* or auricle is thin walled chamber receiving blood from the posterior side The *ventricle* is a thick walled chamber and propels the blood forwards. Two valves are present between atrium and ventricle and do not permit backward flow of blood from ventricle to atrium. A funnel-shaped *sinus venosus* receives blood from veins and communicates with auricle between *sinuauricular valves.* A *conus arteriosus* lies near ventricle and is enlarged at contractile base of the ventral aorta; it contains semilunar valves to prevent backward flow of blood into the ventricles.

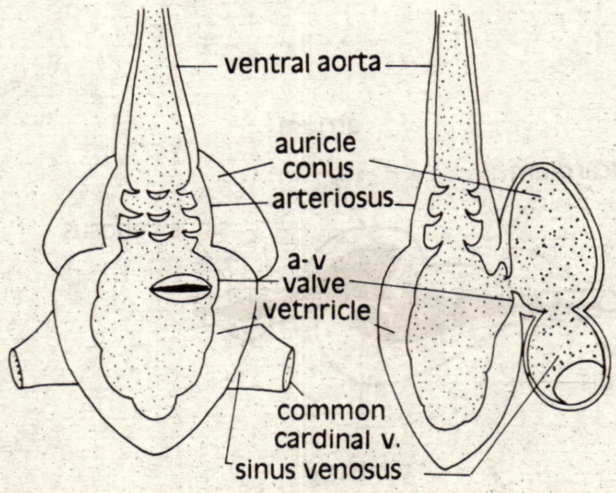

Fig. 7.5 Heart of shark, in frontal and sagittal sections

The pericardial cavity is not separated from the abdominal cavity in clasmobranches and a *pericardioperitoneal canal* communicates with abdominal coelom The bony fishes, the two cavities are separated completely.

SCOLIODON

The heart of *Scoliodon* is enclosed in pericardium and consists of sinous venosus, one auricle, one ventricle and conus arteriosus (Fig. 7.6). It is situated on the ventral side of the body in the region of the gill slits. The *sinous venosus* is thin walled, tubular, highly contractile, chamber from where the beating of the heart starts. It receives two large veins *ductus cuveri* opening on each lateral side and two hepatic sinuses entering posteriorly. The sinous venosus opens into the auricle by *sinuauricular aperture* guarded by two valves. The auricle is thin walled, large triangular chamber on the dorsal side of the ventricle in front of the sinus venosus. It communicates with the ventricle by auriculo-ventricular aperture having two valves. The ventricle has a thick muscular wall with many muscular strands on its inner surface giving it spongy structure. The *conus arteriosus* is a median muscular tube starting from the ventricle and its lumen bears two transverse rows of semilunar valves. The free ends of the valves are attached to the ventricular wall by fine tendinous threads called *chorda tendianae,* which keep the valves in position. The conus arteriosus continues forwards into the *arota*. The heart of *Scoliodon* receives deoxygenated blood from all parts of the body and sends it to gills for respiration. Since its function is to circulate deoxygenated blood, it is called *branchial heart* or venous heart.

LABEO

The heart of *Labeo rohita* is of general teleostean plan (Fig. 7.7). It venous heart made of sinus *venosus*, an *auricle*, and a *ventricle*, thus exhibiting typical piscian condition. The conus arteriosus is represented by a pair of valves. The sinus venosus bears a pair of lateral appendages and is large and spongy chamber.

RANA

The heart of frog (Fig. 7.8) is three chambered conical structure lying in the pericardinal cavity. It is enclosed in two membranes an outer *pericardium* and an inner *epicardium.* The pericardial or serous fluid lies between the two membranes. Two auricles are formed by the division of the atrium and are separated by an *inter-auricular septum* of connective tissue and endothelium. The sinus venosus is thin walled and opens into the right auricle, its opening is controlled by two *semilunar valves.* A pulmonary vein opens into the left auricle and is not guarded by valves. The two auricles open into a ventricle by an auriculo-ventricular aperture having two pairs of auriculoventricular valves, two of these valves are provided with *chorda tendinae* joining them of the wall of the ventricle.

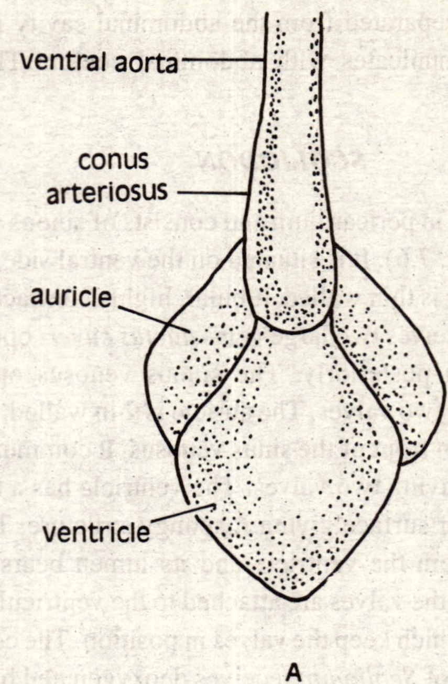

A

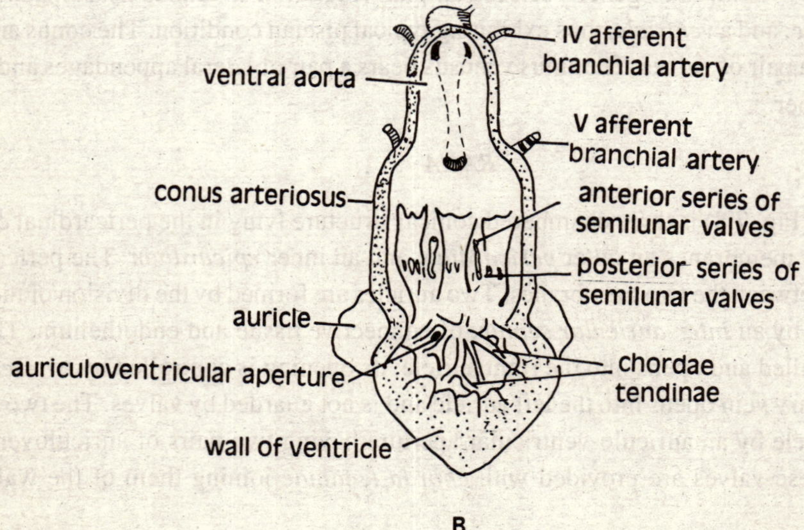

B

Fig. 7.6 A, B Heart of Scoliodon

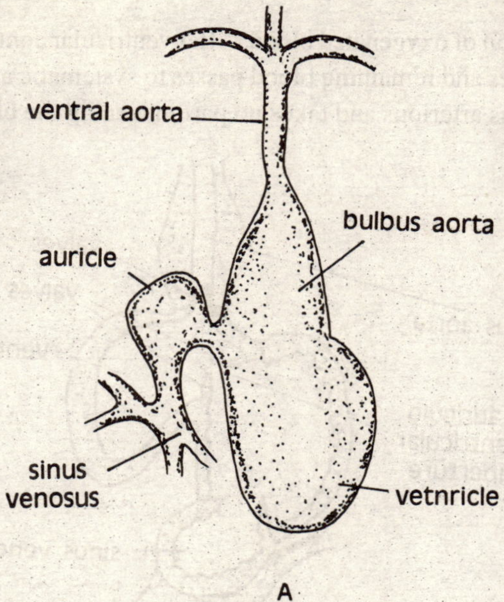

Fig. 7.7 A. Heart of Labeo

These valves prevent a reflex of blood into auricles. The walls of the ventricle are thick and muscular with ridges or *columnae carnae* having deep pockets in between which prevent the mixing of the blood from the auricles to some extent. A *truncus arterious* arises from the upper side of the ventricle and its opening is having three semilunar valves preventing the backward flow of the blood into the ventricle. The truncus arterious has two chambers; a *conus areteriosus* or *pylangium* and a *bulbs arterious* or *synangium* (modification of the ventral aorta of fishes). One set of *semilunar valves* is present at the junction of pylangium and synangium and one of these valves is modified to form a large spirally twisted spiral valves which divides the conus arterious into two passages. These are a *cavum pulmocutaneum*, and a *cavum aortium*. The cavum pulmocutaneum has an aperture leading to two pulmonary arches. The truncus arterious then divides into two halves, each half having a carotid arch, a systematic arch, and a pulmonary arch.

Working of heart: The heart beat is initiated at *sinuaricular node* situated at the wall of the sinus venous. The heart beat passes to the auricles. The circulation of blood is effected by the periodic heart beats. Contraction (systole) and relaxation (diastole) are continually taking place. The blood from sinous venosus comes to the right auricle and from lungs to the left auricle through pulmonary veins. The auricles contract simultaneously and push their blood into the ventricle. When ventricle contracts the blood is pushed into aortic arches through the truncus arterious. In frogs, the respiratory organs are lungs, skin, and buccal cavity. The oxygenated blood from lungs comes to left auricle. The blood coming from musculotataneous vein is mixed and from the buccal cavity is oxygenated; this blood enters the heart through sinus venosus into the right auricle. In this way, the blood received by the heart from different parts of the bodyis not fully deoxygenated and contains a good proportion of oxygenated blood. After ventricular contraction the blood passes first into the pulmonary arches and remaining blood passes to systematic and catorid arches. The spiral valves supports the conus arterious and takes no part in turning the blood.

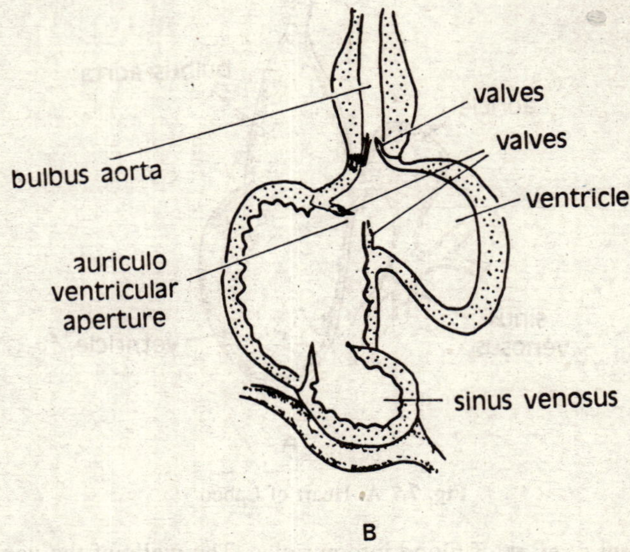

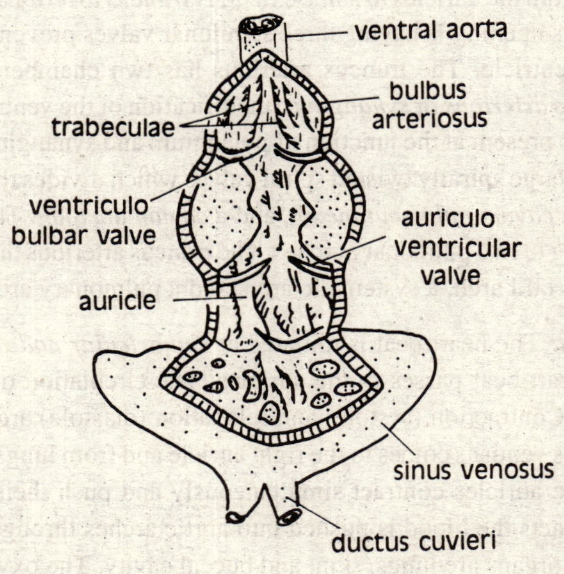

Fig. 7.7 B, C Teleost Heart in Different view

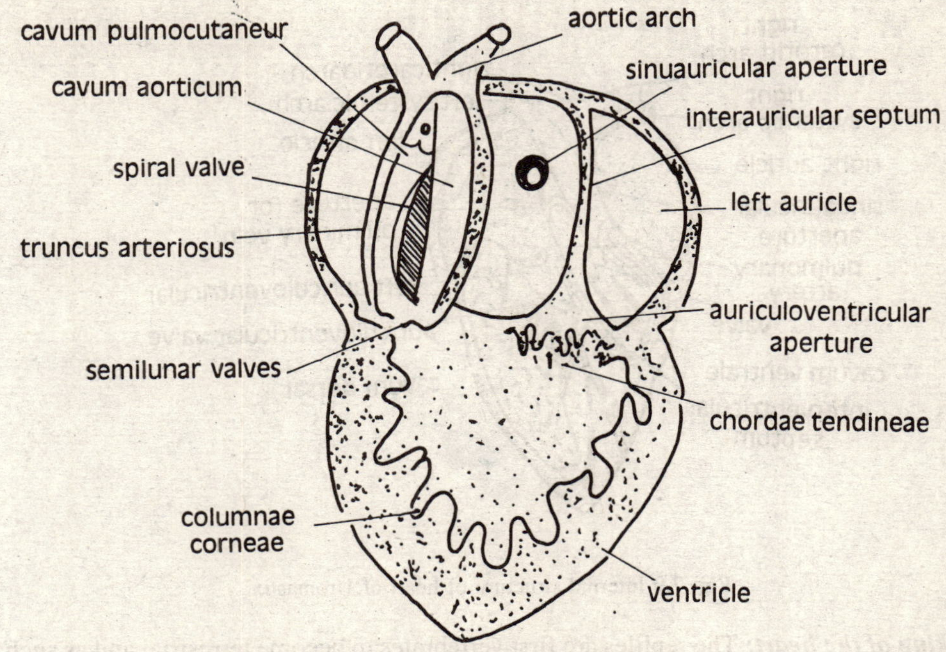

Fig. 7.8 Amphibian Heart Showing Internal Structure

UROMASTIX

The heart of *Uromastix* is a triangular structure enclosed in a thin transparent *pericardium* (Fig. 7.9). It is a chambered structure, more advanced than that of frog. It consists of a large dorsal sinsus venosus, two auricles and a single ventricle. The sinus venosus is a thin walled chamber attached to the wall of the right auricle. Two precaval veins and one postcaval vein open into the sinus venosus. The sinus venosus opens into right auricle by *sinuauricular aperture*, which has no valves and is guarded by folds of the auricular wall. The right auricle is larger than the left and the two auricles are separated by an interauricular septum. The left auricle has an opening of the pulmonary vein, which is without valves. The two auricles communicate with a single ventricle by two openings having an auriculoventricular valve. An interventricular septum extends vertically as a septum from the apex to the centre of the ventricle incompletely and divides it into two unequal chambers, a right *cavum ventrale* (or cavum pulmonale) and a cavum dorsale. The cavum dorsale is subdivided into a left cavum arteriosum and a right cavum venosum. The conus arteriosus present in amphibians is absent in *Uromastix* and the ventral aorta is divided into three trunks. These trunks are *pulmonary aorta* originating from the cavum ventrale, a *right systematic* originating "from the left cavum arteriosum and *left systematic* arch arising from the right cavum venosum. Each of these trunks is provided with a valves at its base. The right and left systematic trunks cross one another after leaving the heart. The point of the crossing of two systematics has a small *foramen of Panizzae* which connects their cavities. The right systematic arch gives off small coronary arteries which supply blood to the heart. The coronary sinus returns blood from the heart to the right auricle.

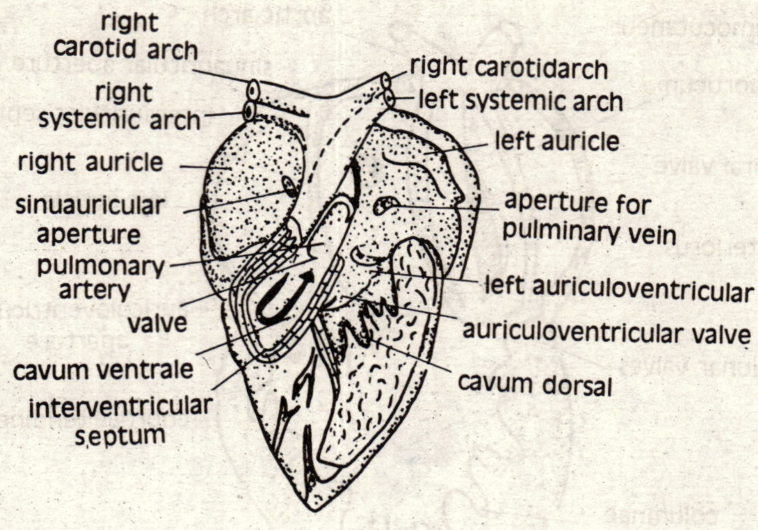

Fig. 7.9 Internal structure of heart of Uromastix

Working of the heart: The reptiles are first vertebrates to become terrestrial and as such, they have well developed pulmonary system where lungs are the only respiratory organs. The oxygenated blood is separated from the deoxygenated one. The sinus venosus receives venous blood which goes to the right auricle. The left auricle receives oxygenated blood from the pulmonary veins. The heart beat is started by sinuauricular node and contraction of auricles continues of ventricle as their muscles are continuous. The interventricular septum keeps the oxygenated and deoxygenated blood separate. When the ventricle contracts the deoxygenated blood from right auricle goes through cavum ventracle and pulmonary aorta to lungs. The mixed blood from cavum venosum goes to left systematic trunk. The oxygenated blood from the left auricle goes through the cavum arteriosum and the right systematic arch to the head region. At the foramen of Panizzae some mixing of oxygenated and deoxygenated blood taken place.

COLUMBA

The heart of pigeon (Fig. 7.10) is oval, large in proportion to the body and placed in the anterior part of the thoracic cavity ventral to the oesophagus. It is four chambered structure covered over by a *pericardium* and lined with *endocardium.* The pericardium is filled with pericardial fluid.

The sinus venosus is absent and the two precavals and a postcaval open into the right auricle. Four pulmonary veins open into the left auricle, which is smaller than the right auricle. The ventricle is completely divided into two parts by an *interventtricular septum.* The left ventricle is a large muscular chamber with circular cavity and great pumping power. The right ventricle is smaller chamber having thin walls with cresentic cavity in section and partly encircles the left ventricle. Longitudinaly arranged bundles of muscle fibres called *columnae carnvue* are present in ventricles. The left auricle opens in the left ventricle and its opening is guarded by a *bicusped valve* made of three membranous flaps which are connected to the wall of the ventricle to chordae tendineae. The right auricle opens into the right ventricle and its opening is guarded by a single large muscular valve. This valve is characteristic of birds and is connected with chordae tendineae. The auriculoventricular valves are having chordae tendineae. The auriculoventricular valves are having chordae tendinae on one side and papillary muscles on the other side attached to the ventricular wall. From the left ventricle a single *right aortic arch* originates and carrys oxygenated blood to different parts of the body. The right ventricle gives rise to pulmonary arch which carries deoxygenated blood to the lungs. The opening of the aortic arches is guarded by three cup-shaped semilunar valves. The heart is supplied with *coronary arteries* arising out of right systemic arches and a *granary sinus* returns blood into the right auricle.

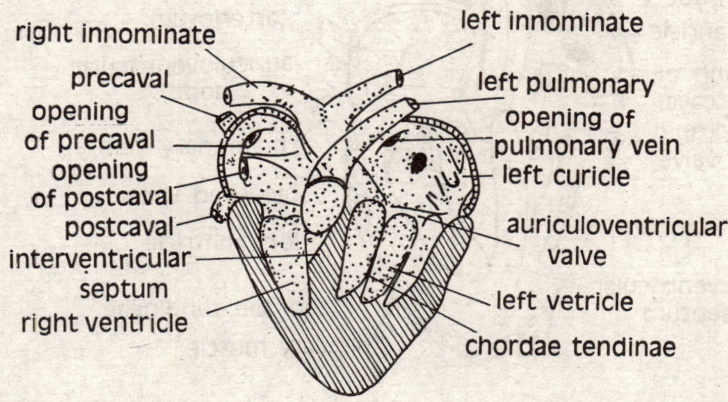

Fig. 7.10 Internal structure of heart of Columba

Working of heart: The blood circulation is completely double and the mixing of oxygenated and deoxygenated blood does not take place. The heart is provided with elaborate nervous system which controls it's working. A pacemaker, *sinu-auricular node* is present on the wall of the right auricle and another pace maker auriculoventricular node, is present around right, auriculoventricular wall. The rate of muscular contractions, systole and diastole is much faster in birds as compared to other vertebrates. When *diastole* starts the heart relaxes and the right auricle receives deoxygenated blood from the venacavae and left auricle gets oxygenated blood from the lungs through pulmonary veins. The *systole* starts from the right auricle at the sinuauricular node and is passed on to the auriculoventricular node from where it spreads to the remaining parts of the heart. During this auricular systole the blood from auricles enters ventricles through auriculoventricular apertures. When ventricles contract the deoxygenated blood from right ventricle goes to lungs through pulmonary arches and oxygenated blood from left ventricle to the different parts of the body through the right aortic arch.

ORYCTOLAGUS

The heart of rabbit (Fig. 7.11) is conical structure situated obliquely in the thoracic cavity between the lobes of the lungs. It is chambered like that of birds with complete double circulation and the deoxygenated blood is not allowed to mix with oxygenated one. The heart is enclosed in a thin walled two layered pericardial cavity. The inner layer is visceral layer and outer is parietal layer. A *pericardial fluid* fills to the diaphragm, which maintains its position.

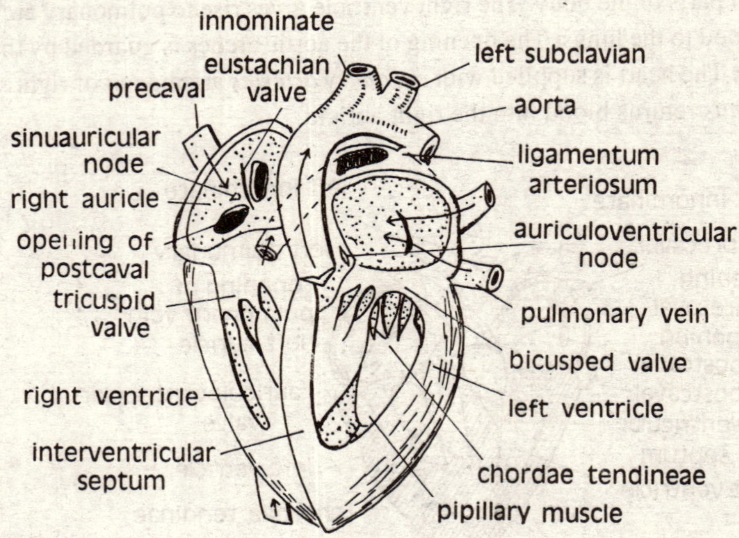

Fig. 7.11 Internal structure of heart of Oryctolagus

There are two *auricles* and two *ventricles.* Sinus venosus is absorbed in the right auricle and truncus arteriosus into the ventricle. The two auricles are completely separated by interauricular septum. A small depression called *fossa ovails* is present, on the interauricular septum. (An aperture called foramen ovale remains present at the region of fossa ovalis in embryos, which is closed before the birth of the animal). The ventricles are thick walled and separated by an interventricular septum into right and left ventricles. The left ventricle is larger and more muscular than the right one. Muscular ridged called *columnae carnae* made of longitudinal bundles of muscle fibres are present" in the inner wall of the ventricles. The opening of left auricle into the left ventricle is guarded by bicusped or *mitral valve* and has two membranous cusps. The opening of left auricle into the left ventricle is large and guided by a *tricusped valve* made of three membranous cusps to prevent the backward flow of blood. The bicusped and tricusped valves have chordae tendine attached to them, which are joined by papillary muscles at their ends. The papillary muscles are extensions of the columnae carnae lining the ventricles. Only *left systemic aorta* is present in mammal and originates from the upper end of the left ventricle. The opening of the left systemic aorta is guarded by three *semilunar valves.* The pulmonary aorta arises from the upper end of the right ventricle and its opening is guarded by three *semilunar valves.* The coronary arteries take their origin from systemic aorta and supply blood to the heart tissues, which is collected by several coronary into a coronary sinus opening into left precaval vein.

Working-of the heart: Alternate contractions (*systole*) and relaxation (*diastole*) of the heart affect efficient circulation of blood in mammals. During systole the two auricles contract simultaneously and their blood is pushed into both the ventricles. The deoxygenated blood goes into the right ventricle and oxygenated blood into the left ventricles. The auriculoventricular valves prevent the backward flow of blood. When the ventricles are filled with blood *ventricular systole* starts and blood from the right ventricle is pushed into pulmonary arch and the blood from the ventricle is pushed into the aortic arch. The whole heart undergoes diastole after the systole with semilunar valves closed. The deoxygenated blood from venae cavae enters the right auricle and oxygenated blood from pulmonary veins enters the left auricle.

7.5. EVOLUTION OF HEART VERTEBRATES

In *Branchiostoma* definitive heart is not present and ventral blood vessel extending between liver diverticulum and gills is contractile and forces the blood to the anterior side.

In *cyclostomes*, the heart has two chambers, atrium and ventricle having valves. A small conus is also present. The straight tube has only constricted in the middle with unequal thickenings.

In fishes (Fig. 7.12A) the heart is more advanced which is called *venous* or *branchial heart.* The blood from the body comes in the heart and then goes to gills for purification and goes to the entire body again. The heart lies posterior to the gills.

In Elasmobranchs the heart has a thin walled *atrium,* a muscular ventricle below the atrium, a small thin walled sac, *sinus venosus* attached to the atrium, and a tubular valvular structure *consus arteriosus* at the antrerior side of the ventricle. The valves are present guarding all the passages and census arteriosus has semilunar valves.

In Teleosts all structures present in clasmobranches are present excepting the conus arteriosus, which is replaced by another tubular muscular structure bulus *arteriosus*, which is an enlargement of the basal portion of ventral aorta.

In Dipnoi major changes take place in the structure of heart due to pulntonary respiration. The *atrium* is partially divided into two auricles. The left auricle receives oxygenated blood from lungs or modified air bladder and sends it to the body directly without passing through the gills. In this way, the circulation is partly double in which the pulmonary circulation is also developed along with the branchial circulation.

In amphibia (Fig. 7.12) further advancement of heart takes place due to air breathing system. The atrium is fully divided into two auricles by a septum. The auricle receives oxygenated blood from lungs and the right auricle receives deoxygenated blood from the entire body. The ventricle has many trabeculae, which prevent the mixing of oxygenated and deoxygenated blood. The *conus arteriosus* is divided into two chambers by a spiral valve to help in the distribution of blood into systemic and pulmonary arches. Thin walled *sinus venosus* receives deoxygenated blood and opens into the right auricle The valves are present in all passages, which guard their opening and allow the flow of blood in one direction only.

In Reptilia (Fig. 7.12) the heart is more developed than in amphibians. The auricles are completely divided by *interauricular septum.* The ventricle is also divided by a septum into two

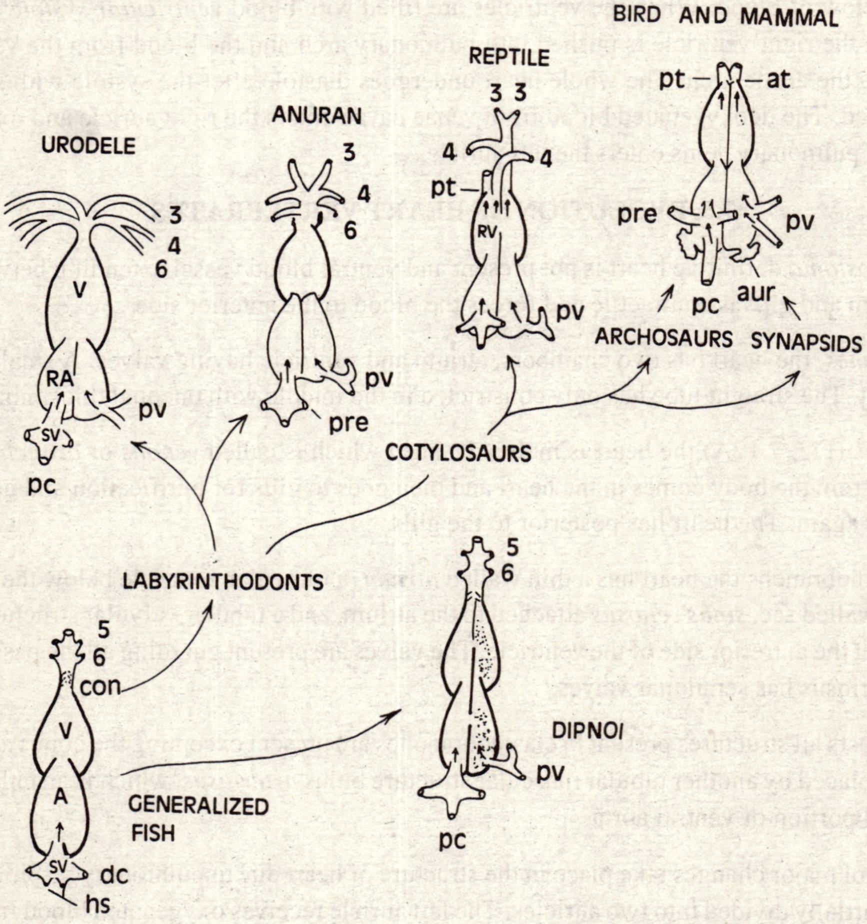

Fig. 7.12 Modifications of the atria and ventricles that result in increased separation of oxygenated and deoxygenated blood. The parts of the heart shown are, **A,** atrium; **RA,** right atrium; V, ventricle; **RV,** right ventricle; **SV,** sinus venosus; **con,** conus arteriousus; **aur,** auricle of mammalian heart. **3** to **6,** Third to sixth aortic arches. Other vessels are, **at,** aortic trunk; **dc,** common cardinal vein; **his,** hepatic sinus; **pc,** postcava; **pre,** precava (common cardinal vein); **pv,** pulmonary veins; **pt,** pulmonary trunk. Gray chambers contain chiefly, or only, oxygenated blood.

chambers partially, which allow the separation of atrial and venosus blood upto some extent. The aorta and conus arteriosus are divided at the base so that the pulmonary artery is a separate vessel starting from the heart. The sinus venosus is reduced and joins the auricle. (The ventricle is completely divided in crocodiles).

In birds and mammals (Fig. 7.12) the two sides of the heart are completely separated. The auricle and ventricle are completely divided by *interauricular* and inventricular septa. The blood enters twice in the heart to make a complete circuit making it *double circulation*. The venous blood enters the heart through the right auricle, goes to right ventricle and is forced out to lungs for oxygenation. The oxygenated blood from lungs is received in the heart and goes to the body from the left ventricle through catorid and systemic arteries. The mixing of oxygenated and deoxygenated blood is completely prevented in this mechanism. The venosus and conus arteriosus are completely absorbed into right auricle and ventricle respectively.

The hearts of birds and mammals are evolved from the reptilian heart and minor structural differences are present. The cavity of right ventricle in birds is narrow and partly encircles the wide and oval cavity of the left ventricle. In mammals the two ventricles are separated situated. In birds only right aortic arch persists while in mammals the left aortic arch is found.

In this way, we find that the heart has undergone gradual evolution and perfection as we proceed from lower groups to the higher group of the vertebrae series.

7.6. THE ARTERIAL SYSTEM

The arterial systems of adult vertebrates appear to be different in structure but the study of their development exhibits similar fundamental plan. The aortic (arterial) arches supply the gills in fishes and amphibians and their later modification in Amniota is connected with the increasing complexity of the heart.

7.6.1. Embryonic Aortic Arches

The anterior end of the ventral aorta divides during early development into two aortic arches, which extend dorsally in the mandibular region above pharynx and make *lateral dorsal aortae* or *radices aortae.* Additional pairs of aortic arches are then formed in anteroposteriorerior direction, which connect the ventral and dorsal aortae on each side. Typically the number of such arches is six and each arch passes through tissue between adjacent pharyngeal pouches. The first arch is called *mandibular aortic arch,* the second *hyoid arch* and remaining are third fourth, fifth and sixth *aortic arches* respectively (Fig. 7.13). Each arch is present anterior to the visceral cleft of the corresponding number. On the posterior side of the pharynx the two dorsal aortae fuse together making one dorsal aorta, which continues into the tail region as the caudal artery. The dorsal aorta supplies different paired and unpaired blood vessels along its entire length and supplies to the tissues of body posterior to the pharynx. The unpaired ventral aorta and paired dorsal aortae continue anteriorly and supply to the head and anterior branchial region.

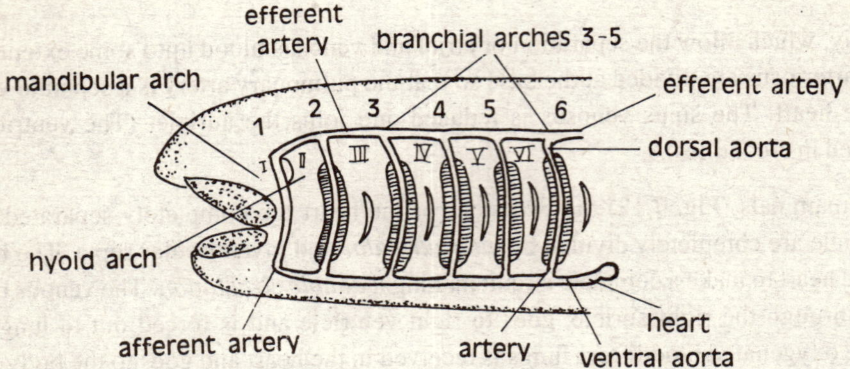

Fig. 7.13 Aortic arches in Vertebrates

The blood from the heart is pumped to the aortic arches through ventral aorta. The aortic arches carry blood to the paired dorsal aortae from where it goes to the head anteriorly and posteriorly to the single dorsal aorta. The venous blood returns to the sinus venosus.

7.6.2. Aortic Arches in Vertebrates

The aortic arches mentioned alongwith embryonic condition undergoes major changes in different groups of vertebrates. A progressive reduction in the number of aortic is found in the vertebrate series.

BRANCHIOSTOMA

The number of the aortic arches in *Branchiostoma* (Fig. 7.14) is much more than higher chordata. The ventral aorta (or heart) is a contractile vessel situated on the ventral side of the gills. About sixty or more afferent branchial arteries or lateral branches are given off on both sides alternatively which extend upto the primary gill bars of the pharynx. Each afferent branchial artery has a contractile *bulbillus* at its base and gives off small branches, which connect with vessels in the secondary gill bars. The arteries present in the primary and secondary gill bars are connected with the paired dorsal aortae. The two aortae join together behind the pharynx and form a *medium dorsal aorta,* which supplies to the parts of body posterior to the pharynx. When the blood passes through gill bars the oxygenation of the colourless blood takes place.

Cyclostomes: In *Petromysfon* the ventral aorta is continued forwards from the heart upto some distance to the level of fourth gill pouch and gives off two branches. Four paired *afferent branchial arches* arise from the unpaired portion of the ventral aorta and four arise each of the paired ventral aortae. *Efferent branchial arteries* collect blood from the gill lamellae and remaining position similar to that of afferent arteries. The afferent arteries are joined with a single dorsal aorta. The dorsal aorta gives off a pair of arteries anteriorly which join to form a cephalic circle and supply to brain, eyes, tongue and various parts of the head.

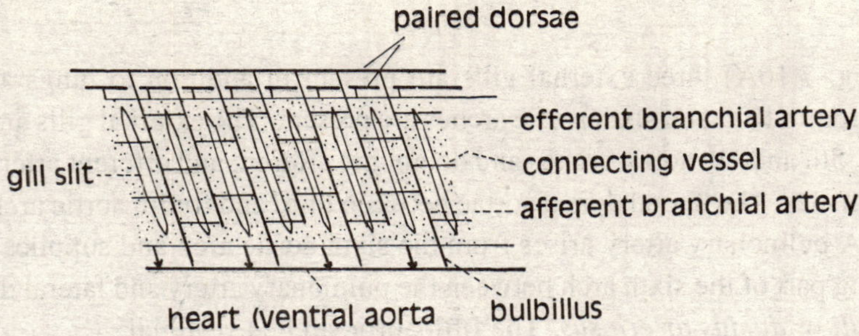

Fig. 7.14 Diagram of portion of pharyngeal region of Amphioxus showing arrangement of aortic arches

FISHES (FIG. 7.15)

In Elasmobranchs the second, third, fourth, fifth and sixth arches persist as five *afferent branchial arteries* and four *epibranchial arteries* formed by nine *efferent branchial arteries.* The first and part of second arch form head arteries. The first arch is modified to form arteries of the head region, as the first gill cleft called the spiracle has no gill filaments. The afferent arteries supply to gills by their branches and efferent arteries collect blood from gills. The afferent efferent arteries are connected together by capillary loops in the gill.

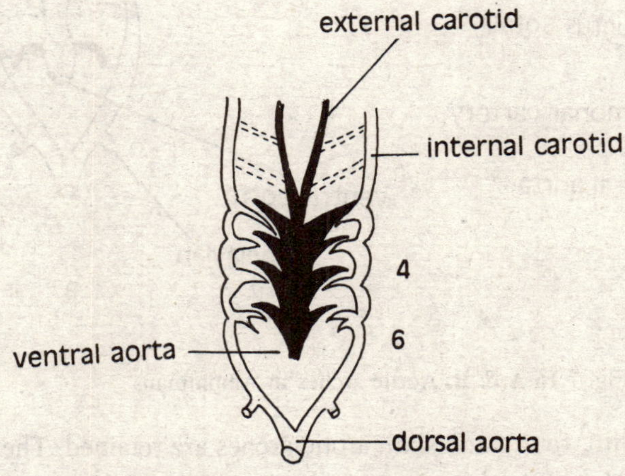

Fig. 7.15 Aortic arches in Elasmobranchs

In bony fishes and Dipnoi, the third, fourth, fifth and sixth arches form four branchial arches constituting afferent and efferent branchial arteries, supplying to four gills. The first and second arches are absent as the hypoid arch does not retain gills. The *external* and *internal carotids* arise from the ventral and dorsal ends of the arch. The afferent and efferent arteries are always one in number in each gill arch in teleosts. but in Dipnoi, each gill carries two efferent arteries and one afferent artery. In Dipnoi, the dorsal ends of each sixth arch give rise to a pulmonary artery to the lungs.

In Urodela (Fig. 7.16A) three external gills are present in addition to lungs as respiratory organs. The third, fourth fifth and sixth aortic arches are present. The external gills are supplied by branches from 4th, 5th and 6th aortic arches and do not gel afferent and efferent arteries as in case of internal gills of fishes. The lateral dorsal aortae between third and fourth aortic arches persist as *ductus caroticus.* A pulmonary artery arises from the sixth aortic arch and supplies blood to the lung. The remaining part of the sixth arch between the pulmonary artery and lateral dorsal aorta is called *ductus botalli* or *ductus arteriosus.* The fifth aortic arch is vestigial.

In Anura (Fig. 7.16 B) only third, fourth sixth arotic arches persist and the fifth arch disappears completely. The third arch and part of ventral aorta modify into a carotid arch. The fourth arch and lateral dorsal aorta modify into *systemic arch.* The ductus caroticus present between third and fourth arches in Urodela also disappears. The sixth arch becomes *pulmocutaneous arch* and ductus arterious disappears completely.

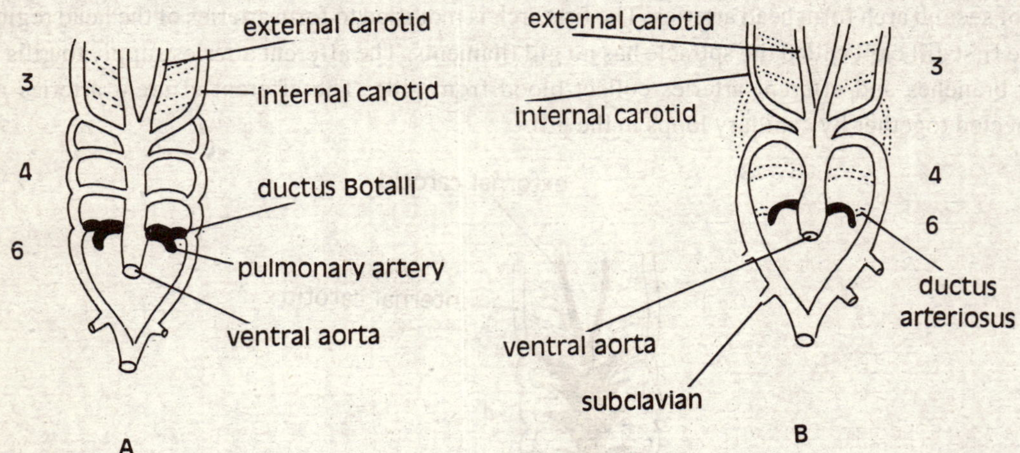

Fig. 7.16 A & B. Aortic arches in Amphibians

In reptiles (Fig. 7.17), third, fourth and sixth arotic arches are retained. The third arotic arch modifies into the *carotid arch* and the common carotids both sides are connected with the *right systemic aorta.* The ductus caroticus persists in some reptiles and ductus arteriosus disappear in most reptiles. The fourth and sixth aortic arches have midified into *systemic* and *pulmonary* aortae respectively. The *right systemic aorta* arises from the left side of the ventricle and left systemic and pulmonary aortae arise from the right side of the ventricle.

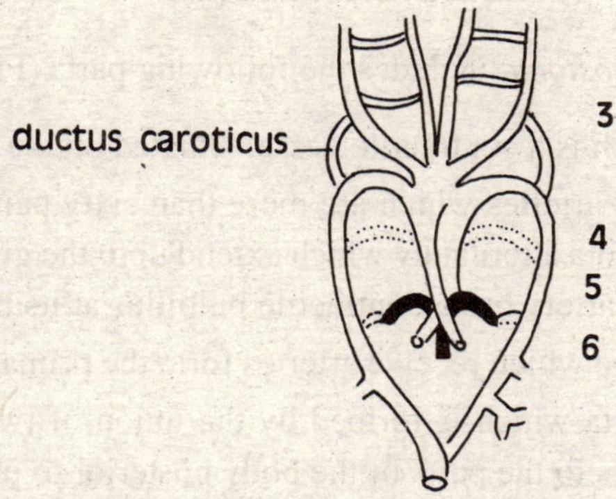

Fig. 7.17 Aortic arches in Uromastix

In birds (Fig. 7.18 A.) minor differences are found with reptilian arrangements of aortic arches. The third arch modifies into the carotid arteries. The fourth arch modifies into the *systemic aorta of right side* only and a part of it on left side forms left *subclavian artery* and remaining part of this arch disappears completely. The *sixth aortic* arch is modified into *pulmonary aorta.* The systemic aorta arises from the left ventricle and pulmonary from the right ventricle.

In mammals (Fig. 7.18 B.) the third, fourth and sixth aortic arches are present. The third arch is modified into *carotid arteries.* The fourth arch forms *systemic aorta* of *left* side only and a part of ft on its proximal right side forms an *innominate* and right *subcalvian artery* with remaining part of this arch and its lateral dorsal aortae disappear completely. The sixth arch modifies into the pulmonary aorta. The ductus arteriosus disappears. The systemic aorta arises from the left ventricle and pulmonary aorta from the right ventricle.

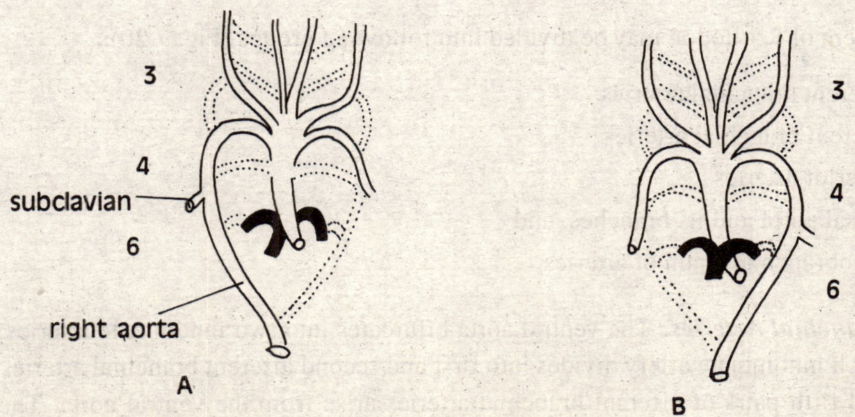

Fig 7.18 Aortic arches in higher vertebrates (a) Birds (b) Mammals

7.6.3. Arterial System in Vertebrate Types

BRANCHIOSTOMA

The arterial system of *Branchiostoma* includes the following parts (Fig. 7.19):

1. Ventral arota which is a contractile vessel situated on the ventral side of the gills.
2. Afferent branchial arteries which are more than sixty paired blood vessels given off form the ventral arota alternately which extend upto the gill bars of the pharynx. Each afferent branchial artery has a contractile bulbillus at its base which pumps blood.
3. Paired dorsal aortae which receive arteries form the primary and secondary gills bars.
4. Median dorsal aorta which is formed by the union of two lateral aortae behind the pharynx. It supplies to the parts of the body posterior to pharynx.

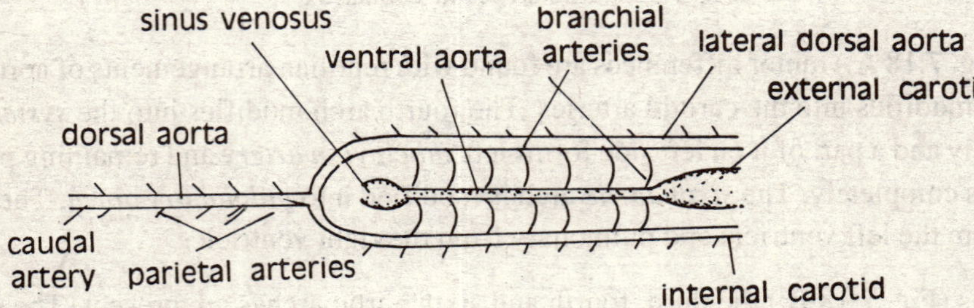

Fig. 7.19 Arterial system of Branchiostoma

The oxygenation of the colouress blood takes place when it passes through the gill region.

SCOLIODON

The arterial system of Scoliodon may be divided into following groups (Fig. 7.20):

(1) Afferent branchial arteries
(2) Efferent branchial arteries
(3) Anterior arteries
(4) Dorsal aorta and its branches, and
(5) Hypobranchial chain of arteries

Afferent Branchial Arteries: The ventral aorta bifurcates into two innominate arteries at its anterior end. Each innominate artery divides into first and second afferent branchial arteries. The third, fourth and fifth pairs of afferent branchial arteries arise from the ventral aorta. The first afferent branchial artery goes to hyoid arch and remaining four run along the outer borders of the first four branchial arches, break up into capillaries and carry blood to gills for oxygenation.

Efferent Branchial Arteries: The blood from gills is collected by nine pairs of efferent branchial arteries. A series of four complete loops is formed around the first four gill slits by the first eight arteries. The ninth efferent branchial artery collects blood from the gill of the fifth gill slit and sends its blood to the fourth loop. All the four loops of the efferent branchial arteries are connected with one another by four short longitudinal connectives and a network of longitudinal *commissural vessels* called the *lateral hypobranchial chain.* Each loop gives rise to an *epibranchial artery* and four pair of epibranchial arteries join in the mid dorsal line to form a dorsal aorta.

Anterior Arteries: The first efferent branchial of each side gives off three arteries. The first is external carotid supplying to the head and lower jaw; the second is afferent spiracular which runs around the spiracle and goes to brain as *spiracular epibranchial artery;* the third is a *hyoidean epibranchial artery* winch supplies to the muscles of the eye and the lower jaw

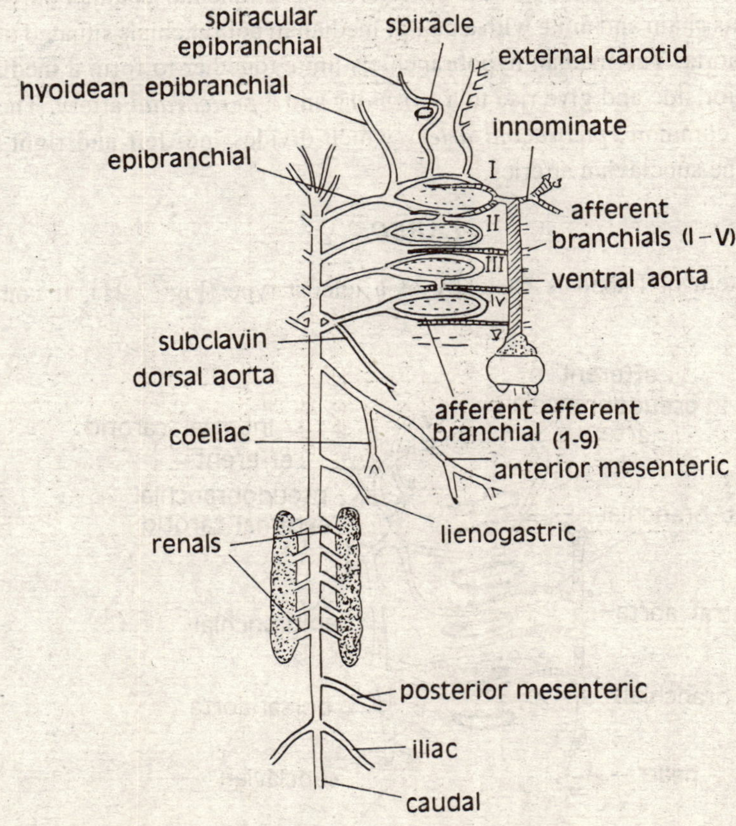

Fig. 7.20 Arterial system of Scoliodon

The Dorsal Aorto: The dorsal aorta formed by the union of epibranchial arteries, runs posteriorly below the vertebral column through the haemal canals and continues upto the tip of the tail as the caudal artery. It gives off the following branches in the anteriposterior direction:

(i) A pair of *subclavian* arteries to pectoral fin.

(ii) Unpaired *coeliacomesenteeric* which divides into two branches, a small *celiac artery* to liver and stomach and a large *anterior mesenteric artery* to pancreas and intestine.

(iii) Unpaired lineogastric artery to stomach, spleen, intestine and gonads.

(iv) Paired *renal arteries* in several pairs to kidneys.

(v) Unpaired, small *posterior mesenteric artery* to the rectal gland.

(vi) Paired *parital arteries* in a series to the vertebral column and its muscles

(vii) Paired *iliac arteries* (1 pair) to the pelvic fins.

Hypobranchial Chain: A network of slender arteries arises from the ventral ends of the loop of the efferent branchial arteries and makes a lateral hypobranchial chain. Four commissural vessels arise from this chain and unite with a pair of median hypobranchials situated on the ventral wall of the ventral aorta. The median hypobranchials unite together to form a median *coracoid artery* on the posterior side and give rise to a *coronary* and a *pericardial* artery. The pericardial artery gives rise to common *epicoracoid artery* which divides into left and right epicoracoid arteries which join the subclavian arteries.

LABEO

The arterial system of Labeo is basically of a teleost type (Fig. 7.21). It consists of the following:

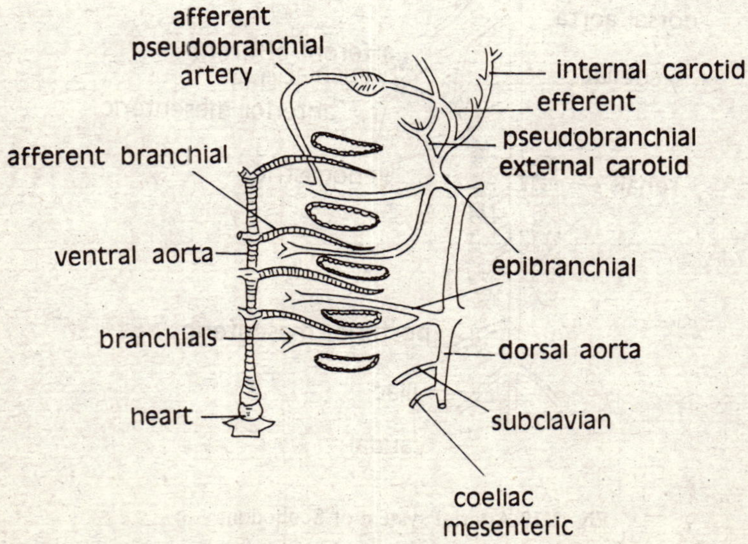

Fig. 7.21 Arterial Circulation in Labeo

Afferent Branchial Arteries: Four pairs of afferent branchial arteries supply deoxygenated blood to the gills. The ventral aorta bifurcates anteriorly into first two afferent branchial arteries. Remaining afferent branchial arteries lake their origin directly from the ventral aorta independently.

Efferent Branchial Arteries: The blood from gills is collected by four pairs of efferent branchial arteries after oxygenation. The efferent arteries join the lateral dorsal aorta of each side. The lateral dorsal aorta continue infront as internal carotid arteries going to head, join together posteriorly to form a *median dorsal aorta* and ring infront of gills called *circulus cephalicus.*

Dorsal Arota: The dorsal aorta continues posteriorly and gives off unpaired arteries to the alimentary canal and paired arteries to the paired fins and kidney

RANA (FIG. 7.22)

The *truncus arteriosus* formed by conus arteriosus and ventral aorta divides into two parts called *right* and *left arches*. Each arch divides further into 3 arches called carotid, systemic and pulmocutaneous arches.

The carotid arch: It gives off the following arteries:

(1) *External carotid* or *lingual* supplying to the tongue and lower jaw.

(2) *Internal carotid* going to orbit and brain. Each internal carotid artery has a swollen carotid body or carotid labyrith having a network of capillaries and chromaffin cells and detects oxygen pressure in the blood so that the rate of respiration may be adjusted by the brain accordingly.

The systemic arch: Each systemic arch gives out three arteries and joins the other systemic to form a dorsal aorta.

(1) *Subclavian artery* which is large and goes to the fore limb.

(2) *Occipitoveriebral artery* which goes to head, vertebral column and spinal cord.

(3) *Oesophageal artery* which goes to oesophagus. The dorsal aorta gives off the following arteries.

(4) Unpaired *coelicomesenteric artery* which divides into a small *celiac artery* and a large *anterior mesenteric artery.* The celiac artery supplies to stomach and liver by gastric and hepatic branches respectively. The anterior mesenteric sends branches to duodenum, spleen and ileum and continues further as posterior mesenteric artery supporting the large intestine.

(5) Four or five pairs of *renal arteries* to kidneys.

(6) A pair of *gonadial* (ovarian or spermatic) arteries to reproductive organs.

(7) A pair of *iliac arteries* which divide into a *femoral* and a *sciatic* arteries which supply to the hind limbs.

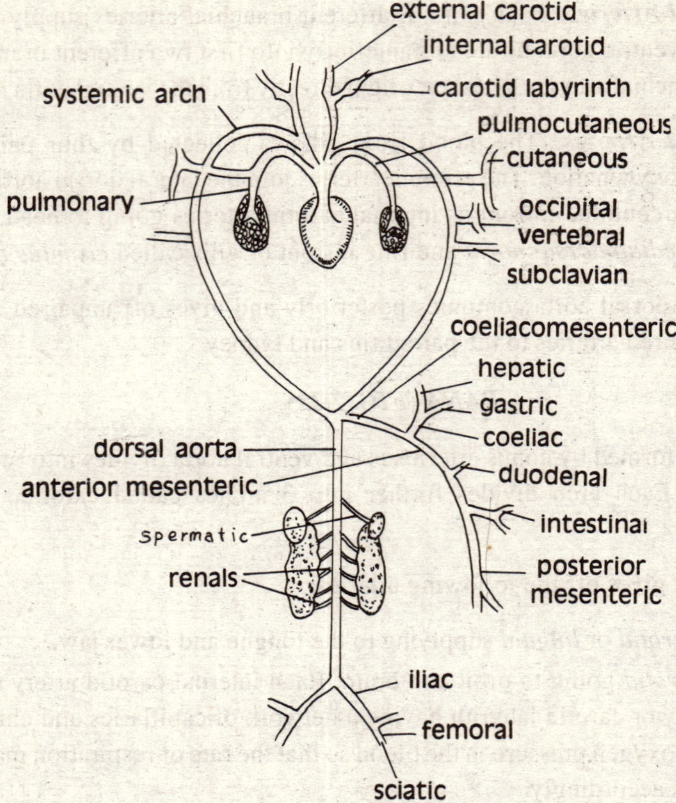

Fig. 7.22 Arterial system of Rana

The pulmonary arch: Each pulmonary arch divides into two arteries.

(1) A pulmonary artery going to lungs.

(2) A cutaneous artery going to the skin and buccal cavity.

UROMASTIX (FIG. 7.23)

One *pulmonary* and two *systemic aortae* lake their origin from the ventricle. The pulmonary aorta arises form the right side of the ventricle and divided to form two pulmonary arteries each one of them going to lung of its side.

The *right systemic aorta* originates from the left side of the ventricle and left systemic aorta from the right side of the ventricle. Two systemic arches cross one another, curve ground and meet below the heart to form a *dorsal aorta.* The cavities of the two systemic aortae are connected together their crossing by a *foramen of Panizzai.* The right systemic aorta gives rise to an innominate artery which divided to form right and left *carotid arteries.* Each of the carotid arteries is joined with systemic artery of its side by a *ductus caroticus.* Before joining the dorsal aorta the right systemic artery gives a common *subclavian artery* and a pair of *parietal arteries* supplying to the body wall. The *subclavian artery* divides into left and right subclavian arteries supplying to left and right fore-limbs respectively.

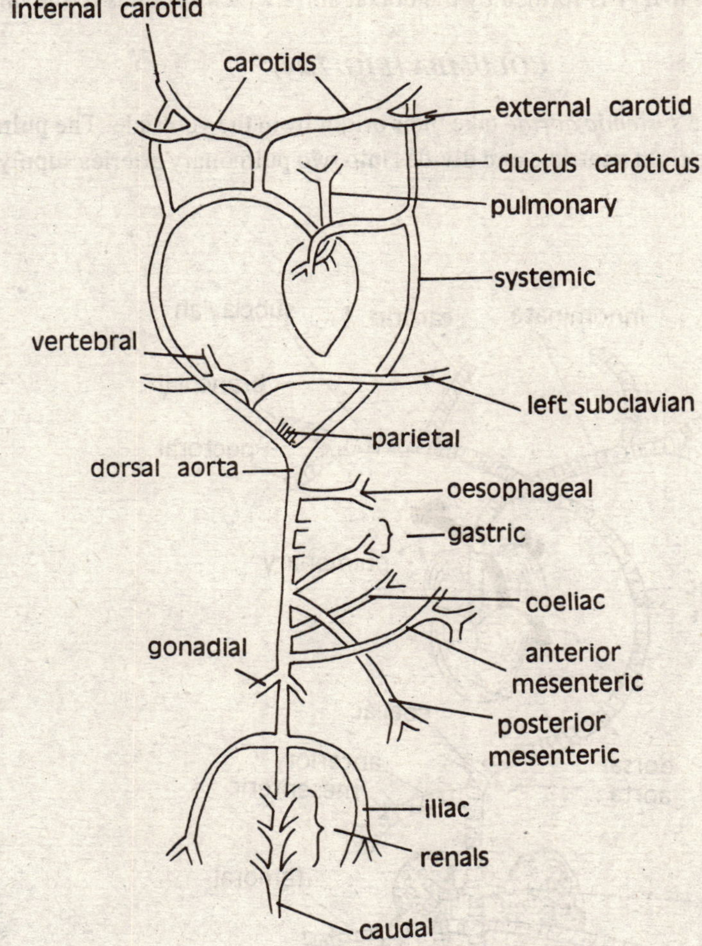

Fig. 7.23 Aterial system of Uromastix

The *dorsal aorta* gives off the following arteries:

1. An *oesophageal* artery to oesophagus.
2. About 15 pairs of *parietal arteries* to the vertebral column.
3. Three unpaired *gastric arteries* to stomach.
4. One unpaired *posterior mesenteric artery* to caesum and colon.
5. One unpaired *celiac artery* to stomach and spleen.
6. One unpaired *anterior mesenteric* artery to the small intestine.
7. One pair of *gonadial arteries* to the gonads.

8. One pair of *iliac arteries* of legs.
9. Three pairs of *renal arteries* to kidneys.
10. The *caudal artery* is formed by the dorsal aorta which supplies to the tail.

COLUMBA (FIG. 7.24)

One *pulmonary* and one *systemic aortae* take their origin from the ventricle. The pulmonary aorta takes its origin from the right ventricle and divides into two pulmonary arteries supplying blood to lungs of each side.

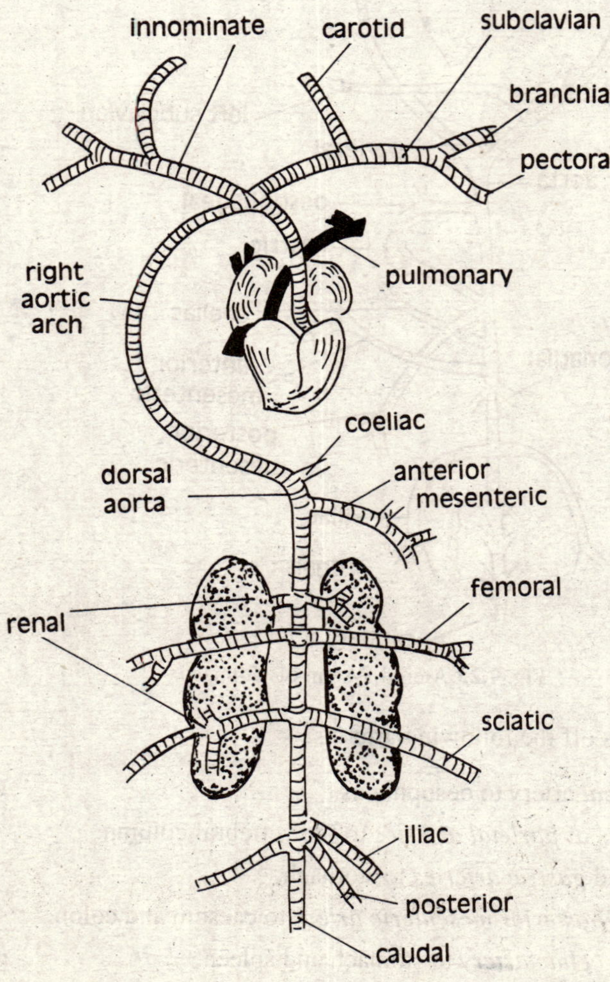

Fig. 7.24 Aterial system of Columba

The ductus caroticus is absent in pigeon and is represented by a cord of connective tissue on each side called *ligamentum anteriosum.* The right systemic aorta or aortic arch is characteristic of birds and originates from the left ventricle. The left aorticarch is absent in adult birds. A pair of *innominate arteries* take their origin from the right systemic arch and divide in to the following branches:

1. A *carotid artery* supplying to the head.
2. A *subclavian artery* which divides into a branchial artery supplying to the wing and a pectoral artery supplying to the pectoral muscles connected with the flight.

The right aortic arch then passes backwards and curves over the head and makes the dorsal aorta which gives off the following branches:

1. A *celiac artery* to stomach and liver.
2. An *anterior mesenteric* artery to the small intestine.
3. Three pairs of *renal arteries* to kidneys.
4. Paired *femoral* and *sciatic* arteries to the legs.
5. One pair of *iliac arteries* to the pelvic region.
6. One unpaired *posterior mesenteric artery* to the large intestine.
7. The dorsal aorta continues to form a small *caudal artery* supplying to the tail.

ORYCTOLAGUS (FIG. 7.25)

Two aortic arches originate from the ventricle. These are *pulmonary aorta* from the right ventricle and *systemic aorta* from the left ventricle. The pulmonary aorta divides to form two pulmonary arteries which supply the lungs.

The systemic aorta or the left aortic arch from the left ventricle and the right arch is absent in adults. This arch curves to left and gives off the following arteries.

1. *Coronary arteries:* Right and left coronary arteries take their origin from the systemic aorta and supply to the muscles of the heart.
2. *Innominate artery* is a short, stout artery which divides to form a *right subclavian,* and right and left *common carotid arteries.* The common carotid arteries run through the neck parallel to trachea and divide into a small *internal carotid* and an *external carotid.* The internal carotid supplies to brain and external carotid to the back of head, tongue, salivary glands ana jaw muscles.
3. *Left subclavian artery:* The aorta makes an arch over the heart and gives rise to a left subclavian artery which divides into three branches, *brachial, vertebral* and *internal mammary.* The branchial artery supplies to fore limb, vertebral artery to vertebroarterial canal of cervical vertebrate, spinal cord and brain, and the internal mammary artery

supplies to the ventral thoracic muscles and then to abdominal muscles as superior epigastric artery.

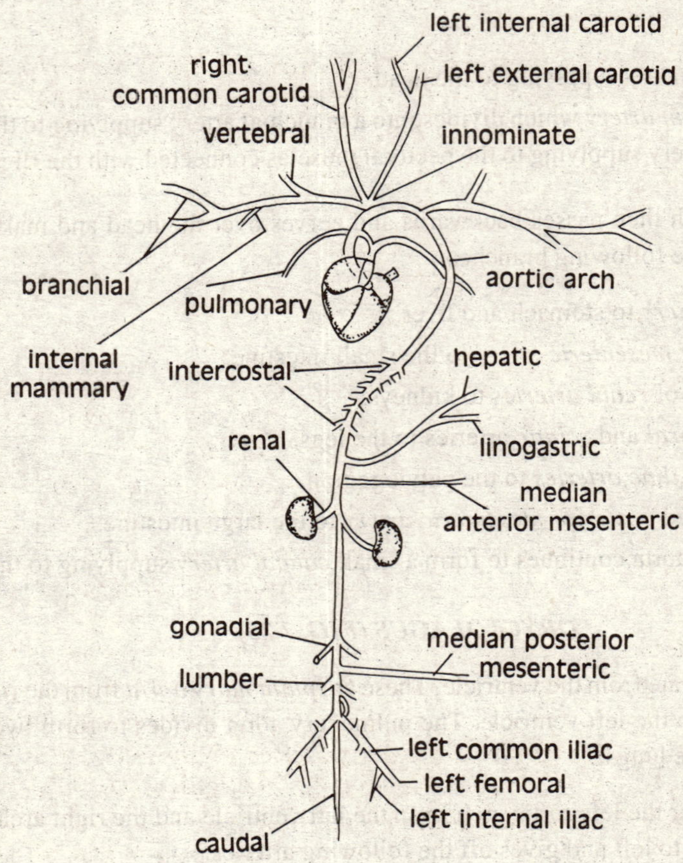

Fig. 7.25 Arterial system of Oryctolagus.

The systemic aorta turns to the left and continues under the vertebral column as dorsal aorta. It gives off the following arteries:

1. *Intercostal arteries* in a series in thorax which supply to intercostals muscles between ribs.
2. *Phrenic arteries* fewer in number and below the intercostals arteries, which supply the diaphragm.
3. A median *celiac artery,* which passes through mesentery and divides into *hepatic* artery to liver and *lineogastric* artery to stomach and spleen.

4. Large *anterior mesenteric artery* which supplies to duodenum, pancreas, ilum, cacum and colon by its branches.
5. A pair of *renal arteries* supplying to kidneys.
6. A *median posterior mesenteric artery* supplying to lower parts of colon and rectum
7. A pair of *ganadial arteries* (spermatic or ovarian) to gonads.
8. Two small *lumber arteries* supplying to the body wall.
9. *Common iliac arteries:* The dorsal aorta divides into two common iliac arteries and its median portion continues in the tall region as *caudal artery.* Each iliac artery gives off an *ilio-lumber artery* near its origin which supplies to muscles of the back and divides into *internal iliac* or *hypogastric* going to organs of pelvis, and an *external iliac* going to hind limbs. The common iliac also gives a small *vesicular* or *umbilical artery* going to urinary bladder and the uterus in females. The external iliac continues into the leg and is called *femoral artery.*

7.7. THE VENOUS SYSTEM

Veins are blood vessels, which return the blood to the heart from tissue. The venous system consists of the following main veins in adults:

1. *Hepatic portal veins* to liver and *hepatic veins* from liver to heart.
2. *Renal poral veins* to kidneys.
3. *Venae cavae* and their connected veins: normally two anterior and one posterior venae cavae are present.
4. The *abdominal vein* which returns blood from the ventral part of the body wall.
5. The *pulmonary vein* which returns blood to the heart, from lungs in air breathing vertebrates.

7.7.1. The Primitive Venous System (Embryonic Veins)

The venous system is originally paired but, due to the introduction of air-breathing habits, it becomes asymmetrical. The venous system of higher forms during the course of its development passes through some stages common to the embryonic stages of the lower forms. The venous system in embryos includes the following veins (Fig. 7.26):

1. A pair of *subintestinal veins* develops in splanchnic mesoderm ventral to gut. This pair fuses to continue as *caudal veins* excepting in the region of anus where they form a loop.
2. A pair of *vitelline veins* joins the posterior side of the heart, which is destined to become sinus venosus. Each vitelline vein is joined by a *subintestinal vein* on the posterior side.
3. Paired *anterior* and *posterior cardinal veins* develop from the dorsal side of the head

region and posterior part of the body respectively and get connected with fused vitelline veins. They meet the sinus venosus on each side by common *cardinal vessel* or *duct of cuvier* or *cuverian sinus.* Anterior cardinal are also called *jugular veins.*

4. An *inferior jugular vein* from the ventrolateral portions of the head joins the common cardinal vein in fishes and salamanders; it is absent in higher groups.
5. A pair of *lateral veins* or *ventral abdominal veins* from the lateral or ventral portions of the body wall also enter common cardinal veins in lower vertebrates. These veins make secondary connections with allantois or umbilical cord in amniotes. They are called *vitelline veins* or *umbilical veins.* After the formations of liver, these vitelline and umbilical veins loose their connection with sinus venosus and get connected with sinus venosus and liver through a large channel ductus venosus.

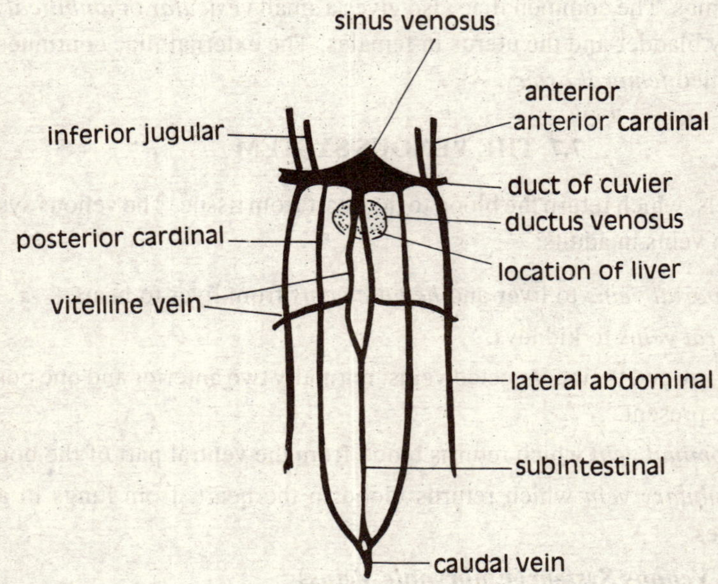

Fig. 7.26 Showing the primitive venous system

7.7.2. Evolution of Venous System in Vertebrates:

The principal veins are derived from the primitive venous complex given above. Variations are found in the venous system of vertebrates but the arrangement of veins is consistent within each vertebrate class. The vitelline and umbilical veins are lost in adult stages but their vestiges are usually found.

BRANCHIOSTOMA (FIG. 7.27)

The simplest venous system of vertebrates is exhibited in *Branchiostoma.* The following veins are present.

1. The *caudal vein,* which continues anteriorly as subintestinal vein. It is connected with postcardinal vein to allow the return of blood to the tail
2. The *subintestinal vein,* which breaks up into capillaries in the intestine.
3. The *hepatic portal vein* which is formed by the capillaries in intestine and goes to the hepatic caecum and breaks up into capillaries forming a true hepatic portal system.
4. The anterior cardinal vein join the postcardinal vein on each side and the veins join the posterior part of the heart, sinus venosus by a common cardinal vein.
5. A hepatic vein, which is contractile, and carrys blood from the hepatic caecum to the heart.

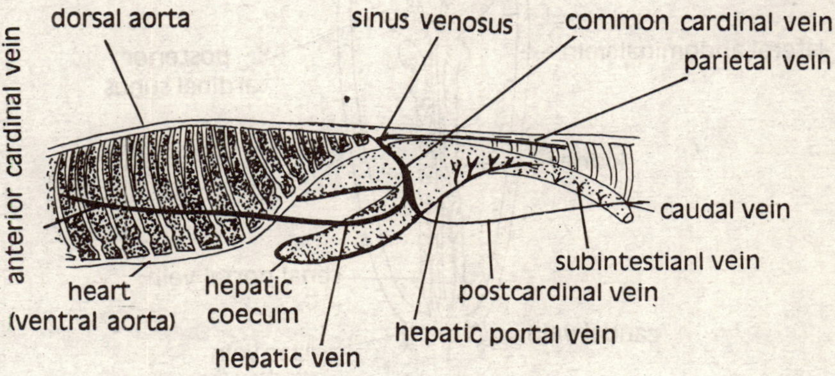

Fig. 7.27 Venosus system of Branchiostoma

CYCLOSTOMES

The cyclostomes the system is similar to that of *Branchiostoma.* The paired anterior and posterior cardinal veins are present and common cardinal vein enters the sinus venosus. Two inferior jugular veins from the anteroventral parts of the body unite to form a single vein, which joins sinus venosus.

1. The *caudal vein* divides at cloacal region and connects the postcardinal veins and runs on the ventrolateral sides of the dorsal aorta on either sides.
2. The anterior part of the *subintestinal vein* is situated in the typhlosole of intestine and posterior part is not connected with caudal vein.
3. The *hepatic portal* vein is formed by the subintestinal vein. It receives a branch from the head region. A contractile " portal heart" is present in this vein.
4. The hepatic vein collects blood from liver and connects the sinus venosus.
5. Segmentally arranged veins from kidneys, gonads and body wall enter the postcardinal vein. Renal portal system is absent in cyclostomes.

SCOLIODON

In *Scoliodon* the venous system is similar to the primitive plan (Fig. 7.28). The veins form large, irregular venous sinuses during their course to the head. The venous sinuses have thin walls. The formation of venous sinuses is characteristic of elasmorbranchs. The veins found in *Scoliodon* are as follows:

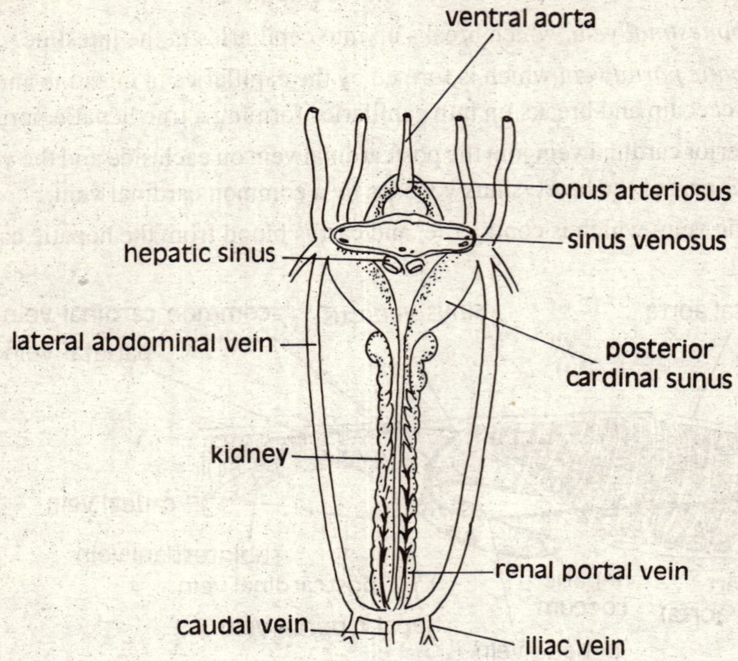

Fig. 7.28 Venous system of Scoliodon

(1) *Cardinal vens:* Paired jugular and anterior cardinal sinuses bearing blood from the anterior region of the body. A pair of posterior cardinal sinuses returns blood from the posterior region. The anterior and posterior cardinal veins unite to form a transverse sinus called *ductus cuveri.*

(2) *Hepatic portal system:* Veins from the alimentary canal and associated glands unite to form the hepatic portal vein. It receives lienogastric vein anterior and posterior gastric veins. It breaks up into capillaries in the liver (The hepatic portal vein is formed by the union of anterior and posterior intestinal veins, found in cyclostomes). The blood from liver is collected by two large hepatic sinuses, which open in the sinus venosus.

(3) Two paired *lateral cutaneous veins,* one dorsal and other ventral are present which join together near the anterior end of the pectoral fin and open into the branchial vein.

(4) Anterior *ventral veins* send blood to the ductus curveri through the *inferior jugular sinuses.* Each inferior jugular sinus is formed by the union of the submental sinus from lower jaw. hyoidean sinus and ventral nutrients from gills.

(5) The posterior veins send blood to ductus cuveri through the subclavian vein.

(6) A small cloacal vein and two femoral veins from pelvic fins join to form two large lateral abdominal veins which are connected together by a commissural vein.

(7) The subclavian vein is formed anteriorly by the joining of lateral abdominal veins with branchial veins. It opens in ductus cuveri.

(8) *Renal portal system:* The caudal vein from the tail region passes through the haemal canals of vertebrae and divides into right and left renal veins. The renal veins divide into numerous afferent renal veins in the kidneys. The blood from kidneys is collected by numerous *efferent renal veins*, which join together forming posterior cardinal sinus.

TELEOSTS

The veins in teleosts are large and wide but do not form sinuses as in clasmobranches. The venous system in teleosts consists of the following (Fig. 7.29):

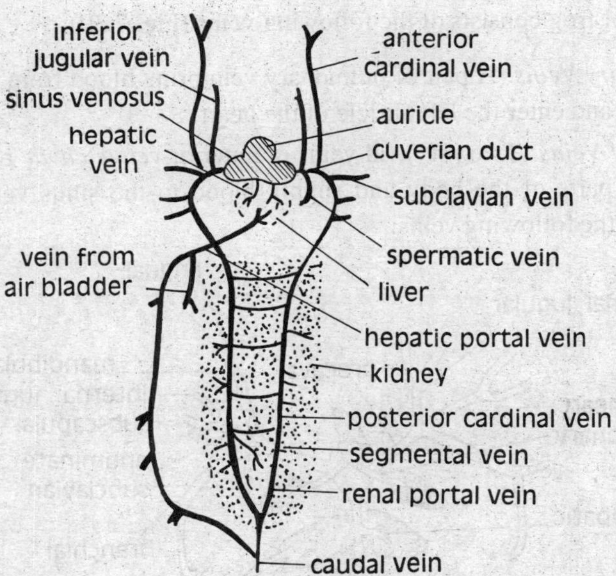

Fig. 7.29 Venous system of Labeo

(1) *Ductus cuveri* are present on right and left side and carry blood to the sinus venosus. Each ductus cuveri receives three veins, an anterior cardinal sinus, a jugular sinus, a posterior cardinal sinus, pectoral vein, pelvic vein and hepatic vein.

(2) The *anterior cardinal sinuses* bring blood from the anterior part of the body.

(3) The *posterior cardinal sinus* brings blood from the posterior part of the body and receives blood from the segmental, renal and genital veins.

(4) The *pectoral* and *pelvic veins* receive blood from the pectoral and pelvic veins respectively.

(5) The *hepatic vein* is slender vein from the liver.

(6) The *caudal vein* receives blood from the tail region and bifurcates into two branches, right and left *posteriorcardinal sinuses* on entry into the trunk. The right posterior cardinal sinus passes through the right kidney and opens into the right ductus cuveri.

(7) The *renal portal system:* The left branch of the caudal vein breaks up into capillaries in the kidney to form portal vein. These capillaries reunite to form a posterior cardinal sinus.

(8) *Hepatic portal system:* The capillaries from the digestive canal and its associated glands form a hepatic portal vein, which enters the liver and breaks up into capillaries. These capillaries reunite to form a hepatic vein supplying to the ductus cuveri.

RANA

The venous system of frog consists of the following veins (Fig. 7.30):

(1) *Pulmonary Veins:* A pair of pulmonary vein bring blood from two lungs. They unite together and enter the left auricle of the heart.

(2) *Precaval Veins:* Two precaval veins or anterior *venae cavae* receive blood from the anterior parts of the body and supply blood to the sinus venosus. Each precaval receives the following veins:

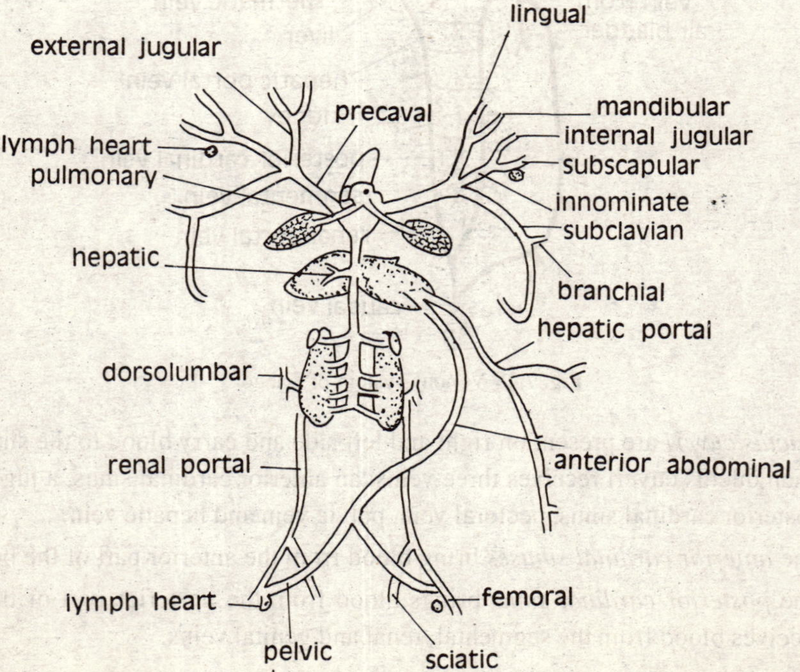

Fig. 7.30 Venous system of Rana

i. An *external jugular,* which receives a lingual vein from the tongue and mouth, and a mandubular vein from the lower jaw.

ii. An *innominate vein,* which receives an internal jugular vein from the brain and orbit and a *subscapular vein* from the shoulder.

iii. A *subclavian vein,* which receives a branchial vein from the and a large *musculocutaneous vein* from skin and muscles of the back and side, head and mucous membrane of the mouth cavity.

(3) *Postcaval Vein:* The renal veins from kidneys form a postcaval vein which receives a pair of *gonadial veins* from gonads and a pair of *hepatic veins* from the liver. It passes forwards and enters the sinus venosus. (Ture posterior cardinal veins found in fishes are replaced by the postcaval vein).

(4) *Renal Portal System:* The venous blood from each leg is collected by a *femoral vein* and a *sciatic vein.* Each vein after entry to the bodycavity divides into two veins, a *dorsal renal portal* and *ventral pelvic vein.* The sciatic vein and a dorsalumber vein from the back muscles join the renal portal vein which passes to the kidney and breaks up there into capillaries to make a renal portal system. The blood from kidneys is collected by five or six renal veins.

(5) *Hepatic Portal System:* An anterior abdominal vein is produced by the union of two pelvic veins. It passes forwards in the mid-ventral body wall and receives an unpaired vesicle vein from the urinary bladder and divided into two small branches which enter the left liver lobe and break up into capillaries there. A hepatic portal vein receives blood from the gastric, intestinal, duodenal, splenic and pancreatic veins and divides into capillaries to from a hepatic portal system.

A part of the blood from the hind limbs has to pass through renal portal system where it filtered in kidneys and sent to the heart through the postcaval vein. The remaining blood from legs is sent to the liver through pelvic and anterior abdominal veins where it undergoes some changes and ultimately reaches the heart through the postcaval vein.

UROMASTIX

The venous system of *Uromastix* includes the following veins (Fig. 7.31):

(1) A pair of *pulmonary veins,* which bring blood from the lungs. They unite together and open into the left auricle.

(2) *Precaval veins:* Two precavals enter a bilobed sinus venosus, which is connected with the right auricle. The right precaval receives the following veins.

(i) An *external jugular* from the neck.

(ii) An *internal jugular* from the head,

(iii) A *subclavian vein* from the arm.

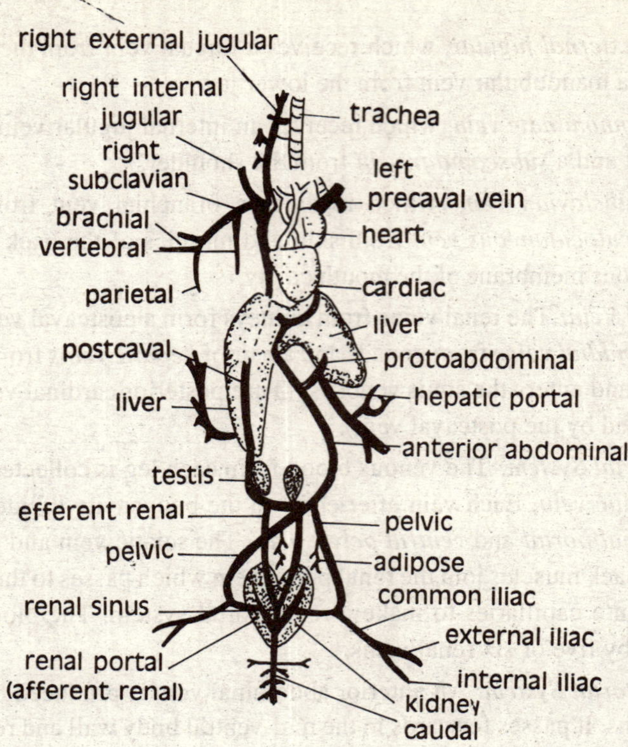

Fig. 7.31 Venous system of Uromastix

(3) *The Postcaval Vein:* The capillaries in kidneys unite to form a renal sinus, which continues infront as two efferent renal veins. The efferent renal veins join each other behind the gonds and form a postcaval vein, which passes through the liver and receives a large *hepatic vein* and enters the sinus venosus.

(4) *The Renal Portal System:* The caudal vein collects blood from the tail and bifurcates into two renal portal veins. Each renal portal vein divides into capillaries to form a renal portal system.

(5) *Hepatic Portal System:* One *femoral* and one *sciatic vein* come from each leg and unite to form an *iliac vein*. Each iliac vein is joined by a small transverse connection to the renal portal vein and continues forward as *pelvic vein.* The two pelvic veins unite in the middle to form an *anterior abdominal* or *epigastric vein,* which enters the left liver lobe and divides into capillaries. (The anterior abdominal vein is formed by incomplete joining of the two lateral abdominal veins of the embryo and is considered a primitive character). A *hepatic portal* vein is formed by small gastric, intestinal, duodenal, and splenic and pancreatic veins, which join the abdominal vein and enter the left lobe of liver and divides to form a hepatic portal system.

The *renal portal system* is reduced in reptiles. Most of the blood from legs and tail passes through the abdominal veins to the liver and a little amount goes to the renal portal system.

COLUMBA

The sinus venousus is absent in pigeon (Fig. 7.32) and the two precavals and one postcanal open directly into the right auricle. The renal portal system is reduced .The venous system of pigeon is given below:

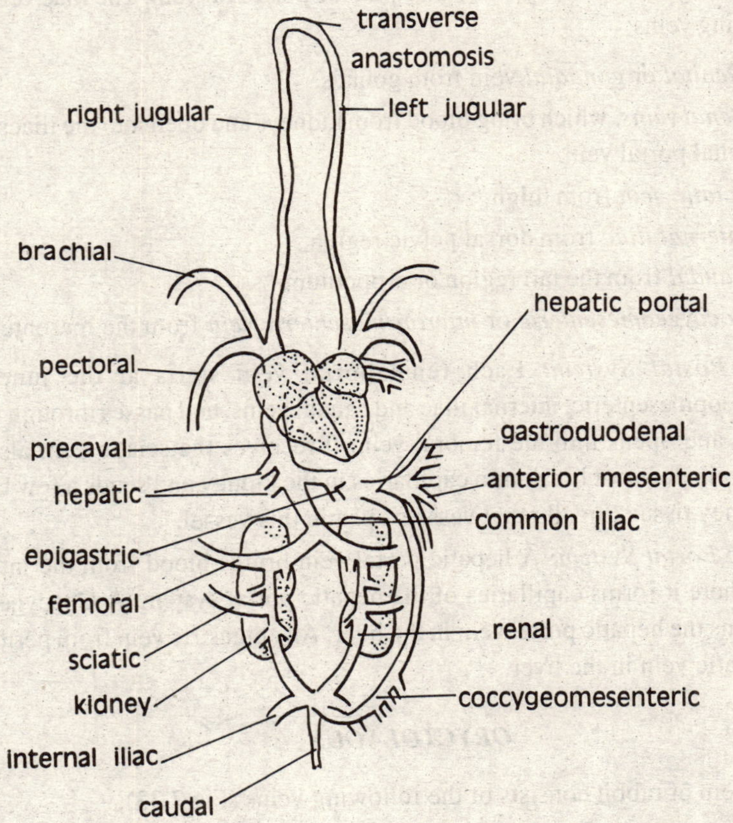

Fig. 7.32 Venous system of Columba

(1) *Pulmonary Veins:* Each lung gives off two pulmonary veins, which carry oxygenated blood to the left auricle of the heart.

(2) *Precaval Veins:* Two precaval veins collect deoxygenated blood from the anterior side of the body. Each precaval receives the following veins.

(i) *Branchial vein* receives blood from the wing and some branches from the shoulder also open into it.

(ii) *Pectoral vein* is formed by the branches of the pectoral region.

(iii) *Internal mammary vein* brings blood from the sternum, coracoid region and the ribs.

(3) P*ostcaval Vein:* The postcaval vein is formed by the union of two iliac veins. The iliac veins are continuation of the femoral veins coming from the region of hind legs. The femoral veins pass through the kidneys. A few hepatic veins from liver and a small vein from the ligament of the gizzard also join the postcaval vein. The iliac veins receive the following veins:

(i) *Genital* or *gonadial* vein from gonads.

(ii) *Renal veins,* which bring blood from kidneys and open into the iliacs and also into renal portal vein.

(iii) *Sciatic vein* from thigh.

(iv) *Internal iliac* from dorsal pelvic region.

(v) *Caudal* from the tail region or uropodium.

(vi) *Coccygeomesenteric* or *inferior mesenteric vein* from the mesentery.

(4) *Renal Postal System:* Each renal portal vein starts at the junction of the coccygeopmesenteric, internal iliac and caudal veins, and passes through the kidney of its side and opens into the femoral vein. It receives the sciatic vein also. The renal portal vein does not break into capillaries in the kidney and sends a few branches into the kidney tissue. Small renal vessels open in this vessel.

(5) *Hepatic Portal System:* A hepatic portal vein brings blood from the intestine to the liver where it forms capillaries of the hepatic portal system. A Coccygeomesenteric vein joins the hepatic portal vein in the liver. An epigastric vein from peritoneum joins the hepatic vein in the liver.

ORYCTOLAGUS

The venous system of rabbit consists of the following veins (Fig 7.33):

(1) *Pulmonary Veins:* A pair of pulmonary veins brings oxygenated blood from the lungs by a common opening into the left auricle.

(2) *Coronary Veins:* A pair of coronary veins takes blood from the muscles of the heart and open into the left precaval vein.

(3) *Precaval Veins:* A pair of precaval veins open into the right auricle by separates openings. Each precaval vein is made of the following veins:

(i) An *external jugular vein,* which brings blood from the tongue, jaws and head muscles. The two external jugulars in rabbit are connected by a transverse *jugular anastomosis.*

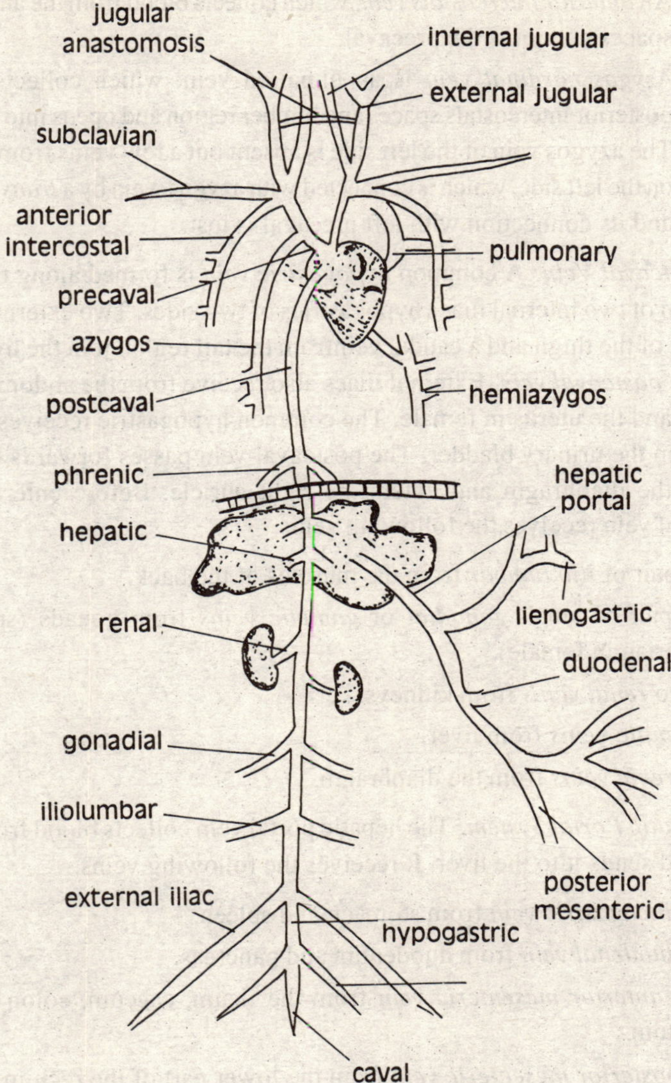

Fig. 7.33 Venous system of Oryctolagus

(ii) An *internal jugular vein* brings blood from the brain and neck joins the external jugular vein.

(iii) A large *subclavian vein* from the shoulder and forelimb is joined by external jugular vein.

The precaval veins formed thus are further joined by the following veins before entering into the right auricle.

(i) An *anterior intercostals vein,* which collects blood from the anterior intercostals, spaces and joins the precaval.

(ii) *Azygos cardinal vein* is an unpaired vein, which collects blood from the posterior intercostals spaces and lumber region and opens into the right precaval. The azygos vein of the left side is absent but a few veins from a *hemizygos vein* on the left side, which is connected with azygos, vein by a *transverse anastomosis* and its connection with left precaval is lost.

(4) *The Postcaval Vein:* A common *hypogastric vein* is formed along the middle line by the union of two internal iliacs hypogastrics of two sides. Two external iliac veins from the back of the thigh and a caudal veinfrom the tail region join the hypogastric vein to form the *postcaval vein.* External iliacs also receive from the abdominal wall, urinary bladder and the uterus in female. The common hypogastric receives a short vesicular vein from the urinary bladder. The postcaval vein passes forwards through the liver, pierces the diaphragm and enters the right auricle. Before entering the heart the postcaval vein receives the following veins.

(i) A pair of *ilio lumbar* from the muscles of the back.

(ii) A pair of small *gonadial* or *gametic veins* from gonads (spermatic in male ovarian in female).

(iii) Two *renal veins* from kidneys.

(iv) *Hepatic veins* from liver.

(v) *Phrenic veins* from the diaphragm.

(5) *The Hepatic Portal System:* The hepatic portal vein collects blood from the alimentary canal and sends it to the liver. It receives the following veins.

(i) A *lineogastric vein* from stomach and spleen.

(ii) A *duodenal vein* from duodenum and pancreas.

(iii) An *anterior mesenteric vein* from the ileum, caecum, colon and parts of the rectum.

(iv) A *posterior mesenteric vein* from the lower part of the rectum.

The renal portal system is absent in rabbit.

7.8. THE LYMPHATIC SYSTEM

In addition to the venous system vertebrates have a second series of vessels returning blood from the tissues to the heart. The plasma and leucocytes escape from the blood capillaries into the tissues. This colourless poation of the blood without erythrocytes and blood protein is called *lymph.* The lymph carrys oxygen and food to cells of the body and takes substances from tissues which reenter blood by means of lymphatic vessels. The lymphatic system includes *lymph vessels lymph sinuses, lymph hearts, lymph nodes,* and *lymphoid organ* such as tonsils and spleen.

7.8.1. The Lymphatic Organs

The *lymph vessels* or lymphatics are very thin walled tubes with a thin coal of muscles and connective tissue. In birds and mammals they are provided with valves to prevent the backward flow. The lymph vessels arise from their capillaries, which are blind at their tips. The lymph capillaries are highly permeable and carry particles like colloids, broken tissue cells and bacteria, which cannot pass through the walls of blood capillaries.

The *lymph sinuses* are present in all vertebrates between skin and muscles and among the muscles next to the skeleton. The lymph sinuses contain many valves arranged in such away that the pressure from the movement of muscles and other organ forces the lymph.

The *lymph hearts* are found in some fishes, amphibians and reptiles, which help in pushing the lymph, by the slow contraction of their smooth muscle walls. These hearts usually have two chambers with valves.

The *lymph nodes* are found scattered in many pans of the body in birds and mammals. The lymphatic vessels meet in spongy clumps of varying sizes to produce such nodes. In mainmasls, lymphocytes are produced in lymph nodes, which circulate freely and check infection by phagocytic action. They destroy bacteria and supply lymphocytes to the lymph.

The *lymphoid organs* are comparable to the lymph nodes. The *tonsils* are present on the sides and posterior wall of the pharynx. The *thymus body* is another lymphoid structure with uncertain functions. The spleen contains pulpy lymphoid tissue with irregular sinuses.

Circulation Of Lymph (Fig. 7.34): The lymph capillaries are smallest vessels, which collect tissue fluid between the cells. The tissue fluid after entry to lymphatic vessels is called *lymph.* The lymph capillaries end blindly at their tips and their bases unite to form larger vessels, which combine with vessels of greater size. The larger lymph vessels mostly open directly into the veins near the heart. The lymphatic system is a closed system but does not form a complete system.

7.8.2. Other Lymphatic Organs

The structures usually associated with the lymphatic system are tonsils, adenoids, thymus gland, Peyer's patches and spleen. Special mention of spleen is necessary owing to its importance in the animal physiology and anatomy.

Spleen: The spleen is found in all vertebrates excepting cyclostomes and Dipnoi. It is largest lymphoid organ of the body. It is a dark red, lymphatic organ, usually found in relation to stomach, sometimes far away from it.

Structure (Fig. 7.35): The spleen is covered over by an outer *capsule.* The trabeculae pass into the substance of the spleen from the capsule and divide it into *lobules* or compartments. The capsule and trabeculae contain dense connective tissue containing clastic fibres and unstriped muscles, nerves, lymph vessels, and blood vessels. The blood vessels pass beyond the trabeculae enters the parenchyma of the spleen and form capillary network of arteries and veins with sinusoids connecting the arteries and veins directly. The lobules of spleen have different sizes of cells, which may be pigmented. The pulp consists of diffuse lymphatic tissue of reticular cells and fibres, blood corpuscles, macrophages, and lymphocytes. The *white pulp* is made of lymphatic tissue around small arteries, which appear as isolated circular masses in section called *splenic nodules*.

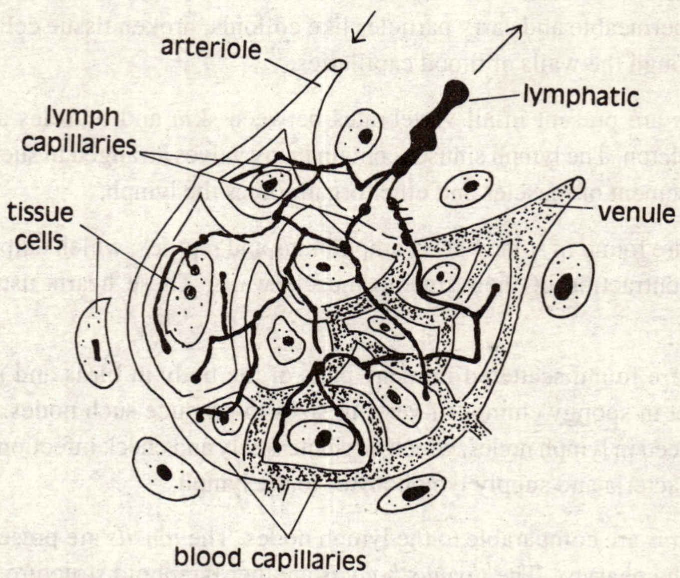

Fig. 7.34 Lymph and blood capillaries

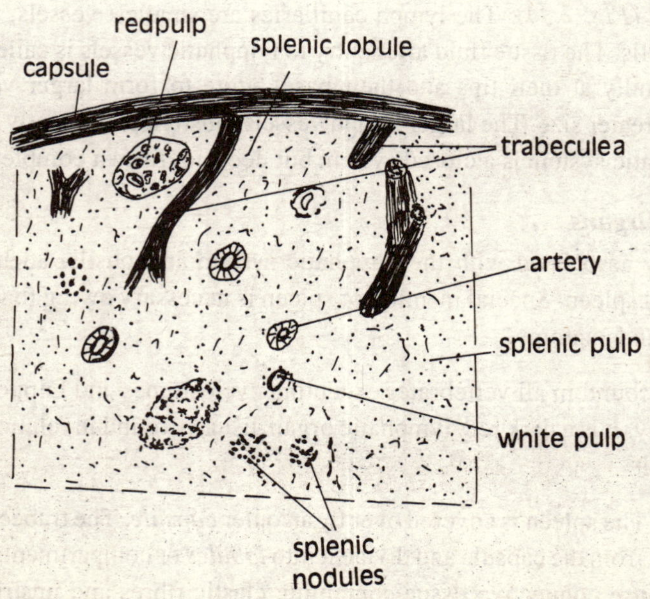

Fig. 7.35 Spleen in a section

In an embryo the spleen is formed as a thickening from mesenchyme cells in the mesentery suspending the stomach. The differentiation of mesenchyme cells to form splenic cells and pulp is brought about by the origin of blood vessels in the spleen.

Function of Spleen: The spleen is considered as a haemolymph gland as ill is interposed in the blood system rather than in lymphoitic vessels. Its function is as follows:

1. The disintegration of red blood carpuscles into fragements takes place in blood, which comes in contact with phagocytes in spleen. The phagocytes engulf the blood fragments.
2. *Storehouse For Erythrocytes:* The spleen serves as the storehouse for erythrocytes in large number. On demand by the blood vascular system the erythrocytes are returned to the blood by the contraction of the spleen. One-sixth of blood may be stored in the spleen.
3. *Blood Formation:* In embryo the spleen acts as blood forming organ (*Haemopoietic organ*). In mammals, the erythrocytes are not produced by the spleen during the adult stage but lymphocytes are produced.
4. *Production of Antibodies:* It produces antibodies serving as defensive mechanism against diseases. The toxin of diphtheria is retained by the splenic cells.

7.8.3. The Lymphatic System in Vertebrates

The comparative anatomy of the lymphatic system is not well known as the small size and delicate structure of vessels restrict their studies. The irregularity of the lymph vessels in different classes of vertebrates do not follow evolutionary sequence.

The protochordates do not have lymphatic system.

The *Cyclostomes* and *Elasmobranchs* do not have lymphatics but thin walled *sinusoids* with little or no arterial connections are present. They probably represent the first stage in the development of the lymphatic system.

In *teleosts* the lymph vessels are well developed. They are present below the skin in the muscles and viscera and extend into head, tail and fins. The lymph vessels open through connections into veins in the anterior, middle and posterior parts of the body. The lymph hearts are present in some fibres. The lymph nodes are not found in teleosts.

In *Amphibia* two sets of lymphatic vessels are present in Urodela. 14 to 20 *lymph hearts* are present in the lymph vessels, which run superficially below the skin and connect to contaneous and postcardinal veins. Deeper lymph channels run parallel to the dorsal aorta and enter the subclavian veins.

In *Rana* the lymph capillaries in tissues unite to form *lymph vessels.* Some lymph vessels are detailed to form large lymph sinuses. Large subcutaneous lymph sinuses are present between skin and muscles, which are separated from each other by fiberous membranes. A *subvertebral sinus* is present around the dorsal aorta and kidneys. Two pairs of *lymph hearts* are present, the first opens into the subscapular vein behind the transverse process of the third vertebra, and the second pair opens into the femoral veins at either side of the end of urostyle. The lymph from lymph sinuses is pumped into veins by lymph hearts.

In *Reptilia* the lymphatic system is well developed. The lymph capillaries from lymph vessels. The main lymphatic vessel is subvertebral trunk, which divides into two branches infront. Each branch enters a precaval vein. One pair of lymph hearts is present which open into the iliac veins.

In *birds* the lymph hearts are absent in adults. The lymph vessels are extensively distributed. The lymph vessels jointly form two thoracic ducts homologous to the subvertebral trunks which open into the precaval veins.

In *mammals* the lymphatic system is extensively developed. The lymph capillaries form lymph vessels, which are interrupted by lymph nodes at different places. The lymph nodes contain lymph capillaries, *lymphocytes*, phagocytes, reticular cells etc., and protect the body from infection. The lymph nodes are present in the head, neck, armpits, groins, and close to large blood vessels. The tonsils and Peyer's patches on intestine are also formed by the lymph nodes. The lymph capillaries from intestinal villi called *lacteals*, carry milky white lymph containing emulsified fats. Lymph hearts are Totally absent in mammals. A thoracic duct collects lymph from all the posterior lymphatic vessels present on the left side of the head, neck, and thoracic regions. It is situated below the vertebral column and passes anteriorly and opens into the subclavian vein near its junction with internal jugular vein. The posterior end of the thoracic duct is produced into an enlarged *cisterna clyili* or *receptaculum.* A right lymphatic duct collects lymph from the right side of the neck, right arm, and right side of the thorax and enters the right subclavian vein near its junction with the internal jugular vein.

CHAPTER SUMMARY

1. The circulatory system of vertebrates includes blood vascular and lymphatic systems. The blood vascular system is composed of heart, arteries, veins and capillaries. The lymphatic system is composed of lymphoid organs of different types, lymph capillaries, vessels, sinuses and nodes.
2. The vertebrate blood consists of plasma, crythrocytes and platelets. The composition of blood varies in different vertebrates. The erthrocytes or red blood corpuscles are nucleated, flattened and oval cells in all groups excepting in mammals. In camels and ilamas the erythrocytes are oval cells in all groups excepting in mammals. In camels and ilamas the crythrocytes are oval. In adult mammals the crythrocytes are nonnucleated. The leucocytes of several types are formed in bone marrow and lymphoid tissue and may pass out of different tissue through the blood They have phagocytic properties and may ingest or destroy bacteria. The platelets are homologus to the thrombocytes of lower animals and help in the clotting of the blood.

3. The blood vascular system consists of arterial, venous and portal systems. The arterial system carries blood away from the heart. The venous system brings blood to the heart. The portal system consists of veins breaking up into a capillary network before reaching the heart.
4. The arteries and veins have three layered walls of varying thickness. These layers are tunica intima, tunica media and tunica externa. The capillaries have single layered wall composed of the endothelium which allows exchanges of materials with tissues.
5. The heart is formed in the embryo by endocardial cells which form two endothelial tubes. The endothelial tubes fuse to form a single tube which divides into chambers called atria and ventricles.
6. The heart is absent in *Branchiostoma.* In cyclostomes and most fishes it is two chambered with one atrium and one ventricle. In dipnoi and amphibians the heart is three chambered with two atria and one ventricle. In most reptiles the heart is three chambered but incomplete partition in ventricle is found in crocodiles it is four chambered. In birds and mammals the heartis four chambered.
7. The arterial system of adult vertebrates appears to be different in structure but its development exhibits similar fundamental plan. The aortic arches supply the gills is fishes and amphibians and their later modifications in amniota are connected with increasing complexity of the heart. There are usually six pairs of aortic arches extending from the ventral aorta to the dorsal side of the pharyngeal region where they fuse to form a vessel on each side: these two veseels combine forming a single dorsal aortaon the posterior side. A progressive reduction in the number of aortic arches takes place from fish to mammals.
8. The venous system is originally paired but due to introduction of air breathing habits it has become asymmetrical. The venous system of higher forms during its development passes through some stage common to embryonic stage of the lower forms. The primitive venous system consists paired subintestinal, vitelline, anterior and posterior cardinal veins, lateral or a ventral abdominal veins and an inferior jugular vein. The vitelline and subintestinal veins give rise to hepatic portal vein, which breaks up into liver and breaks off its connection with sinus venosus. The vitelline veins are lost in the adult. The posterior cardinal veins divide into two portions, one of which makes renal portal vein alongwith the caudual vein. The poscaval of higher forms is derived as an outgrowth of the vitelline veins.
9. The lymphatic system returns blood from the tissues to the heart. The colourless portion of the blood without erythrocytes and blood proteins is called lymph. The lymphatic system includes lymph vessels, lymph sinuses lymph hearts, nodes and lymphoid organs such as tonsils and the spleen.
10. The spleen is largest lymphoid structure and serves as hemolymphatic organ. It also stores and releases red blood corpuscles according to body requirements.

8

Excretory System

8.1. GENERAL ACCOUNT

ɪhe excretion is process which deals with the elimination of waste materials from the body. These waste materials are end products of body metabolism such as carbon dioxide, ammonia, urea, uric acid, different types of pigments, inorganic salts and creatinine. The function of excretion is to remove these wastes from the tissues of the body to outside. The carbon dioxide is mostly executed by skin, lungs or gills and remaining substances mostly soluble in water, are eliminated through kidneys in vertebrates. The kidneys also serve to eliminate excess of water from the body. The kidneys are mesodermal in origin.

An intimate interrelation exists between the excretory and reproductive organs in vertebrates.

The excretory ducts carry the genital products directly to the exterior or provide seat for the partial development of egg. As such, these organs are collectively called *urinogenital* or *urogenital organs.* In the present chapter only excretory organs have been described.

8.1.1. The Vertebrate Kidney

There are two kidneys in vertebrates. Each kidney consists of a mass of coclomoducts opening in a collecting duct (Fig. 8.1). The kidneys are usually situated in the posterior part of the body and develop segmentally as one pair per body somite (pronephros and mesonephros). In higher animals such as reptiles, birds and mammals, the adult kidneys are not segmental. Typically, each kidney consists of a mass of *uriniferous tubules or nephrons* (Fig. 8.2). At one end the nephrons are invaginated to form a cup like *Bowmans capsule* which contains a mass of blood capillaries called *glomerulus.* The Bowman's capsule and glomerulus are called *Malpighian body.* The glomerulus is formed by the capillaries of the renal artery which enters the Bowman's capsule as an *afferent arteriole* and leaves the capsules as *efferent arteriole* which supplies blood to the remaining part of the tubule from where the blood returns to a renal vein. The uriniferous Tubule, in addition to Malpighian body, contains *proximal convoluted tubule, loope of Henle, and distal convoluted tubule.* The distal convoluted tubule joins a *collecting tubule.* A collecting duct, *ureter* carries wastes from each kidney posteriorly. The two ureters discharge into cloaca in amphibians, reptiles and birds. In mammals the two ureters are connected with a urinary bladder and a median duct *urethra* discharges to the exterior.

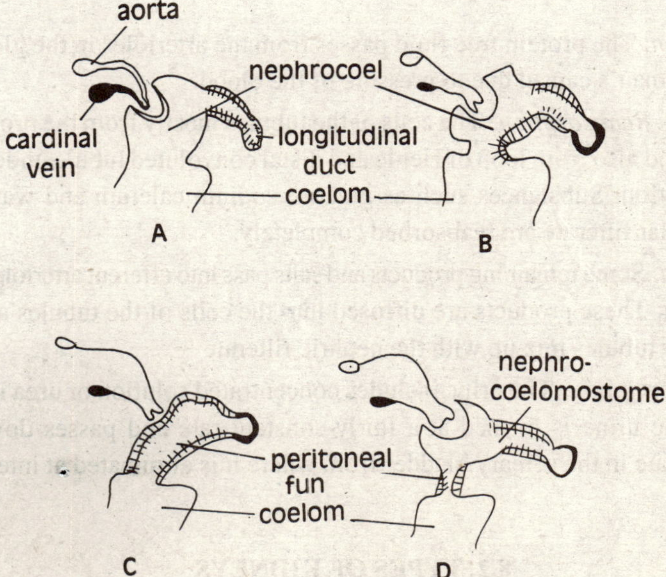

Fig. 8.1 Showing the evolution of vertebrate kidney

The excretory organs develop from the mesomere of the mesoderm which arise as separate units called *nephrotomes* in series with myotomes above them.

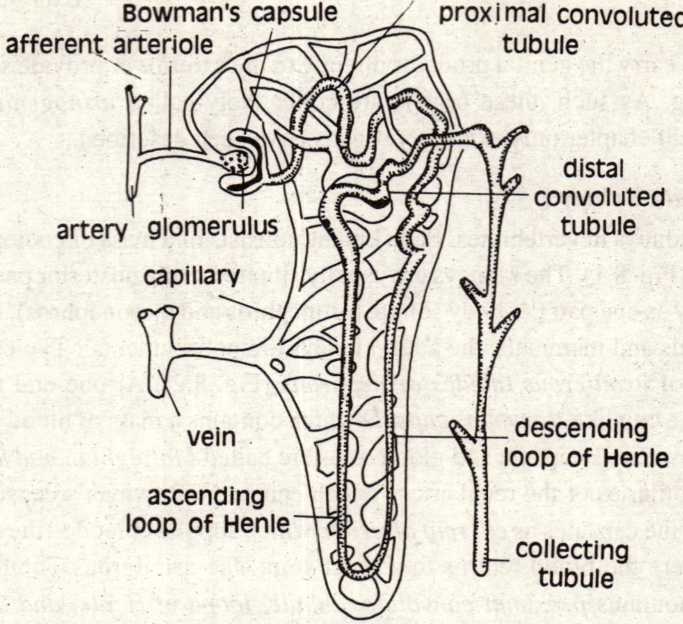

Fig. 8.2 Nephron with vascular supply

8.1.2. Working of Kidney

The waste and other materials are brought in the blood by renal arteries to the arterioles. The kidney function involves the following steps.

1. *Filtration*: The protein free fluid passes from the arterioles in the glomerulus through the Bowman's capsul due to pressure of the blood.
2. *Selection Reabsorption:* The cells of the tubules mostly from the proximal convoluted tubule and also from loop of Henle and distal convoluted tubule undertake differential resbsorption. Substances such as glucose sodium calcium and water present in the glomerular filterate are reabsorbed completely.
3. *Secretion:* Some remaining products and salts pass into efferent arterioles after glomerular filtration. These products are diffused into the cells of the tubules and pass on to the lumen of tubules mix up with the nephric filterate
4. *Urine Formation:* The urine includes concentrated solution of urea in water and some salts. The urine is formed at a fairly-constant rate and passes down the ureters to accumulate in the urinary bladder from where it is eliminated at intervals through the urethra.

8.2. TYPES OF KIDNEYS

The kidneys of different vertebrates have similar structure and function but differe in their position in relation to body cavity and alterations in the number, complexity arrangement and location of glomeruli and tubules. Earlier writers classified kidneys into three types vix. Pronephrons, mesonephros and metanephros on the basis of their structure. Recent writers have considered their true homologies and divided them into three distinct groups which include archinephors or primitive hypothetical kidney, *anamniote kidneys of pronephros* and *pisthonephros* types and amniote kidneys of *mesonephors* and *metanephros* types.

8.2.1. Archinephros (Fig. 8.3.4)

It is presumed that primitive vertebrate ancestors had archinephros or holonephros type of excretory organs. The larval forms of hagfish and larvae of some caecilians have such a kidney. The kidneys of different types have been derived from this kidney. It consists of a pair of archinephric ducts running parallel on the dorsal side of the coelome and extending throughout the length of the animal. These ducts are joined by a series of segmentally arranged tubules, one tubule in each segment. Each tubule opens into the coelom at the other end by means of ciliated nephrostome or funnel. A small bundle of blood vessels called glomerules is associated with each tubule.

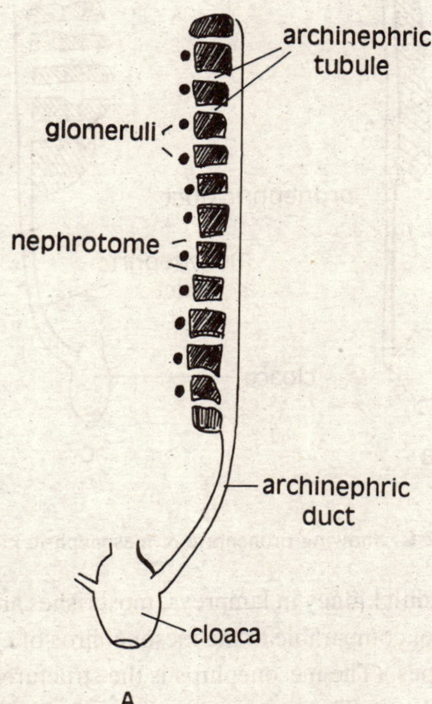

Fig. 8.3 A. The primitive kidney

8.2.2. Anamniote Kidneys

Such paired kidneys are of two types: pronerphric and opisthonephric.

Pronephros. (Fig. 8.3 B): Pronephros are paired kidneys found in vertebrate embryos and cyclostomes. In chondrichthyes, the pronephros degenerates soon after its formation. In many fishes and amphibian larvae it is modified and functional. In hagfishes and a few teleosts it persists in the adult life, Each pronephros contains 1-13 uriniferous tubuless, one pair in each segment. Each tubule contains a glomerules but Bowman's capsule and peritoneal fuunel are absent. When glomerulus projects into coelom without Bowman's capsule it is called external glomerules. (When Bowmen's capsule is present it is called internal glomerulus). The uriniferous tubules of each pronephros open into a common pronephric duct which extends posteriorly to open into the cloaca.

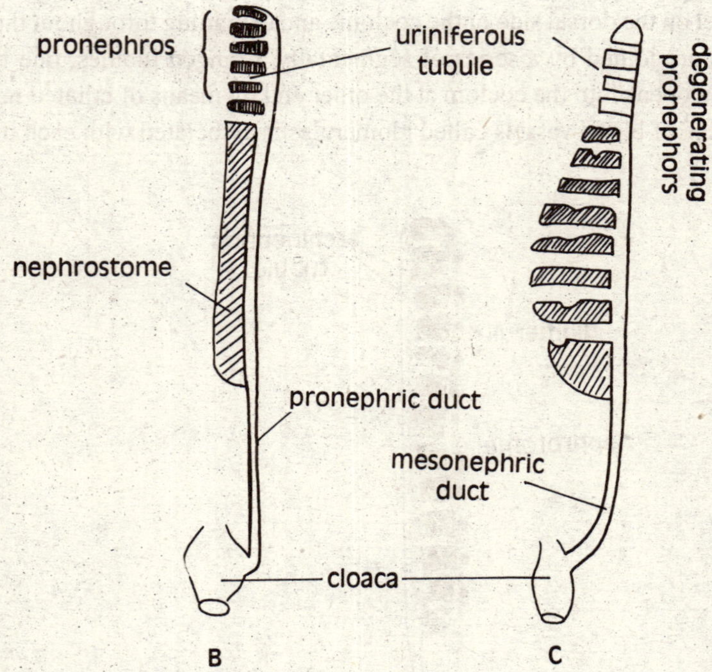

Fig. 8.3 B & C. Showing pronephric & mesonephric kidneys

Opistonephros: It serves as adult kidney in lampreys, most fishes and amphibians. It is usually referred as mesonephros but is not comparable to the mesonephros of embryonic amniotes inspite of similar structure of the two types. (The mesonephros is the structure, which appears during the embryonic development of amniotes). The archinephric duct of opisthonephros is modified into genital ducts and accessory urinary ducts are formed which carry the waste products. The segmental arrangement of urinary tubules is lost and many tubules may lie in a single segment. The connections between uriniferous tubule and coelom are lost and internal glomeruli with Bowman's capsule are present.

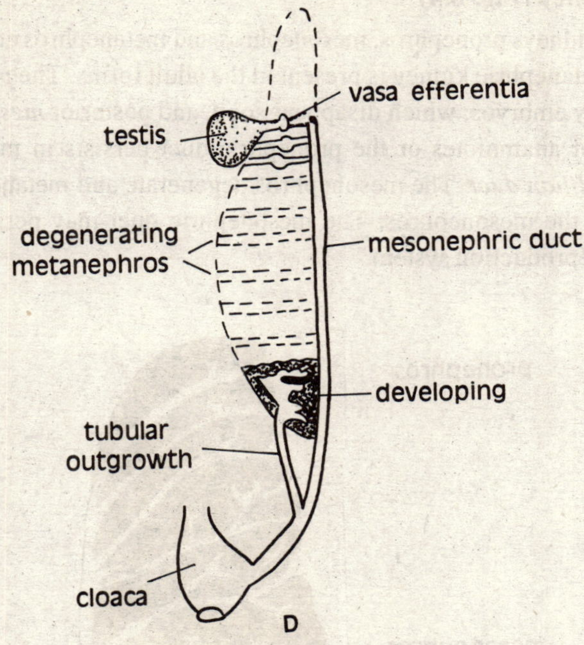

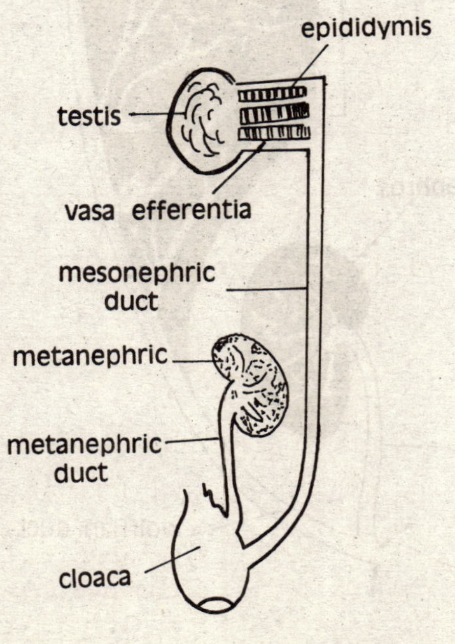

Fig. 8.3 D & E Development of urinary & genital ducts in vertebrates.

8.2.3. Amniote Kidney (Fig. 8.4)

Three types of kidneys pronephros, mesonephros and metanephros are recognized in amniotes; out of them only metanephric kidney is present in the adult forms. The *pronephros* appears on the anterior side of early embryos, which disappear soon, and posterior *mesonephros* developed. The archinephric duct of anamniotes or the pronephric duct persists in mesonephros and is called *mesonephric* or *Walffian duct.* The mesonephros degenerate and metanephros develops from the posterior region of the mesonephros. The mesonephric duct may persist and contribute in the formation of male reproduction system.

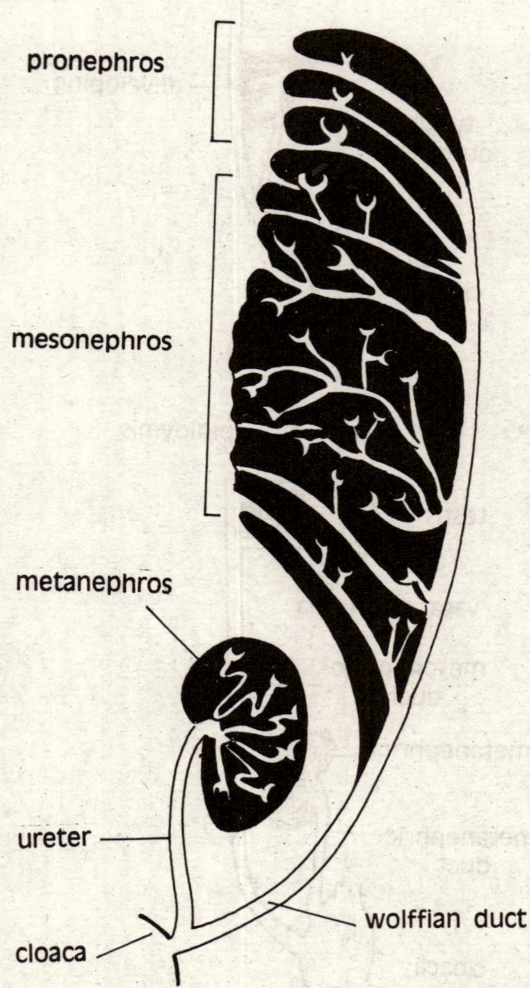

Fig. 8.4 Development of kidney in amniotes

Mesonephros (Fig. 8.3. CE). The mesonephros or Wolffain body arises by the development of mesonephric tubules from the nephrotome behind the pronephros. The mesonephric tubules extend laterally and meet and fuse with the pronephric duct and form a mesonephric or Wolffian duct. The mesonephric tubules form numerous tubules later on by budding, which may result in the loss of their segmental arrangement. The mesonephric tubules have better blood supply and blood capillaries called glomerulus enclosed by the Bowman's capsule make the Malpighian body. The mesonephric kidney may become voluminuous structure temporarily occupying an extensive portion of the coelom, or poorly developed transitional structure (in embryos of guinea pigs and mouse), or may persists even after the birth (Echidna, certain marsupials). When the mesonephros persists, the mesonephric duct or Wolffian duct serves as urinary passage. The mesonephros develops into male reproductive organs including epididymis, vas deferens. seminal vesicles etc. and rudiments of female organs, such as *epoophoron, paraoophoron* and blind Gartner's duct.

Metanephros: The metanephric kidney develops posterior to mesonephros on each side. It is found only in amniotes and is developed from the posterior most part of the nephrotome behind the embryonic mesonephros. The metanephros are functional kidneys of adult amniotes with separated urinary and genital functions and show evolutionary advancement over the mesonephros. The metanephros is more compact than mesonephros and does not show any trace of segmental character. The metanephros corresponds to the posterior portion of the opisthonephros of anamniotes. Nephrostomes are absent and other essential parts of the tubules are present. The metanephrosarise from the caudal end of the embryonic mesonephric duct near cloaca as an anteriorly directed metanephric duct. It grows into the nephrotome and divides into branches producing thousands of uriniferous tubules without segmental arrangement. The proximal part of the tubular outgrowth becomes ureter or matanephric duct and branches form *collecting tubules* and *calyces.* The metanephic tubules become elongated and coiled and contain glomeruli enclosed in Bowman's capsules.

The structure of the kidneys may be summarized by the following table:

TABLE 8.1

Archinephros

Primitive hypothetical kidney with archinephric duct and ciliated nephrostomes
Various vertebrate kidneys may be derived as successive stages in anteroposterior direction.

Anamniote Kidney Found in fishes and amphibia		*Amniote Kidney Found in reptiles birds and mammals Anterior end of archinephric duct modified*		
Pronephros	*Opisthonephros*	*Pronephros*	*Mesonephros*	*Metanephros*
head kidney Found in adult hagfishes and certain teleosts.	Found in fishes and amphibia. Several urini-ferrous tubules	Found in embryos but disappears soon,	Appears after pronephros and functions during	Persists as functional kidney

When persists in adults it is called head kidney. It contains nephrostomes opening from coelome but glomeruli are absent.	may be present in one segment. Its connection with peritoneum is lost. It is comparable to the mesonephric kidneys of. amniotes.	embryonic life and disappears before birth. Its ducts persist as vas deferens in males.	after birth or hatching.

8.3. THE URINARY BLADDER

A urinary bladder is present in most vertebrates. It is usually a distensible, sac-like reservoir where urine is stored. The urinary bladder is absent in cyclostomes. elasmobranchi snakes, crocodile, some lizards and most of the birds. In fishes the bladder are terminal enlargements of the mesonephric ducts called *tubal bladders.* In dipnoi the bladder originates from the dorsal wall of the cloaca as a diverticulum anterior to the opening of the Wolffian ducts. It is homologous with the rectal glands of the clasmobranches. In amphibia the evaginations of the ventral wall of the cloaca give rise to the bladder. In amniotes the base of the allantois or the part proximal to the cloaca in embryos contributes the adult bladder called *allantoic bladder.*

In turtles and most lizards large bladders are present. In turtles accessory bladders used for earning water for nest building are also present. In birds the urinary bladder is present in ostriches.

In amphibians and reptiles the urine goes back to the bladder from the cloaca where urinary ducts open. In birds the urine gets mixed with faeces in the cloaca. In mammals excepting in monotremes the urinary ducts open directly into the bladder, which is drained out by the urethra. In monotremes the ureters open through small papillae near the base of the bladder. Elastic tissue is present at the junction of bladder and urethra and most of the bladder musculature continues down into the urethra in the form of bundles.

The mammalian bladder is highly distensible structure with stout walls and thick coat of smooth musculature. The lining is peculiar calléd transitional epithelium. When the bladder is empty the epithelium appears thick and stratified, when empty it thins down to a layer of flats cells. In males the urethra passes through the *penis* and opens at its tip through external *urethral orifice* or *meatus.* In females the urethra may open indepently to the exterior passing through the clitoris as in rats, or may enter into a vestibule or urinogenital sinus, which forms the terminal part of the urinogenital tract. The flow of urine is caused when the urethra is lengthened reducing the diameter of its lumen.

The urinary bladder stores urine, helps in water conservation and prevents desication under terrestrial conditions.

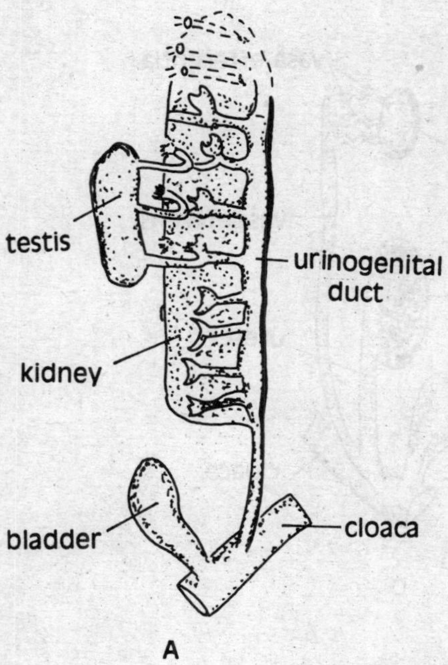

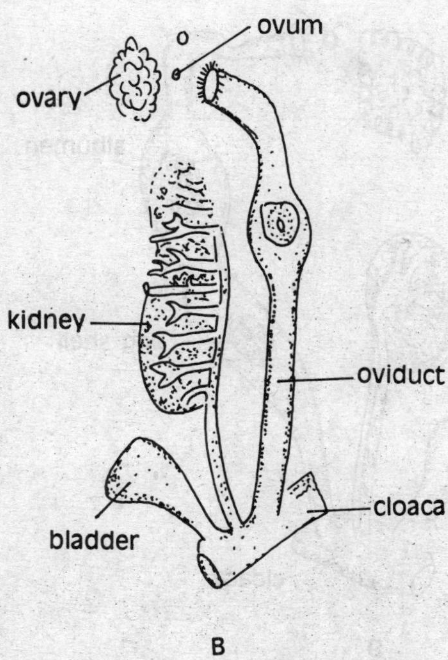

Fig. 8.5 A, B Urinogenital organs in vertebrates

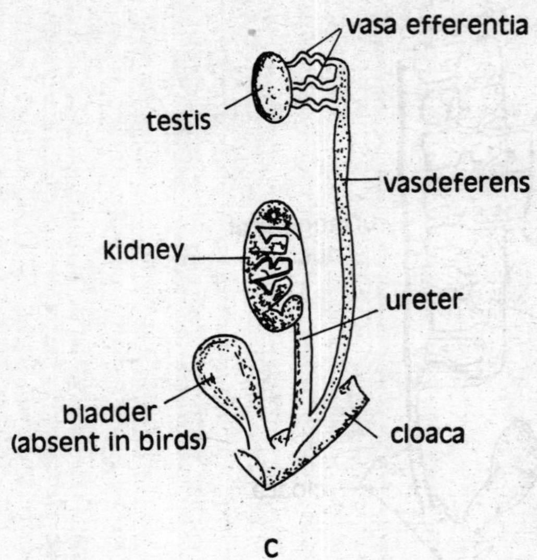

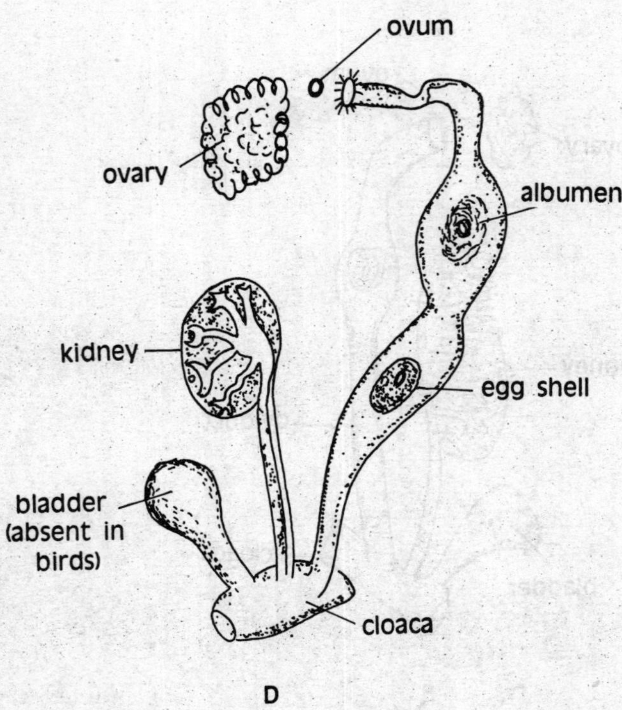

Fig. 8.5 C, D Urinogenital organs in Birds

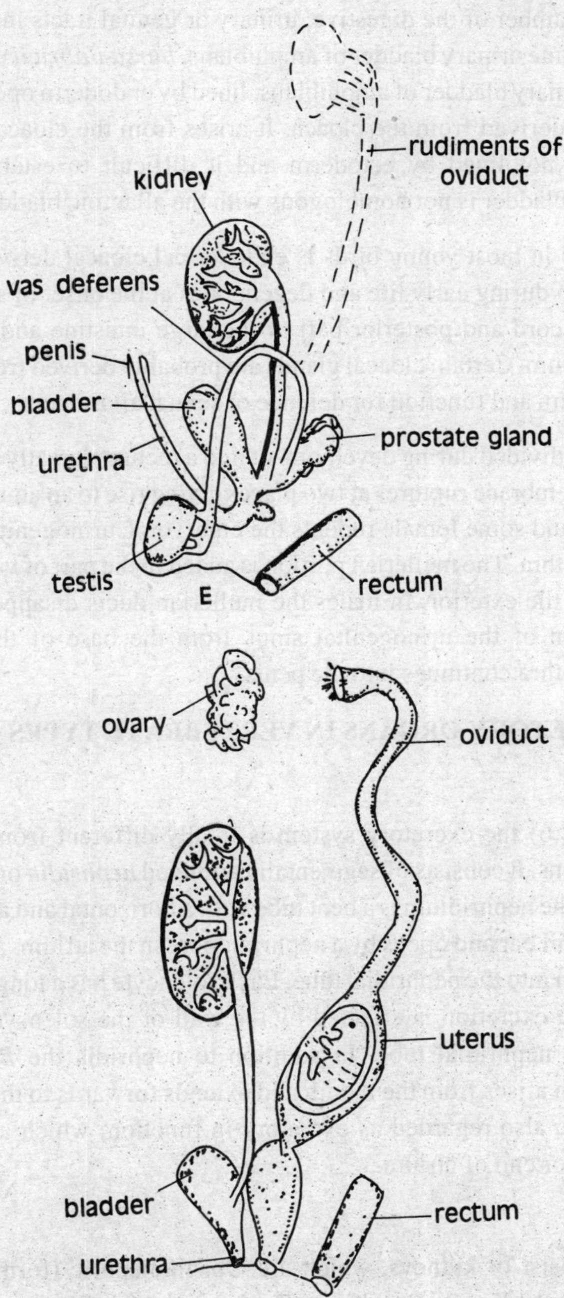

Fig. 8.5 E, F Urinogenital organs in mammals

8.3.1. Cloacal Derivatives

The cloaca is terminal chamber of the digestive, urinary or genital tracts in most vertebrates. The derivatives of cloaca include urinary bladder of amphibians, *bursa Fabricii* of young birds and certain cloacal glands. The urinary bladder of amphibians, lined by endoderm open directly into the cloaca and is believed to be derived from the cloaca. It arises from the cloaca after the cloacal membrane is ruptured. It is not lined by ectoderm and it difficult to establish, endodermal contribution. The amphibian bladder is not homologous with the allantoic bladder of amnoites.

The *bursa Fabricii* found in most young birds is endodermal cloacal derivative. It probably functions as a lymphoid organ during early life and degenerates at the onset of sexual maturity. It develops between the spinal cord and posterior part of the large intestine and forms secondary connection with the proctodaeum. Certain cloacal glands are probably derived from the ectodermal imagination of the proctodaeum and function for defence or sexual allurement.

The cloaca of mammals is divided during development into a rectum dorsally and a urinogenital sinus ventrally. The cloacal membrane ruptures at two places giving rise to an anus and urinogenital aperture. In female primates and some female rodents the embryonic urinogenital sinus is further divided into a vagina and a urethra. The mullerian portion is added to the part of vagina arising from the cloaca with direct exit to the exterior. In males the mullerian ducts disappear and urethra is formed due to the elongation of the urinogenital sinus from the base of the bladder to the urinogenital aperture. The urethra continues into the penis.

8.4. EXCRETORY ORGANS IN VERTEBRATE TYPES

8.4.1. Branchiostoma

In *Branchiostoma* (Fig, 8.6) the excretory system is totally different from vertebrates and resembles with that of flatworms. It consists of segmentally arranged *nephridia* on either side of the pharynx just above gill slits. The nephridium is a bent tube with a horizontal and a vertical limb and corresponds to each primary gill bar and opens by a nephridiopore in the atrium. A large number of *solenocytes* or flame cells open into the nephridial tube. Each solencyte has a long tubular stalk and a terminal rounded body. The excretion is effected by the wall of the solensytes and excretory products pass down into the nephridial tube. In addition to nephridia the *Brachiostoma* has nephridium of Hatschek which arises from the mouth and extends forwards to the right side of the notochord. Brown funnels are also regarded as excretory in function, which are blind saccular structure situated at the anterior end of atrium.

8.4.2. Scoliodon

The excretory organs consist of kidneys, which are Opisthonephric (formerly regarded as mesonephric) and extend from the liver to the cloaca (Fig 9.8 and 9.9). They are covered over by peritoneium. The anterior part of the kidney is non functional and posterior part is well developed and functional. The anterior part is called *organ* of *Leydig* or epididymis and takes part in conveying the genital products. A large number of coiled glandulars uriniferous tubules or *nephrons* are present in the posterior part of the kidney (Fig. 8.7). Each tubule consists of a Bowman's capsule enclosing the glomerulus and a greatly coiled renal tubule. The renal tubules open into a common collecting tubule. The collecting tubules open into a common ureter, which is not homologous to the vertebrate ureter. The two ureters open into the urogenital sinus by separate apertures in males, and unite together before opening into the urogenital sinus in female. The urogenital sinus is mesodernal in origin and serves as the urinary bladder.

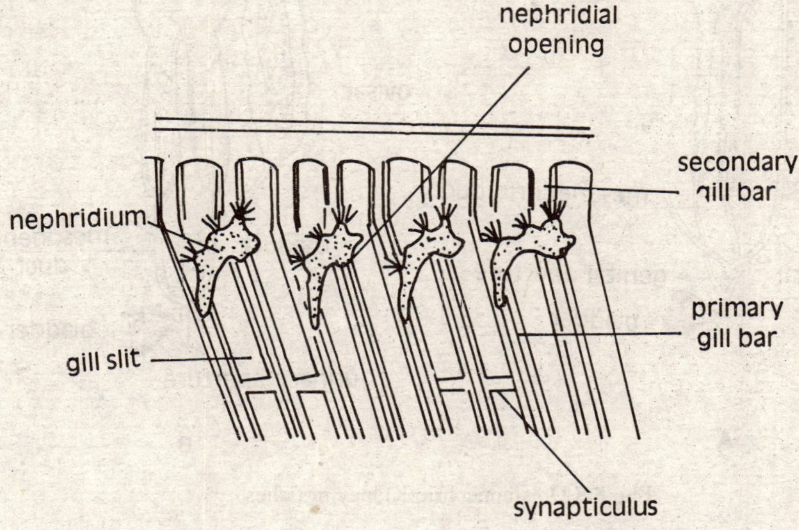

Fig. 8.6 Excretory organs in Branchiostoma

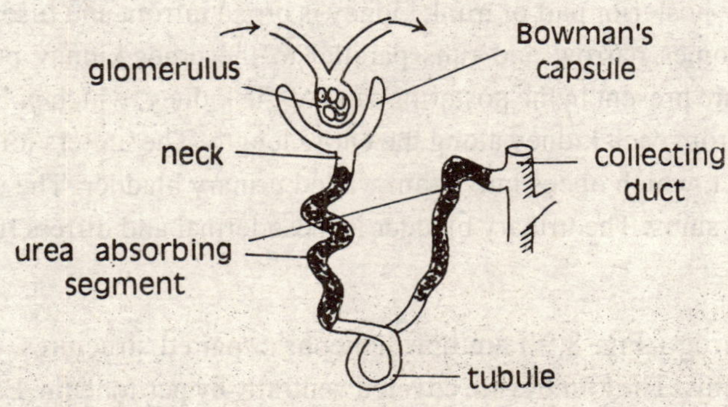

Fig. 8.7 A. Single nephron

The archinephric duct divides into a dorsal and a ven'ral part. The dorsal part of Wolffian du.t becomes vas deferens in males and connects the vasa efferentia from the testis. The ventral part or Mullerian duct becomes oviduct in the female. The 'ureter' of *Scoliodon* is a special accessory urinary duct.

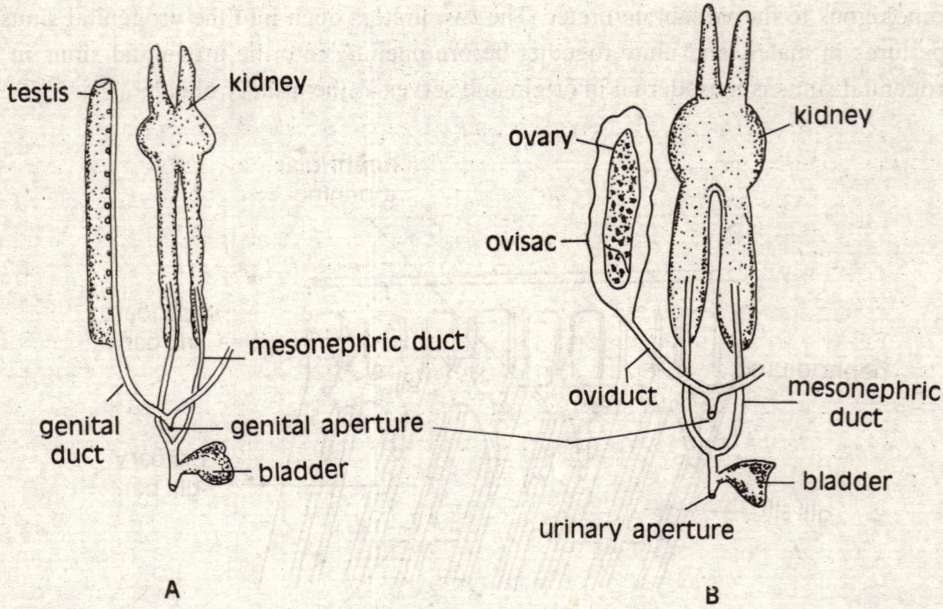

Fig. 8.8 Opisthonephric Kidney in fishes

8.4.3. Labeo

The kidneys of *Labeo* are opisthonephric and extend throughout the length of the visceral cavity (Fig. 8.8). They lie below the spine above the air bladder. Each kidney is made of anterior and posterior parts. The anterior part called "head kidney" is non renal and narrow structure with lymphoid tissue. The posterior part or trunk kidney is broad infront and fuses with other kidney in the middle and becomes narrow and runs parallel to the other kidney posteriorly. Numerous uriniferous tubulus are present in the posterior parts of the kidney, which pour their contents into a pair of ureters, one from each kidney along the entire length. The ureters join together and form a common urinary duct, which opens into a thin walled urinary bladder. The urinary bladder opens into the urinogenital sinus. The urinary bladder is mesodermal and differs from that of tetrapoda.

8.4.4. Rana

The kidneys of frog (Fig. 8.9) are opisthonephric, paired structures, situated dorsal to the coelom in a lymph sinus. The kidneys are covered ventrally by peritoneum. Each kidney consists of a mass of coiled uriniferous tubules held together by connective tissue and richly supplied with blood capillaries. In Anura (hence in *Rana*) a large number of *nephrostomes* or funnels open into the veins of kidney and draw wastes from the coelome in young conditions: in the adult this connection is lost and they open into the branches of afferent renal veins that are present near the ventral surface of the kidney. The nephrostomes are not connected with the uriniferous tubules. Each *uriniferous tubule,* has a double walled cup Bowman's capsule enclosing a *glomerulus*. The glomerulus receives blood from the afferent arteriole of the renal artery and the efferent arteriole carry's away the blood to renal vein. The Bowman's capsule and glomerulus are together called *Malpighian body* or *corpuscle*. The Bowman's capsule is ciliated and leads into a short neck. The neck is continued into a proximal and distal convoluted tubule. The distal convouted tubule opens into a *Bidder's canal,* which is longitudital and opens by a collecting duct into the Wolffian duct. The *Wolffian duct* or ureter passes posteriorly and opens into the cloaca by a papilla (In males it dilates near the kidneys and forms seminal vesicle for the storage of sperms and serves as urogenital duct). The urinary bladder is endodermal structure and arises from the floor of the cloaca as a bilobed and thin-walled sac. It is called *cloacal urinary bladder.* The urine passes first into the cloaca, as there is no direct connection of the ducts with the urinary bladder.

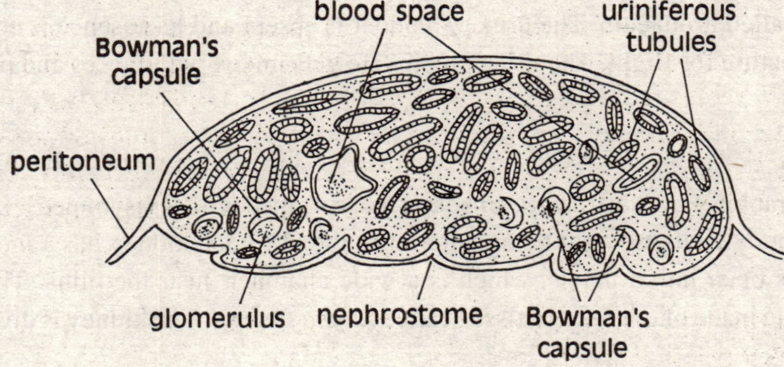

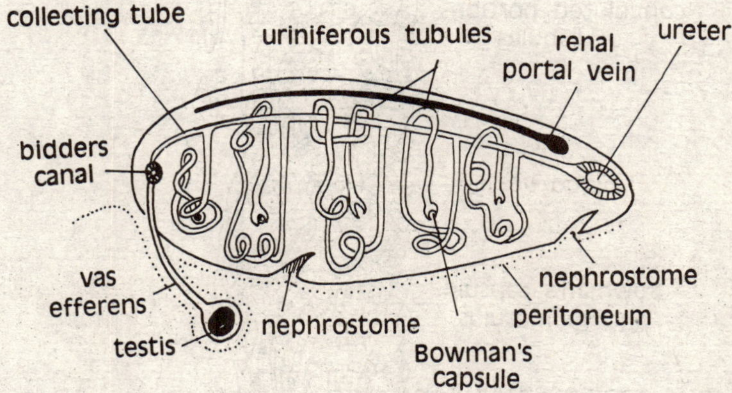

Fig. 8.9 A. T.D. Kidney of frog. **B.** Diagramatic representation of the kidney of frog

8.4.5. Uromastix

The kidneys of *Uromastix* are metanephric paired, small, compact and lobular. They lie on the roof of the posterior part of the coclom. They are covered over with peritoneum on their ventral surfaces. The two kidneys are united posteriorly and appear V-shaped. The posterior parts of the two kidneys are separated from one another by a connective tissue matrix. A short ureter, which is a true metanephric duct, arises from each kidney and opens into the urodacum of the cloaca. The *urinary bladder* is allantoic and attached to the cloaca. The excretory fluid is reabsorbed from the cloaca and bladder and urir e consists of mainly of uric acid.

8.4.6. Columba

The kidneys of pigeon are metanephric, three lobed and flattened structures. They remain closely applied to the dorsal wall of the pelvis. The uriniferous tubules are highly specialized. They have small *glomerulus* and the *loop of Henle* is very extensive. The loop of Henle reabsorbs water from the filterate in glomerulus. The urine contains high concentration of uric acid and precipitate and little water. Each ureter starts from the first and second lobe of the kidney and joins the middle part of the cloaca called urodaeum. The urinary bladder is absent and its absence is considered as an advantageous feature for flight. The urine precipitate gets mixed with faeces and passes out as semisolid.

8.4.7. Oryctolagus

The kidneys of rabbit (Fig. 8.10) are metanephric, bean shaped organs, asymmetrically arranged against the dorsal body wall; the right being anterior to the left Each kidney has a *hilus* or *notch.* Uriniferous tubules open into a pelvis, which is a wide chamber, near the hilus. The kidney is enclosed in a capsule made of connective tissue called *tunica fibrosa.* The kidney is divided into an

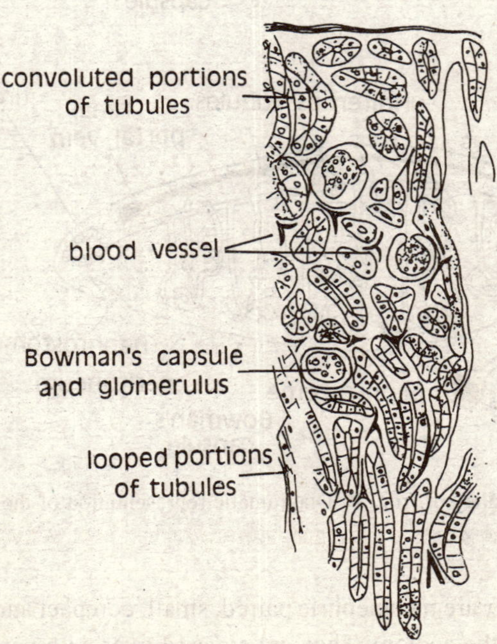

Fig. 8.10 Mammalian kidney through cortex in section

outer *cortex* and inner *medulla.* The cortex and medulla are intimately connected with one another. The medulla is made of many lobes called *pyramids.* The pyramids extend into channels called *calyces.* The calyces open into a pelvis or *renal sinus,* which opens into the ureters. Between the pyramids of medulla, cortex extends as renal *columns of Bertini.* The kidney contains a large number of uriniferous tubules. Each tubule contains a large glomerulus, a proximal convoluted tubule, a narrow U-shaped loop of Henle with discending and asending limb and a distal convoluted tubule, which opens into a collecting tubule. The Malphighian bodies or capsules are situated in the cortex and loop of Henle and collecting tibules lie in the pyramids of the medulla. The pelvis continues as a true ureter, which is metanephric duct. The two ureters pass posteriorly to open into the neck of the urinary bladder. The *urinary bladder* is allantoic and endodermal in origin provided with muscular walls. Its neck is having a sphicter which controls the passage of urine. The neck of bladder continues into a urethra which leads to the exterior. The urethra is also called urinogenital canal as it carries genital products also.

CHAPTER SUMMARY

1. Paired kidneys and their ducts are excretory organs of vertebrates. An intimate relationship exists between the excretory and reproductive organs of vertebrates; they are called urinogenital or urogenital organs.
2. Each kidney consists of a mass of coelmoducts opening in a collecting duct. The kidneys develop segmentally from the mesomere of the mesoderm as one pair per body somite but in higher animals the adult kidneys are not segmental.
3. The kidneys consists masses of uriniferous tubules or nephrons. A cup like Bowman's capsule containing a knot of capillaries called glomerulus, together called Malpighian body is present at one end of the nephron. The uriniferous tubule contains proximal convoluted tubule, loop of Henle and distal convoluted tubule, which join the ureter. The ureters carry waste from the kidneys and discharge into cloaca in amphibians, reptiles and birds, and into the urinary bladder in mammals. The urinary bladder in mammals communicates to the exterior through a median duct called urethra.
4. The kidneys are distinguished into archinephros or primitive hypothetical kidney, anamnoid to kidney (pronephros and metanephros) Various types of kidneys have been considered as successive stages evolved in anteroposterior direction from the original archinephros.
5. It is presumed that primitive vertebrate ancestors had archinephros or holonephros. Such a kidney is found in larval hagfishes and larval caecilians. It consists of a pair of archinephric ducts running parallel on the dorsal side of the coelom extending throughout the length of the animal. These ducts are joined by a series of segmentally arranged tubules, one in each segment. Each tubule opens into the coelom at the outer end by means of ciliated nephrostome or funnel. A small bundle of blood vessels called glomerulus is associated with each tubule.

6. Anamniotic pronephros are paired kidneys found in vertebrate embroyos and cyclostomes. In hagfishes it contains 1-13 uriniferous tubules one pair in each segment, with glomerulus but Bowman's capsule and peritoneal funnel are absent. The uriniferous tubules open into a common pronephric duct which opens into the cloaca.
7. Opisthonephros retains original archniephric duct. It is found in lampreys, most fishes and amphibians. It contains several uriniferous tubules in each segment without peritioneal connections.
8. In amniotes there is a tendency to concentrate kidney tubules towards the posterior side. The anterior end in male becomes ductus deferens and accessory urinary ducts are formed by the fusion of two or more kidney tubules for the discharge of excretory products. The pronephros and mesonephros appear in embryonic development but adults contain only metanephros. Rudiments of pronephric duct persist as Wolffian duct and those of mesonephric duct give rise to epididymis, vas deferens and certain parts of reproductive system in males and rudiments of some female organs.
9. The metanephros is developed from the posteriormost part of the nephrotome behind the embryonic mesonephros .It does not show segmental character. Nephrostomes are absent and thousands of uriniferous tubules are present without segmental arrangement. The proximal part of the tubular outgrowth becomes ureter or metanephric duct
10. The urinary bladder is not homologous structure in all groups of vertebrates. It is absent in cyclostomes, elasmobranchi, snakes, crocodiles, some lizards and most of the birds. It is formed by terminal enlargement of mesonephric duct in some fishes (tubal bladder), from the dorsal wall of the cloaca in dipnoi, ventral wall of the cloaca in amphibians and at the base of allantois in amniotes (allantoic-bladder). The urinary bladder stores urine, help in water conservation and prevention of desiccation under terrestrial conditions.
11. The cloaca is terminal chamber of the digestive, urinary or genital tracts in most vertebrates. Its derivatives include urinary bladder of amphibians, bursa Fabricii of young birds and certain cloaca glands. The cloaca of mammals is divided into a rectum dorsally and a urinogenital sinus ventrally during development. The embryonic urinogenital sinus in primates and some rodents is further divided into a vagina and a urethra.

9

Reproductive System

9.1. GENERAL ACCOUNT

All living being have ability to produce new living individuals for the perpetuation of the species. This process is called reproduction, which takes place in vertebrates, by gametes or germ cells. The germ cells are produced in gonads. The male germ cells, *spermatozoa* are produced in testes and female cells, *ova*, in ovaries. The gonads are called *primary sex organs.* Most of the vertebrates have other parts, such as ducts glands and organs associated with gonads, which aid in the reproduction and constitute the reproductive system. The reproductive systems of different animals show variety in structural details, but the basic pattern remains the same.

The gonads are formed in early stages from a pair of genital ridges between the mesentery and the Wolffian ridge on either side. The primary germ cells arise from the endoderm and migrate through the developing mesoderm to their definitive position in the epithelium of the gentital ridge. The *genital ridges* usually extend along almost the whole length of splanchnocoel in lower animals but become restricted to small region higher animals. The sexual differentiation takes place at later stages and gonads become definite testes and ovries. The folds of peritoneum, which suspend the testes and ovaries in body cavity, are called *mesorchium* and *mesovarium* respectively. In ovaries, the germinal epithelium lies in the outer surface and eggs are discharged directly into the splanchnocoel. In testes, the germinal epithelium lines the seminiferous tubules and the sperms are released into the lumina of these tubules. The tubules are connected to *pronephric duct* by *vasa efferentia.* The mesenchyme lying below the germinal epithelium forms connective tissues and interstitial cells in higher animals. Finger-like primary sex cords are produced from the germinal epithelium and extend into the substance of the gonad. These cords contain germ cells and supporting elements. At this stage the male and female gonads diverge in their formation.

The male reproductive system consists of *testes* (Fig. 9.1 A. 9.2) *vasa deferentia. seminal vesicles* and *accessory glands.* The spermatozoa are produced in seminiferous tubules of the testes and travel through small ducts or *vasa efferentia.* The vasa efferentia join a large duct called *vas deferens.* The vas deferens leads to the exterior either directly or through penis. The posterior part of the vas deferens may be enlarged to form *seminal vesicles* in some forms. The *penis* is a copulatory organ which transfers the sperms into the female reproductive organs. The sperms are sometimes activated by accessory glands, which provide secretions.

The female reproductive system includes *ovaries* (Fig. 9.1 B), oviducts, uterus, and vagina. The ova are produced in the ovaries. In mammals, each ovum is surrounded by a *Graafian follicle,* which ruptures and releases the ovum after its maturity. The follicles produce hormones having important role in reproduction. The ovaries are solid in mammals and saccular in frog.

The *ovum* is released by the ovary after maturity and passes through tube oviduct. The oviduct is lined internally by cilia. The lower part of the oviduct in mammals and viviparous vertebrates is usually enlarged as *uterus,* which retains the eggs for development inside the body. The terminal part of the female genital tract is modified as vagina, which is used during copulation.

A continuity of chromosomes is maintained where each cell of the off springs contains hereditary mechanism established in the zygote after fertilization. The gonads are formed either during embryonic condition or when the animal attains sexual maturity. The germ cells of the gonads divide by mitosis and every chromosome divides longitudinally indentical halves and each daughter cell receives similar set of chromosomes. Each daughter cell thus, contains diploid number of chromosomes. Each chromosomal pair contains two homologous chromosomes, one derived from male parent and the from female parent. The homologous chromosomes in a pair are similar in shape and size but their genetic constitution may be different. The germ cells multiply on sexual maturity and produce *spermatogonia* cells in males and *oogonia* cells in females. These cells undergo maturation or gametogensis and produce gametes (spermatozoa or ova) after meiosis. The meiosis includes two nuclear divisions called first and second maturation divisions. The processes of the formation of sperms and ova are called *spermatogenesis* and *oogenesis* respectively.

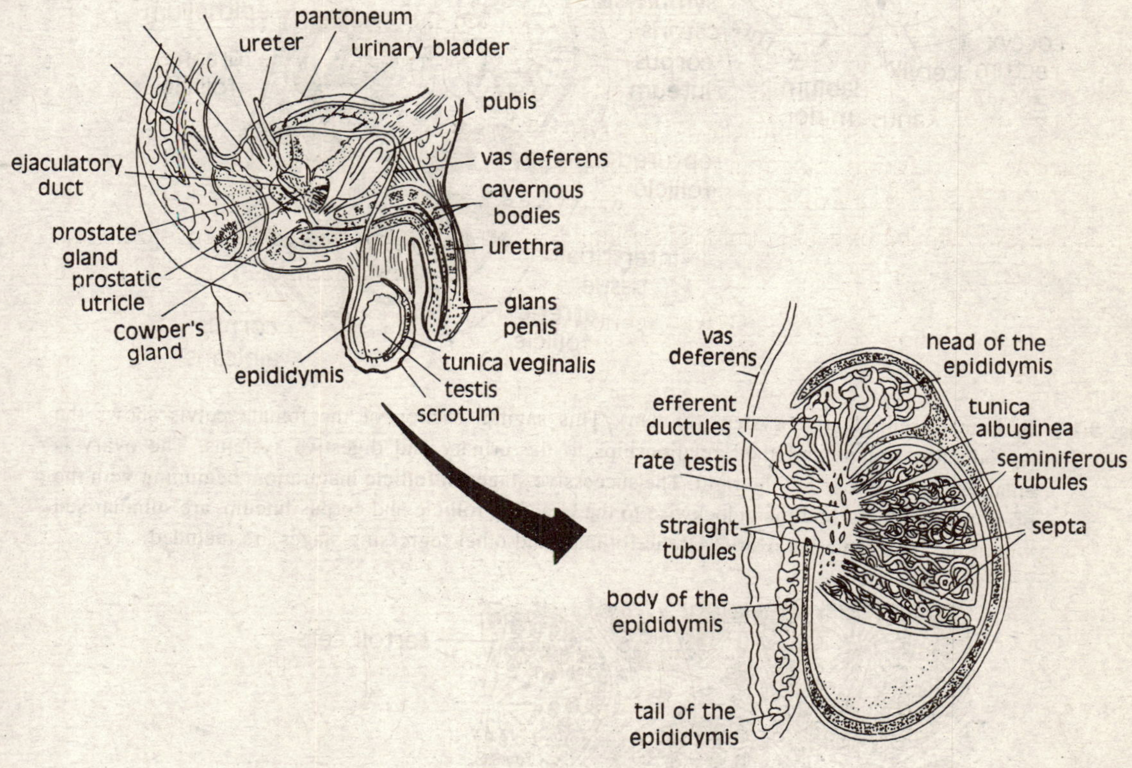

Fig. 9.1 **A.** Male reproductive system (human) This sagittal section of the male pelvis shows the reproductive organs and their relationships to the urinary and digestive systems. The enlarged and cutaway view of the testis and its duct system is shown at the bottom. Spermatozoa produced in the seminiferous tubules eventually pass through the straight tubules into the rete testis and enter the epididymis. Fluid is added as spermatozoa are moved through the vas deferens by contractions of sheets of smooth muscle in its walls.

9.2. MALE REPRODUCTIVE ORGANS

9.2.1. Testes

The testes are compact and regular structures. The early development of testes is similar to that of the ovary. Each testes consists of a mass of *seminiferous tubules* formed by the germinal

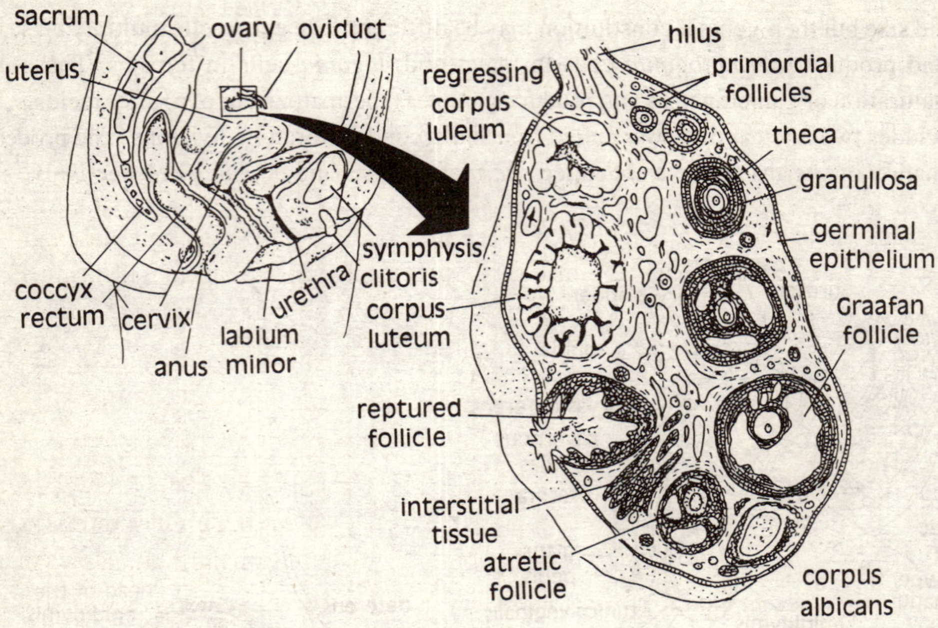

Fig. 9.1 **B.** Female reproductive system (human). This sagittal section of the female pelvis shows the reproductive organs and their relationships to the urinary and digestive systems. The ovary is enlarged and sectioned at the right. The successive stages in follicle maturation, beginning with the primordial follicles and then clockwise to the Graafian follicle and corpus luteum, are summarised within the representative ovary. Atretic follicles and other regressing stages are included.

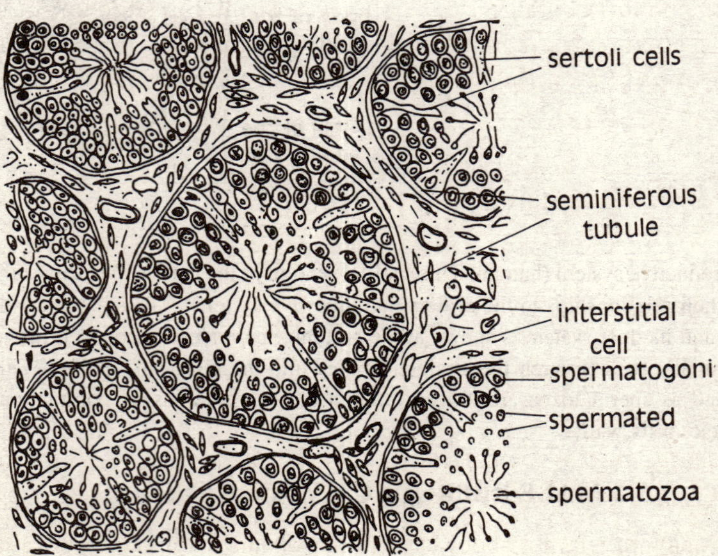

Fig. 9.2 Internal structure of Testis showing spermatogenesis

epithelium (Fig 9.2) The walls of the seminiferous tubules have spermatogonia derived from the large primordial germ cells and *sertoli cells.* The spermatogonia form mature cells after repeated divisions which are released on the surface of the epithelium as spermatozoa or sperms. The seminiferous tubules are connected with measonephric uriniferous tubules by thin tubes called *vasa efferentia.* Loose connective tissue of *interstitial cells* is present between adjacent seminiferous tubules. These interstitial cells produce male hormone *testosterone.* Nerves, blood vessels and lymphatic vessels are also present between seminiferous tubules. In lower classes of vertebrates actual tissue secreting testosterone hormone is not known but in mammals, the interstitial cells or *cells of Leydig* produce this hormone. The sperm production is a cyclic phenomenon in vertebrates. The seminiferous tubules of higher vertebrates are permanent structures producing successive batches of spermatozoa throughout the reproductive age of the animal. The testes are situated in the dorsal part of the coelom but in mammals they undergo descent to a position outside the coelom and come to lie in pouch-like structure called *scrotum* (Fig 9.3). The size of the testes flucturates in annual breeders. They are largest in the breeding season and reduce to a small size in inactive period after they are discharged.

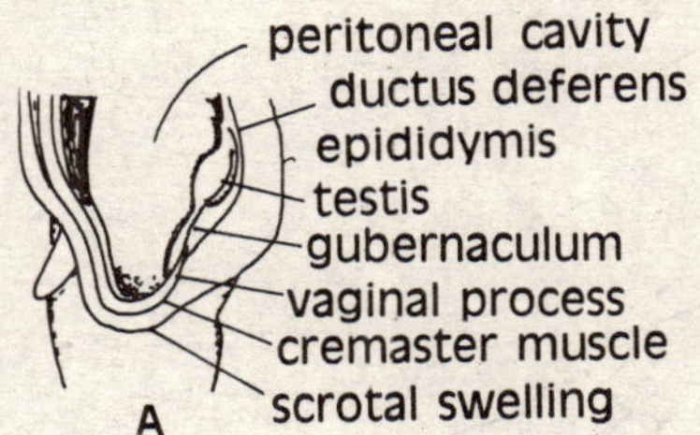

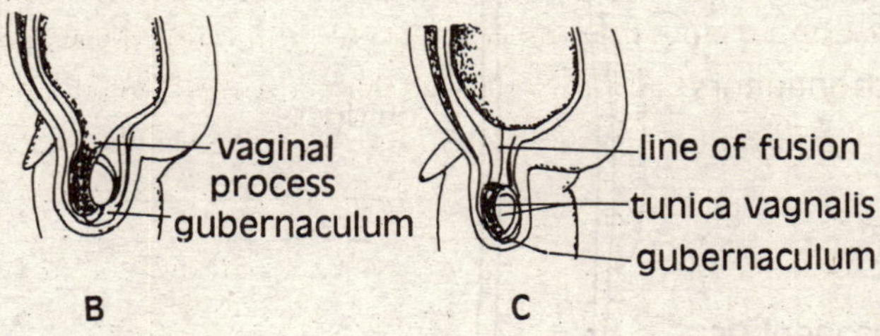

Fig. 9.3 Diagram illustrating the descent of the testes in man.

9.2.2. The Male Ducts (Fig. 9.4)

The transportation of spermatozoa to outside is done by archinnephric ducts or the Wolffian ducts which are developed in connection with the kidneys. The archinephric duct is the kidney duct of anamniotes and Wolffian duct is the duct of amnoites which is formed in connection with the pronephric and mesonephric kidneys. In fishes and amphibians certain modified kidney tubules called *efferent ductules* are used in earning spermatozoa to *archinephric duct.* The archinephric

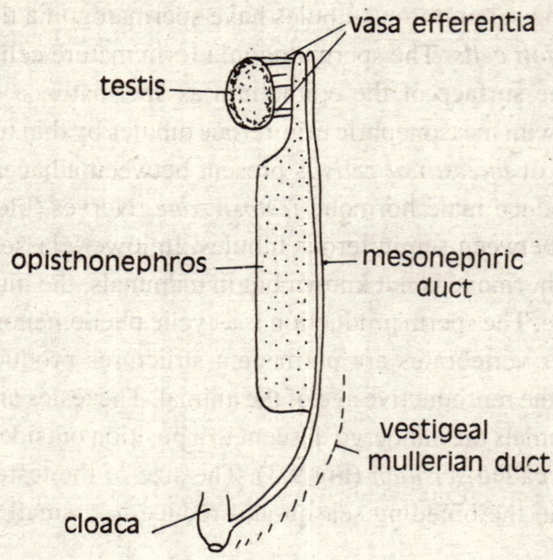

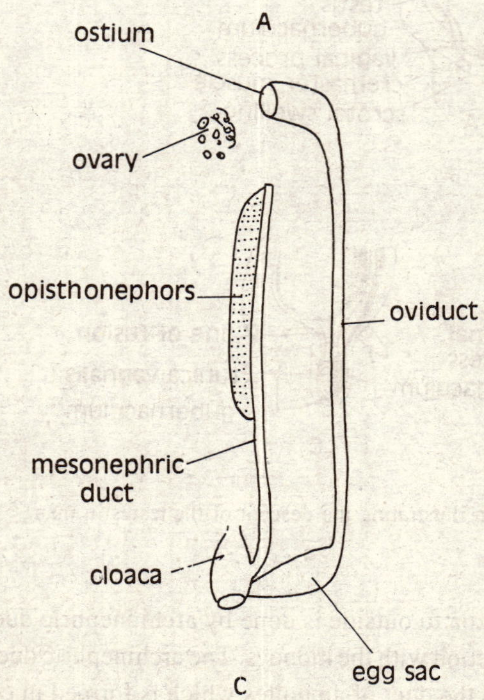

Fig. 9.4 A and C. Reproductive organs in Anamniote

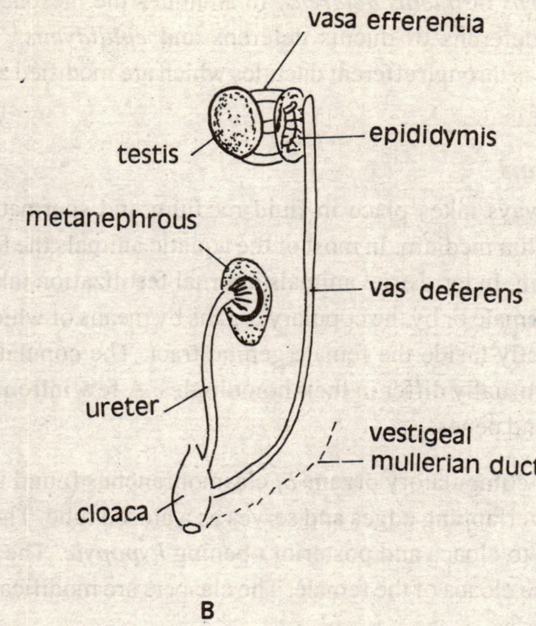

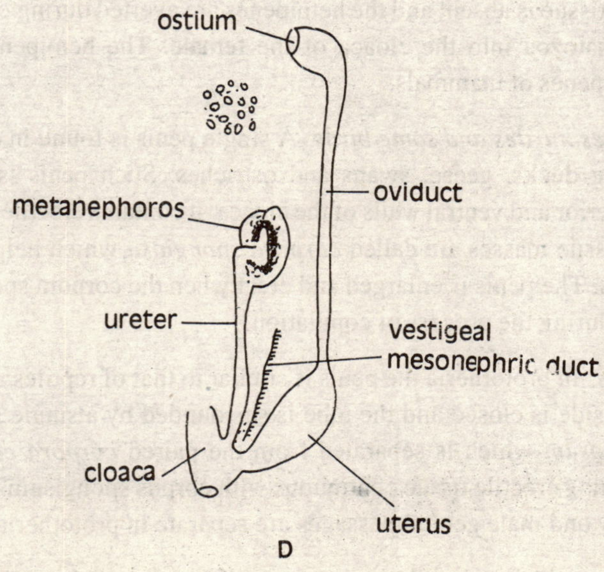

Fig. 9.4 B & D. Reproductive organs in Amniote

duct becomes *vas deferens* or *ductus deferens.* In amnoites the mesonephros degenerates and its duct becomes the vas deferens or ductus deferens and *epididymis.* The epididymis develops connections with the test is through efferent ductules which are modified and persistent mesonephric tubules.

9.2.3. Copulatory Organs

The fertilization always takes place in fluid medium and spermatozoa swim to the ova by flagellar movements in thin medium. In most of the aquatic animals the fertilization is external and water serves as a medium. In terrestrial animals internal fertilization takes place either by cloacal apposition of male and female or by the copulary organs by means of which spermatozoa in seminal fluid are deposited directly inside the female genital tract. The copulatory organs are of various types in vertebrates and usually differ in their homologies. A few intromittent organs are claspers, gonopodia, hemipenes and penes.

Claspers: These are compulatory organs *of* elasmobranches found in males. Each clasper has a median groove with overlapping edges and serves as a closed tube. The clasper tube has anterior opening *appopyle* close to cloaca and posterior opening *hypopyle.* The sperms enter apopyle and clasper is inserted into the cloaca of the female. The claspers are modification of the medial portions of the pelvic fins.

Gonopodium: The anterior border of the anal fin in some teleosts is elongated posteriorly. An intromittent organ is thus formed which serves for internal fertilization.

Hemipenes: In snakes and lizards. sac-like hemipenes are found lying close to the cloaca under the skin. The erectile tissue is absent and the hemipenes are averted during copulation and provide passage to the spermatozoa into the cloaca of the female. The hemipenes of reptiles are not homologous with the penes of mammals.

Penes of crocodiles, turtles and some birds: A single penis is found in crocodiles, turtles and a few birds including ducks, geese, swans and ostriches. Such penis is derived from paired thickenings in the anterior and ventral walls of the cloaca. It contains erectile and connective tissue. The paired erectile tissue masses are called *corpora spongiosa* which help in the extension and retraction of the penis. The penis is enlarged and erect when the corpora spongiosa are filled with blood and distended during the process of copulation.

Penis of mammals: In prototheria the penis is similar to that of reptiles and a few birds but the groove on the dorsal side is closed and the tube is surrounded by a single mass of erectile tissue called *corpus spongiosum* which is separated from the paired *corpora cavernosa.* A sensitive swollen *glans* containing erectile tissue continuous with corpus spongisum is present at the end of the penis. The urinary and male genital passages are separate in prototheria.

In Metatheria the penis does not lie in the cloaca. It is covered by a sheath and opens to the exterior on the anterior side of the anus. It may be protruded or withdrawn. The two corpora cavernosa are separated by a septum and the urethra is surrounded by the corpus spongiosum. In some Metatherians the tip of the penis is bifurcated. There pairs of *bulbourethral glands* are present.

In Eutheria the penis is usually directed forwards and lies at the anterior side or scrotum within a sheath. It may be protruded or retracted. In primates, the sensitive glans is covered over by a fore skin or *prepuce* and is permanently protruded. In some mammals including Rodentia. Carnivora. Cetacea and lower Primates, a penis bone is present in the septum between the corpora cavernosa which helps to increase the rigidity of the penis. The enlarged distal end of the corpus spongiosus is glans penis supplied with sensory nerve endings extremely sensitive to certain stimuli. The urethra passes through penis and opens at the tip by an *external urethral orifice* or meatus.

9.2.4. Accessory Glands

The glands associated with the urethra are as follows:

(i) The *prostate gland* which secretes a fluid for the seminal fluid.

(ii) The *Cowper's gland or bulbo-urethral gland* which secretes a clear, viscous fluid for lubrication during sexual excitement.

(iii) The *Urethral glands of Littre* are small mucus secreting glands.

(iv) The *preputial glands* are situated in the region of prepuce near the distal end of the penis.

The function of the accessory glands is controlled by the hormones secreted by the testes.

9.3. FEMALE REPRODUCTIVE ORGANS

9.3.1. Ovaries

The ovaries are usually paired structures with oval shape but may become distended or irregular in outline during the breeding season. The ovaries are masses of connective tissue with outer germinal epithelium where ova in different stages of development are present. In cyclostomes the ovary is a single median structure formed by the fusion of two gonads. In teleosts the two ovaries are fused into one. In many elasmobranches the right ovary becomes functional. In birds and primitive mammal *Ornythorhynchus* the left ovary matures and right remains undeveloped. In amphibians and reptiles the ovary is hollow with lymph filled cavities in the centre. In birds and mammals the central portion of the ovary is *medulla* made of connective tissue and outer portion cortex having eggs, follicle and germinal epithelium (Fig. 9.5).

9.3.2. Female Ducts

The ovaries are not connected with kidneys or ducts, and ova pass into the coclom after being discharged. In anamniotes a coelomic funnel is formed from the nephrotome and grows to form a groove on each side which is closed to form a *Mullerian duct.* The Mullerian duct lies on the outer side of the mesonephric duct and joins the cloeca posteriorly. The Mullerian duct serves as an *oviduct* which is separate from a mesonephric duct serving as ureter.

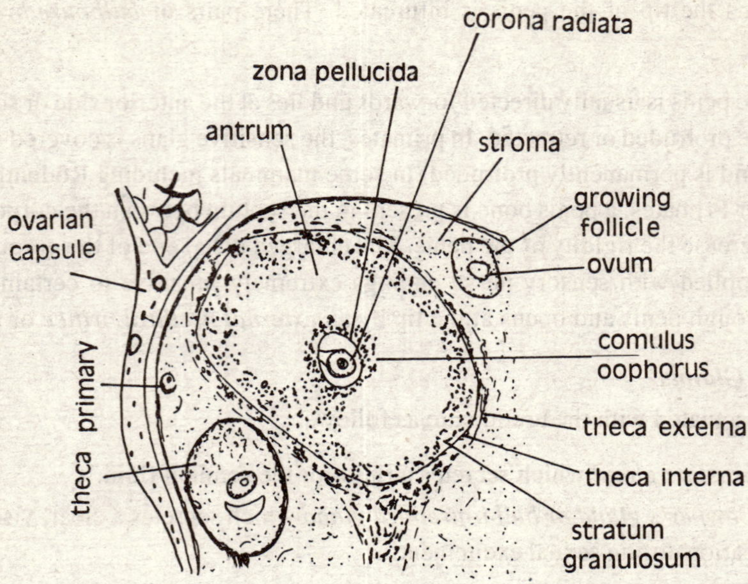

Fig. 9.5 Graafian follicle in Mammals

The two oviducts usually remain separate and open into the cloaca. In mammals, however, each oviduct has a narrow anterior *fallopian tube* and a wide posterior *uterus*. In Prototheria, the vagina is absent, the two oviducts are separate and two separate uteri open into the cloaca (Fig 9.6 A). In Metatheria and Eutheria the urinary and genital systems are separate and cloaca does not participate in reproductive system. In Metatheria two uteri are present and the terminal portion of the oviducts are called *vaginae*. In Eutheria the two vaginae have fused into one tube and the two uteri may fuse partly or completely. Four types of uteri have been classified in Eutheria (Fig 9.6 B-E).

(i) *Duplex type* in which the two uteri enter and open separately in the vagina. It is found in rodents, elephants and a few chiroptera.

(ii) *Bipartite type* when the lower ends of the two uteri fuse together and open into the vagina by a single aperture called *os uterus*. Such uterus is found in carnivora. pigs and cattles.

(iii) *Bicornuate type* where the lower ends of the two uteri fuse completely making a single cavity and no partition wall exists between the fused uteri. Such uterus is found in ungulates, whales and most of the chiroptera.

(iv) *Simplex type* uterus in which complete fusion of the two uteri forming a single body with one cavity. Such uterus is found in primates.

The female reproductive ducts open to the exterior by *vulva*. The primates have two folds of skin called *labia minora* at the margin of the opening of vulva. In some mammals, such as apes and man two additional outer folds are present called *labia majora*. The ventral wall of vulva has a small *clitoris* with a rudimentary *glans clitoritis* at its tip which is comparable to the glans penis of the male.

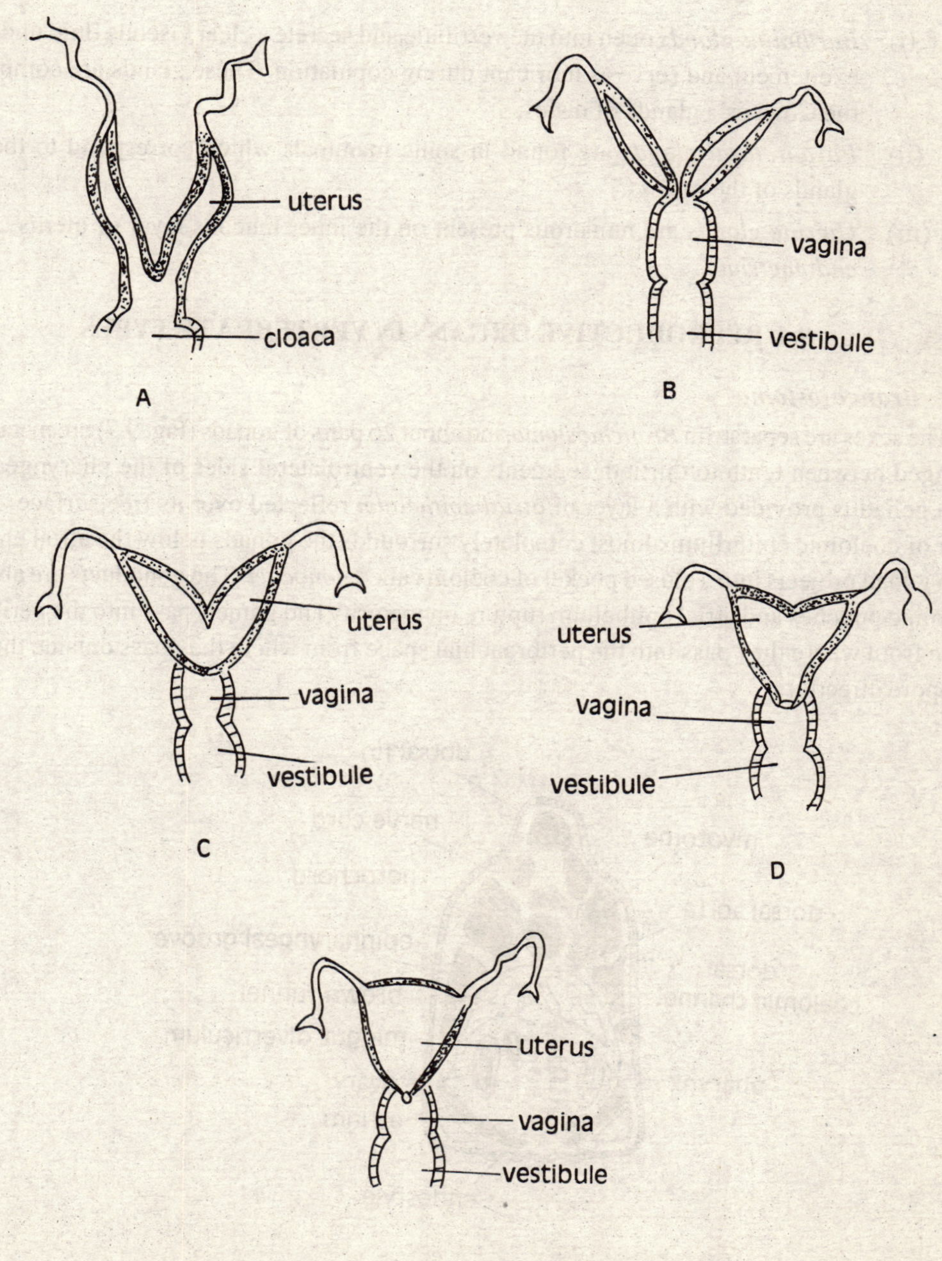

Fig. 9.6 **A, B, C, D.** Female reproductive ducts in different vertebrates, **E** Female reproduction ducts in Mammals

9.3.3. Accessory Glands

Some glands are associated with female reproductive system. These are rudiments of degenerated mesonephros and its duct.

(i) *Bartholins glands* open into the vestibule and secrete a clear viscous fluid under sexual excitement and serve as lubricant during copulation. These glands are comparable to the Cowper's glands of males.

(ii) *Paraurethral glands* are found in sonic mammals which correspond to the postate glands of the males.

(iii) *Uterine glands* are numerous present on the inner mucous layer of uterus called the *endometrium.*

9.4. REPRODUCTIVE ORGANS IN VERTEBRATE TYPES

9.4.1. Branchiostoma

The sexes are separate in *Branchiostoma* and about 26 pairs of gonads (Fig. 9.7) are metamerically arranged between tenth to thirtieth segments on the ventrolateral sides of the pharyngeal region. Each gonad is provided with a layer of *atrial epithelium* reflected over its free surface. A double layer of coelomic epithrlium almost completely surrounds the gonads below the atrial epithelium. Each gonad projects into a closed pocket of coelom called *gonocoel.* The gonoducts are absent. The coelomic pouches and atrial epithelium rupture on maturity and gamets pass into the peribranchial space from where they pass into the peribranchial space from where they pass outside through the atriopore directly.

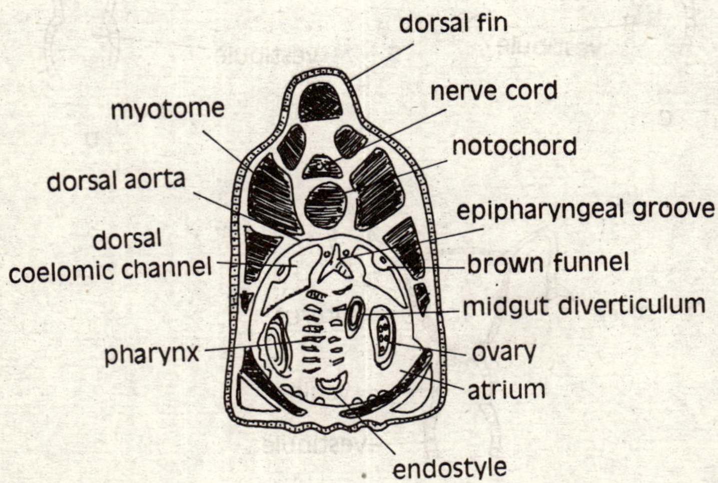

Fig. 9.7 T.S. of Amphioxus through gonads

9.4.2. Scoliodon

Male reproductive system (Fig. 9.8): The testes are two elongated structures extending from liver to rectal (caecal) glands. Each *testis* is attacheds to the dorsal body wall by *mesorchium* and posteriorly with the ordinary tissue of the rectal gland. The testis is made of a large number of *seminiferous tubules* and *interstitial cells*. The interstitial cells secrete a male hormone homologous with testosterone. A large number of thin *vasa efferentia* arise from each testis, enter into the Leydig's gland and join a *vas deferens.* The vas deferens forms a coiled structure called *epididymis* in the anterior part of the kidney and emerges out of it to form a broad *vesicula-seminalis.* The vesiculae seminales of the two sides enter the *urinogenital sinus.* The vas deferns and vesicular seminalis are "mesonephric" or Wolffian ducts. The urinogenital sinus opens into the *cloaca.* The wall of the urinogenital sinus is invaginated to form a tubular *sperm sac.* The cloaca is joined by

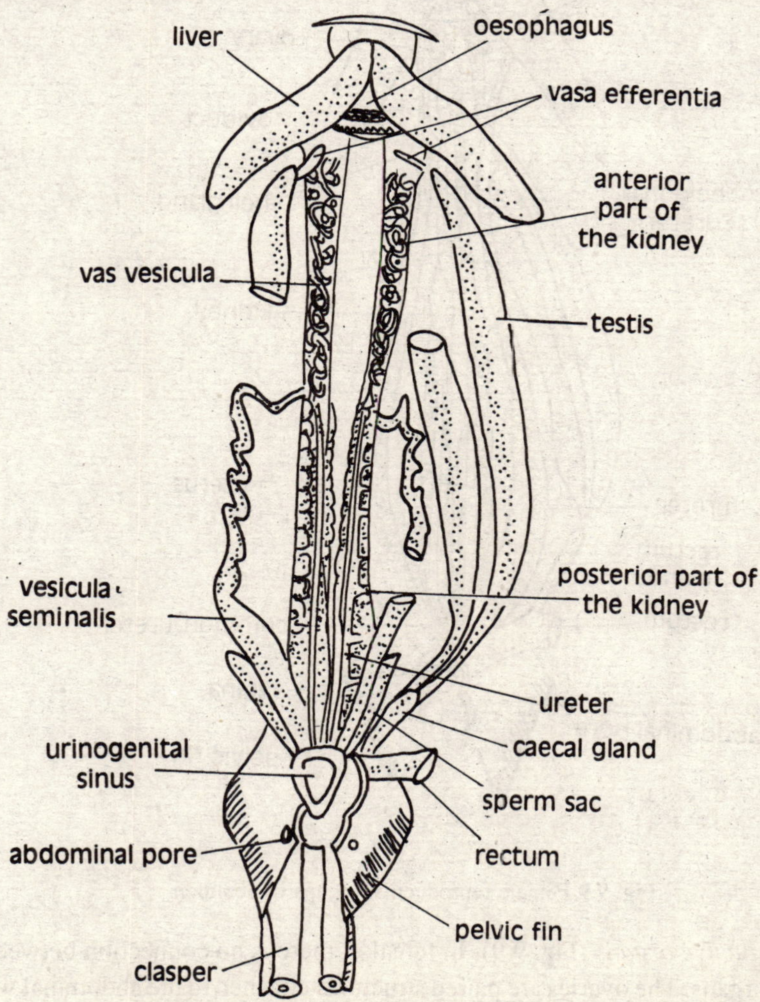

Fig. 9.8 Male reproductive organs in Scoliodon

two grooved erectile *claspers* which serve as intromittent organs during copulation. A pair of elongated spaces called *siphons* are present on the ventral wall of the body between the skin and muscles which open into the grooves of the claspers. The siphon receive sea water which forces semen during copulation along the groove of a clasper.

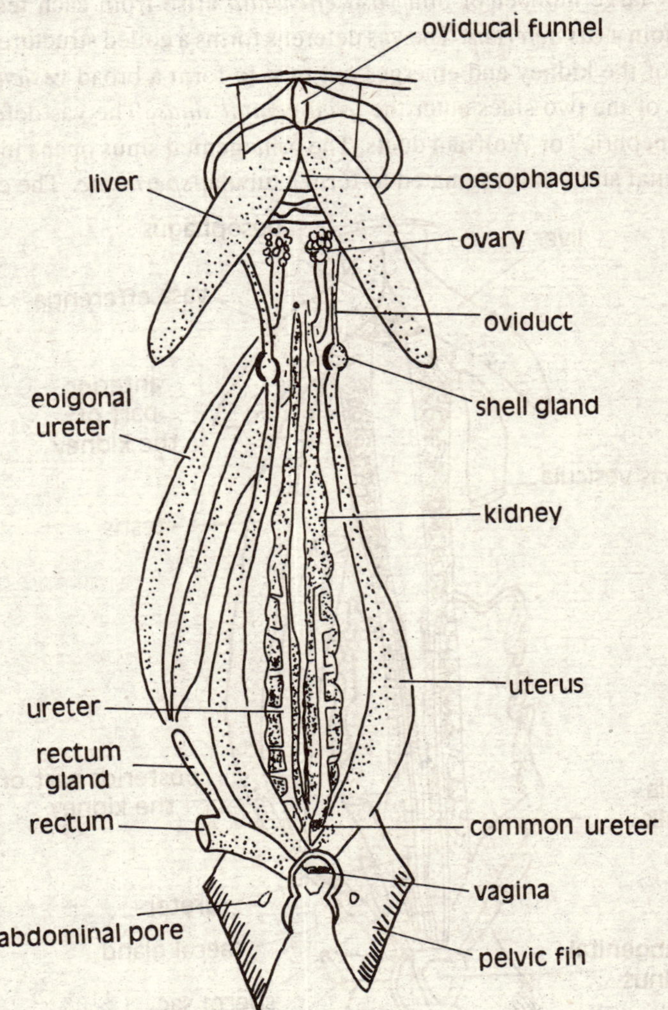

Fig. 9.9 Female reproductive organs in Scolidon

Female reproductive organs (Fig. 9.9): In females, there is no connection between kidney of and reproductive organs. The ovaries are paired structures attached to the abdominal wall by a fold of peritoneum called *mesovarium.* The *ovary* has round follicles projecting from it. Each follicle contains one ovum and follicle cells produce a female sex hormone called *estrogen.* A long tubular *epigonal organ* extends between each ovary and *rectal gland,* which is lymphoid structure concerned with the manufacture of red and while blood corpuscles. The *oviducts* are long lubes united with one another at their anterior and posterior ends. The oviducts open anteriorly into the coelomic cavity by a longitudinal slit-like opening called *oviducal funnel* or *oviducal aperture* near the liver. A *shell gland* is present at the anterior portion of the oviduct in the form of an enlargement. The shell gland serves to store spermatozoa after copulation. The oviduct dilates to form a *uterus* or *uterine chamber.* The two uteri unite to form a vagina which opens into a cloaca The ova are released by the bursting of the follicles and move in the coclom by the action of the epithelial cilia. They enter the oviducal funnel and move downwards by the action of the cilia of the oviducts. The fertilization takes place in the shell gland and fertilized ova undergo further development in the uterus.

The *Scoliodon* is viviparous and three to seven embryos develop in each uterus. Each embryo lies in its own compartment and its intestine receives the nourishment through a yolk stalk attached to the yolk sac. The yolk-sac gels embedded in the uterine wall and forms *yok sac placenta* or *omphaloidean placenta.* The yolk stalk becomes *umbilical cord* after the development of vitalline veins and the food passes into the embryonic circulation through this vein. When the embryos are fully formed, they are given birth by the mother.

9.4.3. Labeo

The urinary and genital organs are separate unlike other fishes where they are united together. The glands become considerably enlarged during the breading season.

In males the testes (Fig. 9.10 A) are two elongated structures extending throughout the length of abdominal cavity suspended from the body wall by *mesorchium.* Each testis has a vas deferens arising from its posterior end. The vasa deferentia open info a urinogenital sinus.

In female the ovaries (Fig. 9.10 B) are paired structures much larger than testes and suspended by *mesovarium.* They are hollow, bag like structures enclosed in an *ovisac.* Each ovisac continues behind into an *oviduct.* This oviduct is not homologous to Mullerian duct and is formed by the peritoneum, which surrounds the ovary. (According to some authors the oviducts are absent in *Labeo*). The two oviducts meet each other posteriorly and open to the exterior by a genital aperture situated between the anus and urinary aperture. On maturity the eggs are released into the cavity of the ovisac formed by the peritoneum. According to some authors the eggs are liberated into the body cavity after the rupture of the ovary. The eggs are laid in water through the oviduct and genital aperture. The fertilization takes place in water when the male discharges spermatic fluid or milt.

9.4.4. Rana

The sexes are separate in frog. The male has *copulatory* or *amplexusary* pads which are swelling on the inner side of the palm. In addition to this the *articular pads* are present below the joints of the digits of hand and foot. Such swellings are absent in females.

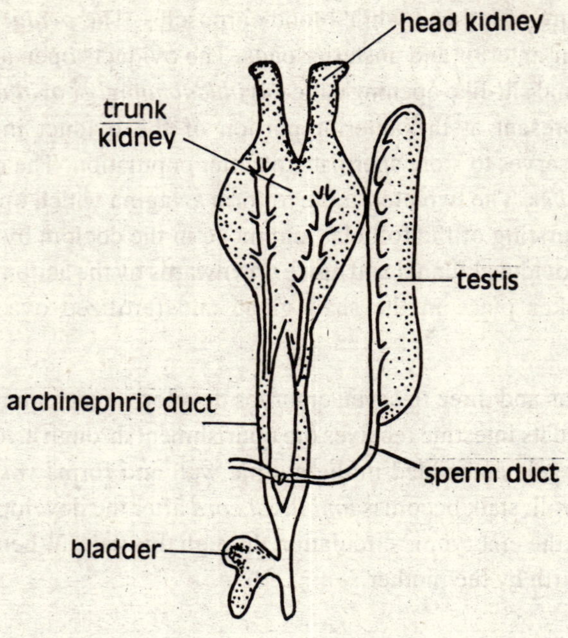

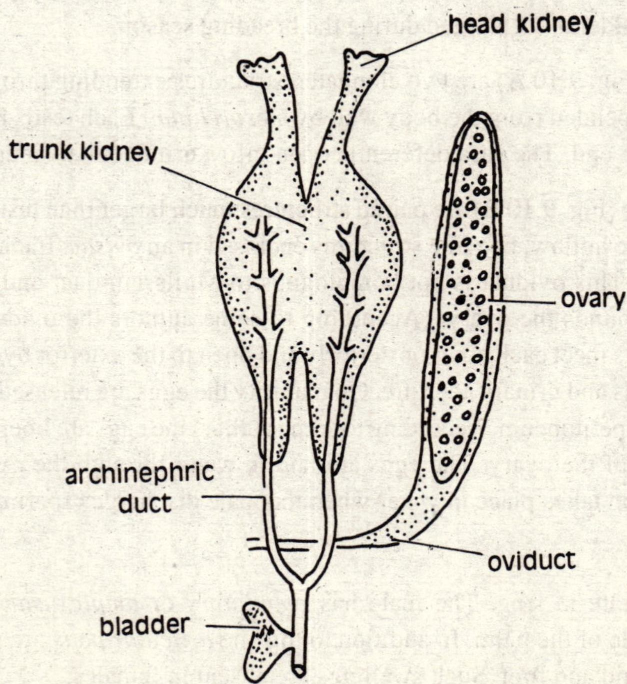

Fig. 9.10 A, B Reproductive organs in a bony fish

Male reproductive organs (Fig. 9.11): The male reproductive organs include *testes, vasa efferentia.* Bidder's canal collecting tubules, urinogenital duct and cloaca. The testes are paired structure attached to the ventral side of the kidneys by *mesorchium.* Each testis is a yellow ovoid body having a length of about 1.25 cms. A large number of seminiferous tubules or crypts containing spermatozoa are present in each test is (Fig. 9.12). The space between seminiferous tubules is a filled with connective tissue containing some blood capillaries, lymph and groups of small interstitial cells. The interstitial cells secrete a male hormone. Each seminiferous tubule is lined by *germinal epithelium* containing primordial germinal cells, which produce spermatozoa by the process of spermatogenesis. Each spermatozoon has a conical head with a nucleus, a middle part with one or two centrioles and long thin vibratile tail. The sperms are motile and have chemical specificity for eggs of their species. About six vasa efferntia originate form the inner margin of each testis as dedicate ductules from the network of thin tubes called *rete-testis.* The spermatozoa produced in the crypts of testis are conveyed to the kidney tubules by vasa efferentia and to the ureters, which serve as urinogenital ducts. The Bidder's canal is a long tubule, which maintains the relation between vasa efferentia, and kidney tubules. The collecting tubules or seminal vesicles or vesicular seminalis are bagilike swellings present posterior to the kidneys and serve to store the spermatic fluid.

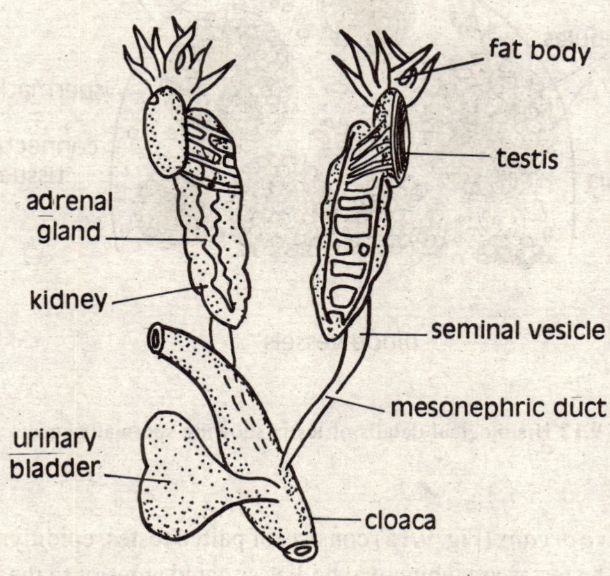

Fig, 9.11 Male reproductive organs in frog

Female reproductive organs (Fig. 9.13): They include ovaries, oviducts, ovisacs and cloaca. Ovaries are paired structures attached by mesovarium on the ventral side of the kidneys. The position of ovaries in females corresponds to those of testes in males, but the connection between reproductive organs and kidneys is absent in female frogs and ureters serve as renal ducts only. Each ovary is greatly folded sac, which is enlarged during the breeding season. The inner surface of the ovary is produced into numerous rounded projections containing ova at different stages of the development. The wall of the ovary consists of germinal epithelium, which divides during the breeding season to form groups of cells, which produce large *ova* and *follicle* by the process of *oogenesis* (Fig. 9.14). Each ovum is a large rounded cell covered with a *vitelline membrane* and contains large nucleus and many nucleoli inside the follicle. On the dorsal side of each ovary long oviduct is situated in the body cavity. Each oviduct is long convoluted tube opening into the body cavity at the base of the lung by a small aperture called *ostium*. The cavity of the oviduct is lined by ciliated cells and glandular cells. The glandular cells secrete a slimy albumen which makes a jelley after coming in contact with water. On maturity the eggs are discharged into the body cavity from where they enter the oviducts by ostia. The oviduct forms a thin walled wide chamber *ovisas* or *uterus* on the posterior side. The ovisacs communicate with the cloaca. The ova pass through oviduct, ovisac and cloaca. The fertilization is external and takes place in water after copulation.

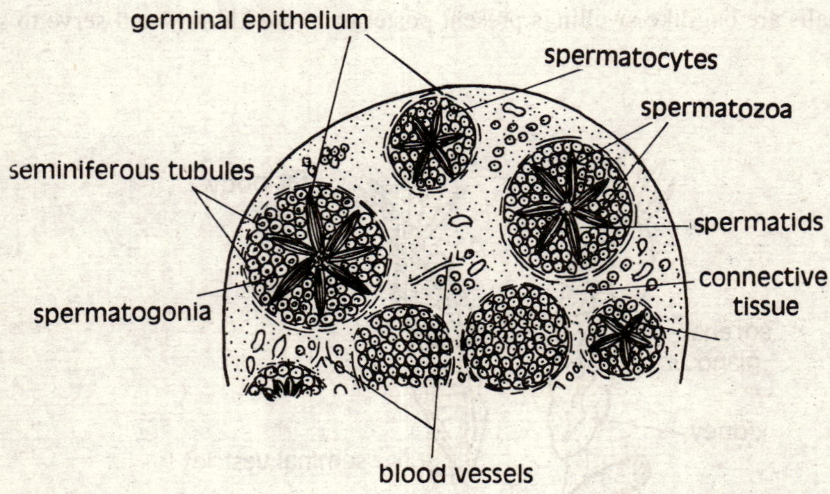

Fig. 9.12 Histological details of testis showing spermatogenesis

9.4.5. Uromastix

The *male reproductive organs* (Fig. 9.15) consists of paired testes, epididymes. vasa deferentia and paired hemipenes. The testes are white oval bodies situated anterior to the kidneys in the body cavity and suspended by *mesorchium* The right testis is placed more anteriorly than the left. Each testis gives several delicate *vasa efferentia* which pass through the mesorchium and enter the *epididymis*. Vasa efferentia are outgrowths of mesonephric tubules. The epididymis is closely applied to the testis. The epididymis continues posteriorly as a genital duct *vas deferens* which is

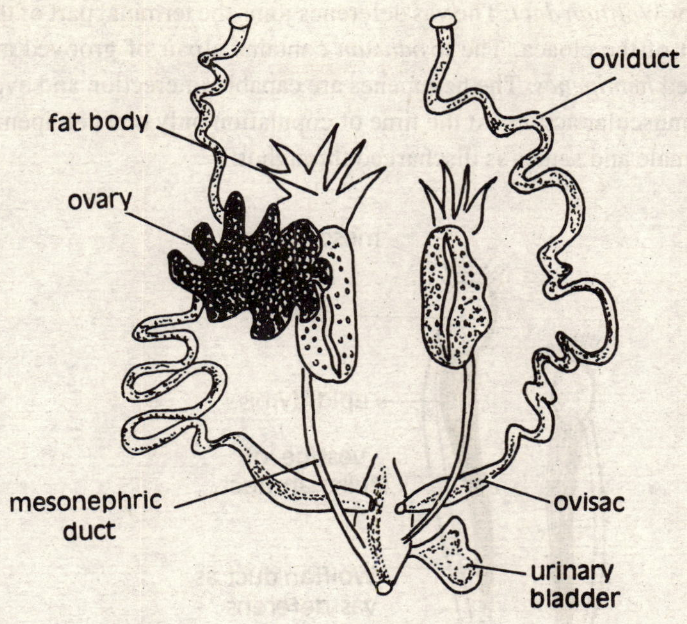

Fig. 9.13 Female reproductive organs of Rana

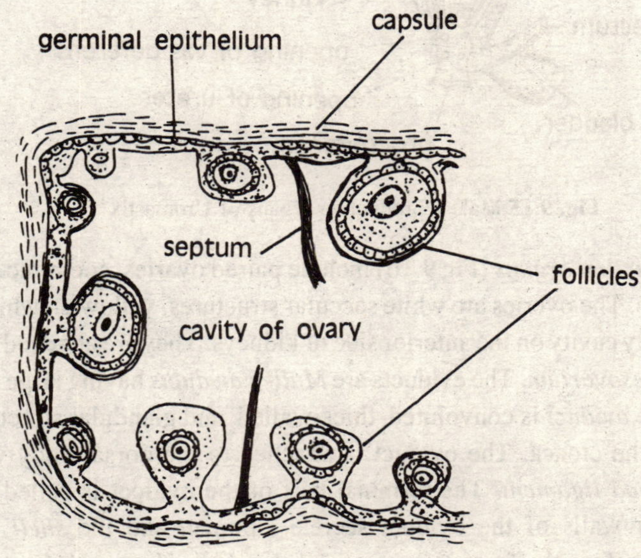

Fig. 9.14 Amphibian Ovary in section

persistent mesonephric or *Wolffian duct.* The vas deferencs joins the terminal part of the ureter and enter the *urodaeum* part of the cloaca. The *urodaeum* contains a pair of grooved and eversible copulatory pouches called *hemipenes.* The hemipenes are capable of erection and evertion due to the filling of blood and muscular action. At the time of copulation only one hemipenis is inserted into the cloaca of the female and semen is discharged through it.

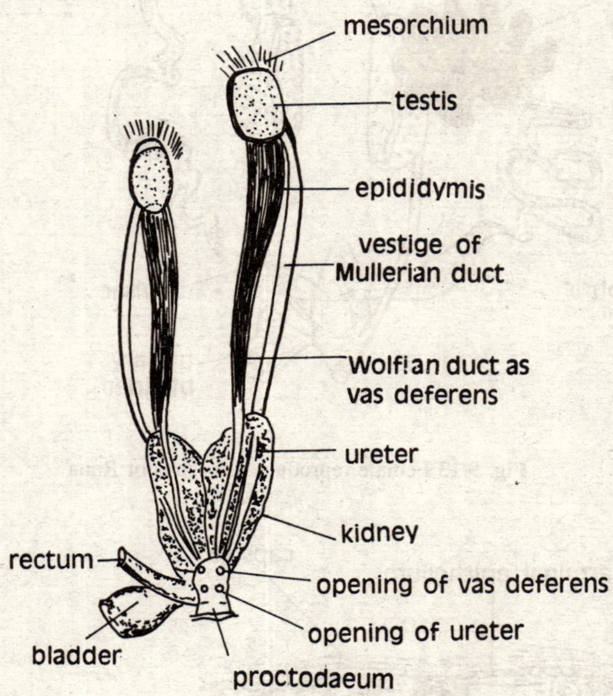

Fig. 9.15 Male reproductive organs of Uromastix

The *femaie reproduction organs* (Fig 9.16) include paired ovaries, coelomic funnels, oviducts and unpaired urodaeum. The overies are white saccular structures, yellowish white, situated in the posterior part of the body cavity on the anterior side of kidneys. They are attached to the dorsal wall of the body cavity by *mesovarium.* The eviducts are *Mullerian ducts* having large coelomic funnels on the anterior side The *oviduct* is convoluted, thick walled, and glandular structure opening in to the urodaeum part of the cloaca. The oviduct is attached to the dorsal body wall by a fold of peritoneium called *broad ligament.* The terminal part of the oviduct is called *vagina* by some authors. The glandular walls of the oviduct secrete albumen and the *shell* of the egg. The fertilization is internal before the formation of egg of the shell. *Uromastix* is oviparous and lays about 15 eggs in burrows where further development takes place.

9.4.6. Columba

The *male reproductive organs* (Fig 9.1/) include paired testes epididymes, vasa deferentia.

vesicula seminalis, and unpaired urodaeum. The testes are ovoid bodies attached to the anteroventral end of the kidneys by *mesorchium.* The size of the testes is variable according to the season. In breeding season they increase many times in size. Each testis consistes of masses of coiled *seminiferous tubules* with *interstitial cells* in between. From each testis a coiled mesonephric or *Wolffian duct* arises. The anterior part of the mesonephric duct is an epididymis which continues posteriorly as a *vas deferens.* The epididymis is covered with the seminiferous tubules. The distal part of the vas deferens is dilated into a *vesicula seminalis* which open into the *urodaeum* by a small *erectile papilla* posterior to the ureter. The copulatory organ is not present in pigeon and transfer of sperm cells to female takes place by the close apposition of the two cloaca.

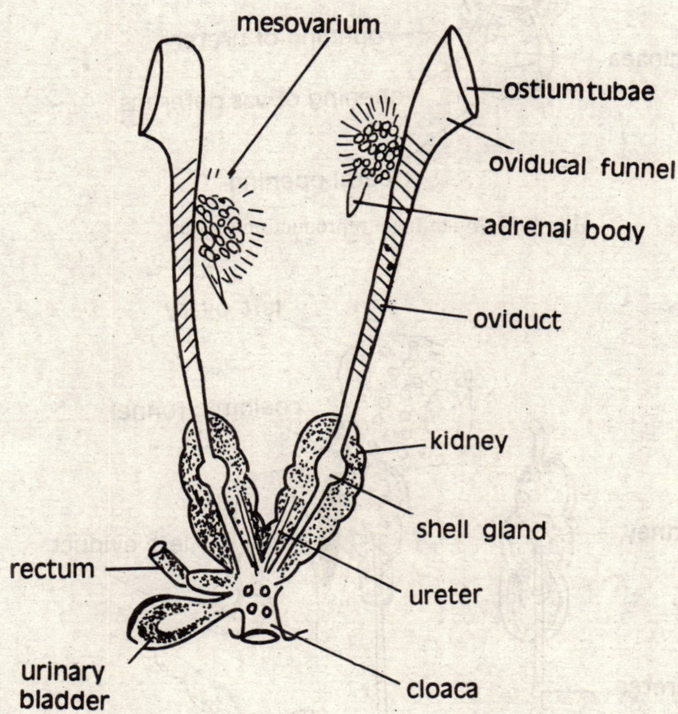

Fig. 9.16 Female reproductive organs of Uromastix

The *female reproductive organs* (Fig 9.18) include left overy, left oviduct and urodaeum. The ovary is situated on the left side of the body cavity suspended from the body wall by *mesovarium.* The right ovary and oviduct are atrophied in adult birds. The ovary is large and contains numerous follicles with eggs of different sizes. The overy covers the anterior part of the left kidney. The left *oviduct* is large tubular structure attached to the dorsal body wall by board ligament or *mesotuberium.* The anterior end of the oviduct communicated with coclom by *ostium* or *oviducal funnel* or *infundibulum.* The posterior end of oviduct is expanded to form uterus which opens into the

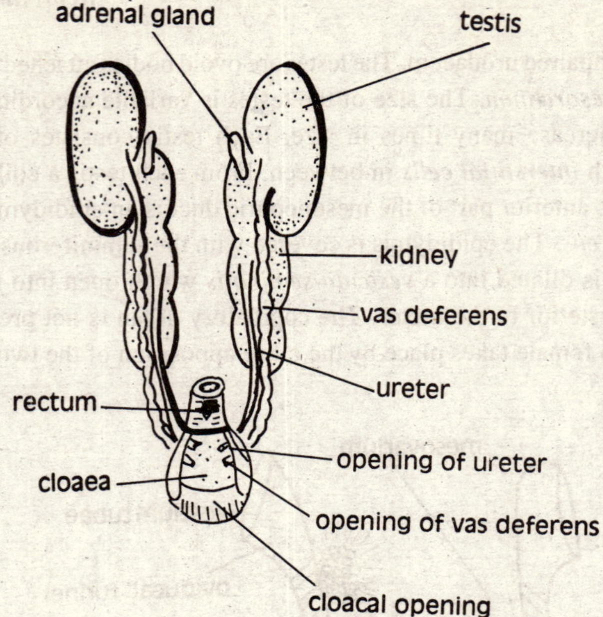

Fig 9.17 Avian male reproduction organs

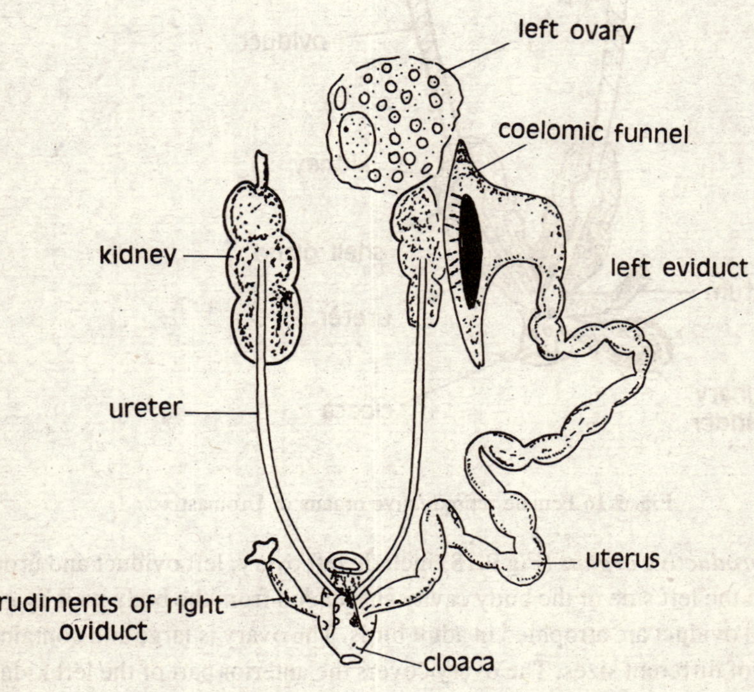

Fig 9.18 Avian female reproductive organs

urodaeum. The rudiments of right oviduct are attached to the right of the urodaeum. The inner linning of the oviduct contains various glands. On maturity the ovarian follicle burst and ovum liberated into coelomic cavity from where it enters the oviduct through *oviducal funnal.* The ovum gathers the secretion of various glands of the oviduct while passing through it. The fertilization is internal and ovum is fertilized in the upper end of the oviduct. The middle part of the uterus secretes albumen and the posterior part of the uttrus called *vagina* secretes shell and shell membranes. Two eggs are usually laid by the pigeon at one time.

9.4.7. Oryctolagus

The *male reproductive organs* (Fig 9.19) include a pair of tests a pair of vasa deferentia unpaired *uterus masculinus,* urethra and penis. Some gland and rectal glands are associated with male genital system.

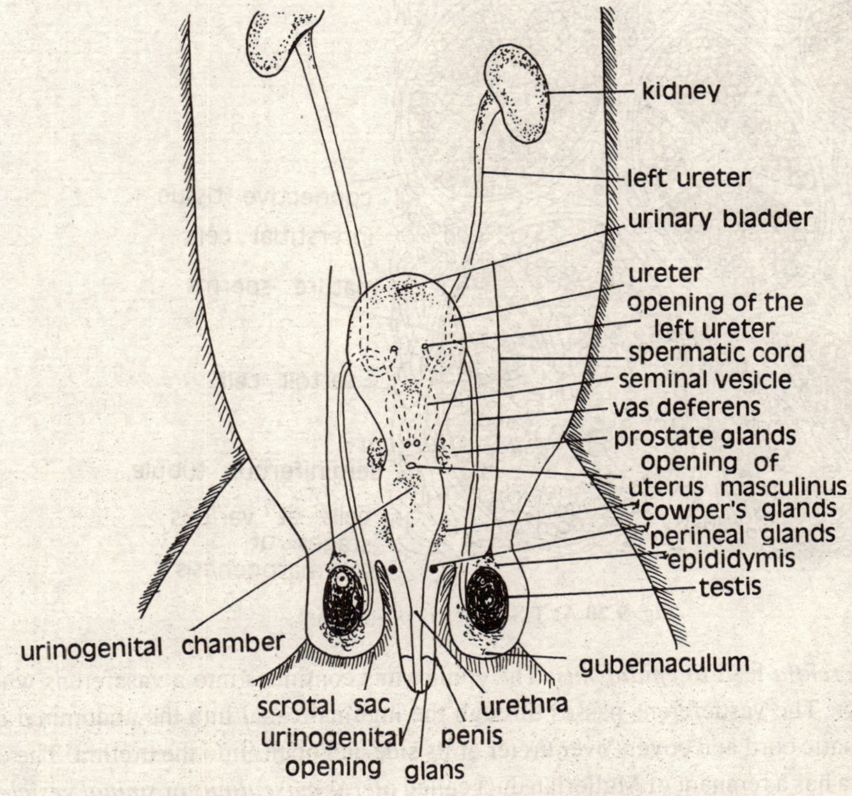

Fig. 9.19 Urogenital system of male rabbit (ventral)

The *testes* are pink oval bodies, each about 15 mm. long which develop in the abdominal coelom but descend downwards outside the coelom just before birth to lie in bag of skin called *scrotal sac.* The cavity of each sac is connected with abdominal cavity by an *inguinal canal.* The tests is attached to the wall of the scrotal sac by a fiberous *gubernaculum.* The scrotal sacs are situated on the sides of the penis. The testis remains connected to its original position by a *spermatic cord* made up of an artery a vein and a nerve. The inner side of each testis is half circled by an *epididymis* which is product of memsonephros. The body of the epididysis is called *corpus epididymis* which remains connected with two enlargements *caput epidiymis* and *cauda epididymis* situated at the anterior and posterior ends of the testis respectively The tests is enclosed in a fibrerous connective tissue called *tunica albuginea.* The seminiferous tubules are present in testes and open into a system of spaces called the *reti-testses* giving rise to delicate vasa-efferentia. Between the tubules some connective tissue with fibers, blood vessals and interstitial cells of Leydig are present. The interstitial cells produce a male sex harmone testosterone. The seminiferous tubules have lining of germinal epithelium which produces spermatozoa. Various stages of developing spermatozoa are found in the seminiferous tubules (Fig 9.20). In between the germinal epithelial cells *sertoli cells* are present which provide nourishment to the developing sperms.

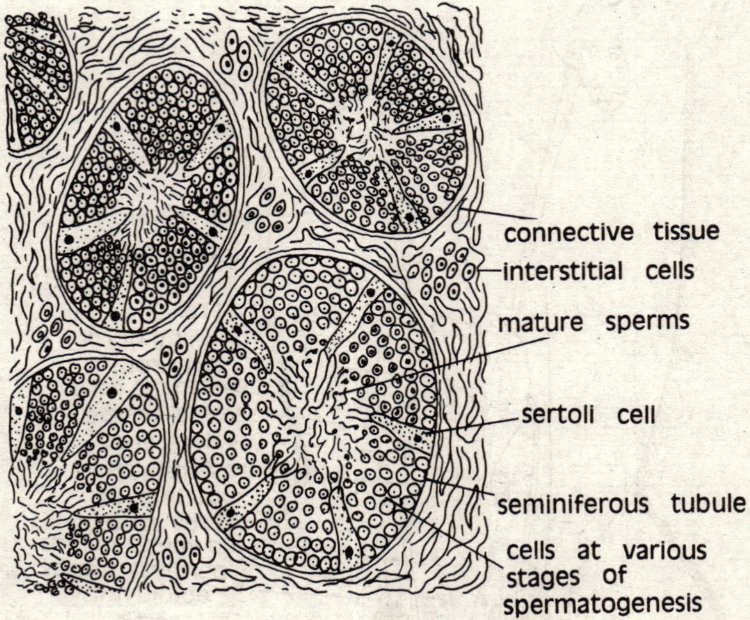

Fig. 9.20 A. T.S. of testis of mammal

The *vasa efferentia* lead to *epididymis.* The epididymis continues into a vaseferens which is mesonephric duct. The vasdeferens passes through the inguinal canal into the abdominal cavity along with spermatic cord and covers over ureter of its side and opens into the urethra. The dorsal side of the urethra has a remnant of Mullerian duct called *uterus masculinus* or *sminal vesicle.* The urethra passes through a male intromittent organ *penis* which opens at its tip by *urinogenital aperture.* The penis hangs infront of anus from the ventral wall of the abdomen. The dorsal wall of penis is made of *corpus spongiosum*, a spongy tissue and anterior wall has two hard structure called *corpora covernosa.* A loose fold of skin called *prepuce* covers the anterior end of the penis.

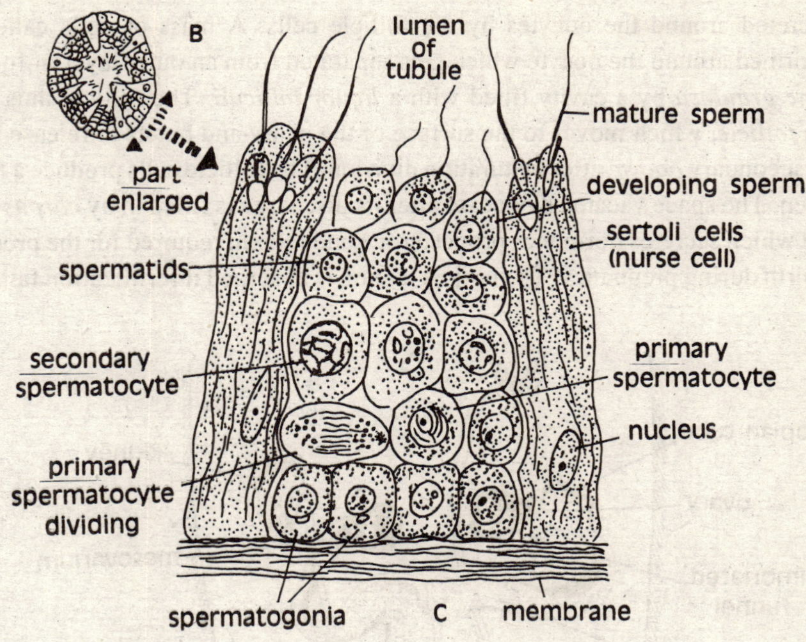

Fig. 9.20 B, C. Part of seminiferous tubule (enlarged)

The *prostate gland* is made up of 4 or 5 lobes. It is situated on the dorsal and lateral sides of the uterus masculinus and communicates with the urethra by small ducts. Its secretion is whitish and activates the spermatozoa. The Cowper's glands are ovoid glands situated behind prostate gland on the dorsal wall of the urethra and open by small ducts. Its secretion is whitish and activates the spermatozoa. Its secretion is supposed to protect the spermatozoa agains. the acidity inside urerthra and vagina. Behind Cowper's glands at the sides of the base of penis are two *perineal glands* which open to the exterior by ducts into the hairless perineal depression of the skin. The secretion of perineal glands gives characterstic odour to the animal. The *rectal glands* of unknown function are situated on the sides of the return in both sexes.

The *female reproductive organs* (Fig 9.21) include a pair of ovaries, a pair of oviducts, unpaired uterus, vagina and vestibule. The associate glands are *Bartholin's glands,* perineal glands and rectal glands. The urethra is short and serves for urinary function only. The *ureters* open in the main body of the urinary bladder. The ovaries are two small oval bodies situated in the body cavity attached to the dorsal body wall by *mesovarium.* The ovary is a solid structure made of a mass of germinal epithelium lying externally and a mass of connective tissue called *stroma.* The stroma contains ova in different stages of development in follicles. The *follicles* are formed by groups of germinal epithelial cells. One cell of the follicle is enlarge to form an oocyte surrounded by a single layer of cells. At this stage tne follicles enter into the stroma and a striated membrance called *zona pellucida* is secreted around the oocytes by the follicle cells. A mass of cells called *discuss proligerous* is formed around the oocyte which gets separated from an outer layer of follicle cells called *membrane graulosa* by a cavity filled with a *liquor folliculi.* The follicle thus formed is called *Graafina follicle,* which moves to the surface of the ovary and bursts to release its *oocyte,* which becomes secondary *oocyte* after maturation division. The follicle cells produce a female sex hormone estrogen. The space vacated by oocyte after its discharge is filled in by *corpus luteum* an endocrine gland which secretes hormones progesterone and relaxin required for the production of milk and child birth during pregnancy. The corpus luteum is absorbed if fertilization falls to occur.

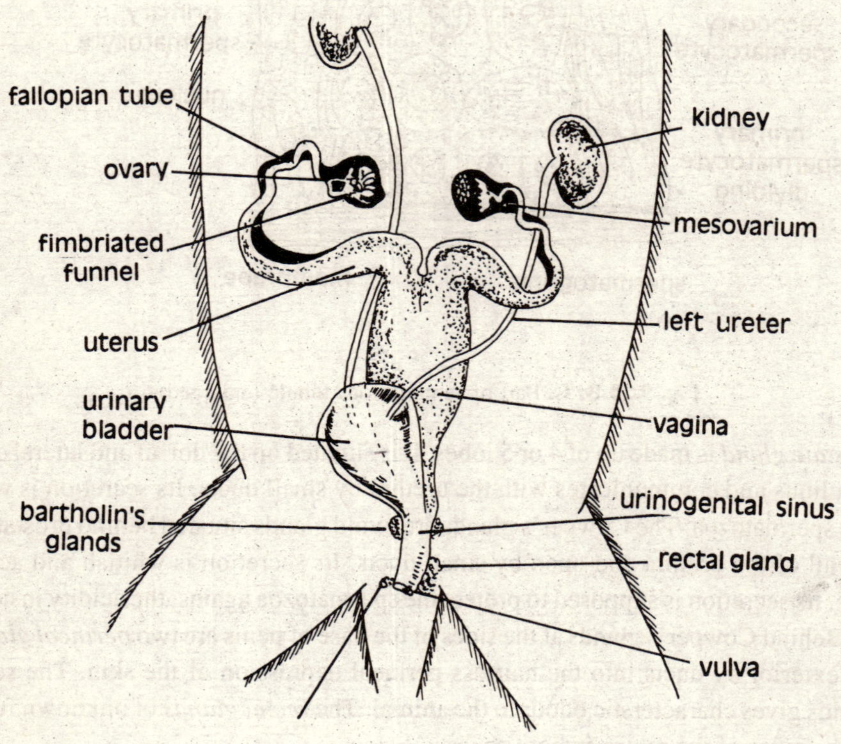

Fig. 9.21 Urogenital system of female rabbit.

The *oviducts* are formed by Mullerian ducts. Each oviduct contains a broad *coelomic funnel* closely applied to ovary. The coelomic funnel joins the *Fallopian tube* (Fig. 9.22) which form anterior narrow part of the oviduct. The middle part of the oviduct is dilated to form a thick walled *uterus* where the embryo develops. The posterior part of the uterus fuses with its fellow of the opposite side to form a median structure *vagina.* The vagina and the urethra joined in a urinogenital canal or *vestibule* opening the exerior by *vulva.* The vulvae are guarded laterally by two folds called *labia majora* and a small rod-like *clitoris.* Homologous to penis of male, projects into the vestibule. The *Bartholins glands* are comparable to the Cowper's glands of male and open into the dorsal wall of the vestibule. The secretion of Bartholin's glands provide lubrication to the vestibule during copulation. The fertilization takes place into the uterus where the secondary oocytes descend through the fallopian tube. Prostate glands are absent in females. *Perineal glands* and *rectal glands* are similar to those found in males.

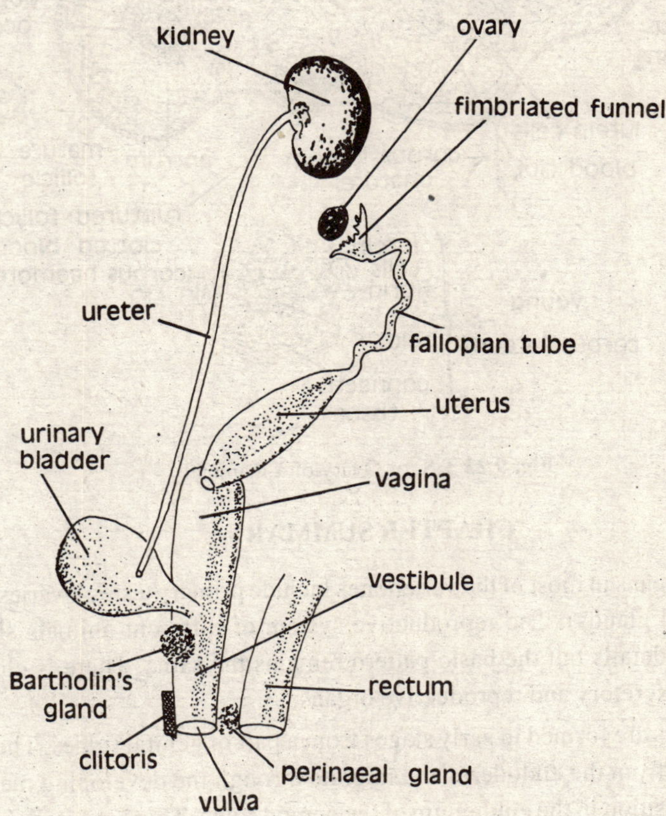

Fig. 9.22 Female Urogenital organs (lateral view).

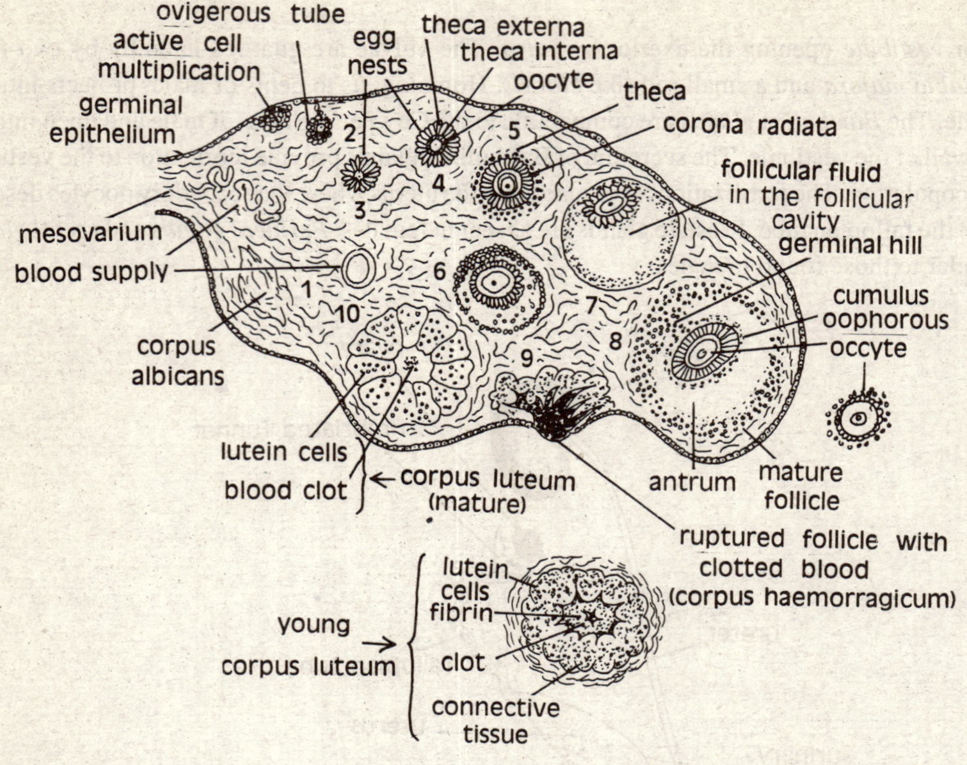

Fig. 9.23 T.S. of Ovary of a mammal

CHAPTER SUMMARY

1. The sex organs in most of the vertebrates include primary testes, ovaries) and secondary (ducts and glands). The reproductive system of different animals shows variety in structural details but the basic pattern remains the same. There is close relationship between excretory and reproductive organs.
2. The gonads are formed in early stages from a pair of genital ridges. The primary- germ cells arise from the endoderm and migrate through the developing mesoderm to their definite position in the epithelium of the genital ridge. There is parallel development of male and female reproductive organs.
3. The male reproductive system consists of testes, vasa deferentia, seminal vesicles and accessory glands. The spermatozoa are produced in seminiferous tubules of the testes and travel and through small ducts vasa efferentia. The vasa efferentia join a larger duct vasa deferens which opens to the exterior directly or through the penis.
4. The female reproductive system includes ovaries, uterus and vagina. The ova are produced in ovaries which are sacular in frog and solid in mammals. The ovum passes through a tube oviduct. The oviduct is enlarged to form uterus in mammals where the

egg is retained for development in viviparous and ovoviviparous species. The vagina is terminal part of the female genital tract in amniotes.

5. The oviducts are paired and separate in most animals excepting in, mammals. Each oviduct is divided into fallopian tube, uterus and vagina in mammals. The uteri may be duplex, bipartite, bicornuate or simplex types. The female accessory glands such as Bartholin's paraurethral and uterine glands are rudiments of degenerated mesonephros and its duets. In anamniotes a coelomic funnel grows to form a Mullerian duct which serves as an oviduct; it is separate from the mesonephric duct seving as ureter.
6. The transportation of spermatozoa to outside is done by archinephros or Wolffian duets developed in connection with the pronephric and matanephric kidneys. The mesonephros degenerates and becomes vas deferens or ducts deferens and epididymis.
7. The testes are abdominal in most cases and extra abdominal in a number of mammals. In mammals the testes are temporarily or permanently located in scrotal sacs which serve as temperature regulators. The spermatozoa are suspended in seminal fluid secreted by accessory glands.
8. The copulatory organs used for transfer of spermalozoa from male to female are claspers (modification of the pelvic fin) in clasmobranchs, gonopodium (anterior border of anal fin) in some teleosts, hemipenes in snakes and lizards, penes of crocodiles, turtles and some birds and penis of mammals. The urethra in mammals passes through the penis (excepting in monotremes) and serves for passage of both urinary and seminal fluids.

10

Endocrine Organs

10.1. GENERAL ACCOUNT

Sonic body cells, groups of cells or organs specialize in structure and functions to produce secretions necessary for body, activities. Such structure are called glands which discharge their products through ducts. Liver, salivary glands etc. are examples of this type called *exocrine glands.* Some glands do not have ducts and their secretion are carried by the blood stream to different parts of the body. Such glands are called *endocrine-glands,* glands of internal secretions, or ductless glands. The secretions of these glands are hormones which exercise considerable regulatory influence over many physiological activities and behaviour of the animals. Most of the endocrine glands work together and constitute an endocrine system. Anatomically it is not possible to deal with endocrine glands in a comparative manner but they show common features physiologically. The endocrine system works with the nervous system for the regulation of physiological activities. Certain glands like pancreas and gonads share double role in their mode of secretion. They produce a secretion which passes through a duct or ducts and give endocrine secretion in addition to it.

A hormone is a complex organic compound usually of low molecular weight with high potency even in smallest quantity. The exact mechanism of endocrinal influence on physiological activities is not known. They may participate in chemical reaction or act as catalysts. They diffuse freely and get oxidized or destroyed easily. The effects of hormones are not permanent and, as such, their continuous supply should be maintained in the body. A hormone shows uniformity of nature throughout the series of animals, where it is found. It influences only specific tissues or activities without effecting other tissues.

The science of endocrine glands is called *endocrinology* and has important applications on human medicine family planning and some bearing on the production of domesticated animals.

The endocrine glands have been classified under the following categories:

1. *Endocrine glands derived from the Ectoderm*
   Pituitary gland or hypophysis
   Uropophysis
   Adrenal medulla and Chromaffin tissue
2. *Endocrine glands derived from the Mesoderm*
   Interenal bodies Adrenal cortex and Gonads
3. *Endocrine glands derived from the Endoderm*
   Islets of Langerhans (Pancreatic Islets)
   Thyroid gland
   Parathyroid glands
   Thymus
   Ultimobranchial bodies

10.2. THYROID GLAND

It develops as a hollow outgrowth from the floor of the buccopharyngeal cavity in all vertebrates and separates from it as an independent gland. In mammalian embryo it arises as a median ventral diverticwlum of the pharynx, at the level of the second pharyngeal pouches. It consists of a mass of glandular tissue, subdivided into minute *follicles* or *lobules* below the pharynx. The *thyroid follicles* are separated from one another by delicate strands of connective tissue (Fig. 10:2). The thyroid gland is homologous to the endostyle of lower chordates. Each thryriod follicle is lined with a layer of cuboidal epitheĥal cells enclosing a mass of colloid which is a viscous product of secretary epithelial cells. The amount of colloid inside follicles fluctuates alongwith changes in the physiological activities of the gland.

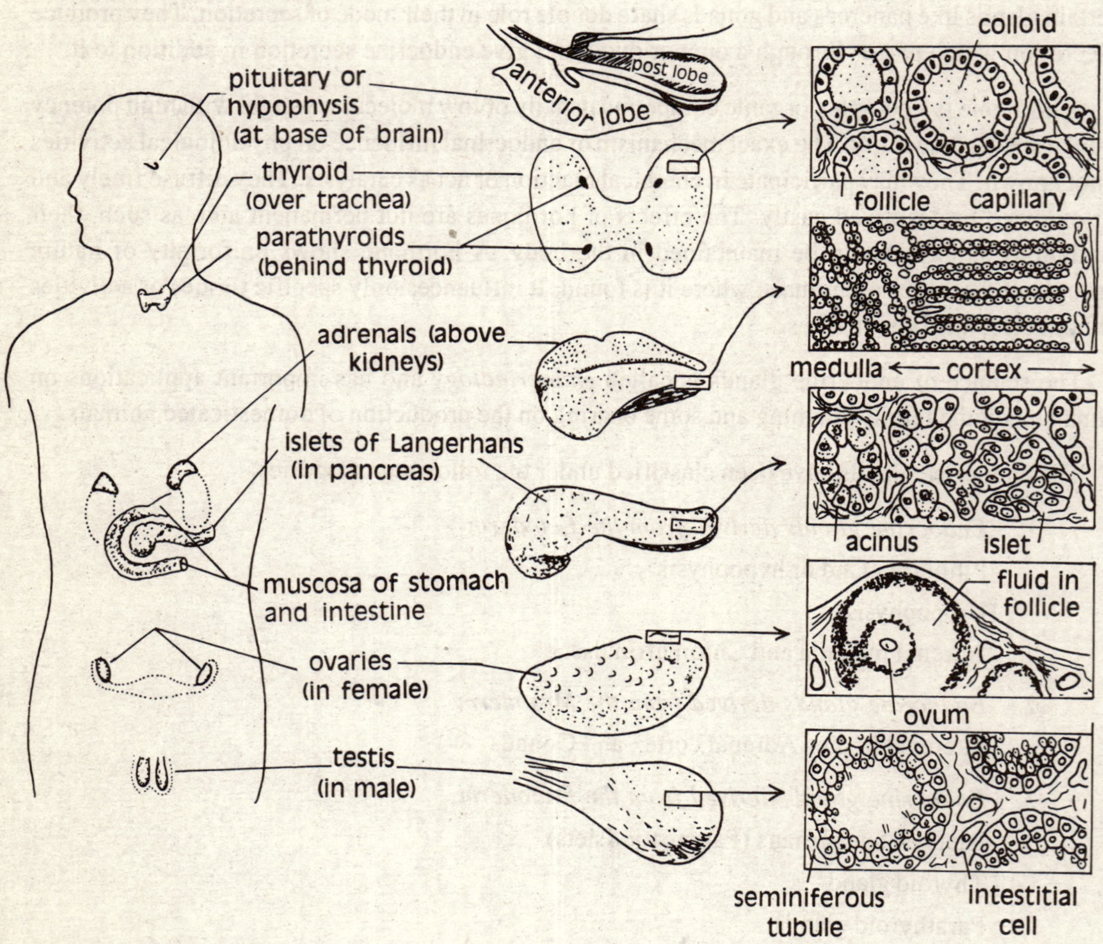

Fig. 10.1 The human endocrine glands. *Left.* Position in body. *Center.* Form of each (not to same scale) *Right.* Enlarged details of cellular structure.

The thyroid secretes *thyroxin* which is an aminoacid containing 65% of iodine. The thyroxin combines with other amino-acids present in the thyroid gland and forms a complex protein molecule *thyroglobulin* which is stored in follicle as colloid. The thyroxin enters the body fluid after its separated from other aminoacids of the tlnroglobulin by enzyme action. The thyroxin is responsible for the control of the rate of metabolism of the body. It increases the rate of metabolism for the release of energy. The thyroid is controlled by the hypothalamus (pituitary) by a *thyroid stimulating hormone* (TSH). It influences other endocrine glands including adrenals and gonads. It activates metamorphosis in young animals like tadpoles. In human beings the severe thyroid deficiency in children results in *cretinism* which causes stunted growth, low intelligence, bent legs, protruding tongue, protrusion of abdomen and immaturity of sex organs. In adult humans the severe thyroid deficiency results into *myxedema* characterized by sluggishness, swelling of subcutaneous tissue mainly of face and hands, low body temperature and dull mind. When the thyroid produces more thyroxine, *exophthalmic goiter* results which is characterized by enlarged thyroid gland, protrusible eyes, high metabolic rate, rapid heart beat and wasting away of the body. Deficiency is iodine in thyroxin causes *simple goiter* in which the thyroid is enlarged.

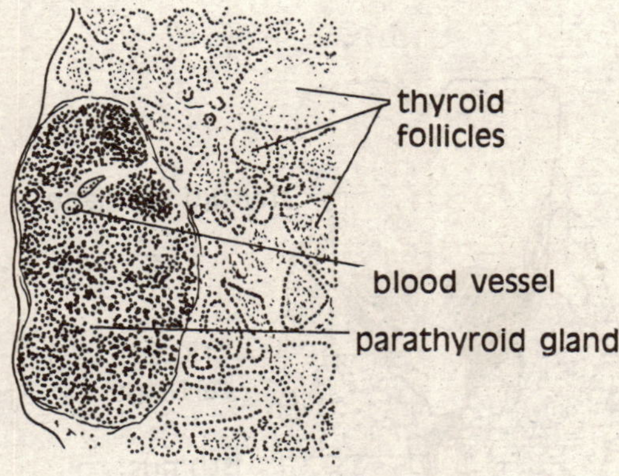

Fig. 10.2 Section through parathyroid and thyroid glands of rat, showing parathyroid partially embedded in thyroid issue.

In *Protochordates* the thyroid gland is not found but a homologous structure called *endostyle* aids in getting food particles and there is no evidence of its endocrine role.

In *cyclostomes* a subpharngeal gland or *endostyle* of uncertain functions is found in ammocoetes larvae of the lamprey.

In *fishes* the thyroid gland makes its appearance in early embryo and differs in its location in adult. In cartilaginous fishes it is a single, compact structure and lies at the anterior end of the ventral aorta. In most telcosts it is paired structure situated on either side of the anterior visceral arches. In some teleosts like perch it consists of diffuse masses present under the ventral aorta and bases of afferent branchial arteries. In Dipnoi the thyroid gland has two lateral lobes connected by a constricted central portion and lies under the tongue epithelium.

In *amphibia* the urodeles have two lobes situated on either side of the external jugular vein. In anura, the thyroid gland consists of two lobes, each situated on the lateral side of the hyoid apparatus. The thyroid hormone plays important role in the metamorphosis of tadpoles and ecdysis.

In *reptilies* the thyroid gland of snakes, turtles, and crocodiles is unpaired and that of lizards is paired in adult condition. In lizards it is situated on the ventral side of the middle parts of trachea but in other reptilies it is posteriorly situated infront of pericardium. The thyroid hormone partially controls ecdysis in reptilies excepting in snakes where the removal of thyroid increases the ecdysis. The phenomenon of reverse effect of thyroid in snakes has not been understood properly

In *birds* the thyroid gland consists of paired structures on either side of the posterior most part of the trachea.

In *mammals* the thyroid gland typically consists of two lobes connected across the ventral side of the trachea by a narrow isthmus (Fig. 10.3).

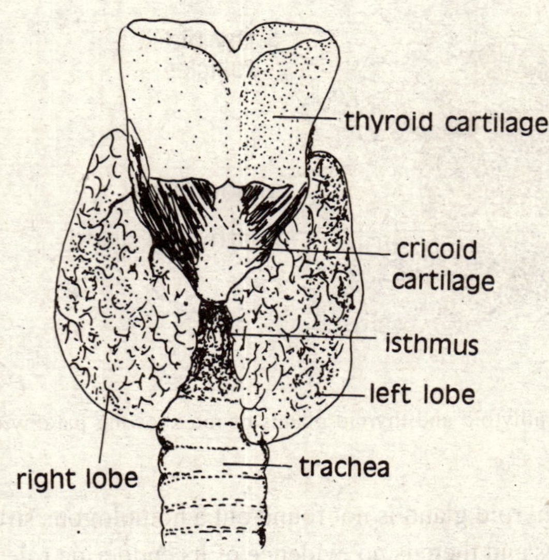

Fig 10.3 Ventral view human thyroid gland shown in relation to trachea and larynx.

10.3. PARATHYROID GLANDS

The parathyroid glands are frequently embedded in the thyroid or thymus glands. They arise from small masses of cells which proliferate from the dorsal and ventral ends of the pharyngeal pouches. In man, the parathyroids show lobules like those of thyroids. The parathyroids originate from the parts of embryonic gill pouches and occur generally in all tetrapods. The parathyroids secrete a hormone called *parathyromone* which regulates the concentrations of calcium and phosphate in blood plasma and their metabolian. Removal of parathyroids results into muscular thickenings and spasmodic contractions, spasm (tetany), and ultimate death. Such effects may be slopped by the injection of parathyroid extract. Overproduction of parathyroid secretion raises the level of calcium in the blood and withdrawal of calcium from bones. In man there are two pairs of parathyroids.

In *cyclostomes* and *fishes* the parathyroids have not been found as compact glands. They are probably represented by diffuse cells posterior to the last pair of visceral clefts. However, the presence of diffuse cells as parathyroid gland has not been established properly.

In *amphibians* the parathyroids have been identified for the first time as derivatives of the third and fourth pair of pharyngeal pouches together with thymus gland. In Anura the parathyroids arise from the second pouches. In Urodela they are 1 to 3 pairs present on the ventral side of the thymus gland and lateral to aortic arches. The parathyroids are widely separated from thyroids in urodela. In Anura they are small rounded bodies, out of them two are present at the posterior part of the hyoid cartilage on either side.

In *reptilies* the parathyroids develop from the ventral ends of the and III and IV pharyngeal pouches like in amphibia. There are usually two pairs of parathyroids situated on the side of the neck on the posterior and lateral sides of the thyroids.

In *birds* the parathyroids are derivatives of III and IV pharyngeal pouches as in amphibibia and reptilies. There may be one or two pairs of parathyroid glands on the posterior side or on the dorsal surface of the throid.

In *mammals* the parathyroid glands have been well studied. They develop from the ends of the III and IV pharyngeal pouches as usually two paired structures on either side. They are usually embedded in thyroid tissue. There are two types of cells, principal or *chief cells* and *colloid* or cells. The chief cells are connected with the secretion of parathyromone.

10.4. GASTRIC AND INTESTINAL MUCOSAE

The lining of the stomach and small intestine produces several hormones which control the secretion of digestive enzymes. The stomach *mucosa* secretes a hormone *gastrin* which enters the blood and return to stomach to stimulate the secretion of gastric juice. The mucosa of the duodenum secretes a hormone *secretin* which enters the pancreas through blood and stimulates the secretion of pancreatic juice. One more hormone *cholecystokinin* is produced by the mucous membrane of the duodenum which activates the gill bladder for the release of stored bile.

The gastrointestinal hormones are not affected by other hormones and do not arise from specific endocrine gland. The mucous lining of stomach and intestine serve other functions also. These hormones have been studied in mammals only.

10.5. ISLETS OF LANGERHANS

The pancreas serves as exocrine and endocrine gland. There are many small groups of cells the *islets of Langerhans* without connections with ducts. These cells are of two types *alpha cells* and *beta cells.* The alpha cells produce a hormone *glucagons* and beta cells produce *insulin.* The insulin controls the entry of sugar into muscle cells, connective tissues, and fat synthesizing tissue and reduces the blood sugar level. The glucagons has opposite effect on the blood sugar level. Insulin is a protein of two peptide chains with 21 and 30 aminoacids respectively, joined by disulphide bonds. Its molecular weight is 12,000. Damage to islets of Langerhans or removal of pancreas causes the increase of sugar in blood and urine which causes a disease called *diabetes mellitus.* Excess of unused sugar in the blood due to diabetes is exreted with urine and the body proteins and fats form increased amount of carbohydrates and the body is weakened ultimately causing coma and death. This condition is controlled by daily injection of insulin extracted from the pancreas of cattle and sheep.

The pancreas is not found in *cyclostomes.* A few small cellular masses are found in the liver and intestinal wall may represent endocrine pancreatic tissue. In *teleosts,* the pancreas is diffuse tissue having a few large islets of Langerhans. In some *bony fishes* the islets of langerhans are grouped together and bound by a capsule outside the pancreas. In amphibians, reptiles, birds and mammals the islets of Langerbans are scattered throughout the pancreas.

10.6. ADRENAL GLANDS

The *adrenal* or *suprarenal glands* are two small bodies situated anterior to kidneys in their close proximity in tetrapods. Each adrenal consists of a *cortex* and a *medualla* of different microscopic structures. The adrenal of mammals has one inner medualla derived from ectoderm and an outer-cortex derived from the mesoderm. The medulla is controlled by the sympathetic nervous system, and produces, *adrenalin* (epinephrin). The cortex produces several endocrinal substances including *dexycorticosterone, cortisone* and *corticosterone.* These bormones are steroids and help the body to face the cold, infection, or low oxygen supply. The complete removal of both adrenals may result into the loss of appetite, weakness, reduction in body temperature and metabolism, loss of water and sodium chloride from blood, and death in 10-15 days. Destruction of adrenal cortex in man results into Addision's desease in which the skin becomes bronze, and gradual wasting away lakes place which results into death. The removal of cortex may cause rise of sodium in the urine and corresponding decrease in the blood and increase the sexual character in some way. The adrenal medualla secretes *adrenalin* which is sent into the blood stream in emergency and acts upon parts innervated by sympathetic nervous system. The adrenalin increases the rate of heart beats, acts upon liver to convert glycogen into the sugar, upon blood vessels, intestine, and skin etc.

The adrenal glands have been observed in *cyclostomes* for the first time in vertebrates. There are two series of such bodies called the *suprarenals* and *internals* which remain separated from one another. The interrenal or cortical series consist of small, irregular structures and extend throughout the length of the trunk and tail. The *suprarenal* or *chromaffin bodies* or medullary series are paired structures situated the course of dorsal aorta and its branches.

In *cartilaginous fishes* the paired suprarenals and interrenals extend throughout the length of the trunk and are independent of each other (Fig. 10.4). The suprarenals are independent of each other but paired interrenals are connected transversely. In *Dipnoi* the suprarenals are associated with intercostal arteries. In teleosts the suprarenals are associated with the cardinal vein and interrenal is related to the kidney.

In *Amphibia* the suprarenals and internals are combined to form the *adrenal.* In Anura the adrenal is yellow, elogated, rod-like structure situated on the ventral surface of the kidney. In Urodeles the adrenal glands are not compact and the intermingled strips of tissue extend throughout the entire length of the kidney.

In Reptilia the interrenal and suprarenal parts are intermixed. The adrenals of crocodiles and chelonia are similar to those of birds.

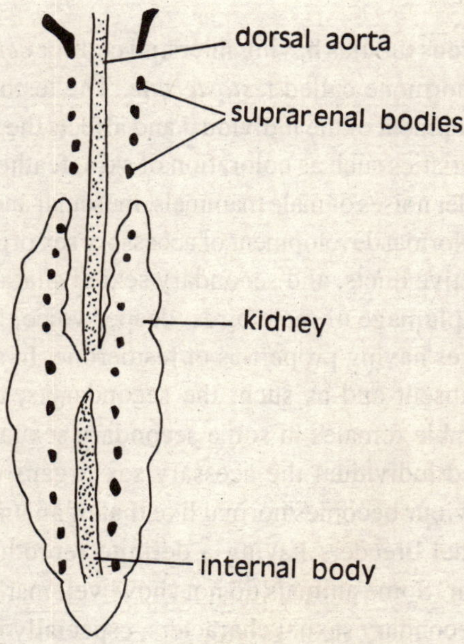

Fig. 10.4 Supra-renal and inter-renal bodies.

In *Birds* the adrenal glands are yellowish bodies present on either side of the postcaval vein just anterior to kidneys near the gonads.

In *Mammals* the adrenals are present at the cranial end of the kidneys. They are clearly distinguished into the a *cortex* and a *medulla.* The adrenal corted is divisible into 3 zones on the basis of histological character and arrangement of cells. These are:

i. Outer zone or zona glomerulosa.

ii. A middle zone fasciculate.

iii. An inner zona reticularis.

The epithelial cells of the adrenal cortex are rich in vitamin "C" (ascorbic acid) and bound with connective tissue. The adrenal medulla has irregular or polyhedral cells held in a loose network of connective tissue. The cells are stained brown by chromic acid and chromium salts and called chromaffin cells.

10.7. THE GONADS

The gonads arise from the coelomic mesothelium as a pair of gonadal ridges close to pronophros and produce hormones (steroids) in addition to the production of eggs and sperms. The gonadal hormones effect the accessory organs and sexual characterstics. The sex organs and functions are also influence by other endocrine glands such as pituitary and thyroid glands.

10.7.1. The Testes

The *testes* are made of seminiferous tubules having interstial cells or *cells of Leydig* in between them. These cells produce a male hormone called *testosterone*. The testosterone in produced in increasing amounts during the development of the individual and affects the various part of the body to produce distinctive male characteristics such as coloration of skin, feathers and hairs, vocal sacs of male frog, syrinx of male birds louder noises of male mammals, muscular and skeletal characteristics and peculiarities of male behaviour. Normal development of accessory sex organs such as intromittent organs, associated glands, reproductive ducts, and secondary sexual characters such as antlers of stags, mane of male lions, brighter plumage of male birds, deeper voice, large body size etc. are produced by *androgens* or substances having properties of testoerone. In some lower vertebrates the glandular interstilal cells are absent and as such, the secondary sexual characteristics are absent. The castrated animals resemble females in some secondary sexual characteristics. If the hormone is injected into a castrated individual the acessary sex organs enlarge and secondary sexual characters develop and behaviour becomes normal like that of an uncastrated animal. Most of the lower vertebrates are seasonal breeders having a definite reproductive cycle where the breeding takes place only once a year. Some animals do not show well marked sexual dimorphism while others have well developed secondary sexual characters, especially in the breeding period.

In male *fishes,* during breeding season the color patterns are altered and fins are modified In some amphibians the thumb pads are swollen, colour patterns are changed and cloacal glands are enlarged. In *reptilies,* the sexual dimorphism is not well marked and seasonal skin colour change is observed. In *birds* the changes of beak color and plumage are observed seasonal. The sexual dimorphism is well pronounced in mammals. Castration of some domestic animals is done to make them more docile. If the castration is done before the age of puberty the typical masculine characters do not develop. If castration is done after puberty some accessory sex organs such as seminal vesicles, prostate and bulbourethral glands are regressed.

10.7.2. The Ovary

The ovaries serve as cytogenic and endocrine gland. It produces three female sex hormones *estradiol progesterone* and *relaxin.* The follicles of ovary produce estradiol, which is responsible for the phenomenon of estrus or "heat" in female mammals. Chemical substances having properties of estradiol are called *estrogens,* which are responsible for the development of secondary sexual characters in female. If ovaries are removed from an immature female the sexual maturity is prevented, accessory sex organs remain immature and sexual instincts are not developed. The injection of estradiol into a castrated female will correct such effects. Another hormone *progesterone* or *progestin* is produced in the ovary by the *corpus luteum* which is formed late in the pregnancy after the discharge of the ovum. The progesterone and estradiol prepare the uterus for receiving a fertilized ovum and implantation of the embryo and induce directly or indirectly. The enlargement of the mammary glands (A lectogenic hormone from pituitary stimulates milk secretion at a later stsate). A third hormone *relaxin* releases the ligaments of the pelvic girdle to facilitate childbirth.

Periodic fluctuations in the levels of estrogen and progesterone secreted by the ovaries are correlated with the cyclic activity of the female reproductive organs including menstruation in higher primates. The production of female hormones is stopped at middle age in mammals and ovaries cease to work as endocrine glands. Certain changes in secondary sexual characters occur at this time resulting into atrophy of gential organs and stoppage of periodic cycles.

10.8. PITUITARY GLAND

The *pituitary gland* or *hypopysis cerebri* (Fig. 10.5) lies below the diencephalons of brain. It connected with the brain by a hypophyseal or infundibular stalk. In its structure and functions the pituitary is not one gland but two glands consisting of a single compact mass. In higher vertebrates the pituitary is divided into two parts called *anterior lobe* or *adenohypopysis* derived from the roof of the stomodaeum and posterior lobe or neurohypophysis derived from the floor of the diencephalon. In higher forms the pituitary gland is situated in depression present on the upper surface of the basisphenoid bone called *sella turcica.* The anterior lobe of the pituitary is developed from an imagination of ectoderm the roof of the stomodacum called *Ratlike's pouch.* The pouch lies under the diencephalon in early stages and soon becomes separated from the roof of the mouth developing into the adenohypophysis (Fig. 10.6). The anterior lobe consists typically of larger and lower *pars-distalis,* an upper *pars tuberalis* surrounds the stem of the posterior lobe and *pars-intermedia* or *intermediate lobe* between pars distalis and posterior lobe. The neurohypophysis or posterior lobe is developed as an outgrowth from the floor of the hypothalamus of the brain below the third ventricle in the diencephalon. The glandular part of the neurohypophysis called *pars-nervosa.* A fraction of hypothalamus also forms an integral part of its mechanism.

The pituitary gland secretes many hormones which influence endocrine tissues, tissues other than endocrine glands and is also influenced by other endocrine glands by reciprocal action. It has been regraded as *'master of the endocrinal orchestra'.* The anterior lobe of the pituitary adenohypophysis consists of masses and cords of secretary epithelial cells supported by a loose framework of connective tissue separated by sionusoids. At least two types of gland cells are present in pars distales but the association of specific cell types with the production of specific hormones is not established completely. The intermediate lobe consists of polygonal secretary cells and pars-tuberalis consists of cells perhaps without secretary activities. The posterior lobe has branched cells formerly regarded as secretory cells but presently as supporting cells. The posterior lobe stores hormones which many be passed on into the small blood vessels which remify here. The latest opinion is that the secretions found in this party pituitary gland are not produced here but are neurosecretions received from the true brain substance of the hypothalamus.

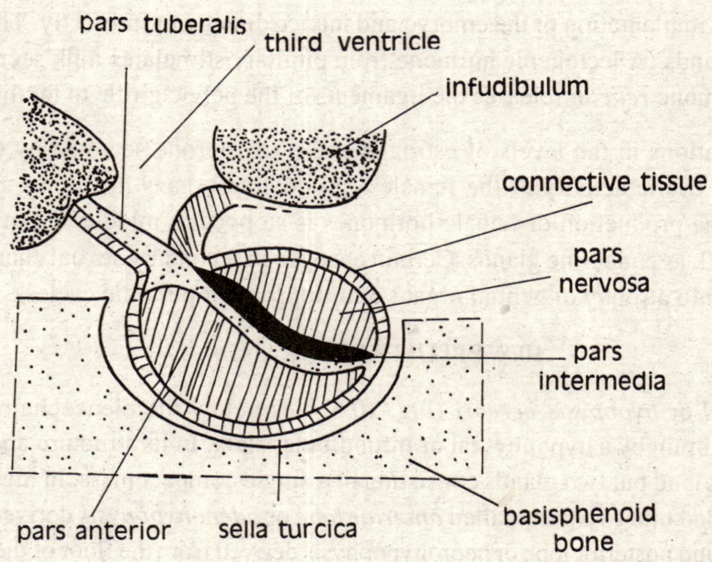

Fig. 10.5 Structure of Pituitary Gland

10.8.1. Functions of Pituitary Gland

The anterior lobe secretes about 10 hormones which are given below:

i. *Growth hormone or somatotropic hormone* (STH) is connected with normal growth of the body and bones and probably stimulates thyroid gland. It possibly stimulates alpha cells of the islets of Lagerhans to secrete glycogen.

ii. *Thyrotropic* or *thyroid stimulating hormone* (TSH): stimulates thyroid glands to secrete thyroxin.

iii. *Adrenocorticotropic hormone* (ACTH) stimulates cells of the adrenal cortex to secrete adrenocorticoids by which body responds to fight the harmful effects of starvation, burns and haemorrhage.

iv. *Lactogenic hormone* or *prolactin* causes the development of mammary glands in female mammals in conjunction with gonadial hormones, maintains secretion of milk by mammary glands, causes development and secretion of crop glands in pigeons, and stimulates corpus luteum to secrete progesterone.

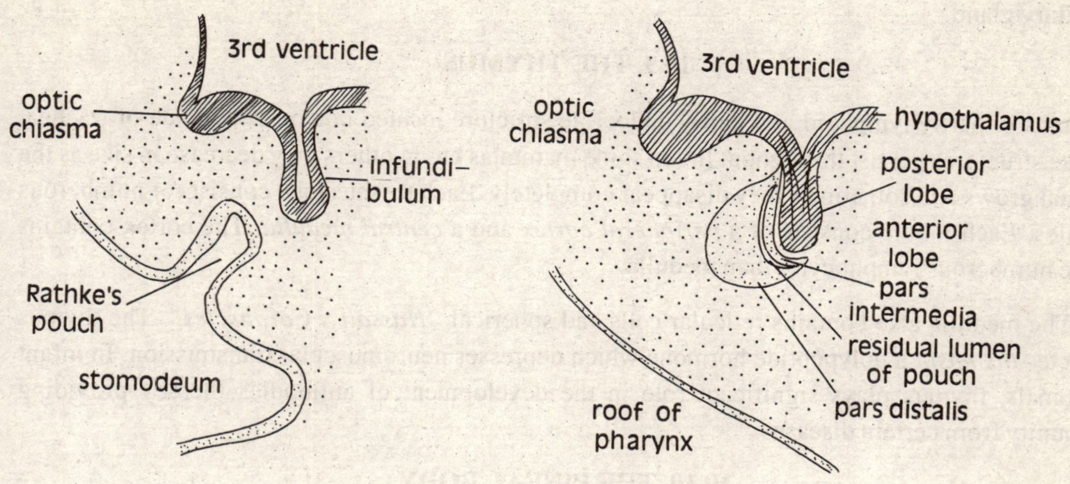

Fig. 10.6 Diagram of pituitary gland (right) and its embryonic development from Rathke's pouch and the infundibulum (left) in higher vertebrates

v. *Gonadotropic hormones* are of two types: The *follicle stimulating hormone* (FSH) controls the endocrine function of the ovarian follicles in females and plays a role in the development of mature sperm cells in seminiferous tubules. The *luteinizing hormone* (LH) is another gonadotropic hormone controlling the development of *corpora lutea* in the ovary and activities of the interstitial cells of the testis. A *luteotropic hormone* influencing the secretion of progesterone by corpora lutea has also been clained.

vi. A separate *interstitial cell stimulating hormone* (ICSH) has also been reported by some authorities.

vii. *ACTH-hypophysiotropic hormone* present in the hypothalamic portion of the brain is connected with the stimulation of adrenocorticotropic hormone. (ACTH).

viii. The intermediate lobe secretes *intermedin* which is responsible for the dispersal of pigment granules in integumentary chromatophores.

ix. The posterior lobe stores two endocrine products called *vasopression or pitressin* and *oxytocin or pitosin.* The vasopressin increases blood pressure and has antidiuretic effect upon kidneys. Oxytocin is connected with the contration of the smooth muscles of the uterus

10.8.2. Pituitary Gland in Vertebrate Series

The origin of pituitary gland is constant throughout the vertebrate series in its development and basic structure. The mammalian pituitary gland is typical of vertebrates in general. In *Branchiostoma* a preoral pit present in front of the mouth of a young individual has been regarded as possible homologous structure. The pituitary gland is typical in cyclostomes in its development and structure from cyclostomes onwards, there is no much of remarkable difference in the structure of the pituitary gland.

10.9. THE THYMUS

Thymus in partly lymphoid and partly endocrine structure located one on either side of trachea. These structures persist throughout life in some mammlas but in others they decrease in size as the animal grows and sometimes they disappear completely. Each thymus lobe consists of numberous lobules. Each lobule consists of a *peripheral cortex* and a *central medulla.* The cortex contains more numberous lymphocytes than medulla.

The medulla also contains reticular cells and spherical *"Hassall's Corpuscles".* The thymus secretes *thymosin* a polypeptide hormone which depresses neuromuscular transmission. In infant mammals, thymus plays significant role in the development of antibodies thereby providing immunity from certain diseases.

10.10. THE PINEAL BODY

It is pea-sized small gland situated deep in the groove between the cerebellum and the cerebral hemispheres in rabbit. It consists of parenchyma cells and neuroglia cells. In lower vertebrates it produces a hormone *melation* in which causes marked lightening of the skin in frogs, thus acting against melanophore stimulating hormone (MSH). In humans it atrophies at the age of seven years and its function is uncertain.

10.11. OTHER ENDOCRINE GLANDS

The Urohypophysis: In a large number of fishes a caudal neurosecretory system called urohypophysis is found associated with the posterior end of the spinal cord. It is highly developed in teleosts. The secretions from urohypophysis probably regulate the transport of sodium ions into or out of the organism in relation to the osmolarity of surrounding water and the organism. About eight neurosecretions have been identified from urohypophysis but their exact physiological effect is not established.

Chromaffin Tissue: In lower vertebrates the homology of medulla and cortex are usually separate from each other. Those homologous to medulla of higher chordates are called *chromaffin tissue* or *chromaffin bodies.* Masses of chromaffin cells are found closely associated with sympathetic ganglia, ovary, testes, heart, kidney and other visceral organs. The physiological significance of chromafin tissue has not been understood properly although they are well distributed.

Interrenal Bodies: The cortical tissue of adrenals of higher vertebrates is homologous to the interrenal bodies of fishes. They are derived from mesoderm. The interrenal cells are found kidneys. In fishes, they may form serially arranged clusters on wo major masses of different shapes. They are also called cortical tissue These are corticoid producing endocrine tissue of lower vertebrates.

Ultimobranchial bodies: These are structures growing from the caudal walls of the last pharyngeal pouches. They are seem to have lost their function.

10.12. ENDOCRINE GLANDS IN VERTEBRATE TYPES

Comparative account of the endocrine glands in vertebrate types has been given in table 10.1.

TABLE 10.1

Gland	*Scoliodon*	*Rana*	*Uromastix*	*Columba*	*Oryctolagus*
Thyroid	Single thyroid present below anterior end of ventral aorts	Paired on lateral sides of hyoids, controls growth, and periodic ecdysis	Paired controls ecdysis	Paired one on each side of trachea	Paired lobed gland below larynx
Para-thyroids	Absent	2 pairs of small red, round glands	2 pairs ner aortic arches posterior to thyroid	Paired present posterior to thyroid	Paired small glands embedded in thyroid gland
Gastric and intestinal mucosae	Not well under stood as endocrines	Not well under-stood as endocrine	Endoerine function not well understood	Endoerine function not well under-stood.	Endocrine Gastrin, secretion & Cholecy-stokinin.
Islets of Langer-Hans	Prism-shaped cells present around a cavity	Typical condition	Typical condition	Typical condition	Typical condition
Adrenals	Two widely-separated sets of tissues suprarenal bodies and interrenal bodies present	Two component are inter-mixed and form a yellow hand on the ventral surface of each kidney	Lie close gonads	Yellow structures close to gonads, the suprarenal tissue lies between masses of interrenal tissue	Present near kidneys
Gonads	Testes have interstitial cells producing testosterone and ovarian follicles producing estrogen and progesterone	Testes have no interstit-ial colls. Ovary has a few follicle cells hut corporalutea absent Endocrine function gonads not	Intersti-tial cells of testes and folli cle cells and corpo-ralulea of ovary present.	Testes have very few interstit-ial cells. Follicle cells present in ovary but of the presence of corporalutea	Typical condition

Gland	*Scoliodon*	*Rana*	*Uromastix*	*Columba*	*Oryctolagus*
		well under stood.		doubtful.	
Pituitary	Present but its function are not fully under-stood. It stimu-lates and keeps the glands healthy & causes expansion of melanophores	Relationship of different lobes differs from typical condition Pars anter-ior and pars tuberais do not form an anterior lobe and pars anterior lies posterior to other lobes, pituitary hormones control skin colour changes and expansion of melanophores	In-fundi-stalk long pituitary in posterior position The anter-ior lobe is loosely connected with other lobes infront.	intenn-ediale lobe is absent but inter medin is secreted b, anterior lobe.	Typical condition

10.13. SUMMARY OF ENDOCRINE FUNCTIONS

The hormones secreted by the endocrine organs and their functions have been summarized below:

TABLE 10.2

Summary of Endocrine Functions

S.N.	*Name of Gland*	*Hormones Secreted*	*Function of hormone*
1.	Thyroid	Thyroxin	1. Controls the rates of body metabolism for release of energy. 2. Activates metamorphosis in young animals. 3. Influences adrenals and gonads.
2.	Parathyroids	Parathyromone	Regulates the concentration of calcium and phosphates in blood plasma and their metabolism.
3.	Gastric and Intestinal Mucosae	1. Gastrin 2. Secretin	Stimulates secretion of gastric juice. Stimulates secretion of Pancreatic juice.

S.N.	Name of Gland	Hormones Secreted	Function of hormone
		3. Cholecystokinin	Activates gall bladder for the release of bile
4.	Islets of Langerhans	1. Insulin (Beta cells)	Controls the metabolism of carbohydrates and reduces the Sugar level in blood.
		2. Glucagon (Alpha cells) level.	Opposite effect of insulin and increase blood sugar.
5.	Adrenals	1. Dexycarticosterone 2. Corticosterone 3. Cortisone	Make the body to cope with cold, low oxygen supply.
		4. Adrenalin Epinephrin)	i. Increase the rate of heart beat, ii. Acts upon liver to convert glycogen into sugar, iii. Acts upon blood vessels, intestine and skin etc.
6.	Gonads Testis	Testosterone	Produces distinctive male characteristics and peculiarities of male behaviour. Develop accessory sex organs and produce secondary sexual characters.
	Ovary	i. Estradiol, (Estrogens)	Development of female secondary sexual characters; productions of estrus or 'heat' in females.
		ii. Progesterone (In corpusluteum)	Prepares uterus for pregnancy.
		iii. Relaxin	Release the ligaments of the pelvic girdle to facilitate child birth.
7.	Pituitary Anterior lobe	Somatotropic Hormone (STH)	i. Maintains normal growth of body and bones. ii. Possibly stimulates alpha cells of Islets of Langerhans to produce glucagons.
		Thyroid stimulating Hormone (TSH)	Stimulating thyroid glands to secrete thyroxin.
		Adrenocorticotropic Hormone (ACTH)	Stimulating adrenal cortex to produce adrenocortioids.
		Lactogenic hormone or prolaction.	i. Development of mammary glands and crop glands in pigeons.

S.N.	*Name of Gland*	*Hormones Secreted*	*Function of hormone*
			ii. Stimulating corpus luteum to secrete progesterone.
		Gonadotropic Hormones	
		i. Follicle stimulating hormone (FSH)	Controls endocrine function of ovarian follicles and plays a role in the development of mature sperms in seminiferous tubules of male.
		ii. Leutenizing Hormone (LH)	Controls the development of corpora lutea in ovaries and activities of interstitial cells of testis.
		iii. Leuteotropic hormone.	Possibly influences secretion of progesterone by corpora lutea.
		iv. Interstitial cells Stimulating Hormone (ICSH)	Possibly stimulates interstitial cells of testis.
	Hypothalamus	Adrenocorticotropic Hypophysiotopic Hormone (ACTH).	Stimulates the secretion
	Intermediate Lobe	Intermedian	Dispersal of pigment granules in integumentary chromstophores.
	Posterior lobe	Vasopression or Pitressin.	Increase blood pressure and has antidiuretic effect upon kidneys.
		Oxytocin or Piotosin	Contraction of smooth muscles of the uterus.
8.	Thymus	Thymosin	Depression of neuromuscular transmission.
9.	Pineal body	Melatonin	Action against MSH.

CHAPTER SUMMARY

1. The endocrine or ductless glands secrete hormones into blood or lymph for circulation to all parts of the body. The hormones are complex organic compounds which bring about specific or general changes in the body.
2. The origin of the endocrine glands may be ectodermal (pituitary, urohypophysis, adrenal medulla and chromaffin tissue), mesodermal (interrenal bodies, adrenal cortex, gonad) or endodermal (Islets of langerhans, thyroid, parathyroid, thymus and ultimobranchial bodies).

3. The thyroid develops as an outgrowth from the floor of the buccopharyngeal cavity. It consists of a mass of glandular tissue containing follicles. The thyroid gland is homologous to the endostyle of lower vertebrates. The thyroid gland secretes thyroxin responsible for the control, of metabolic rate in body or release of energy, activation of metamorphosis in young animals and influences adrenals and gonads.
4. The parathyroid glands are frequently embedded in the thyroid or thymus glands. Parathormone hormone secreted by the parathroids regulates the concentration of calcium and phosphates in blood plasma and their metabolism.
5. The lining of stomach and intestine produces several hormones which control the secretion of some digestive enzymes.
6. The adrenal or suprarenal glands situated near kidneys consists of cortex (derived from mesoderm) and medulla (derived from ectoderm). Adrenal medulla is controlled by sympathetic nervous system and secretes a hormone adrenalin or epinephrine which increases the rate of heart beat, acts upon liver to convert glycogen into sugar and acts upon blood vassels, helping the body against cold, infection or low supply of oxygen.
7. The gondal hormones affect the accessory organs and sexual charactestics. The testes produce testosterone which produce distinctive male characteristic, peculiarities of male behavior, accessory sex organs and secondary sexual characters. The ovary produces estrogenas (estradiol) responsible for female sexual characteristic, production of estrus or heat in female; progestrone from the corpus luteum prepares uterus for pregnancy and relaxin for release of the ligaments of the pelvic girdle to facilities child birth.
8. The pituitary gland lies below the diencephalon of the brain. In higher vertebrates it consists of anterior lobe or adenohypophysis, middle lobe and posterior lobe neurohypophysis. It secretes several hormones and plays a dominant role in endocrine system and controls many other endocrine glands.
9. The thymus gland is considered endocrine in function. It may persist throughout life or decrease in size as the animal grows or disappear completely. The thymus secretes thymosin responsible for the depression of neuromuscular transmission.
10. The pineal body produces melatonin in lower animals, which works against melanophere stimulating hormone. It atrophies in humans at the age of seven years.
11. Other endocrine glands include urohypopysis of teleosts and other fishes, chromaffin tissue of lower vertebrates homologous to the medulla of higher chordates, and interrenal bodies in fishes homologous to the cortex of adrenals of higher chordates.
12. The endocrine glands of all vertebrates are similar in structure and functions with minor variations.

11

The Nervous System

11.1. GENERAL ACCOUNT

The nervous system coordinates and controls various activities including learning, intelligence and memory. The main function of nervous system is to inform the animal about environmental conditions and their correlation for its advantage. The outermost layer of body ectoderm comes in contact with environment and the information is carried to the internal organs by nerves or conducting tracts. The body of the nerve cell where the incoming and outgoing tracts meet serves for the correlation of messages. The *neurons* are units of nervous system of vertebrates. The nervous system is always dorsal to digestive tract in vertebrates and may be divided into the following parts for convenience.

1. The *central nervous system* including anterior brain connected with spinal cord.

2. The *peripheral nervous* system including nerves originating from the brain and a pair of spinal nerves from spinal cord for each primitives body segment.
3. The *sympathetic nervous system* or *automatic nervous system* which controls various involuntary activities of the body.

11.2. HISTOLOGY OF THE NERVOUS SYSTEM

The nerve cells *neurons* (Fig. 11.1) constitute the nervous system. There may be different types of neurons in the different parts of the nervous system or in different animals. Each neuron has a large cell body with a conspicuous nucleus and two or more protoplasmic processes called *dendrites* and *axon.* The dendrite is usually short and branched structure near the cell body but axon is usually short or long unbranched structure. The dendrite transmits stimuli to the cell body (*receptor function*) and axon carries impulses away from it (*motor function*). A neuron may be a very long structure measuring several feet in large animals. Neurons with one dendrite and one axon are *bipolar,* with several dendrites and one axon are multipolar. The neuron cell body has a nucleus and several basophilic *Nissl's granules.* Fine threads or *neurofibrils* form a network inside the cell Body. The sensory neurons (*afferent*) conduct impulses from receptors towards central nervous system. The motor neurons (*efferent*) conduct from the central nervous system to various effectors. Adjuster neurons are present in the brain and spinal cord and join variously between sensory and motor neurons.

When the nerve cell bodies group together outside the central nervous system they form a *ganglion.*

A *nerve* is a group of fibres or processes held together in parallel bundles by connective tissue. *Nonmyelinated* or *nonmedullated nerve fibres* do not have any surrounding sheaths and have gray appearance. Such fibres are found in invertebrates and also in vertebrates where they are found in sympathetic system, certain internal fibre tracts of spinal cord and external fibre tracts of brain. *Myelinated* or *medullated fibres* have axon surrounded by a sheath of mylin containing fatty white material giving white appearance to nerves and outside of spinal cord. The myelin substance is constricted at intervals and forms *nodes of Rainier.*

The *neurilemma* or *Schwman sheath* is a delicate membrane surrounding both types of fibres and plays Important part in the regenration of damaged nerve fibres.

The *gray matter* is formed in the brain and spinal cord by nerve cell bodies, their dendrites and proximal unmyelinated portions of axons.

The *while matter* of brain and spinal cord is formed by bundles of myelinated nerve fibres.

Some cells called *neuroglia* are also present in the brain and spinal cord besides the neurons and serve as packing cells between the neurons. True connective tissue is absent in the brain and spinal cord. The cavities of the brain and spinal cord are lined with some *ependymal cells.*

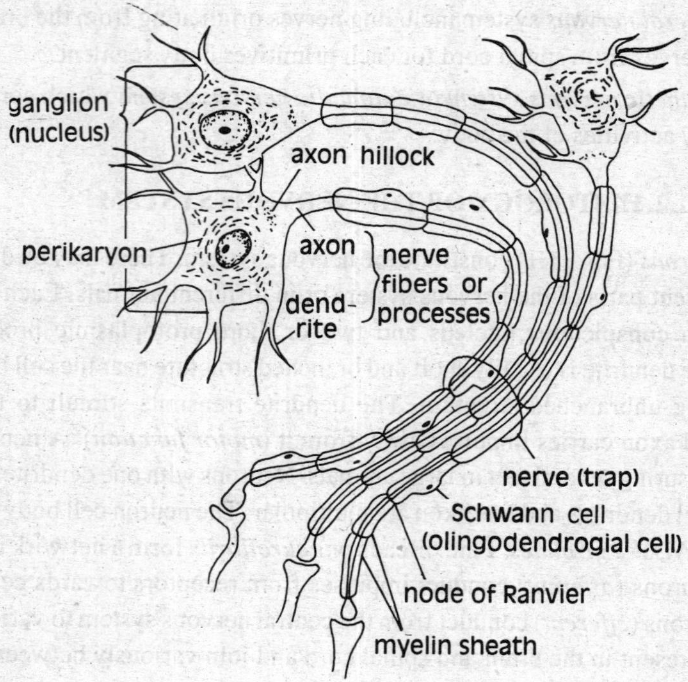

Fig, 11.1 Structure of a neuron. The cell nucleus and surrounding cytoplasm form the cell body of nuron (Perikaryon). Nerve fibers or process are cytocusmic extensions form the perkiaryon. Axons carry impulses away from the perikaryon and dendrites carry impulses toward it. The same structures are given different names in the peripheral and the central nervous systems. Central nervous system terms are given in parentheses.

In vertebrates the impulse always travels in one direction in nerve cells. The impulse passes through dendrites to cell body and then along the axon. The fibre can transmit impulse in both directions.

The neurons are not fused but arranged in chain-like manner so that impulse may be transmitted from one to another in scries. The terminal carbonization of cell comes in contact with dendrites of one or more neurons. The point of junction of two cells is called a *synapse,* which is physiological valve. The impulses are transmitted from one part of the body to another through synapse, which passes nerve impulses from the axon of one neuron to the dendrite of the other in one direction only. The nerve impulse passing through a fibre involves chemical and electrical change and maintains uniform speed with same intensity throughout.

11.3. CENTRAL NERVOUS SYSTEM

11.3.1. Development of the Central Nervous System

In all vertebrates the origin of nervous system is comparatively parallel. The brain and spinal cord develop from the dorsal hollow neural tube (Fig. 11.2). The anterior end of the neural tube enlarges to form the brain after the appearance of two constrictions which form three primary brain vesicles. These vesicles are *prosencephalon* (fore brain), *mesencephalon* (hid brain) and *rhombencephalon* (hid brain). The remaining part of the neural tube makes the *spinal cord.* The prosencephalon divides into an anterior *telencephalon* destined to become *cerebral hemispheres* and a posterior diencephalon by a constriction. The *olfactory lobes* are formed from the anteroventral portion of the *telencephalon.* The mesencephalon gives rise to optic lobes and does not undergo division. The anterior end of rhombencephalon (metencephalon) gives rise to the *cerebellum* and *medulla oblongata* or *myelencephalon* is formed from the remaining portion of the posterior portion of the rhombencephalon which continues into the spinal cord.

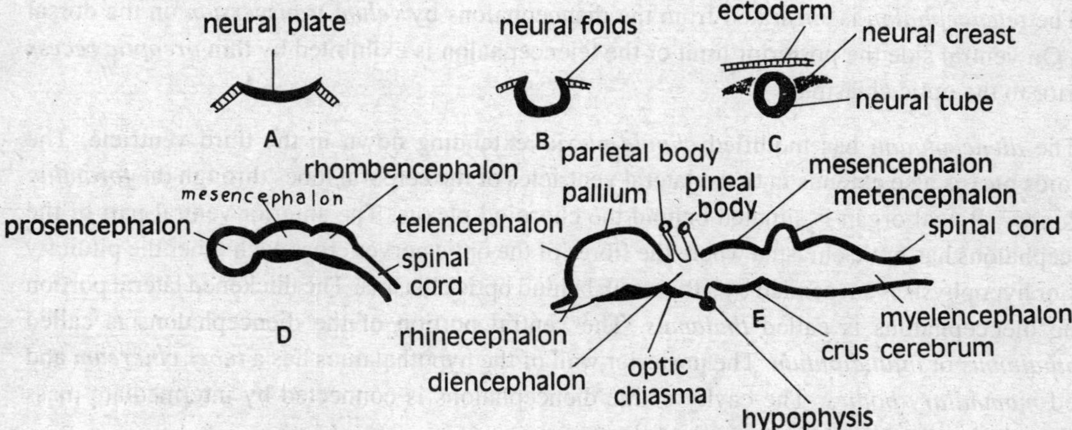

Fig. 11.2 Development of brain.

11.3.2. Typical Structure of the Brain

The structure of the brain may be summarised by the following table:

TABLE 11.1
Brain Structure

	Prosencephalon (Fore brain)	Telencephalon: Includes cerebral hemishpheres, olfactory lobes, corpus striatum, cerebral cortex (pallium) and olfactory bulb.
Brain		Diencephalon: Includes epithalamus thalamus, Hypothalamus and its appendages.
	Mesencephalon (Mid brain)	Tectum: Includes optic lobes, tegumentum; and crura cerebri.
	Rhombencephalon (Hind brain)	Metancephalon: Includes part of medulla oblongata, cerebellum and Pons in mammals. Myelencephalon: Part of medulla oblongata.

The brain is enclosed in connective tissue envelopes called *meninges* which are highly vascular and share *spinomeningeal* or *cerebrospinal fluid.* The central cavity of brain and spinal cord and spaces between the meanings are filled with this fluid. The *cerebral lobes* or *hemispheres* are large structures which fill the cranial cavity in higher mammals. Each cerebral lobe has a *ventricle* connecting the third ventricle or *prosencoel* by a *foramen* of Monroe. The ventricles of the cerebral lobes extend into olfactory bulbs. The two cerebral lobes are united across the mid line above the level of diencephalons by *corpus callosum* of transverse fibres. Bilateral tracts of fibres called *fornix* hang down from the corpus callosum, which are suspended by a *septum lucidum.* In the anterior wall of the brain belew the corpus callosum *anterior commissure* is present. The olfactory bulbs and olfactory nerves are situated below the anterior commissure anterior to it.

The *telencephalon* is spearated from the diencephalons by *velum transversum* on the dorsal side. On ventral side the posterior limit or the telencephalon is exihibited by thin *preoptic recess* anterior to the optic chiasma.

The *diencephalon* has modified *chooid-plexus* extending down in the third ventricle. The choroids plexus also extends in to the lateral ventricles of the cerebral lobes through the *foramina of Monroe.* Pineal organ is situated behind the choroiod plexus. The anterior ventral part of the diencephalons has optic chiasma where the fibres of the optic nerves cross each other the pituitary body or hypophysis is suspended by a thin stalk behind optic chiasma. The thickened lateral portion of the diencephalons is called *thalamus.* The ventral portion of the diencephalons is called *hypothalamus* or *infundibulum.* The posterior wall of the hypothalamus has a *tuber cinereum* and paired *mammilary bodies.* The cavity of the diencephalons is connected by intermediate mass developed from the bilateral in growth of the thalami.

The *mesencephalon* has *corpora quadigemina* on the dorsal side which are two pairs of lump-like structures called *anterior* or *superior colliculi* (visual reflex centres) and *posterior* or *inferior colliculi* (auditory reflex centres) The floor and side wall of the mesencephalon called *tegmentum* have the nuclei or ganglia for the third and fourth nerves supplying to the eyes muscles. Fibre-tracts called the *cerebral peduncles* connect fore brain and hind brain.

The *metencephalon* has *cerebellum* on the dorsal side and *pons* on the ventral side.

The *myelencephalon* has dorsally situated choroids plexus which is membranous roof and a thick ventral portion having bilateral *reticular bodies, olivary nuclei* and *fibre tracts.*

11.3.3. Structure of Brian in Vertebrate Types

In *Branchiosotoma* the anterior part of the nerve tube is slightly dilated to from a cerebral vesicle. The first two anterior dosal nerves emerge out of the cerebral vesicles and convey implses from the oral hood and buccal cirri. Ventral nerves do not come out of the cerebral vesicle.

In *cyclostomes* (Fig. 11.3) the orain is narrow and elongated structure. The *olfactory* and *optic lobes* are well developed. The roof of telencephalon *pallium* is thin and does not share in the formation of cerebral hemispheres. A vertical pouch *infundibulum* is present below the diencephalon and forms posterior lobe of the pituitary gland. The cerebellum is small and the brain does not take part in motor coordination.

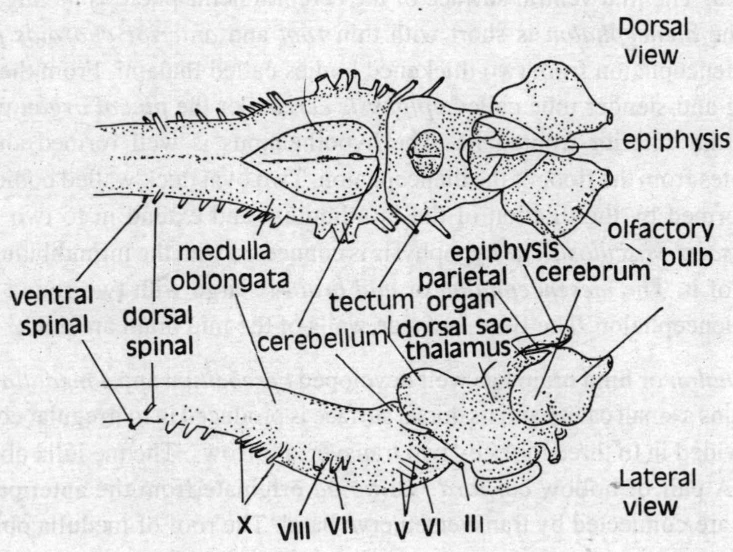

Fig. 11.3 B*rain* in cyclostomes

In elasmobranchis (Fig. 11.4) the brian is large well developed structure. The brian of *Scolidon* is divided in to three primary parts: prosencephalon mesoncephalon and rhombencephelon.

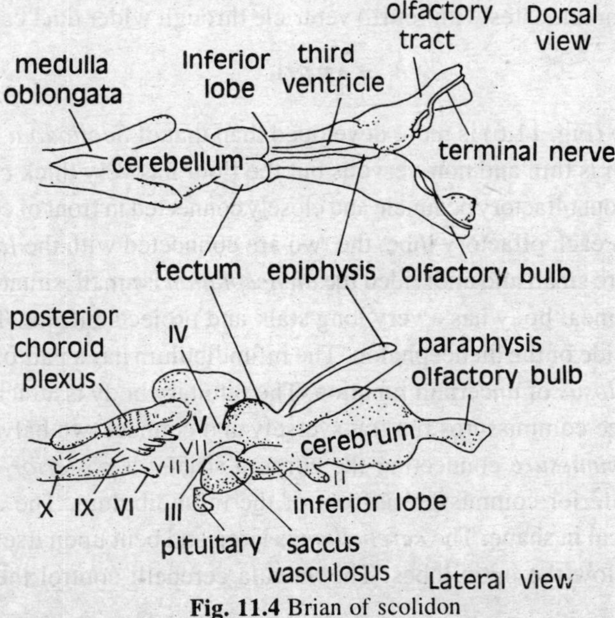

Fig. 11.4 Brian of scolidon

The *procencephalon* consists of an undivided massive *cerebral hemisphere.* Two stout *olfactory peduncles* arise from the anterior end of the cerebral hemispheres and terminate in to larger bilobed *olfactory lobes* near the olfactory capsules. The cerebral hemisphere is having a smooth surface and its walls are thickened. The mid ventral surface of the cerebral hemisphere is having an opening called *neuropore.* The *diencephalon* is short with thin *roof* and *anterior choroids plexus.* The lateral walls of the diencephalon from two thickened bodies called thalami. From the roof of the diencephalons a long and slender tube called *epiphysis cerebri* or the *pineal organ* projects to a membrane covering the anterior frontanella. The hypothalamus is well formed and a hellow infundibulum originates from the floor of the diencephalon. Two oval thick walled bodies called the *lobi-inferiores* are formed by the dilation of the infudibulum and extend in to two thin walled glandular sacs called *sacci-vasculosi.* The hypophysis is connected with the infundibulum and optic chiasma lies infornt of it. The *mesencephalon* or *mid brain* is large with two round *optic lobes* situated behind the diencephalon. The floor and side walls of the mid brain are thick.

The *rhombencephalon* or hind brain has well developed *cerebellum* and a *medulla oblongata.* The cerebellum contains a small cavity and its dorsal surface is produced in to irregular convolutions. The cerebellum is divided in to three lobes by two transverse furrows. The medulla oblongata is a triangular structure. A pair of hollow *corpora restiforma* originate from the anterior end of the medulla and the two are connected by transverse nerve band. The roof of medulla oblongata has *posterior choroid plexus.*

The *cerebral hemispheres* have two narrow lateral ventricles (Fig. 11.5) the lateral ventricles continue to *rhinocoels* which extend in to the olfactory lobes. The *third ventricles* extends forwards about half of the length of cerebral hemispheres. The fourth ventricle is large with thickened floor and extends dorsally in to the cerebellum and continues posteriory in to the cavity of the spinal cord. The third ventricle communicates with fourth ventricle through wider duct called *iter.*

LABEO

The brain of *Labeo* (Fig. 11,6) is more developed than that of *Scoliodon.* The roof of cerebral hemispheres or *pallium* is thin and non-vervous but the floor has very thick *corpus striatum.* The olfactory lobes are without olfactory peduncle and closely connected in front of cerebral hemispheres. *Rhinocoel* is present in each olfactory lobe, the two are connected with the *lateral ventricle.* The cerebral hemispheres are small and undivided the *diencephalon* is small, situated on the dorsal side of the midbrain. The pineal body has a very long stalk and projects infront. The infundibulum is present on the ventral side of the diencephalon. The infundibulum has a pair of *lobi-inferiores* and a median *saccus vasculosus* of uncertain function. The pituitary body is attached to the free tip of the infundibulum. Three commissures run transversely and connect two halves of the fore brain. These are *anterior commissure* connecting the corpora striata; a *posterior commissure* behind pineal body: and an inferior commissure infront of the infundibulum. The optic lobes are well developed and sphereical in shape. The *cerebellum* is large and bent upon itself and forms *valvula cerebelli* extending below the optic lobes. The valvula cerebelli control the active movements characteristic of teleosts. The *medulla oblongata* is well developed with facial and vegal lobes for the entry of lateral line nerves.

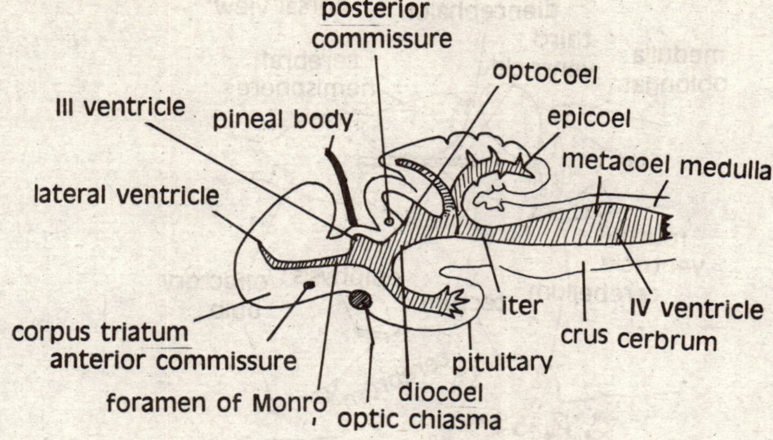

Fig. 11.5 Brain of Scoliodon in side view

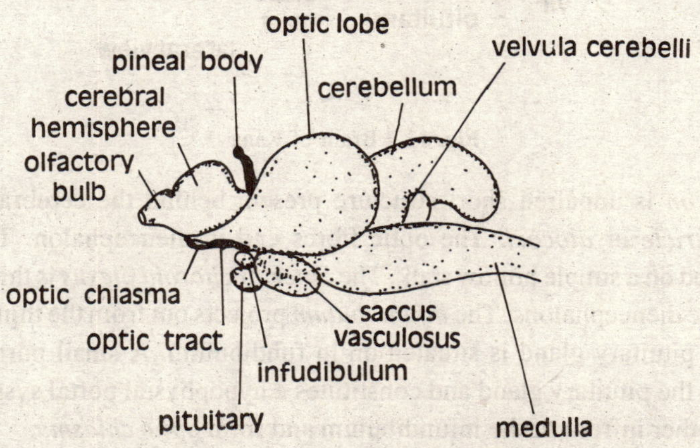

Fig. 11.6 Brain of Labeo in lateral view

RANA

The brain of frog (Fig. 11.7) is bilaterally symmetrical with gray matter on outer surface and white matter inside. The brian is situated in the cranium and has two membrances called *duramater* and *piamater.* The cerebrospinal fluid fills in the space between these membranes. The olfactory lobes are elongations of cerebral hemispheres in the form of a pair of large elongated lobes having olfactory nerves. The two lobes are fused together in the median plane. The cerebral hemispheres are separated by a *longitudinal fissure.* They are having a thick roof or *pallium* and a thick floor made of *corpoa striata.* Each cerebral hemisphere is having a cavity called *lateral ventricle* extending up to the olfactory lobe. Around lateral ventricle masses of cell bodies are situated in layers. The cerebral hemispheres act as centres of correlation; such function is unknown in fishes.

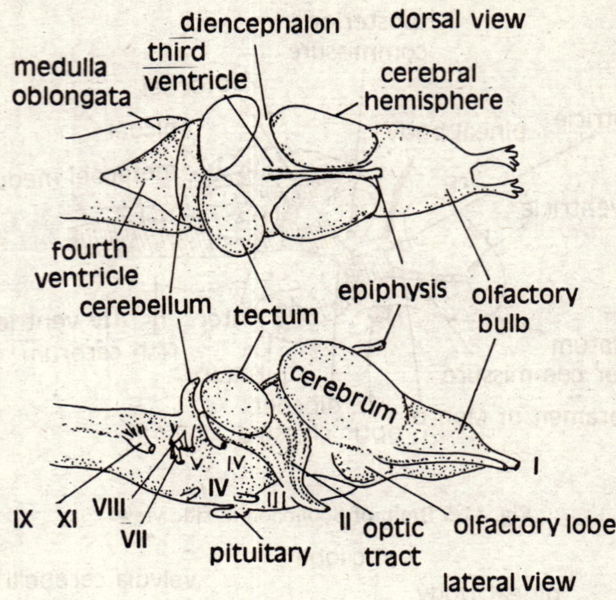

Fig. 11.7 Brain of Rana

The *diencephalon* is unpaired short structure present behind the cerebral hemispheres. Its cavity is *third ventricle* or *diocoel.* The optic fibres end in diencephalon. The pineal body is vestigeal and situated on a simple hollow stalk. The *anterior choroid plexus* is thick and invaginated in to the cavity of the diencephalons. The *infundibulum* projects out from the thin ventral wall of the diencephalon. The pituitary gland is situated on in fundibulum. A small portal vein formed by capillaries supplies the pituitary gland and constitutes a hypophysial portal system. The two optic nerves cross each other infront of the infundibulum and form *optic chiasma.*

The *mid brain* constitutes the larger part of the brain and functions for learning and other complicated action. Two rounded *optic lobes* or *corpora begemina* with optic ventricles are present on dorsolateral sides. A band of nerve fibres called *posterior commissure* lies transversely between diencephalons. Fibrous tracts called *crura cerebri* are present on the floor of the mid brain. The *iter* or *aqueduct of Sylvius* connects with third ventricle and extends in to the optic lobes as *optocoels.*

The hind brain includes *cerebellum* and *medulla.* The cerebellum is very small and the *medulla oblongata* is also short. The fourth ventricles or *metacoel* is present in hind brain. A thin epithelial layer covers the roof of the medulla oblongata to which richly vascularized piamater is applied. This double layer of membranes projects in to the ventricle as tufts called *posterior choroid plexus.*

UROMASTIX

The brain of *Uromastix* (Fig. 11.8) has some advancement over that of the frog. The *cerebral hemispheres* are swollen, elongated, closely applied to other and separated by a *longitudinal fissure*. Each cerebral hemisphere extends in to an olfactory-peduncle which is swollen infront to from olfactory bulb. The olfactory bulb gives origin to olfactory nerves. The *corpora striata* are highly developed and *pallium* has a few nerve cells on the surface. The corprora striata are joined by an *anterior commissure*. A *hippocampal commissure* lies transversely in the posterior part of the cerebral hemispheres. Two *lateral ventricles* are present which continue in to the olfactory lobes as *rhinocoels*.

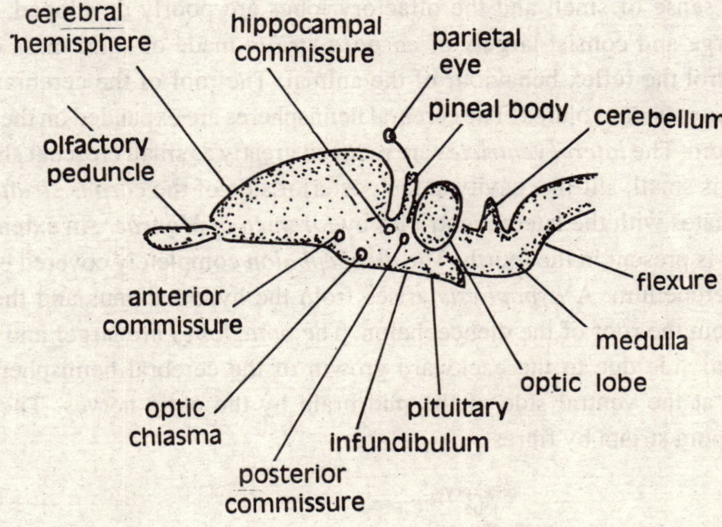

Fig. 11.8 Brain of Uromastix in lateral view.

The diencephalon is a small area invisible from the dorsal side. The anterior choroids plexus situated in diencephalons above a narrow third ventricle and extends in to the lateral ventricles through paired *foramina of Monroe*. The *thalamus* is a thick structure and reduces the cavity of the third ventricle and receives impulses from cerebral hemispheres, medulla, and spinal cord. The *hypothalamus* forms an *infundibulum* with pituitary gland attached on it. The hypophysial portal system is present as in case of frog. The hypothalamus is connected with metabolic and visceral activities. A pineal apparatus or *epithalamus* is present on the roof of the diencephalon. It consists of an anterior *parietal body* and a posterior *pineal body*. A *parietal organ* has a terminal eye-like structure and a stalk. It has been regarded as a third eye which registers radiations of sunlight and influences animals behaviors including reproduction.

The *mid brain* has *optic lobes* receiving fibers of optic nerves and probably acts as coordinating center. The *cerebellum* is small structure. The *medulla oblongata* is having a fourth ventricle and *posterior choroids plexus* and shows a marked ventral flexure.

Due to change habits and connected requirements of the animal, the brain of *Uromastix* shows higher development of cerebral hemispheres, well developed epithalamus and less developed cerebellum.

COLUMBA

The brain of pigeon (fig 11.9) is short and rounded, but built on the same structural plan as that of other vertebrates. Two meninges, an inner *piamater* and an outer *duramater* cover the brain. The pigeons have poor sense of smell and the olfactory lobes are poorly developed. The *cerebral hemispheres* are large and consist largely of *corpora striata* made of solid mass of tissue. The corpora striata control the reflex behaviour of the animal. The roof of the cerebral hemispheres called the *neopallium* is unconvoluted. The cerebral hemispheres are expanded on the posterior side to meet the cerebellum. The *lateral ventricles* are reduced greatly as small crescent shaped cavities. The *third ventricle* is small, slit-like cavity at the posterior side of the *corpus striatum.* The *third ventricle* communicates with the lateral ventricles by *foramen of Monroe.* An extension from the *aqueduct of Sylvius* is present in the fourth. The *diencephalon* completely covered by the cerebral hemispheres and cerebellum. A *hypophysis* arises from the hypothalamus and the pineal body projects dorsally from the roof of the diencephalon. The *optic lobes* are larger and spherical and pushed to the lateral side due to the backward growth of the cerebral hemispheres. The *optic chaisma* is formed at the ventral side of the mid brain by the optic nerves. The mid brain is connected with corpora striata by fibres.

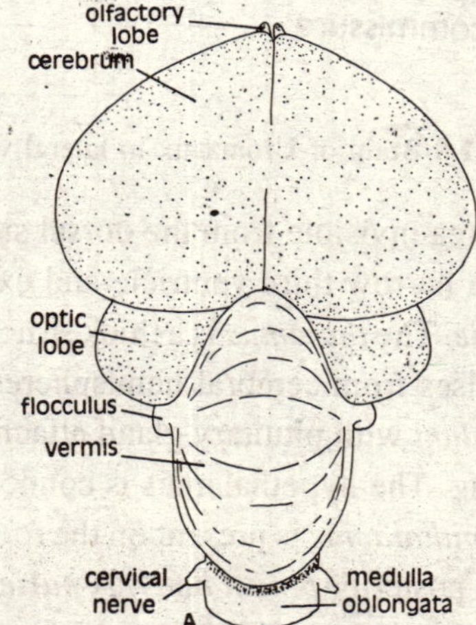

Fig. 11.9 A. Brain of Columba dorsal view.

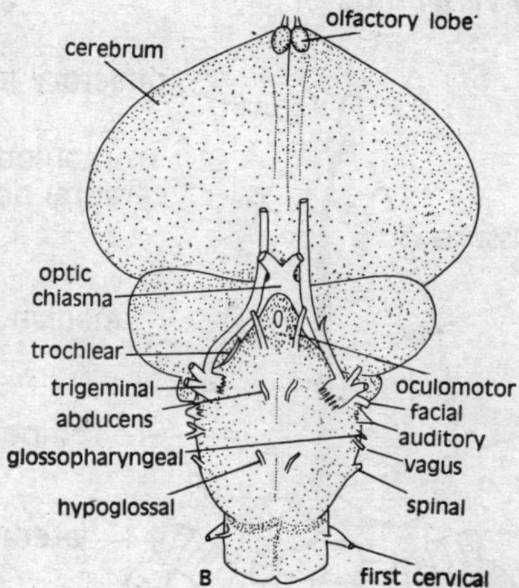

Fig. 11.9 B. Brain of Columba, ventral view.

The *cerebellum* is greatly developed. It consists of larger central *vermis* and two small *lateral lobes,* the *floculi*. The vermis is provided with transverse grooves. The cerebellum is solid structure and the fourth ventricle does not extend in to it. The cerebellum is greatly developed to control the equilibrium, movements and timings during flight. Large tracts of nerve fibres extend from the cerebellum to the spinal cord and parts of the brain.

The *medulla oblongata* has a conspicuous *ventral fissure.* It has a posterior *choroid plexus* and a *fourth ventricle,* which does not extend in to the cerebellum. The fourth ventricle is a small cavity between the cerebellum and medulla oblongata.

ORYCTOLAGUS

The brain of rabbit (Fig. 11.10) and other mammals shows consider advancement over that of other vertebrates. The general pattern of brain is similar to other vertebrates. The description given below deals with the special points about the mammalian brain. The meanings covering the brain have three layers. The outer layer is *duramater*, middle *arachnoids*, and inner *piamater*. The size of brain is much larger than other vertebrates with convolutions to increase the area. The *olfactory lobes* are small, and club shaped. The *cerebral hemispheres* are enlarged and cover the diencephalon and mid brain. There are for lobes in each cerebral hemisphere. These lobes are *frontal, parietal* and marked by grooves called *Sylvian fissure.* The two cerebral hemispheres are separated by a *median fissure.* The large size of the cerebrael hemispheres is due to its overgrown roof *neopallium.* The cerebrael hemispheres are connected by transverse bands of nerve fibres called *corpus callossum.*

The *hypothalamus* present on the ventral side the diencephalon contains *optic chiasma, pituitary body* and pair of round masses called *mammilary bodies.* The *pineal body* or *epiphysis* is present on the dorsal side of the diencephalon. *Anterior choroid plexus* is also situated on the dorsal side of the diencephalon. Four optic lobes called *corpora quadrigemina* are present in the mid brain.

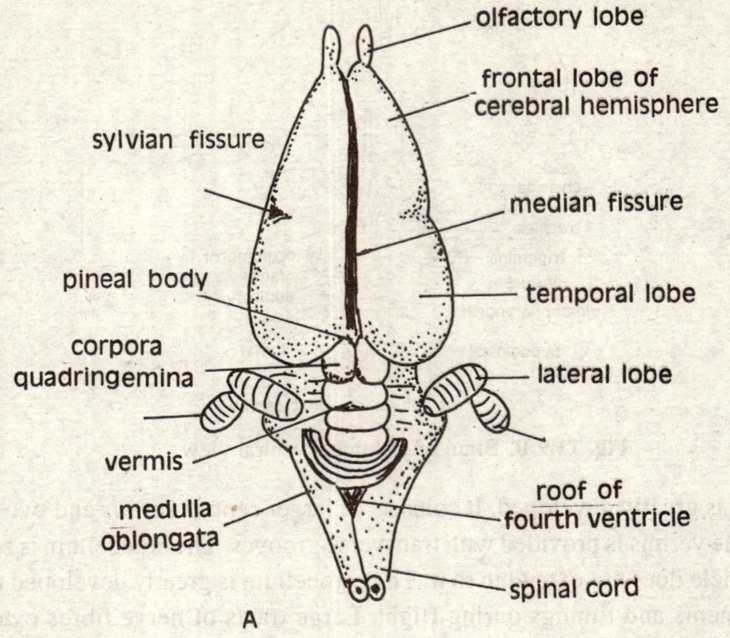

Fig. 11.10 A. Brain of Rabbit (dorsal view)

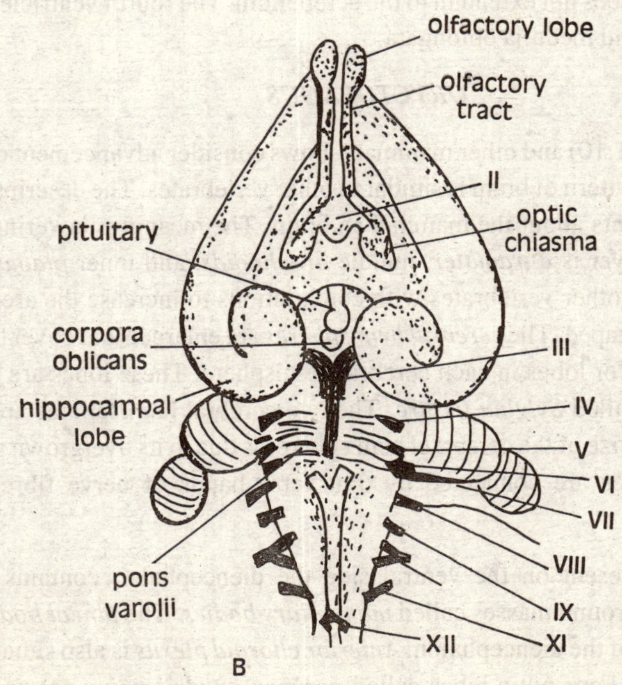

Fig. 11.10 B. Brain of Rabbit (ventral view)

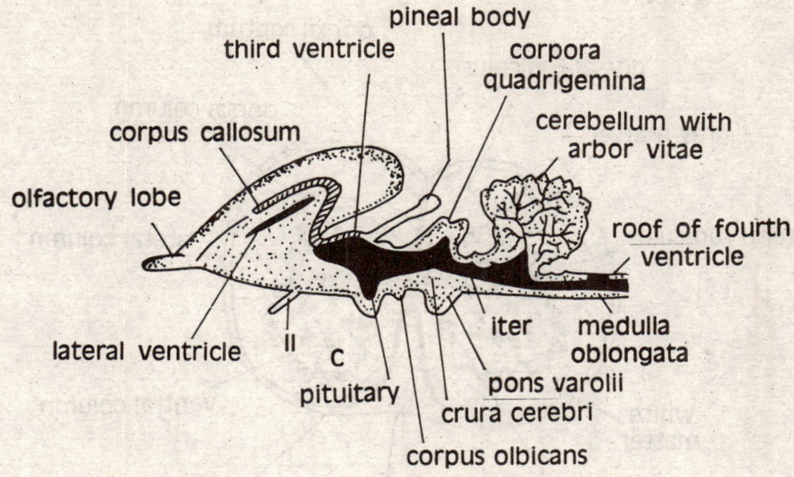

Fig. 11.10 C. Brain of Rabbit in lateral view

The *cerebellum* is enlarged and has a median *vermis* and two *lateral lobes.* Each lateral lobe has a small *flocculus.* The ventral side of the cerebellum has a broad band called the *pons varolii.* The medulla oblongata has a vascularised posterior *choroid plexus* on its roof.

The *lateral ventricles* present in the cerebral hemispheres are small and communicate with the third ventricle present in the diencephalon by a *foramen of Monroe.* The ventricle is absent in the cerebellum. A small cavity called *fourth ventricle* is present on the medulla oblongata, which communicates with the third ventricle by *iter* situated between optic lobes and crura cerebri.

11.3.4. The Spinal Cord

The spinal cord is a circular or oval elongated tube which runs most of the length of the body. It is formed from the neural tube on the posterior side of the brain. It is somewhat flattened dorsoventrally. Its anterior end in continuation with medulla is wide and the posterior end usually tapers in to a fine thread called *filum terminale.*

The spinal cord bears a *dorsal* and a *ventral fissure* in tetrapoda and chondrichthyes (Fig. 11.1 1) The dorsal fissure is a septum composed of neuroglial tissue and ventral fissure is real fissure. In Osteichthyes, the ventral fissure is absent. The cavity of the spinal cord, which is continuation of the fourth ventricle of the medulla is a narrow canal called *central canal* which is lined by ciliated epidermal cells. It lies in the middle of the narrow bridge, which connects the two halves of the spinal cord separated by the dorsal and ventral fissures.

The *gray matter* lies in the central position and *white matter* in the peripheral position. In amniotes in the gray matter is arranged somewhat in the form of the letter H and makes dorsal and ventral columns connected by a connecting bar made of dorsal and ventral gray commissures. The gray matter consists mainly of cell bodies and the white matter contains numerous myelinated fibers. The white matter is divided by the gray matter in to *dorsal, lateral* and *ventral, funiculi.*

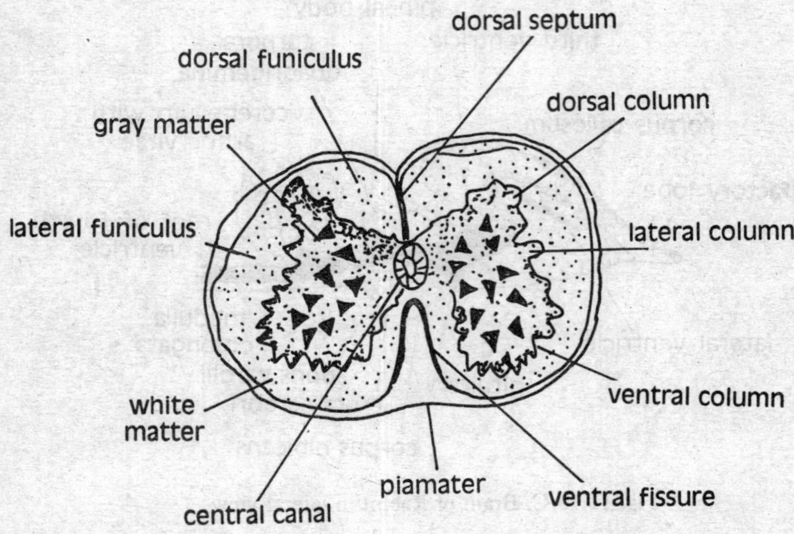

Fig. 11.11 T.S. of spinal cord

11.3.5. Spinal Cord in Vertebrate Types

In *Branchiostoma* the dorsal hollow spinal cord is situated above the notochord. It gives paired nerves to each segment of the body. It has a narrow central lumen. The gray matter surrounds the canal and white matter is situated on the outer side.

In *Cyclostomes* the spinal cord is dorsoventrally flattened band-like structure. The nerve cell bodies are present towards the centre and network of nerve fibres is present towards the periphery. The gray matter and white matter are not sharply marked in cyclostomes and dorsal and ventral fissures are lacking.

In *Scoliodon* the spinal cord is cylindrical rod extending from the medulla oblongata to the end of the tail. The central canal is narrow. The dorsal fissure is small but the ventral fissure is conspicuous. A single menix *pimater* or *menix primitiva* is closely applied to the spinal cord.

In *Labeo* the spinal cord is having a typical structure.

In *Rana* the spinal cord has an outer white matter and inner gray matter which is quadrangular. The dosal and ventral horns are well marked.

In *Uromastix* the gray matter is H-shaped in section with projecting paired dorsal and ventral horns.

In *Columba* the spinal cord extends throughout the vertebral column. The filum-terminale is absent. Large brachial and lumbar enlargements are found. In the enlargement the central canal forms a wide, diamond-shaped cavity called *sinus-rhomboidalis* and dorsal horns of the gray matter are diverged outwards.

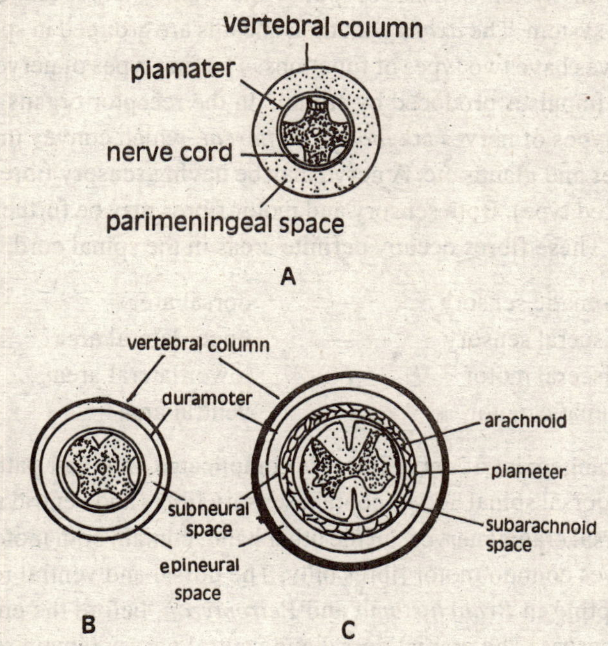

Fig. 11.12 Spinal cord in different vertebrates **A.** Elasmobranch, **B.** Amphibian **C.** Mammal.

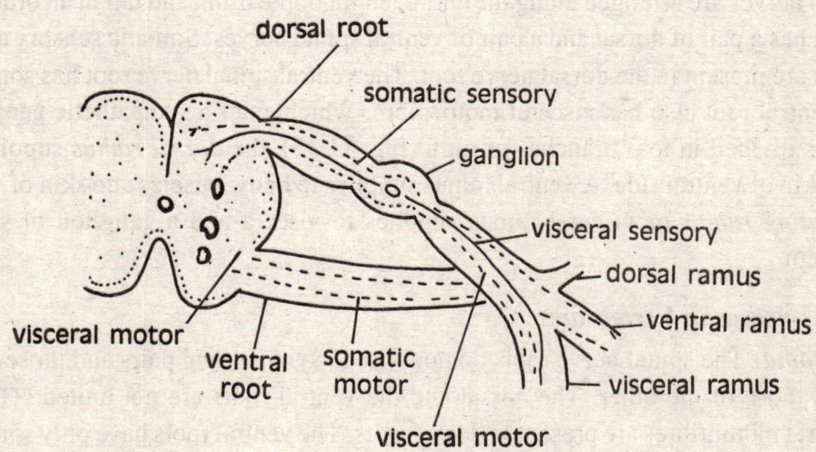

Fig. 11.13 Nerves Originality from Spinal Cord

In *oryctolagus* the spinal cord shows typical structure.

11.4. THE PERIPHERAL NERVOUS SYSTEM

The peripheral nervous system consists of *spinal,* and *cranial nerves* and ganglia, which arise from the central nervous system. The *axons* with their sheaths are grouped in spinal or cranial ganglia or the brain. Such nerves have two types of functions. The first types of nerves are *sensory* or *afferent* which conduct the impulses produced by stimuli on the receptor organs to the spinal cord or the brain. The second types of nerves are *motor* or *efferent,* which convey impulses from spinal cord and brain to muscles and glands etc. A nerve may be having sensory fibres or motor fibres or both types of fibres (mixed type). Both sensory and motor fibres may be further classified in to *somatic* and *visceral* types. These fibres occupy definite areas in the spinal cord, which are as follows:

Somatic sensory	—	dorsal area
Visceral sensory	—	upper lateral area
Visceral motor	—	lower lateral area
Somatic motor	—	ventral area

In brain, the arrangement of nerve fibres is complicated and their pattern found in spinal cord is absent here. The dorsal spinal nerves contain sensory fibres and ventral spinal nerves have motor fibres only. The dorsal cranial nerves, on the other hand, contain both motor and sensory fibres and ventral cranial nerves contain motor fibres only. The dorsal and ventral roots of the spinal nerves join together, excepting in *Branchistoma* and *Petromyzon,* before the emergence of dorsal aorta and ventral spinal nerves. The cranial dorsal and ventral nerves remain separate.

11.4.1. Spinal Nerves

The spinal nerves are arranged along the entire length of the trunk and tail in an orderly matter. Each segment has a pair of dorsal and a pair of ventral spinal nerves. Somatic sensory and visceral sensory fibres are present in the dorsal nerve root. The ventral spinal nerve root has somatic motor fibres. The ventral root also has visceral motor fibres which enter a sympathetic ganglion. Each spinal nerve is divided in to 3 branches or rami (Fig. 11.13), the *dorsal ramus* supplies to body muscles and skin of ventral side. A ventral ramus supplies to body muscles and skin of ventral side A *communicating ramus* or visceral ramus supplies to viscera and a ganglion of sympathetic nerveous system.

11.4.2. Spinal Nerves in Vertebrates

Branchistoma: The spinal nerve roots do not arise in symmetrical pairs and those of one side alternate with those of the other. The dorsal and the ventral roots are not united. The somatic-sensory, visceral motor fibres are present in dorsal roots. The ventral roots have only somatic motor fibres.

Cyclostomes: The roots of spinal nerves are not united in lampreys and united in hagfishes excepting in caudal region. One sensory and one motore nerve is formed in the gill region of the lampreys from dorsal and ventral roots respectively and form hypobranchial nerve which supplies to the central part of the gill region.

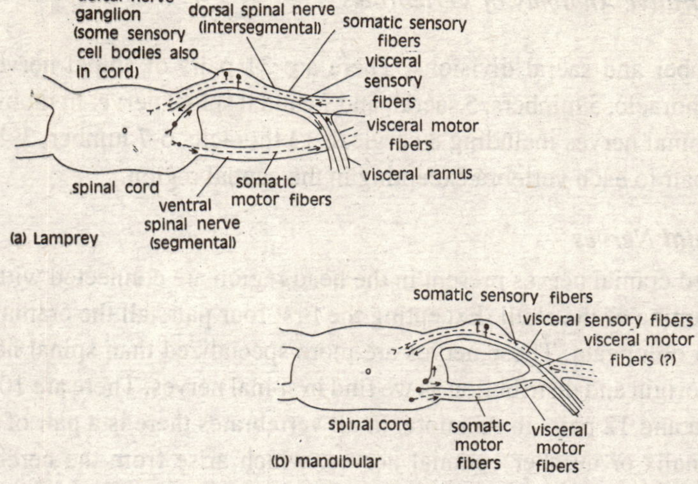

Fig. 11.14 *Somatic and visceral circuits in anamniotes.* (a) Lamprey. (b) Amphibian. It is unclear whether the amphibian dorsal root carries visceral motor output.

Fishes: The union of dorsal and ventral roots occurs in fishes outside the vertebral column. In some of the anterior spinal nerves only the ventral roots persist as *spino- occipital nerves* and dorsal roots have disappeared. Fibres from spino occipital nerves contribute to the formation of hypobranchial *nerve* made of motor fibres alone.

Amphibia: The roots join after passing through the intervertebral foramen. The dorsal root ganglion is present at the joint of junction. A small motor nerve called *sub occipital nerve,* which represents the hypoglossal cranial nerve of higher animals, emerges between the occipital condyles and the first vertebra from the ventral roots. Anura have a comparatively small number of spinal nerves. The first function as *hypoglossal*, the second and third join to form a *brachial plexus* which supplies nerve fibres to the muscles and skin of the anterior limb. The seventh to tenth join to form a *lumbosacral plexus* which supplies to the hind limb.

Reptiles: The spinal nerves of reptiles have dorsal roots with sensory fibres, visceral motor and somatic motor fibres are present in the ventral roots only. In certain snakes and limbless lizards a poorly developed *lumbosacral plexus* is present showing the origin of limbless forms from ancestors with limbs.

Birds: In pigeon and most birds the cervical spinal nerves pass out to neck region and last three cervical nerves and first thoracic nerves from a *brachial plexus,* which supplies to the wing. Next five spinal nerves pass between the ribs. The *lumbosacral plexus* is usually divided into separate *lumbar, sacral* and *pudendal* plexuses supplying to thigh, lower legs, cloacal and pudendal plexuses supplying to thigh, lower legs, cloacal and tail region respectively.

Mammals: The number of spinal nerves in mammals corresponds to the number of vertebrae from where they emerge. The spinal nerves are thus called *cervical, thoracic, lumber, sacral* and *caudal or coccygeal.* The limb plexuses are usually very complicated and divided into cervical, brachial, lumber and sacral divisions. There are 31 pairs of spinal nerves in man including 8 cervical. 12 thoracic, 5 lumbers, 5 sacral, and 1 caudal spinal nerve. In rabbit and squirrel there are 40 pairs of spinal nerves including 8 cervical, 13 thoracic, 6-7 lumber, 3-4 sacral pairs of caudal nerves, one pair to each vertebra excepting in the caudal region.

11.4.3. Cranial Nerves

The paired cranial nerves present in the head region are connected with the brain and emerge from the foramina of the skull. Excepting the first four pairs all the cranial nerves are connected with medulla oblongata. These nerves are more specialized than spinal nerves and do not show symmetry in origin and distribution as we find in spinal nerves. There are 10 pairs of cranial nerves in Anamniota and 12 pairs in Amniota. In all vertebrates there is a pair of terminal nerves called nerves terminalis or number ocranial nerves, which arise from the cerebral *hemispheres.* The cranial nerves are numbered from I to XII. Their origin, distribution and function may be summarized by the table given below:

TABLE 11.2

Origin, Distribution and Function of Cranial Nerves

No. of cranial nerve	*Name*	*Place of origin from brain*	*Distribution*	*Nature*	*Function*
I.	Olfactory	Olfactory lobe	Olfactory lining of the nasal cavity	Sensory	Smell
II.	Optic	Optic lobe	Retina	Sensory	Sight
III.	Oculomotor	Base of mid brain	Eye, iris, lens. upper lid, anterior superior inferior recti and inferior oblique eye muscles.	Motor	Movement of eye hull, iris lens eye lid
IV.	Trochlear	Base of mid brain	Superior oblique eye muscles	Motor	Rotation of eye balls.
V.	Trigeminal	Sides of medulla	A large nerve with three branches supplying to head, jaw muscles and teeth.	Sensory and motor	Movement of tongue and jaws.
VI.	Abducens	sides of medulla	Posterior rectus muscle of eye kill	Motor	Rotation of eye ball.
VII.	Facial	Side and floor of	Anterior part of tongue,	Sensory motor	Taste, Chewing,

No. of cranial nerve	Name	Place of origin from brain	Distribution	Nature	Function
		medulla	muscles of face and neck		neck movement
VIII.	Auditory	Side of medulla	Internal ear	Sensory	Hearing and equilibrium
IX.	Glosso-pharyngeal	Side of medulla	Posterior part of tongue pharynx	Mixed	Taste and touch Movement in pharynx
X.	Vagus	Side and floor of medulla	Pharynx, vocal cords, lungs, heart oesophagus stomach and intestine	Mixed	Muscle movement of connected organs
XI.	Spinal Accessory	Floor medulla	Palate - muscles, larynx, vocal-cords and neck	Motor	Muscle movements pharynx larynx and neck
XII.	Hypoglossal	Floor of Medulla	Muscles of tongue	Motor	Movement of tongue

In addition to somatic sensory, somatic motor, visceral sensory and visceral motor components of spinal nerves, the cranial nerves have two motor components. These are *lateralis system of neuromasts* and the nerves of special sense as *tactile sense organs.* In this way, there are 6 different kinds of conducting tracts in cranial nerves. According to their components and nature the cranial nerves may be grouped as follows:

A. Nerves of special sense including olfactory (I), optic (II), and auditory (VIII).

B. Eye muscle nerves including oculomotor (III), trochlear (IV) and abducens (IV).

C. Acoustico lateral is system including acoustic nerve with portions of VII. IX and X.

D. The branchial nerves including trigeminal (V), facial (VII), glassopharyngeal (IX) and vagus (X).

E. The terminal nerve (O) which originates in diencephalon and supplies to the olfactory mucous membrane.

11.4.4. Cranial Nerves in Vertebrate Types

In *Branchiostoma* the brain is not marked and cranial nerves are not identified. There is a system of spinal nerves, which arise from the tube in segmental pairs. Each spinal nerve has a dorsal root with efferent sensory fibres and a ventral root with efferent or major fibres. The vagus nerve (X) originates as a single outgrowth and becomes branched subsequently. (In other animals, it originates with many fibres, which unite to form a single bundle.)

In *Petromyzon* there are 10 pairs of cranial nerves, which arise from the different parts of the brain. The general arrangement of cranial nerves is similar to that of gnathostomata with some variations. The olfactory nerves (1) have many nerve fibres and optic nerves (II) make simple crossing and do not from optic chiasma. The VI nerve emerges along with trigeminal (V) nerve and its existence is questioned by many workers. The V and VII nerves are intimately associated and their branches to lateral line system are unidentifiable. The IX and X nerves are similar to those of fishes. The vagus nerve (X) originates as a single outgrowth and becomes branched subsequently. (In other animals, it originates with many fibres, which unite to form a single bundle).

SCOLIODON

In *Scoliodon* (Fig. 11.15) there are 10 pairs of cranial nerves along with an extra pair of terminal nerves in *Scoliodon*. The *nervous terminalis* or terminal or preolfactory nerve arises near the lamina terminalis and goes to olfactory septum. The *olfactory nerve* (I) arises as a number of fibres from the cells of the olfactory lobe and its fibres enter in the olfactory lobe. The *optic nerves* (II) arise from the optic thalami of the lobes, from optic chiasma and supply to eyes. The *oculomoter nerve* (III) originates from the ventral surface of the mid brain and supplies to the anterior, superior and inferior recti musclues and inferior oblique muscles of each eyeball. The *trochlear* or *pathetic nerve* (IV) arises from the dorsal surface of the mid brain and supplies to the Superior oblique eye muscles. The trigeminal nerve (V) has three branches. The *ophthalmicus superficialis* branch supplies to skin of the snout. The *maxillaries* branch is further divided into a *maxillaries superior* to skin of upper jaw and a *maxillaris inferior* supplying to the posterior part of upper jaw. The *mandibularis branch* supplies to the muscles of the lower jaw. The *trigeminal* nerve gets secondarily associated with *ophthalmicus* of VII cranial nerve and supplies to eyeball and dorsal surface of the snout. The *abducens nerve* (VI) supplies to the posterior rectus muscles of the eve ball. The *facial nerve* (VII) has two branches. These are *ophthalmicus* superficialis branch like that of V nerve and a bundle of mixed nerve which subdivides into 3 branches: (i) *Ramus-buccalis* supplying to the infraorbital canal of the snout; (ii) *Ramus hyomandibularis* supplying to the lower jaw and throat: (iii) *Ramus palatinus* supplying to roof of the buccal cavity and pharynx. The *auditory nerve* (VIII) gives vestibular and saccular branches to the internal car. The *glossopharyngeal nerve* (IX) divides in the region of first gill cleft into a small *pretrematic* and a large *postrematic nerve.* These supply to the pharynx, pharyngeal muscles and mucous membrane surrounding the first gill cleft. The *vagus nerve* (X) arises from multiple roots and gives off the following branches:

i. Branchial nerve supplying to the gills.
ii. Visceral nerve supplying to the viscera including heart, liver and lungs etc.
iii. Lateralis nerve supplying to the lateral line sense organs.

LABEO

The number of cranial nerve in Labeo is 10 and their arrangement is typical (Fig. 11.16)

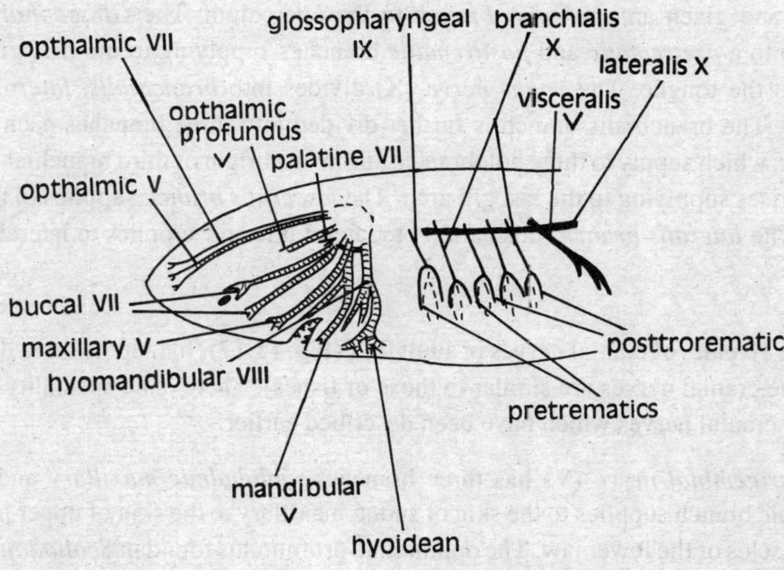

Fig. 11.15 Nerve in Scoliodon

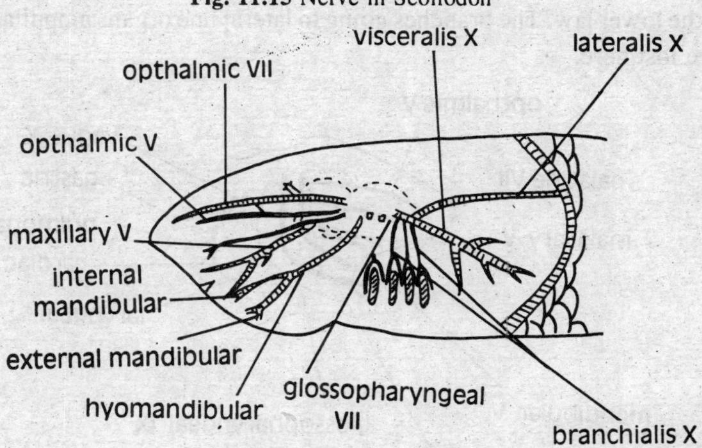

Fig. 11.16 Nerves in Labeo

The *olfactory nerve* (I) shows no speciality. The optic nerves (11) form a true chiasma and fuse incompletely at the crossing. The third (*oculomotor*), fourth (*trochlear*) and sixth (*abducens*) nerves supply to eye muscles. The eighth nerve (auditory) gives many fibres which spread like a fan and supply to internal ear. The fifth and seventh nerves from a *trigeminofacial complex.* The fifth divides into three branches. These are *ophthalmicus superficialis, maxillaris* and *mandibularis.* The VII nerve divides into three branches, *opthalmicus superficialis, buccalis* and *mandibularis.* The opthalmicus superficialis Branches of V and VII nerves runs together. The mandibular branch supplies to lower jaw and divides into two branches, *mandibularis externus* and *mandibularis internus* and given an *opercular branch* to the operculum. The *Glossopharyngeal nerve* (IX) divides into a *pretrematic* and *posttrematic* branches supplying to the first gill arch and gives a branch to the tongue. The *vagus nerve* (X) divides into *branchialis, lateralis,* and *visceralis,* branches. The branchialis branch is further divided into three branches each with *pre* and *post trematics*, which supply to three holobranchs. From the origin of third branchial nerve an additional branch arises supplying to the last gill arch. The *visceralis branch* supplies to the various visceral organs. The *lateralis branch* runs parallel to lateral line and supplies to lateral line sense organs.

RANA

There are 10 pairs of cranial nerves in adult frog (Fig. 11.17) (In tadpole larva due to aquatic mode of life, the cranial nerves are similar to those of fishes). There is no speciality in I, II, III, IV, VI, VII, VIII cranial nerves which have been described earlier.

The *trigeminal nerve* (V) has three branches, *ophthalmic maxillary* and *mandibular.* The ophthalmic branch supplies to the skin of snout, maxillary to the skin of upper jaw and mandibular to the muscles of the lower jaw. The ophthalmic profoundus found in *Scoliodon* in lost in frog. The *facial nerve* (VII) divides into two branches, a *palatine* branch and a hyomandibular branch. The palatine branch supplies to the roof of the buccal cavity and the hyomandibular branch to the tongue and muscles of the lower jaw. The branches going to lateral line organs ampullae of Lorenzini found in *Scoliodon* are lost here.

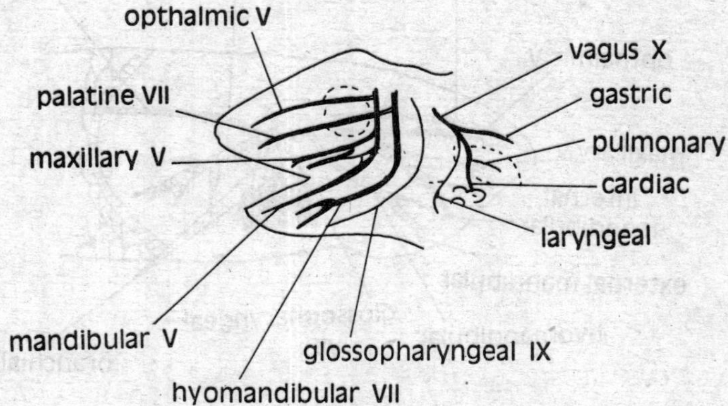

Fig. 11.17 Cranial Nerves of frog

The *glossopharyngeal nerve* (IX) goes to the tongue and pharynx. The pretrmatic and post-trematics of fishes are lost here. The *vagus nerve* (X) has lost its lateralis and branchialis branches. Visceral branch persists which supplies to the larynx. oesophagus, stomach. heal and lungs.

UROMASTIX

There are 12 pairs of cranial nerves in *Uromastix* (Fig. 11.18). The origin and distribution of first ten cranial nerves is similar to that of frog.

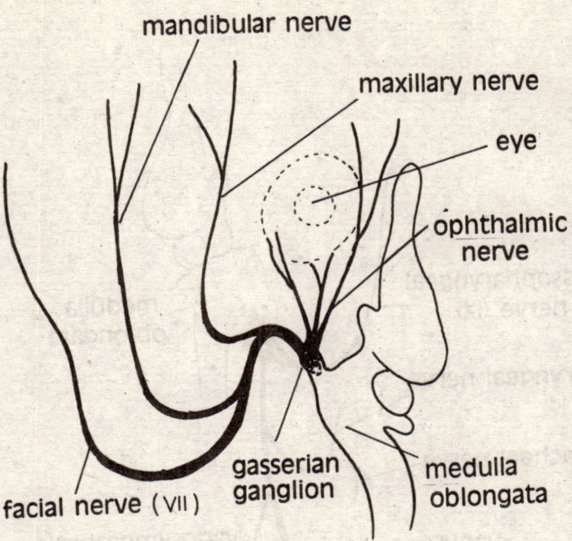

Fig. 11.18 A. Schematic diagram showing the origin and distribution of the fifth and seventh cranial nerves in *calotes.*

The *Spinal accessory nerve* (XI) starts from the posterior part of the medulla and joins the vagus nerve. It is purely a motor nerve and innervates the larynx, pharynx, and cervical region. It represents the dorsal root. The *hypologlossal nerve* (XII) also orginates from the medulla and innervates the muscles of the tongue. It is a motor nerve and represents the ventral root.

COLUMBA

There are 12 pairs of cranial nerves emerging out of the brain (Fig. 11.19). The terminal cranial nerve number zero is absent in birds. The *olfactory* (I) is a somatic sensory nerve and supplies to the nasal capsule. The *optic* (II) is also somatic sensory nerve emerging from optic chiasma. The *oculomotor* (III) is somatic motor nerve innervating the orbit. The *trochlear* or *pathetic* (IV) is somatic motor nerve and arises from the side of the brain between the optic lobe and cerebellum and innervates the brain and contains and some sensory fibres also.

The *trigeminal nerve* (V) is a large nerve and arises on the sides of the medulla below the optic lobe and contains sensory and motor fibres. It continues forwards into *Gasserian ganglion* where it divides into three branches, *ophthalmic, maxillary* and *mandibular.* The ophthalmic branch passes into the orbit through the floor of the skull and reaches the upper portion of the orbit. The maxillary nerve enters the orbit, passes through the floor of orbit and supplies the upper jaw. The mandibular branch enters into the orbit and further divides into 2 branches, one to temporal muscles and second to the region of lower mandible.

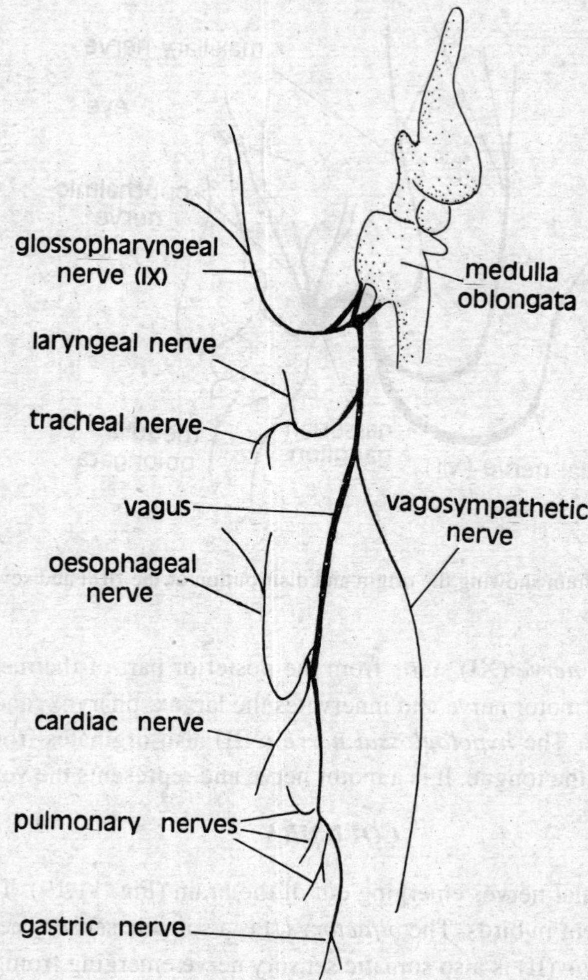

Fig. 11.18 B. Schematic diagram showing the origin and distribution of the ninth and tenth cranial nerves of *calotes.*

The *abducens* (VI) is a small nerve containing somatic motor fibres. It arises from the ventral surface of the medulla and reaches the orbit.

The *facialis nerve* (VII) is a mixed nerve arising from the lateral sides of the medulla. It divides into two branches, which innervate the base of the head.

The *auditory nerve* (VIII) is somatic-sensory nerve and supplies to the ear region. It arises from the medulla on the posterior side of the VII nerve.

The *glossopharyngeal* (IX) is a mixed nerve from the side of the medulla close to the auditory nerve. It supplies to the base of skull and tongue.

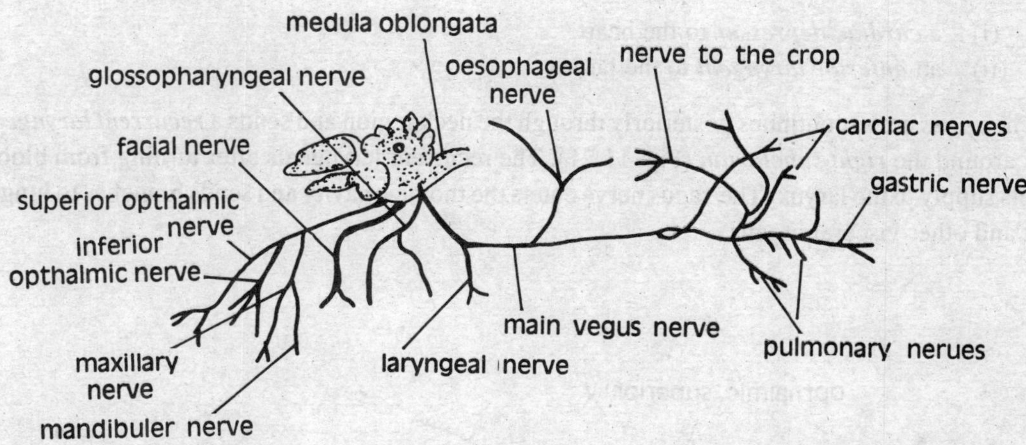

Fig. 11.19 Cranial Nerves of Columba

The *vagus nerve* (X) is a mixed nerve arising from the lateral side of the medulla. It supplied to trachea, lungs, heart, stomach and gizzard.

The *spinal accessory nerve* (XI) is a motor nerve arising from the sides of spinal cord by a number of roots. It enters the cranial cavity through the foraman magnum and passes out again alongwith IX and X cranial nerves and supplies to the necle muscles. The presence of this nerve birds has been disputed by authorities.

The *hypoglossal nerve* (XII) is a motor nerve arising from the ventral side of the medulla. It passes out through a foraman of the skull and supplies to the muscles of the neck.

ORYCTOLAGUS

There are 12 pairs of cranial nerves in rabbit (Fig. 11.20). The first ten crania] nerves are similar to those of reptiles and birds with minor differences.

The *olfactory* (I) arises from the olfactory lobe and supplies to the olfactory membranes of the nasal sacs.

The *optic* (II), *oculomotor* (III), *trochlear* (IV), *trigeminal* (V), *and abducens* (VI) nerves do not show any special features.

The *facialis* (VII) does not have ophthalmic branch and supplies mainly to the highly developed facial muscles. The *auditory* (VIII) and *glossopharyngeal* (IX) are similar to those of other amniotes.

The *vages* (X) forms a ganglion, runs outside the carotid artery and sends the following branches:

(i) a *cardiac depresser* to the heart.
(ii) an *anterior laryngeal* to the larynx.

The *vagus* nerve continues posteriorly through the neck region and sends a *recurrent laryngeal* turns around the *right subclavian* (Fig. 11.21). The recurrent laryngeals after turning from blood vessels supply to the larynx. The vagus nerve enters the thoracic cavity and sends branches to lungs, heart and other visceral organs.

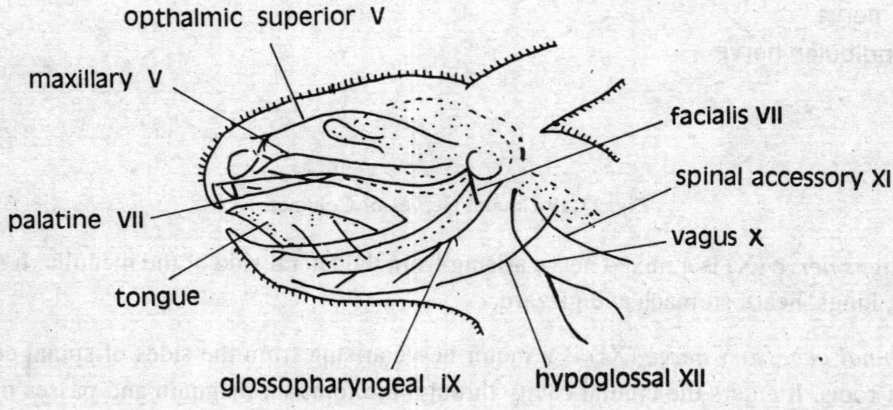

Fig. 11.20 Cranial nerves of Oryctolagus

The *spinal accessory* (XI) is a motor nerve originating from the lateral side of the medulla and supplies to the neck muscles. It sends a branch called ramus interims to the muscles of the pharynx and larynx.

The *hypoglossed* (XII) is a motor nerve supplying to the tongue muscles. It originates from the mid ventral region of the medulla.

1.1.5. THE AUTONOMOUS NERVOUS SYSTEM

The autonomous nerves system controls the functions of the body such as metabolism, maintenance of constancy of blood, lymph and tissue fluids and action and tone of internal muscles. The regulation of temperature in birds and mammals is also done by the sympathetic system. This system controls the function of organs which are not under the control of the will. The nerve centres controlling autonomous system are situated in the hypothalamus of the brain through which the system forms connections with other nervous tissue.

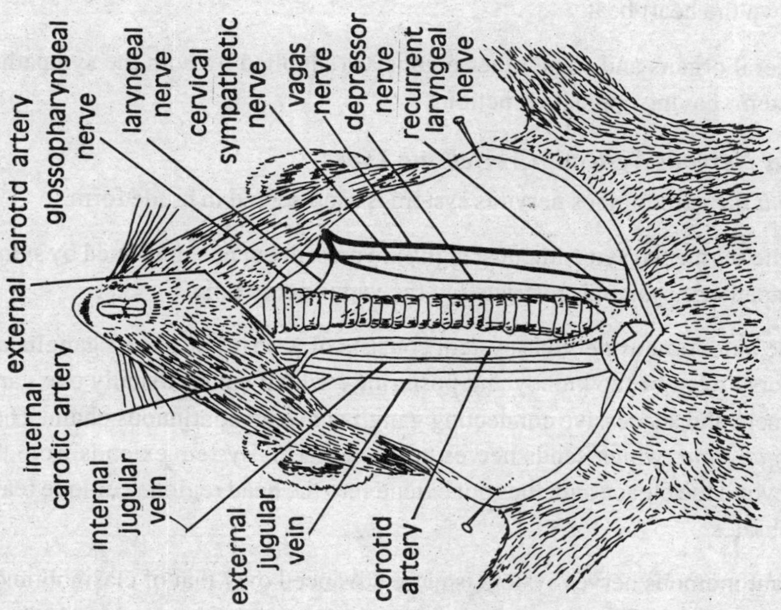

Fig. 11.21 Dissection of the neck region of *Cavia*, showing the distribution of the main neck nerves (left side) and main blood vessels (right side).

The *sympathetic system* or *thoracolumbar outflow* consists of two long chains of connected ganglia parallel to trunk vertebrae and aorta. Efferent fibres from spinal cord called *preganglionic fibres* enter the sympathetic ganglia and leave the ganglion as *post ganglionic fibres.* The post-ganglionic fibres from each group unite to form a plexus and then distribute to different organs such as liver, stomach etc. Fibres from organs called *afferent -fibres* pass directly to the dorsal root of the spinal cord. Some different fibres connect the sweat glands, small blood vessels and erector muscles of the hairs. The function of the sympathetic system is to increase the heartbeat, slowing down of gastrointestinal action and dilation of the bronchioles. It also organises the body resources by increasing the secretion of epinephrin from the adrenal gland during emergencies like injury, fight etc. resulting into the constricition of blood vessels of skin and viscera, dilation of blood vessels from heart and skeletal muscles, release of glucose from liver for muscle metabolism and acceleration of the clotting of blood.

The *parasympathetic system* or *craniosacral outflow* consists of fibres of certain cranial and sacral spinal nerves involved in it (Fig. 11.22). The oculomtor (III), facial (VIII) glossopharyngeal (IX), and vagus (X) cranial nerves send parasympathetic fibres to ganglia, which are situated close to, or in the organs supplies by the system. These fibres supply to organs like iris of the eye, glands and mucous membranes of mouth, upper small intestine and organs of lower abdomen. The preganglionic fibres are long and the posyganglionic fibres are very short. The parasympathetic system promotes the secretion of saliva and digestive juices, increases muscular activities of intestine, constricts the bronchioles of the lungs, constricts the pupils of eye, adjusts the eye for vision and slows down the heart beat.

Most of the visceral organs and some other organs are supplied by both the sympathetic and parasympathetic systems having opposite function.

11.5.1. Autonomous Nervous System in Vertebrate Types

In *Branchiostoma* the autonomous nervous system is represented in crude form.

In *Cyclostemes* the segmental sympathetic ganglia are present but not connected by sympathetic trunks. The parasympathetic system is restricted to the vagus nerve only.

In *Scoliodon*, the autonomous nervous system consists of a scries of paired ganglia arranged irregularly on the dorsal wall of the kidney and posterior cardinal sinus. Usually one ganglion is present in each segment and successive connecting ganglia have no continuous chain. The largest ganglion is *gastric ganglion,* which sends nerves to viscera. This system extends from the sinus venosus to a little beyond the cloaca and does not extend into the head region, a unique feature, not found in other vertebrates.

In *telecosts* the autonomous nerves system is more advanced over that of clasmobranches and shows similarity with tetrapods. The *sympathetic system* is represented by two longitudinal chains having ganglia extending form the anterior end of the medulla to the end of the tail. The *parasympathetic system* is represented by fibres of oculomotor and vagus nerves only.

In *Rana* (Fig. 11.23) there are two longitudinal sympathetic chains extending from the brain to the end of urostyle, one on each side of vertebral column and ventral to it. Segmentally arranged ten sympathetic ganglia are present. The sympathetic chain of each side between first and second ganglia divides to form a loop around the subclavian artery called *annulus of Vieussens.* From the ganglion of one sympathetic chain to another ganglion minute nerves extend and break up to form plexuses in the viscera. *Solar plexus* near the coelico-mesenteric artery and a *cardiac plexus* on the heart are such plexuses.

Preganglionic fibres gray matter of the spinal cord pass through spinal nerves and white rami-communicantes and reach corresponding sympathetic ganglia. Some preganglionic fibres pass through the sympathetic chain and enter into the head to communicate with some cranial nerves. The postganglionic fibres from the sympathetic ganglia pass to heart, lungs, blood vessels, alimentary canal, pancreas, liver, spleen kidneys, adrenals, gonads and bladder. Postganglionic fibres are not found in fishes and lower vertebrates. Some postganglionic fibres enter the spinal nerves through rami communicantes and reach skin to control chromotophores.

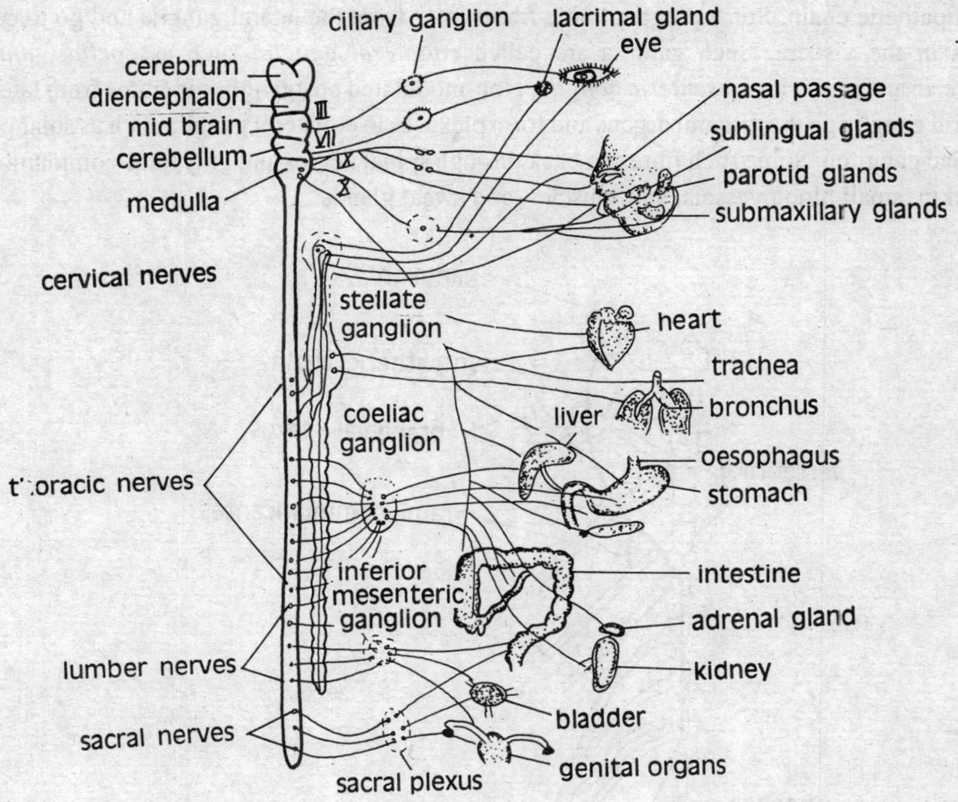

Fig. 11.22 Autonomic Nervous System

The *parasympathetic system* consists of preganglionic and postganglionic fibres in association with oculomotor, facial, glossopharyngeal, and vagus cranial nerves and sacral spinal nerves and enter the various organs. The *sacral flow* has appeared in amphibians for the first time.

In *Amniota:* The autonomous system shows remarkable similarity among its classes. The autonomous system of mammals has been understood properly and consists of sympathetic and parasympathetic systems.

Each *thoracolumber outflow* (sympathetic chain) consists of several lateral or *chain ganglia.* In neck region, there are three ganglia called *superior middle* and *inferior cervical ganglia.* The inferior cervical and first thoracic ganglia usually fuse together to form a *stellate ganglion.* After stellate ganglion, there is a scries of *prevertebral ganglia* in the thoracic and lumber regions. These ganglia are connected to spinal nerves through rami-communicantes. The preganglionic fibres from spinal cord pass through ventral roots of the spinal nerves and white rami-communicantes to the lateral ganglia. Some *preganglionic fibres* communicate with cranial nerves after passing through the sympathetic chain. Some preganglionic fibres arise from the lateral ganglia and go to certain ganglia in the viscera. Such ganglia are called *collateral ganglia* such as *coeliac, anterior mesenteric* and *posterior mesenteric ganglia.* Non-modulated postganglionic fibres from lateral or collateral ganglia go the visceral organs and form plexuses in collateral ganglia such as solar plexus of coeliac ganglion. Some such fibres go back through spinal nerves and gray rami-communicantes to the skin, small blood vessels, hair muscles, and sweat glands.

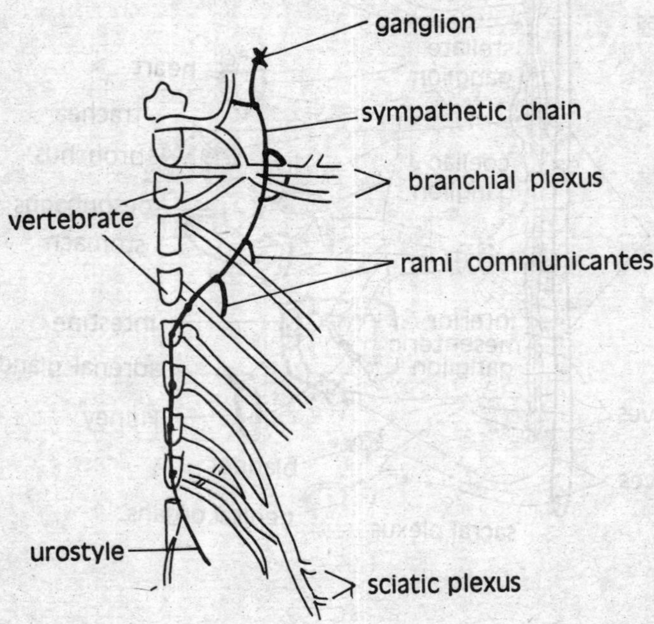

Fig. 11.23 Symapathetic Nerves of Frog

The *craniosacral outflow* (parasympathetic system) consists of ganglia, preganglionic and postganglionic fibres. The III, VII, IX, and X cranial nerves and II, III and IV sacral spinal nerves are joined by preganglionic fibres the preganglionic fibres of cranial and spinal nerves are very long and terminate in ganglia situated in or close to the organs supplied by them. The postganglionic fibres arise from ganglia and go to various aorgans and such fibres are very short.

CHAPTER SUMMARY

1. The nervous system consists of central, peripheral and sympathetic nervous systems. All parts of the nervous systems are derived from the ectoderm. The neurons are units of the nervous system in vertebrates. The nervous system of vertebrates is always dorsal to the digestive system.

2. Each neuron has a long cell body with conspicuous nucleus and two or more protoplasmic processes, dendrites and axon. The sensory or afferent neuron conduct impulses from receptors to central nervous system and motor of afferent neuron from central nervous system to various effectors. Adjuster neurons join between sensory and motor neurons.
3. When cell bodies group together outside the central nervous system they form a ganglion. A nerve is a group of fibers or processes held together in parallel bundles by connective tissue.
4. Gray matter and white matter form nervous material in brain and spinal cord. Neuroglia cells serve as packing cells between the neurons in brain and spinal cord.
5. The brain is divided into prosencephalon or forebrain, mescenphalon or midbrain and rhombencephalon or hindbrain. The forebrain contains telencephalon with cerebral hemispheres, olfactory lobes corpus striatum, cerebral cortex and olfactory bulb; and diencephalon containing epithalamus, hypothalamus and its appendages. The midbrain contains optic lobes tegumentum and crura cerebri. The hindbrain contains metencephalon with part of medulla oblangata, cerebellum and pons in mammals; and myelencephalon with part of medulla oblongata.
6. The brain is enclosed in connective tissue envelopes called meninges. The central cavity of spinal cord and spaces between meninges are filled with cerebrospinal fluid.
7. The spinal cord is circular or oval elongated tube in continuation with the medulla anteriorly and tapering into a fine thread filum terminate posteriorly. It bears a dorsal and a ventral and fissure and a central canal.
8. The peripheral nervous system consists of spinal and cranial nerves and ganglia, which arise from the central nervous system.
9. The spinal nerves are segmentally arranged paired structures. Each spinal nerve is divisible into three branches or rami.
10. The cranial nerves are paired nerves arising from brain and emerging out from the foramina of the skull. There are ten pairs of cranial nerves in anamniota and 12 pairs in amniota I, II and VIII cranial nerves are sensory; III and IV are motor and remaining mixed sensory and motor nerves.
11. The automatic nervous system consists of sympathetic and parasympathetic components. The sympathetic system or thoracolumbar outflow consists of two long chains of connected ganglia parallel to trunk vertebrate and aorta. The parasympathetic system or craniosacral outflow consists of fibers of III, VII, IX, X cranial and sacral spinal nerves. Most of the visceral organs and some other organs are supplied by both the sympathetic and parasympathetic systems having opposite functions.

12

Receptor Organs

12.1. GENERAL ACCOUNT

The receptor organs are closely associated with nervous system and yield conscious sensations to meet the environme • A receptor organ responds to a particular stimulus by starting an impulse, which is transmitted, to the central nerves system by sensory nerve fibres. Such impulses are converted into sensations in the higher centres of the brain such as thalamus or celebral cortex. The function of receptor organs is well known in man but similarity in the structure of the sense organs of man and other vertebrates leads us to assume similar functions. The receptor organs are ectodermal is most likely to be influenced by the environment.

The receptor organs serve to provide access to the nervous system but do not perceive anything by themselves.

The receptor organs of different kinds are named according to the nature of stimulus affecting them. These include the following:

(1) Organs of chemical sense (chemoreceptors) including taste buds, olfactory organs and Jacobson's organs.

(2) Organs detecting movements of the medium and pressure changes such as ear, laterosensory organs, pit organs, ampullae of Lorenzini.

(3) Organs of vision including median and paired eyes.

The receptor organs have classified into *exteroreceptors* and *internal receptors.* The exteroreceptors or external receptors includes *photoreceptors* (vision) *olfactoreceptors* (smell), *phonoreceptors* (hearing), *gustatoreceptors* (taste) *tangoreceptors* (touch and presense), *thermoreceptors* (heat and cold), *algesireceptor* (pain), and *rheoreceptors* (current of water). The internal receptors include *proprioceptors* (hunger, thirst etc.).

The classification of receptors adopted in the text is as follows:

Receptors

1.	2.	3.	4.
General somatic Receptors	Special somatic Receptors	General visercal Receptors	Special viscera; Receptors
	- Visual		— Olfactory
— Cutaneous	— Neuromasts		
— Propriceptors	— Membranous-labyrinth		— Gustatory — Jacobson's organ

12.2. GENERAL SOMATIC RECEPTORS

These receptors include cutaneous and proprioceptors.

12.2.1. Cutaneous Receptors

The external sensory organs are found in the integument covering the entire surface of the body. These organs produce cutaneous sensations including touch, heat and cold. The cutaneous receptors may consist of *free nerve fibre endings* or encapsulated nerve endings. The free nerve endings (fig 12.1 are probably the oldest external receptors in the vertebrate body. They are responsible for the sense of touch (*tangoreceptors*) or pain (*algesireceptors*). The branches of sensory nerves terminate in the epidermis, cornea of eye, hair follicles and muscous membrane. These nerve endings do not reach the outermost layer of epidermis. The stimulus travels through the outer cells before it reaches the nerve endings. The encapsulated nerve endings or corpuscles may be tractile, sensitive to pressure, heat or cold. The sensory nerve terminations are enclosed in a capsule here, which is made of connective tissue. The corpuscles are stimulated by light touch, a range of non-injurious temperatures, and pressures on the skin. Birds and mammals are richly provided with varieties of such corpuscles. Specialized corpuscles are present on the beak, at the tip of the snout and on genitals. The corpuscles present around genitals assist in stimulating centres of brain necessary for sexual reproduction. *Meissner's corpuscles* are present is skin of primates especially in areas undergoing friction.

12.2.2. Proprioceptors

Sensory nerve endings responsible for location of muscles position are present in striated muscle cells, the *bursas of joints,* the *tendinous ending* of *muscles, connective tissues* and *skeletal tissues.* They are stimulated mechanically when a muscle cell contacts, a tendon is stretched, or a joint is under pressure from flexion. These organs do not produce well define sensations but help to coordinate the position of limbs. The impulse go partly to the cerebellum and assist in reflexy maintenance of posture.

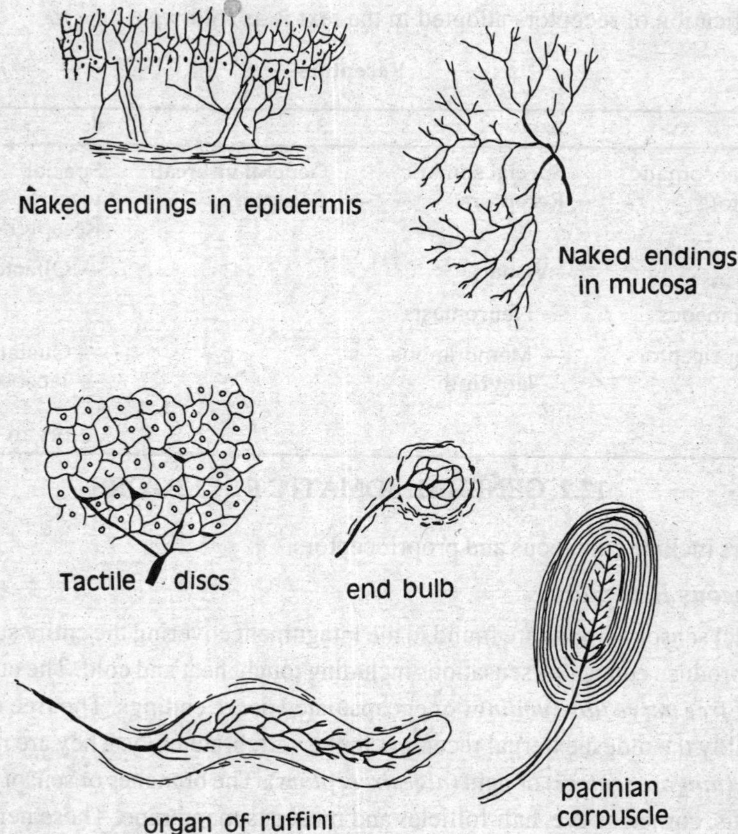

Fig. 12.1 General somatic and visceral receptors. Encapsulated endings include pacinian corpuscles for touch, end bulbs (of Krause) for cold and organs of Ruffini for heat.

12.3. SPECIAL SOMATIC RECEPTORS

They include photoreceptors. neuromast organs and internal ear.

12.3.1. Photoreceptors

The paired eyes are sensitive to radiations in a narrow range or wave lengths. The limits of visible spectrum vary in different species. The chief runctional photoreceptors is *retina,* which is a nervous membrance present behind the lens at the posterior end of a series of fluid filled chambers. The retina, fluid filled chambers, and lenses are surrounded by a protective and light absorbing called *sclerotic* and choroids coats of the eye ball. The structure of paired eyes in vertebrate is remarkably constant with minor differences.

Development of the eye (Fig. 12.2): During development the eye originates from the embryonic forebrain as primary *optic vesicle.* The distal wall of the vesicle is folded inwards to form a double walled secondary optic vesicle, or *optic cup.* The outer wall of the optic cup acquires pigment granules and becomes pigment layer of the retina. The inner wall of the optic cup thickens and forms a sensory layer of the retina with *cones, rods* and *neurons.* The two layers of retina fuse together and form, retina The axons from the neurons of the retina to the brain foom *optic nerve* through *optic stalks.* The optic stalks are connection of the optic cups with forebrain. The ectoderm present outside the optic cup thickens to form a *lens, placode* which is separated from the ectoderm to from a solid lens. The *lens* sinks inwards and comes to lie within the optic cup. The optic cup grows in size but the lens does not grow. The epidermis covering the cornea becomes transparent to form the *conjunctiva.* The mesenchyme present outside the optic cup from outer tough *sclerotic* and *cornea,* and an inner *vascular choroid.* The rim of the optic cup together with a part of the developing choroid becomes *cilliary body* and *iris.* The opening of the optic cup makes the *pupil.* The development of eyes, therefore, takes place from three sources (1) As outgrowth of the forebrain forming iris, retina, ciliary body and optic nerve: (2) from ectoderm which forms the lens and conjuctiva: and (3) from mesenchyme which forms the sclerotic and choroid.

Typical structure of eyes in Vertehrates: The eyes of vertebrates show remarkable similarity in structure excepting some minor variations to suit the particular needs of the animal concerned. The typical structure of the vertebrate eye as illustrated by mammalian eye is given below. The eye consists of the following parts (Fig. 12.3)

(1) Three distinct layers *sclerotic, choroid* and *retina,* on the back.

(2) A spherical biconvex *lens* in front to focus light on retina for the formation of image.

The *sclerotic* layer forms the outermost coast is made of connective tissue, cartilage, and in some reptiles and birds, of bones. It makes firm socket for eye ball and extends infront of the lens as a transparent layer called *cornea.* The cornea is covered by a thin transparent vascular *conjuntiva* formed by modified epidermis and is continuous with lining of eye lid. The *choroid* is highly vascular pigmented middle layer present next to scletotic. It contains blood vessels and lymph spaces and supplies nourishment to the eye. The pigment absorbs the light and allows most

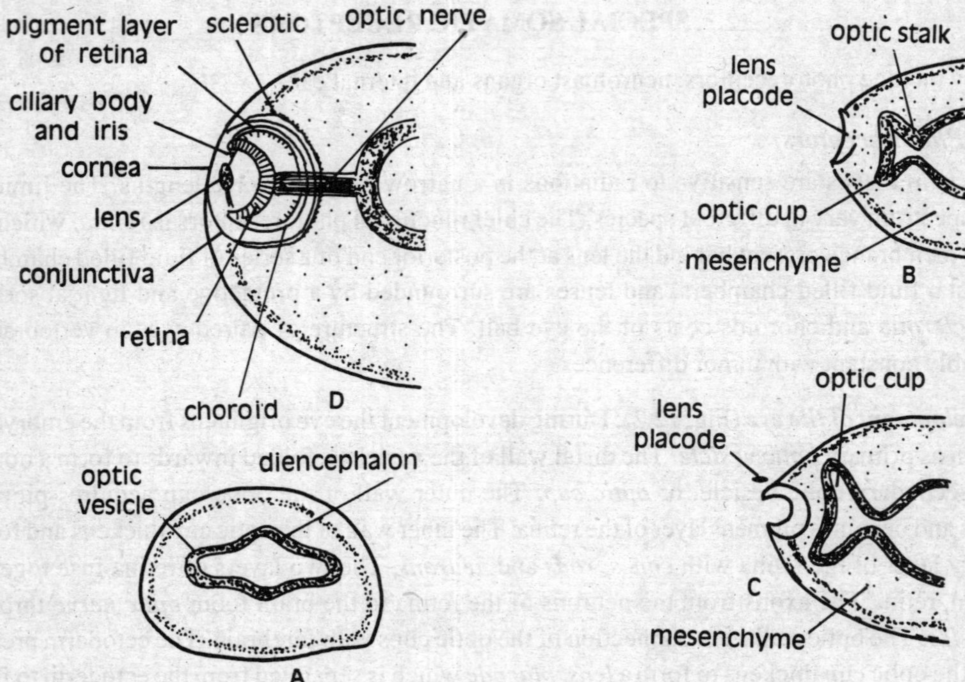

Fig. 12.2 A. Development of eye in Vertebrates

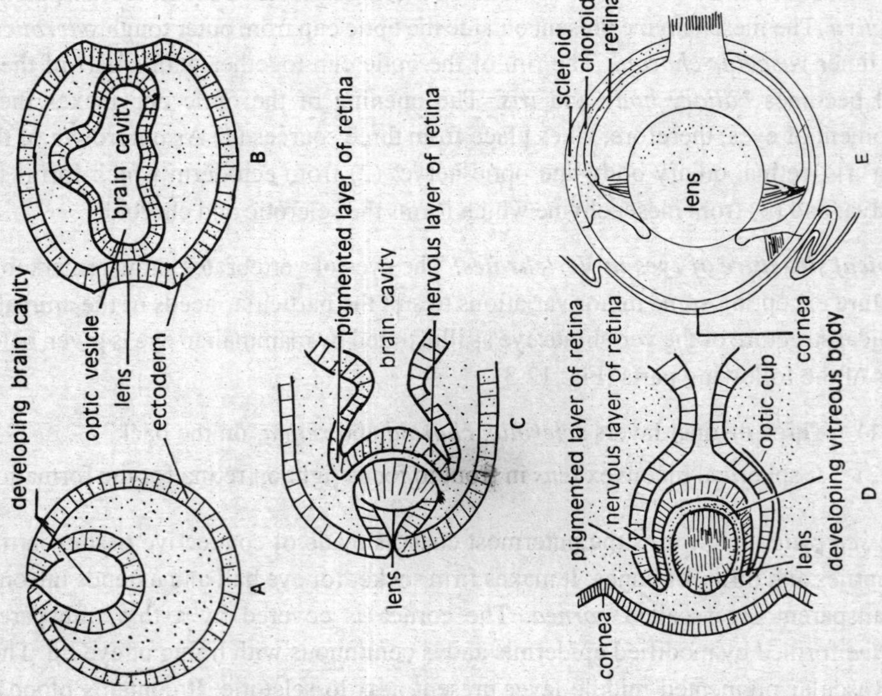

Fig. 12.2 B. Development of eye in Frog

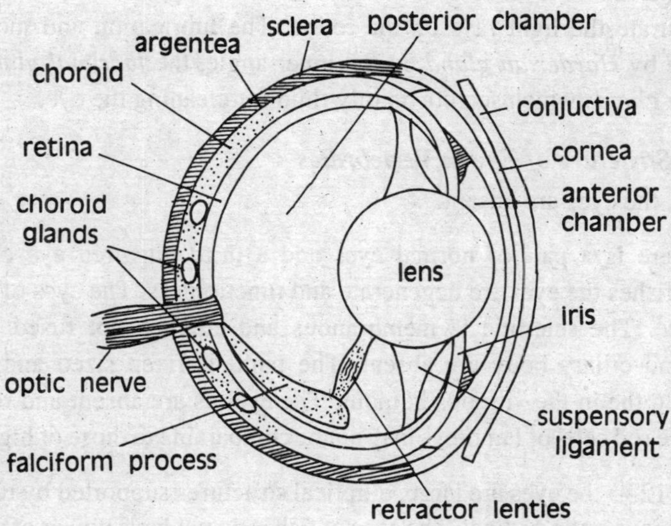

Fig. 12.3 A Typical Structure of Eye

of the light to fall on retina. The choroid extends forwards to form *iris diaphragm* enclosing a circular slit-like opening, the *pupil.* The iris is having instrinsic, circular and radial smooth muscles for adjustment of the size of aperture according to the animals requirements. The peripheral margin or iris is having a ciliary body, which is thick folded band containing smooth ciliary muscles, ciliary processes, blood vessels glands. Sympathetic fibres control the circular and radial muscles of the iris and ciliary muscles. The *retina* is innermost sensitive layer. It is made of two layers, an outer pigment layer and an inner layer is transparent and contains outermost photosensitive layer of cells called *rods* and *cones.* The rods Have long thin cylinder with purple pigment *rhodopsin.* The cones bear short tapering processes. Both rods and cones are nucleated cells with their inner ends continuous with slender nerve fibres. The rods detect difference in the intensity of light and cones detect acute details in bright light and colour differences. In nocturnal animals the retina has rods only. The rods and cones are stimulated by light rays and thereby perceive the image and transmit it to the brain by optic nerve where the image is visualized. The point where the optic nerve enters the retina rods and cones are absent and no image is formed. The point is called *blind spot.* Just near this point there is another point called *yellow spot* containing abundance of rods and cones where the brightest image is formed.

The *lens* is crystalline, spherical or biconvex structure suspended at the back of the iris. It is meant for focusing the light rays on retinal surface. The lens is attached to the ciliary body by a fiberous *suspensory ligament.* The space between the cornea and iris is occupied by an *anterior chamber.* A *posterior chamber* is situated in the narrow space between iris and lens. The anterior and posterior chambers are filled by a watery *aqueous humour.* Behind the lend the large cavity of the eye ball, the *vitreous chamber* is filled with a transparent jelly-like *vitreous humour.*

The *cornea, aqueous humour,* the *lens,* and *vitreous, humour* form a scries of refracting media which help to concentrate the light rays on the retina. The lubrication and moistening of the vertebrate eye is done by *Harderian gland* on the inner angle, the *lacrimal gland* on the outer angle, and *Meibomian glands* which secrete on oily fluid for cleaning the eye.

12.3.2. Comparative Structure of Eye in Vertebrates

In *Branchiostoma*, the eyes are absent.

In *Petromyzon* there is a pair of normal eyes and a third unpaired eye called *pineal* or *epiphysial eye*. In hagfishes the eyes are degenerate and functionless. The eyes of lampreys show flattened outer surface. The sclerotic is membranous and corenea not fused with skin. The suspensory ligment and ciliary body are absent. The pupil is fixed sized and a permanently spherical lens is held in the in the vitreous humour. The eyelids are absent and rods are lager in number than cones. The rod cells of lampreys may not be comparable to those of higher vertebrates.

In *Scoliodon* (Fig. 12.4) the eyes are large, elliptical structures supported by recti and oblique muscles and a cartilaginous optic *pedical.* The eyes of fishes do not have power of accommodation for observing distant objects. Movable eye lids and tear glands are absent. The *Scoliodon* has monocular vision and is color blind. The eye ball is made of usual three layers (Fig. 12.5), the *sclerotic*, the *choroids* and the *retina*. The sclerotic layer is cartilaginous. The pupil is vertical slit, which cannot be dilated or contracted. The retina lacks cones and is made of rods only.

A non-pigmented area centralize is present on the posterior side of the retina where acute vision is affected. A large number of plates made of guanine, called *tapetum lucidum* are present on the inner surface of the choroid and act as reflector. The *crystalline lens* is spherical and hard and pushes the iris forwards. The *ciliary body* has no intrinsic muscles. A dorsal *suspensory ligament* forms a *gelatinous zonule* attached to the ciliary body at the equator of the lens. Slight accommodation of lens for near vision is possible by a small protractor lentis muscle present on the ventral side of the lens on the gelatinous zonule. The eye lids are small folds of skin. The lower eye lid is slightly movable. Glands are absent in eyes.

In *Labeo* (Fig. 12.6) the eye shows typical structure with three layers. The outer sclerotic layer contains cartilage. The median choroid layer is vascular and the innermost retina is sensitive to light. The choroids layer contains choroids glands surrounding the optic nerve. The cornea is flat and eyes have no eye lids. The lens is rounded and in contact with cornea. The anterior chamber of the eye is reduced. A silvery reflecting layer called *argenta* is present between the choroids and sclerotic layers. A peculiar *falciform process,* which is a vascular fold of choroid, is present in the posterior chamber of the eye and helps in changing the position of the lens. The falciform process pierces the retina near the entrance of the optic nerve. It is attached to the back of the by a *retractor lentis* or *companula helleri* muscle. The eye of *Labeo* has monocular vision and accommodation is possible by shifting the position of the lens.

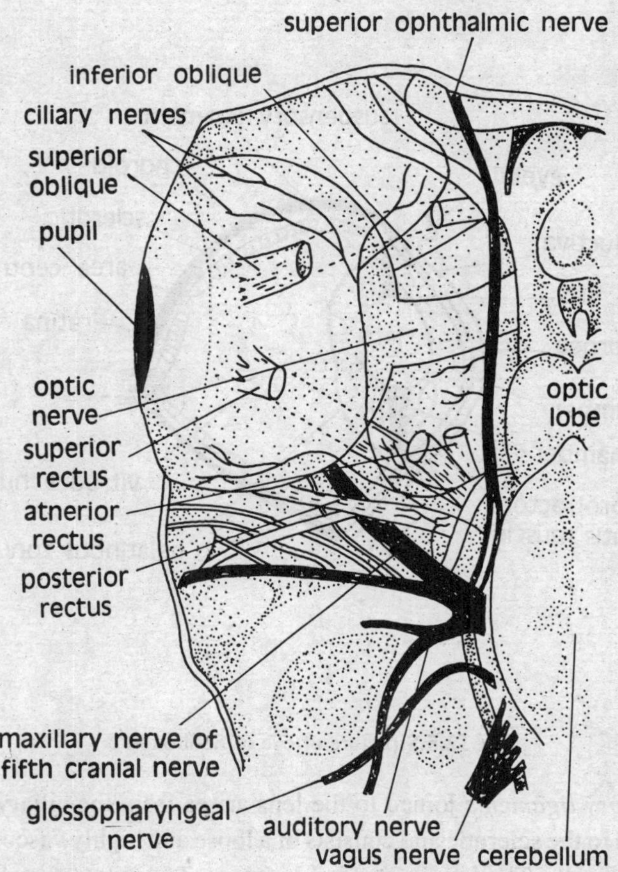

Fig. 12.4 Relationship of the eye with the brain, muscles, nerve in scaliodon.

In *Rana* (Fig 12.7) the eyes are paired globular structures present in the large orbits on lateral sides of the head. Each eye is placed in such a way that it may have its own field of vision. Since amphibians are also adapted for aerial life, sight has acquired a dominant sense and visualization of distant objects has become possible.

There are 3 layers in the eye like in other vertebrates. The *sclerotic* is made of dense connective tissue with some cartilage and its anterior part *cornea* curved outwards to make the anterior chamber large. The cornea is covered with delicate transparent membranes *conjunctiva* formed by epidermis and is continuous with the inner lining of the eye lids. The conjunctiva is supplied by the secretions of *harderian glands* situated in the orbit below the eye ball and keep it moist. The circular and radial muscles of the iris are well developed and help in the dilation or reduction of the pupil. The *ciliary body* without ciliary muscles is attached to the sclerotic and posterior surface of the iris.

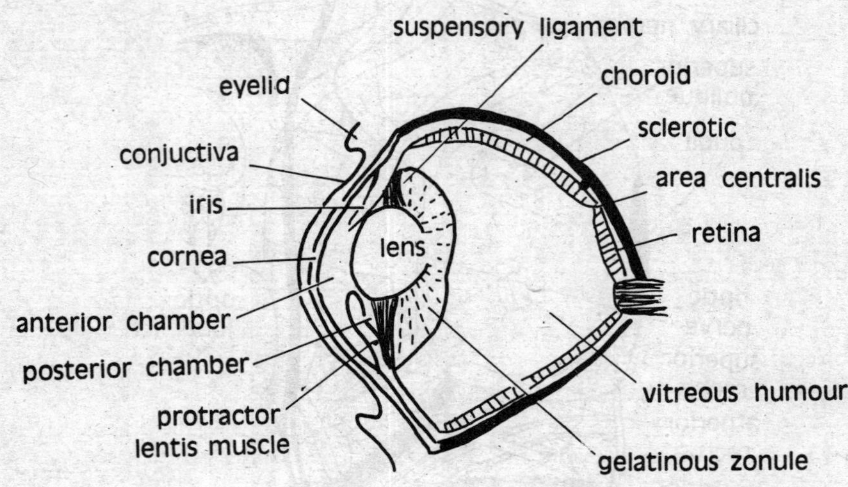

Fig. 12.5 Structure of eye of Scoliodon

A ring of *suspensory ligaments* joined to the lens arises from the ciliary body. The *choroid layer* is closely applied to the sclerotic and consists of a loose and highly vascular connective tissue with numerous pigment cells. The *lens* is rounded structure. Two *protractor lentis muscles* run on the dorsal and ventral side from the cornea to the ciliary body and suspensory ligaments. These muscles bring about the accommodation by moving the lens forwards towards the cornea. The lower lids are well developed in frog. The lower eye lid is movable with its upper border forming a transparent *nictitating membrane.* The nictitating membrane is not homologus with that of higher vertebrates. It protects and cleans the eye and may be pulled radidly over the cornea *Lacrimal-glands* are present below the lower eye lid and secrete lacrimal fluid or tears which keep the eye moist and clean. The *harderian gland* is situated at the inner angles of the eye and lubricates the nictitating membrane.

In *Uromastix* (Fig. 12.8) the eyes are suited to a purely terrestrial life. The *sclerotic* is having cartilaginous structure with a ring of sclerotic bones on the outer side. The number of cones in retina has increased and lizards are able to have color vision. A vascular projection called *ectodermal conus* arises from the *blind spot* region of the retina and nourishes the retina. The *choroid* is made of an inner pigmented layer and an outer vascular later. The circular and radial muscles are absent in iris and the shape of pupil cannot be changed. Striated *intrinsic ciliary muscles* are present on the ciliary body for the first time in reptiles. The lens has an *annular pad.* The accommodation for near objects is possible in reptiles by ciliary muscles and their process squeezing the periphery of the lens. In this way, the curvature of the outer surface of the lens and cornea is increased. The lower eye lid is more developed than the upper one. A true *nictitating membrane* lies between the upper and lower eye lids. The *lacrimal gland* is situated on the outer angle of the lower eye lid and secretes tears which are drained into the nasal passage through nasolacrimal duct. The *Harderian gland* lies at the inner angle of the eye and lubricates the nictitating membrane

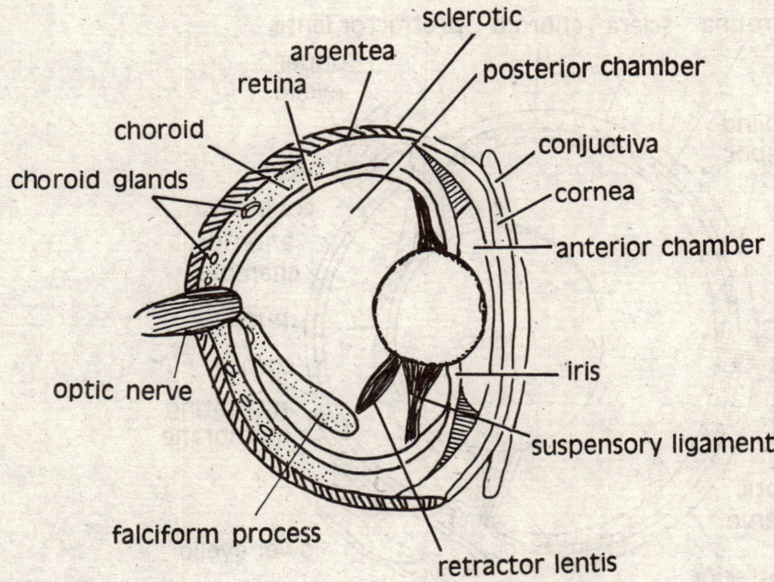

Fig. 12.6 Structure of eye of Labeo

In *Columba* (Fig. 12.9) the eyes are large and well developed. They are built on the typical vertebrate plan suited for precise vision over long distances. The shape of the eye ball is biconvex and lens and cornea protrude outwards. The eye ball is surrounded by a ring of sclerotic bones in the region of the ciliary body Striated sphincter muscles are present in the iris for contraction. The retina lies in the broad posterior region of the eye and distant objects may be sharply focused on it. The vision is binocular in birds and color vision is highly developed. An area centralis with *fovea centralis* is present in retina for visual acuteness. The lens is soft and biconvex. The ciliary body is well developed and ciliary processes are attached to the lens. The striated cells of the ciliary body are divided into anterior *Crampton muscles* and posterior *Brucke muscles,* which are responsible for accommodation. A highly vascular and thick dark pigmented structure called *pectin* is present at the blind spot. This structure is plate like, folded fanwise and looks like a comb. It is made up of a vascular network supported by pigmented glial cells. Its functions are not fully ascertained. It is presumed that (1) it helps in accommodation by pressing the lens forwards; (2) regulate the pressure of fluids in the eye; (3) its shadow falls on the retina and helps in the perception of movements; (4) it nourishes the retina by diffusion through the vitreous humor. The eye lids are well developed and closē down during the sleep. The *nictitating membrane* is present below the two eye lids and covers the cornea during the flight. A *hareian gland* is present at the inner angle of the eye. The *lachrymal glands* are present below the outer angle of the lower eve lid.

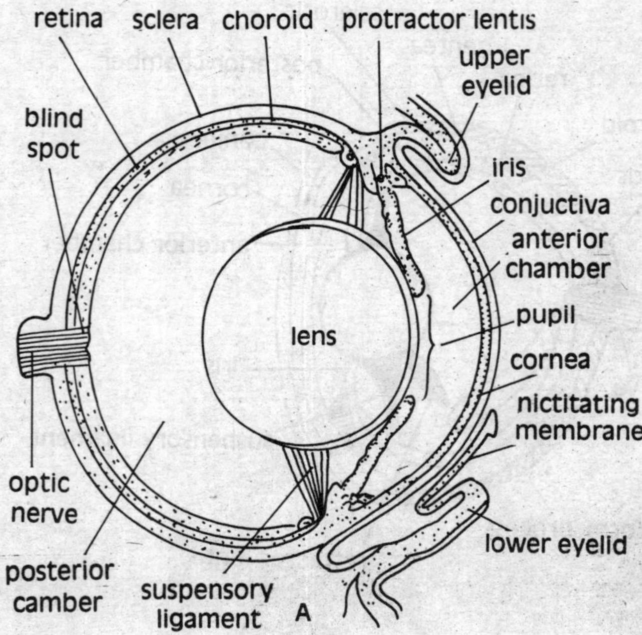

Fig. 12.7 Structure of Amphibian eye

In Orycotolagus the eyes have typical structure as described earlier. 12.3.2.

12.3.3. The Neuromast Organs (Fig. 12.10):

A group of epithelial receptor cells called *neurmasts* are found in cyclostomes, fishes, aquatic larval stages of some amphibians and some adult amphibians. The neuromasts have morphological variations such as *lateral line canal system, cephalic canals,* the *ampullae of Lorenzini*, and *pit organs.* All neuromast organs are fluid filled pits, cups, or ampullae, which open to the surface of the body or occur under the skin in the wall of a sunken, closed canal having pores along its length. The nemomasts develop as epidermal structures and their epithelium has two types of cells.

neuroepithelial cells and *sustentacular cells.* The neuroepithelial cell has one or more sensory hairs projecting from the apical tip of the cell into the fluid in the neuromast chamber. The sustentacular or supporting cell rests on the basement membrane of the epithelium. The neuromast organs are supplied with branches of VII, IX and X cranial nerves. Each neuromast receives a small nerve twig. The exact function of neuromasts has not been confirmed. The sensory fibres of the neuromast system terminate in a portion of the medulla, which contains a continuation of a layer of the cerebellum. The sensory hairs of the neuroepithelial cells are stimulated by mechanical changes in the fluid medium. The lateral-line system may have *kinesthetic function,* as proprioceptors are absent in aquatic animals. They may detect movements of water and currents and furnish this information through the membranous labyrinth. The body wall movements or aquatic medium, or both, may send information to the cerebellum which may be used reflexly to regulate speed and direction of turning. The modified neuromast organs situated superficially may have a tactile function also. However, the evidence is lacking in support of these assumptions.

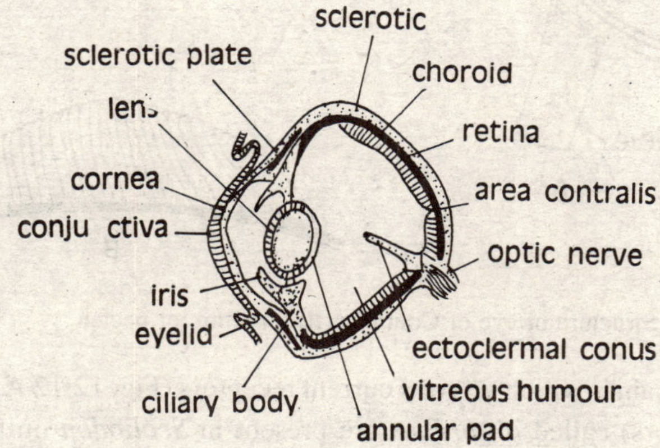

Fig. 12.8 Structure of eye of Uromastix

The lateral line system in Scoliodon: The laterosensory system has two long lateral line canals each running on the laterals side of the entire body in the dermis. (Fig. 12.11). Each canal bends dorsally on reaching the head and joins its fellow from opposite side by a transverse commissure called *occipital canal* situated behind the opening of endolymphatic duct. Beyond occipital canal each lateral line canal runs forwards as a *postorbital canal,* which divides, into two branches, a *supraorbital canal* and an *infraorbital canal.* The supraorbital canal runs above the orbit and infraorbital canal below the orbit and run up to the snout. A *jugal-canal* arises from the infraorbital canal below the eye and extends backwards upto the first gill cleft and gives a mandibular canal to lower jaw. The laterensensory canals are provided with epithelial lining with nocuous gland cells. The canals open to the surface by vertical tubes and are filled with a fluid and mucus.

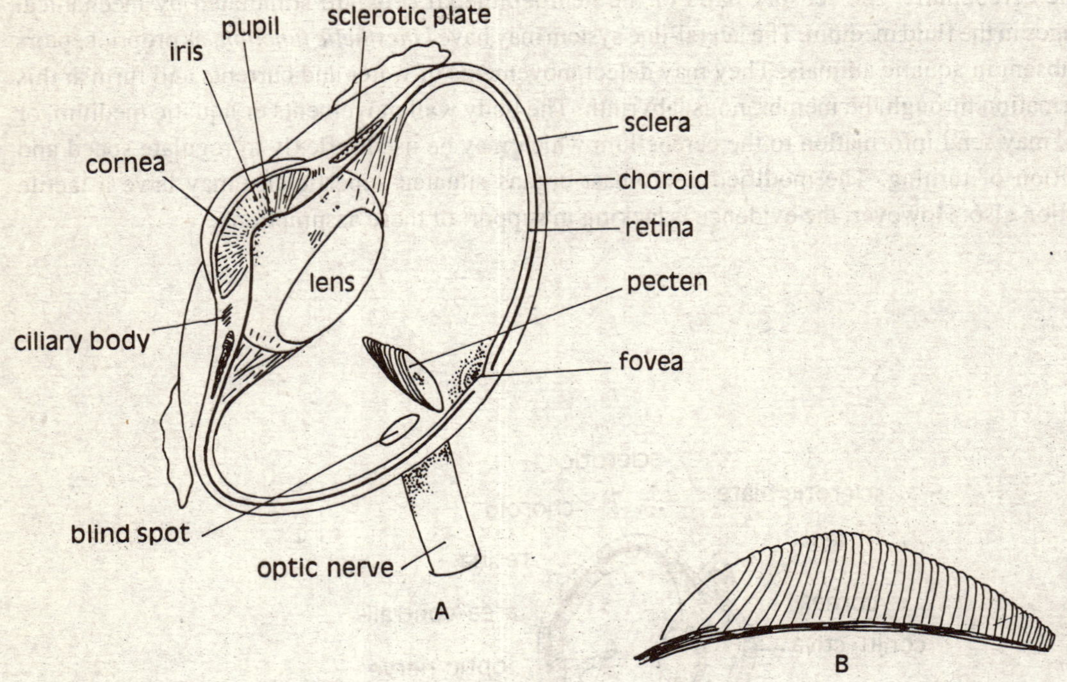

Fig. 12.9 A. Structure of eye of Columba, **B.** Structure of pecten.

Neuromasts are present in canals which serve as current receptors (Fig. 12.12 A). Independent *rheoreceptors* (current receptors) called *pit-organs* are present in *Scoliodon* on the dorsal and lateral surfaces the head. Each pit-organ consists of ectodermal pit having groups of receptor cells supplied by nerve fibres of VII cranial nerves. The pit-organs probably serve for controlling the orientation of body in relation to the currents and waves.

Ampullae of Lorenzini (Fig. 12.12 B, C) are present on the dorsal and ventral surfaces of the head scattered in definite groups. They open to the outside by pores on the skin. The pore of each ampullae leads into a duct or canal filled with mucus. The canal ends below an *ampullary sac* containing 8 or 9 vertical chambers arranged radially around a central core or centrum Each ampullae has pear-shaped mucous gland cells and pyramidal *sensory hair* cells. The ampullae are innervated by VII cranial nerve. They are named according to their location on the head. The *supra-opthalmic* are situated around the supraorbital canal. The outer *buccal* are present between supraorbital and infraobital canals. The *Inner buccal* lie below the infraorbital canal. The real function of the ampullae of Lorenzini is uncertain. They have been regarded as thermoreceptors, weak tactile receptors. or hydrostatic pressure receptors

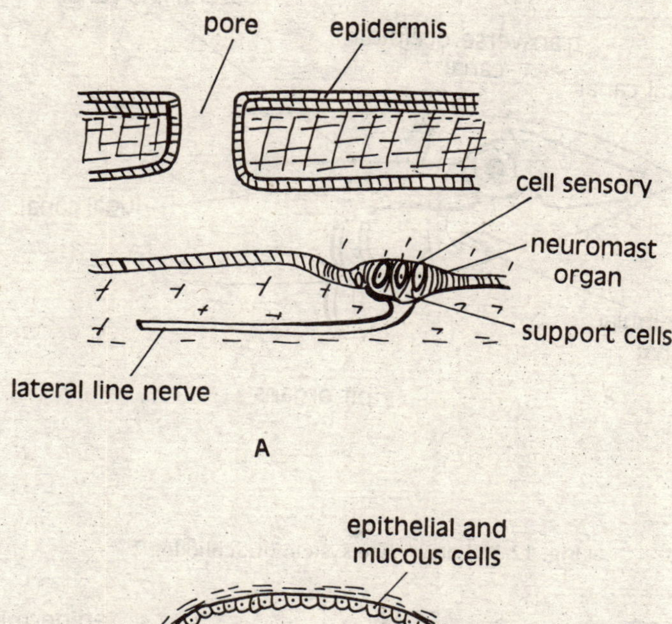

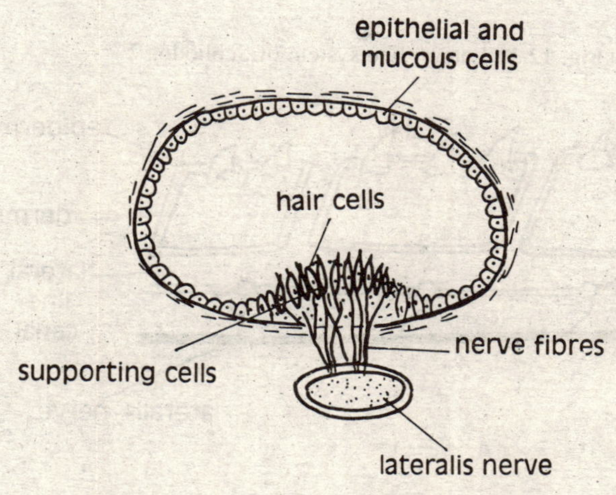

Fig. 12.10 A & B. The Neuromast Organs

Lateral Line System In Teleosts (Fig. 12.13): The lateral line system of teleosts is well developed and consists of a large number of deep seated neuromasts. Each neuromast is situated in a pit and communicates with surrounding water by tubes, which pierce the scales and open by pores. Water circulates within canals in contact with the neuromasts.

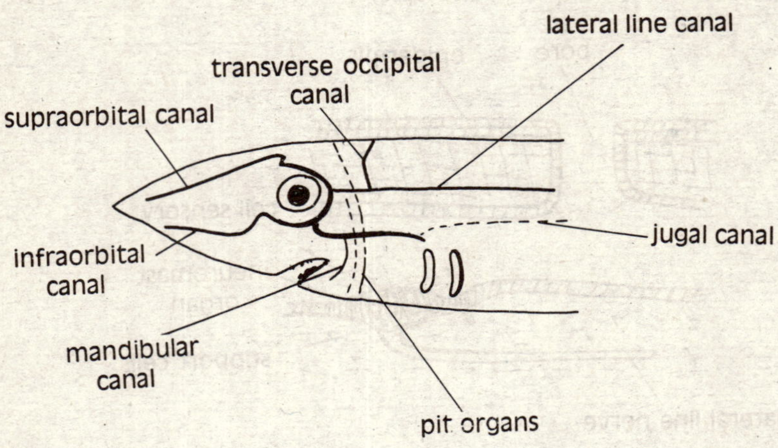

Fig. 12.11 Lateral line system of Scoliodon

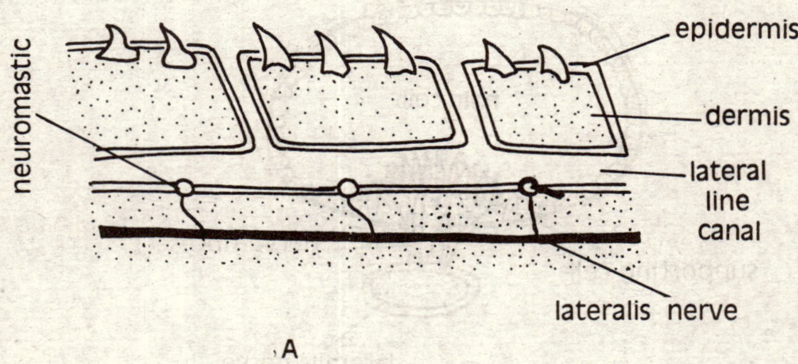

Fig. 12.12 A. Lateral line canals with nervous supply

One lateral-line canal runs the length of the body on each side perforating one row of scales. The laterosensory canals in telcosts show a remarkable constancy in pattern. They consist of the *main lateral canal* running back from the head region where one *supraorbital canal* above eyes and one temporal canal behind otic region joins it. A *post-orbital* canal from behind eye, *infraorbital* from below the eye, a *mandibular* from lower jaw and a *preopercular* branch also join the main lateral line canal. An *occipital* or *supratemporal* canal near the back of the head crosses the head dorsally.

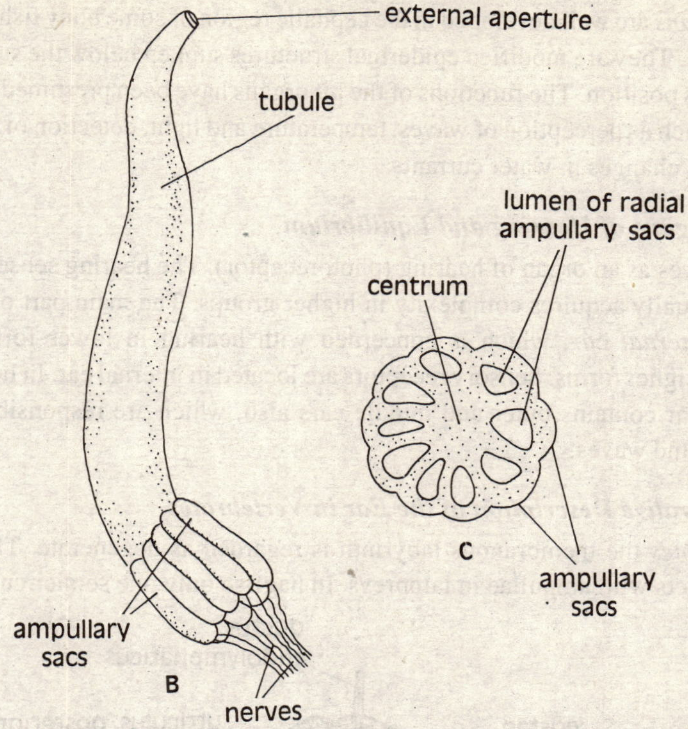

Fig. 12.12 B&C. Ampullae of Lorenzini

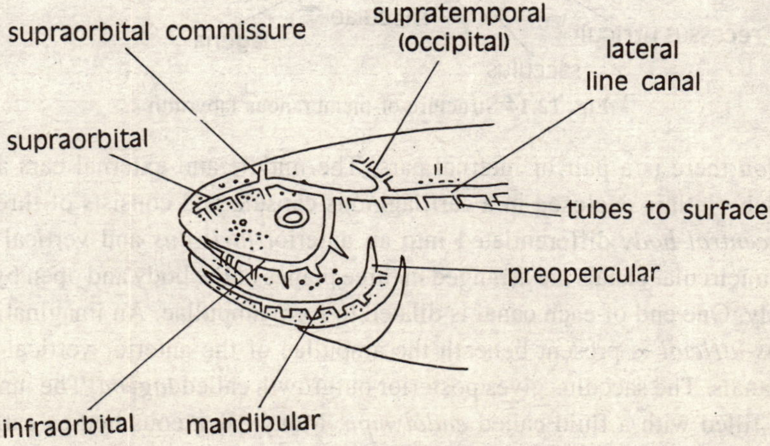

Fig. 12.13 Lateral line system in Teleosts

The *pit-organs* are well developed in the cephalic region of some bony fishes on both dorsal and ventral surfaces. They are modified epidermal structures sunken below the surrounding epidermis in subcutaneous position. The functions of the pit organs have been presumed similar to lateral line sense organs, such as perception of waves, temperature and light, detection of electric currents, and physiochemical changes in water currants.

12.3.4. The Organs of Hearing and Equilibrium

The car serves as an organ of hearing (photoreceptor). The hearing sense is first developed in fishes and gradually acquires complexity in higher groups. The main part of ear is *membranous labyrinth* or *internal car*, which is concerned with hearing in lower forms and hearing and equilibrium in higher forms. Sensory receptors are located in internal ear. In higher animals such as mammals the car contains outer and middle ears also, which are responsible for receiving and transmitting sound waves.

12.3.5. Comparative Description of the Ear in Vertebrates

In *cyclostomes* the membranous labyrinth is regarded as degenerate. There are two vertical semicircular ducts with ampullae in lampreys. In hagfish only one semicircular canal is present.

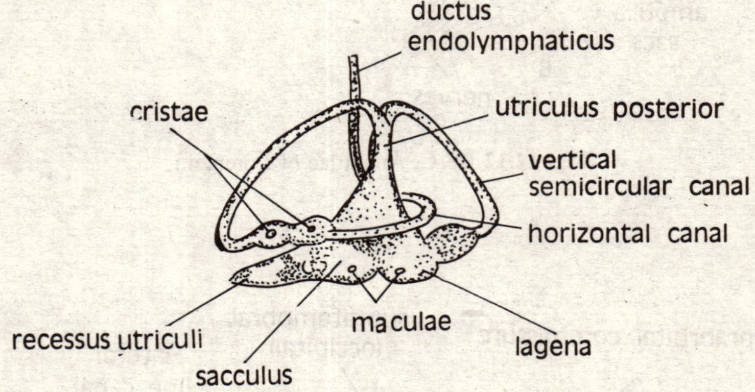

Fig. 12.14 Structure of membranous labyrinth

In *Scoliodon* there is a pair of internal ears. The middle and external ears are absent. The membranous labyrinth is enclosed in a cartilaginous capsule and consists of three *semicircular canals* and a *central body* differentiated into an anterior *utriculus* and vertical *sacculus* (Fig. 12.14). The semicircular canals are arranged in three planes of the body and open by both ends into the central body. One end of each canal is dilated to form ampullae. An imagination of utriculus called *recessus utriculi* is present beneath the ampullae of the anterior vertical and horizontal semicircular canals. The sacculus gives posterior outgrowth called *lagena.* The inner cavity of the internal car is filled with a fluid called *endolymph.* Many calcareous bodies called *otoliths* are present in the endolymph. The space between internal car and cartilaginous capsule (auditory capsule) is filled with *perilymph.* A long duct called *ductus endolymphaticus* from sacculus communicates to the exterior through a small opening. The ductus endolymphaticus of elasmobranches is not homologus with that of higher vertebrates and should be called an invaginated canal. The membranous labyrinth is supplied by auditory nerve (VII cranial nerve). Groups of delicate receptor cells with stiff hairs or *maculae* are found various parts of the internal ear. The movements of fluid in any or all semicircular canals are perceived by appropriate receptors. The utriculus and semicircular canals serve for orientation and accleration (statoreceptors). The sacculus. lagena and cochlea serve for hearing (acoustic receptors).

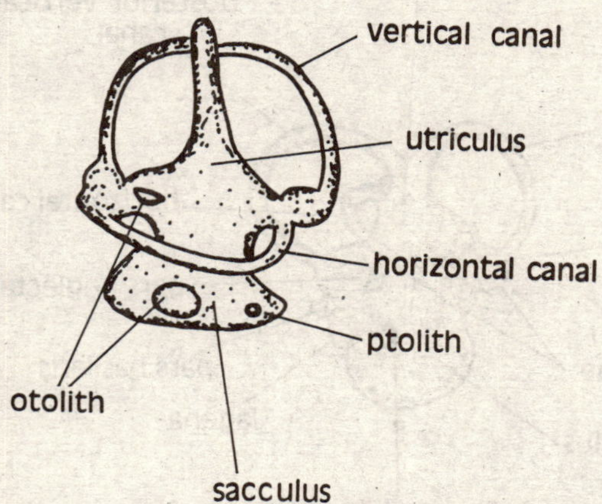

Fig. 12.15 Membranous labyrinth of Labeo

In *Labeo* the ear is composed of utriculus, sacculus, lagena and three semicircular canals (Fig. 12.15). The otoliths called *sagitta* are present inside the sacculus, *asterisus* in lagena, and *lapillus* in the utriculus. The perilymphatic space of internal car is connected with the air bladder by a chain of bony *weberian ossicles* which probably help to intensify sound vibrations and convey them to the ear or help to understand the air pressure in the bladder and transmit changes of such pressure to the perilymph to start a reflex action.

In *Rana* the ear consists of an internal and a middle car on each side of the head (Fig 12.16). The internal car lies in a bony auditory capsule. The connection between sacculus and utriculus is partly constricted and utriculus is large than sacculus. Both sacculus and utriculus have one macula each.

The large macula of utriculus is called *pars neglecta.* The upper part of the sacculus is provided with a special sensory *papilla amphibiorum,* which probably plays an important role in the reception of sound. The endolymphatic duct aries from the sacculus, extends dorsally and expands above medulla in to an endolymphatic sac. A *lagena* projects from the posterior end of the sacculus and extends into another sac a *basilar papilla.* The lagena and basilar papilla are *phonoreceptors.* A gelatinous membrane called *otolithic membrane* contains hairs of the sensory cells of the maculae. The endolymph fills the membranous labyrinth and cerebrospinal perilymph is present between the auditory capsule and membranous labyrinth.

The *middle ear* (Fig. 12.17) consists of an outer tympanic membrane stretched on a tympanic ring or *annulus tympanicus,* an air filled tympanic cavity and *eustachian tube.* The middle ear communicates with the auditory capsule by an aperture called *fenestra ovalis* with a ring-like bony operculum closed by a membrane. The inner surface of the tympanic membrane is connected with fenestra ovalis by a cartilaginous *columella auris.* The head of columella called *extra columella* is joined to the tympanic membrane and its foot is expanded to form a cartilaginous stapedial plate. The *stapedial plate* is fused with the fenestra ovlis.

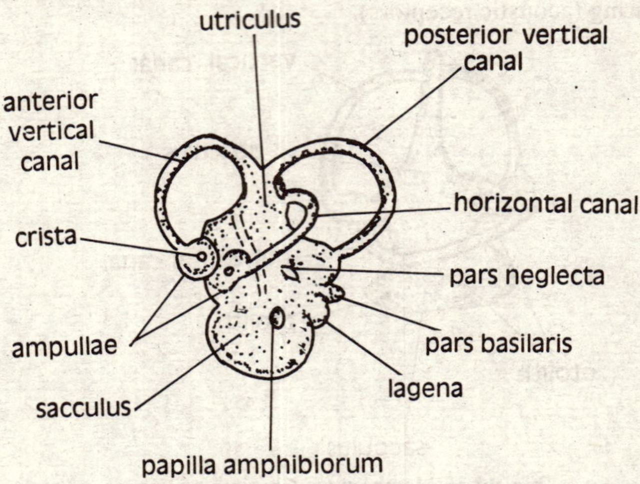

Fig. 12.16 Membranous labyrinth of Frog

The sound waves induce vibration in tympanum which are conveyed to the membrane of fenestra ovalis through columella. These vibrations are passed to the perilymph, endolymph, maculae of langena, basilar papilla and papilla amphibiorm and finally to the brain through the auditory nerve. The car of frog is primarily an organ of equilibrium but serves secondarily for hearing also.

In *Uromastix* the ear consists of an internal car and a middle ear like in frog. The membranous labyrinth is enclosed in the auditory capsule (Fig. 12.18). The *utriculus* and *sacculus* are connected by a narrow *sacculoutricular canal.* The utriculus present in upper position has a cylindrical tube bent sharply. Three *semicircular canals* arise frome the utriculus. A common tube called *crus-commune* is formed by the union of the lower parts of the anterior and posterior cannals. The endolympatic duct starts from the sacculus and ends blindly in to the duramater as a swollen sac. The *legena* or rudimentary cochlea arises frome the sacculus and makes a flat cochlear duct with large *macula* or *basilar papilla.* The basilar is an organ of hearing formed the first time in reptiles.

The internal car communicates with the middle car through *fenestra ovalis.* The tympanic membrane is situated behind the eye. The tympanic cavity lies below the tympanic membrane and communicates with pharynx by a small eustachian tube. The *columella auris* is bony and runs across the tympanic cavity. The head of columella called *extra columella* joins the tympanic membrane and its foot called *stapes* or *plectrum* is attached to the membrane covering the fenestra ovalis.

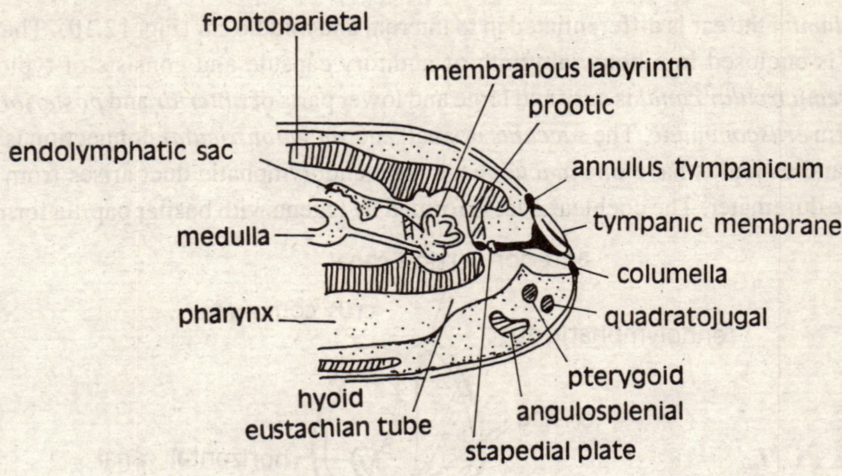

Fig. 12.17 Lateral view of ear showing columela auris

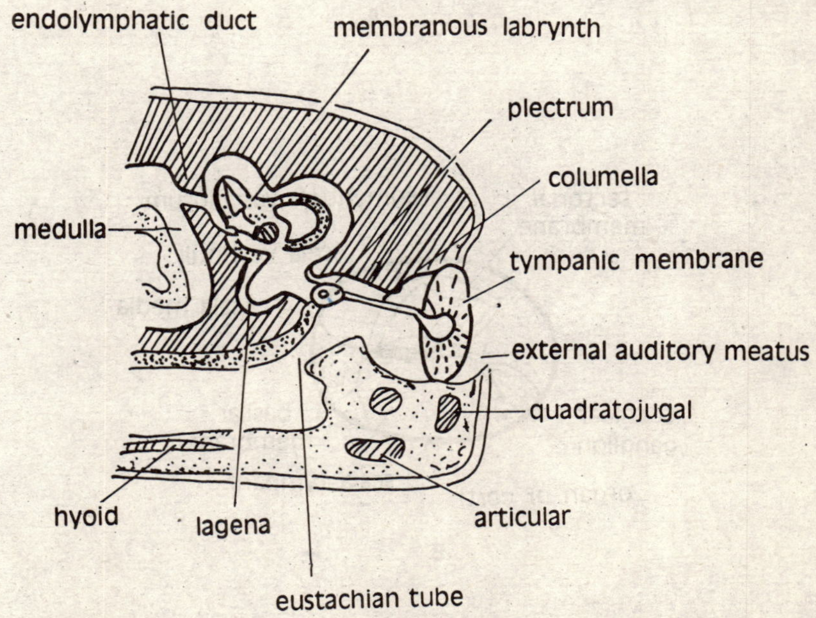

Fig. 12.18 Middle and Inner ear of Uromastix

The membranous labyrinth mainly serves for equilibrium. The lagena and basilar papilla serve for hearing, which is well developed in reptiles.

In *Columba* the ear is differentiated in to internal and middle car (Fig. 12.19). The membranous labyrinth is enclosed in a bony labyrinth of auditory capsule and consists of typical parts. The *anterior semicircular canal* is very and large and lower parts of *anterior* and *posterior semicircular canals* form *cruscommune*. The *sacculus* is small and *sacguloutricular* connection is narrow. Both sacculus and utriculus have one *macula* each. The endolymphatic duct arises from sacculus and ends in the duramater. The cochlear duct is formed by lagena with basilar papilla forming a *basilar*

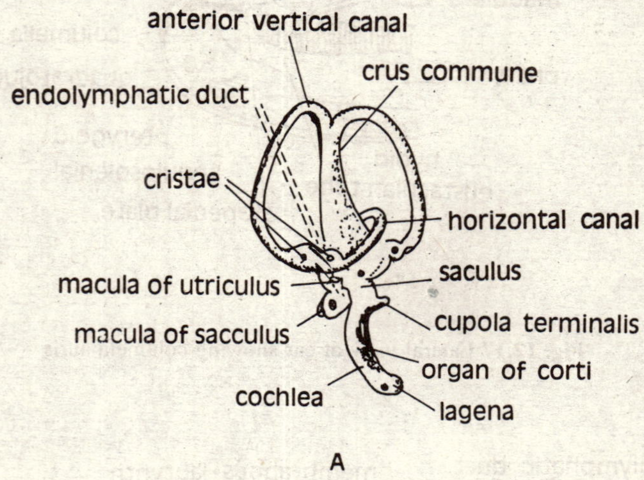

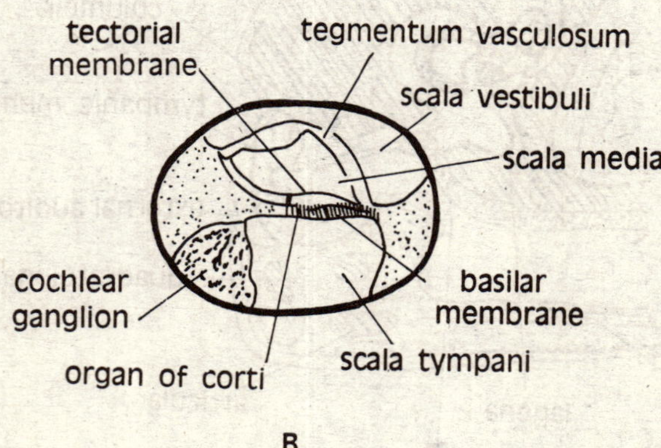

Fig. 12.19 A. Avian membranous labyrinth, **B.** Section of cochlea

membrane. This membrane contains hair cells, which make *organ of corti,* which is sensitive to high frequency vibration. The fibres of hair cells are embedded in a *tectorial membrane* present above the organ of corti. Another set of hair cells with their fibres embedded in a gelationous *cupola terminals* is provided with minute calcareous *otliths*. The cupola-terminalis is situated at the apical end of the cochlear duct. The *lagena* with a macula is present at tip of the cochlear duct and is sensory to low frequency vibrations. The cochlea has three passages; a *scala vestibuli* on the upper side a *scala media* in the middle and a *scala-tympani* on the lower side. The scala media is filled with endolymph and remaining two have perilymph.

The *tympanic membrane* is situated behind an *external auditory meatus* hidden under feathers (Fig. 12.20). The eustachean tube arise from the tympanic cavity, meets it's fellow from opposite side and opens in the pharynx by a common aperture. The *columella auris* has three processes at its outer end, which are connected with the tympanic membrane. The inner end of columella called *stapes* joins the membrane of the fenestra ovalis. The sense of hearing and equilibrium are both well developed in birds.

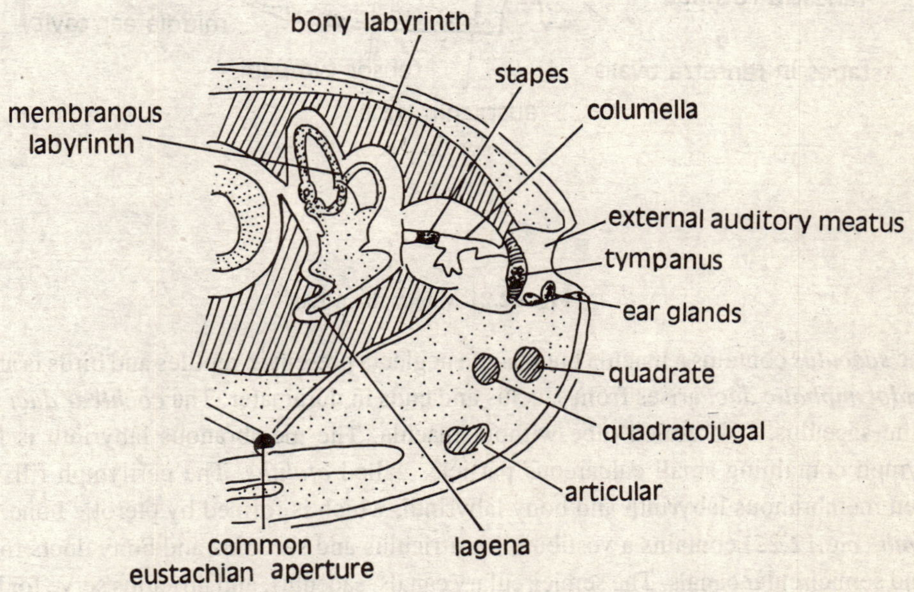

Fig. 12.20 Middle car of Columba

In *Oryctolagus* the ear is having three parts external, middle and internal (Fig. 12.21). The cochlear duct of membranous labyrinth is more developed than in birds. The anterior and posterior semicircular canals a *cruscommune, Sacculus* and *utriculus* are small with a narrow connection.

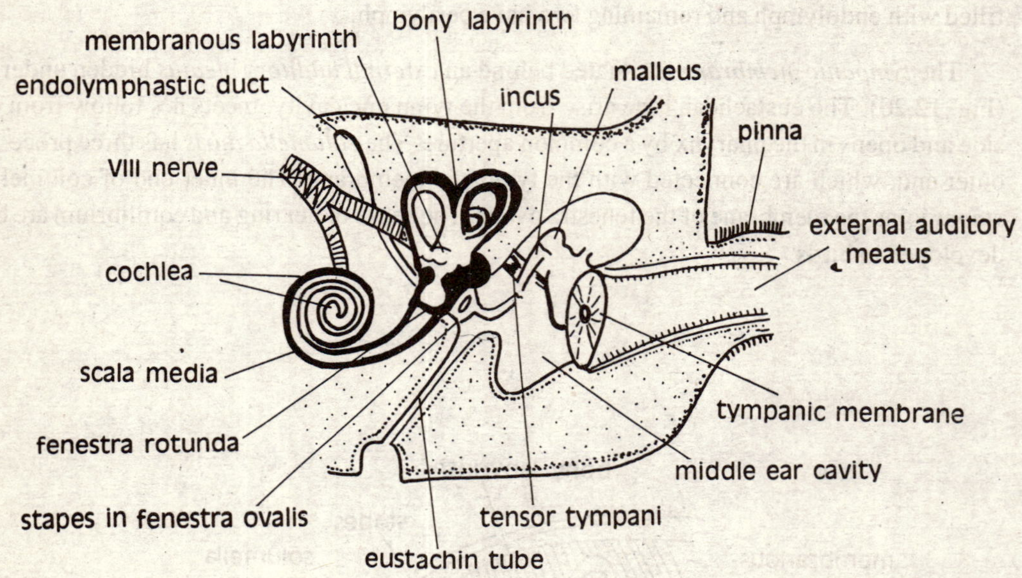

Fig. 12.21 Mammaliam Ear

The *sacculus* contains a macula but macula neglecta present in reptiles and birds is absent here. The *endolymphatic duct* arises from sacclus and ends in duramater. The *cochlear duct* also arises from the sacculus. It is coiled tube without macula. The membranous labyrinth is filled with endolymph containing small calcareous particles called *otoliths.* The perilymph fills the space between membranous labyrinth and bony labyrinth, which is formed by pterotic bone. The bony *labyrinth* (Fig.12.22) contains a vestibule for utriculus and sacculus and bony ducts for cochlear duct and semicircular canals. The semicirculary canals, sacculus, and utriculus serve for balancing. The bony cochlear canal encloses three cavities, *scala-vestibuli*, *scala-tympani* filled with perilymph and *scala-media* filled with endolymph. The scala-vestibuli is situated above the cochlear duct and the scala tympani below it. The scala media (Fig. 12.23) forms the cavity of the cochlear duct. The demarcation between the three cavities is done by *basilar membrane* or *Reissner's membrane.* The receptor apparatus for hearing, the *organ of corti* is supported on basilar membrane inside the cochlear duct. The organ of corti is made of supporting cells and sensory hair cells.

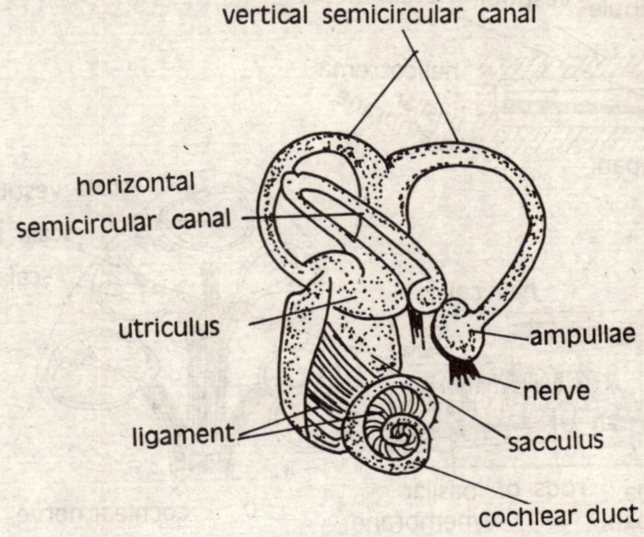

Fig. 12.22 Membranous Labyrinth of rabbit.

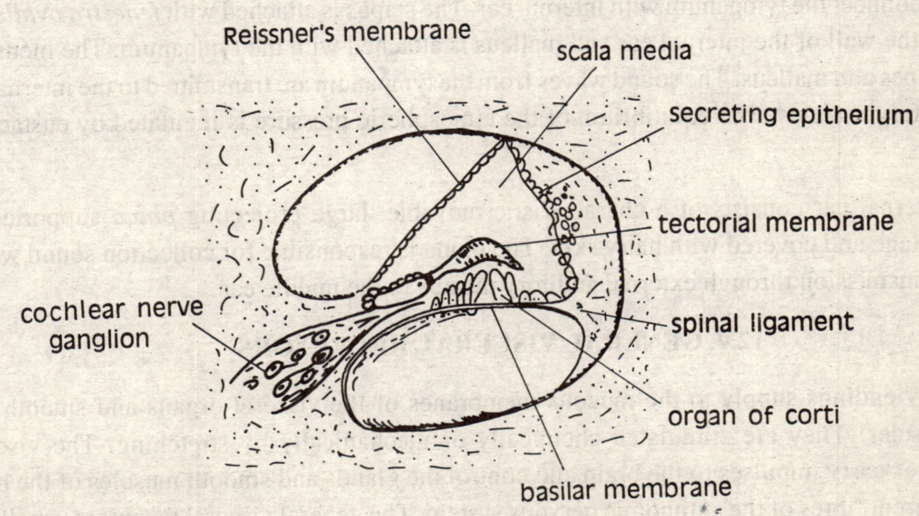

Fig. 12.23 A. Section of cochlear duct

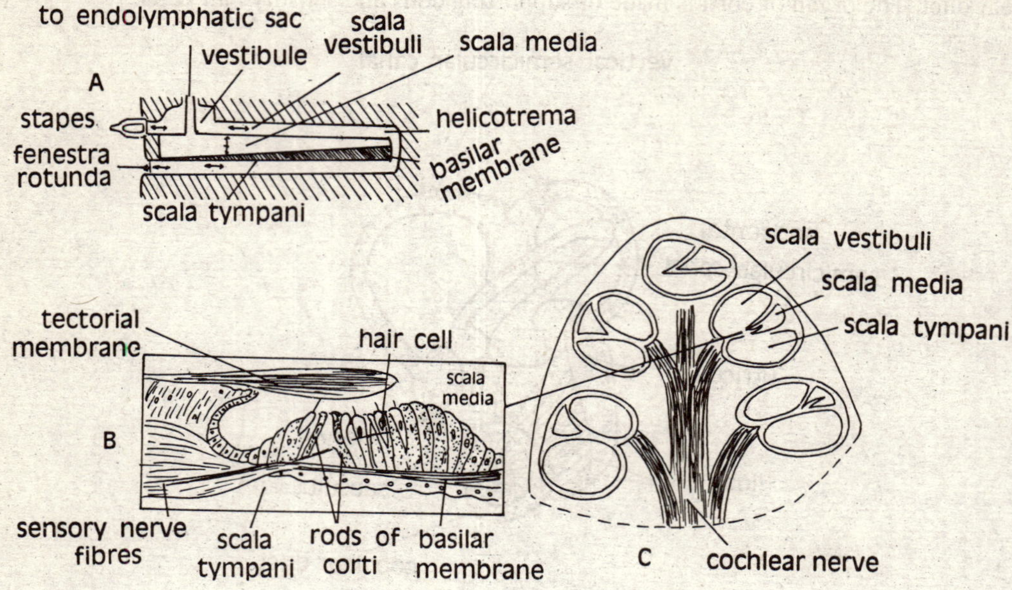

Fig. 12.23 B. (i) Diagram showing transmission of vibrations. (ii) Organs of corti (iii) Section of Cochlea

The *middle ear* is separated from the external ear by *tympanum* and communicates with buccal cavity by *eustachian tube.* Three ear ossicles, elongated *malleus*, slightly curved *incus,* and ring-like *stapes* connect the tympanum with internal ear. The stapes is attached with *fenestra ovalis,* the opening in the wall of the internal ear and malleus is attached with the tympanum. The incus lies between stapes and malleus. The sound waves from the tympanum are transmitted to the internal car by the auditary ossicles. The equilibrium of the atmospheric pressure is regulated by eustachian tube.

The external ear consists of a characteristic movable, large projecting *pinna* supported by elastic cartilage and covered with hairy skin. The pinna is responsible for collection sound waves and their transmission through external auditory meatus to the middle ear.

12.4. GENERAL VISCERAL RECEPTORS

Free sensory endings supply to the mucous membranes of the visceral organs and smooth and cardiac muscles. They are stimulated chemically or mechanically by stretching. The visceral afferent fibres carry impulses to the brain and control the glands and smooth muscles of the body through efferent fibres of the autonomic nervous system. The general visceral receptors usually do not create sensation but play important role in homeostasis. The knowledge about such receptors is very limited.

12.5. SPECIAL VISCERAL RECEPTORS

They include chemoreceptors such as *olfactory* and *gustatory* organs. They serve to provide information about the environment by detecting substances like food, males and enemies.

12.5.1. Olfactoreceptors (Fig. 12.24)

They consist of individual *bipolar neurons* derived from placodes in the ectoderm. Such receptors are situated in the olfactory epithelium or mucous membranes, which consists of *basal cells* and *supporting cells.* The olfactory organs form olfactory sacs in fishes or lining of the nasal sac in tetrapoda.

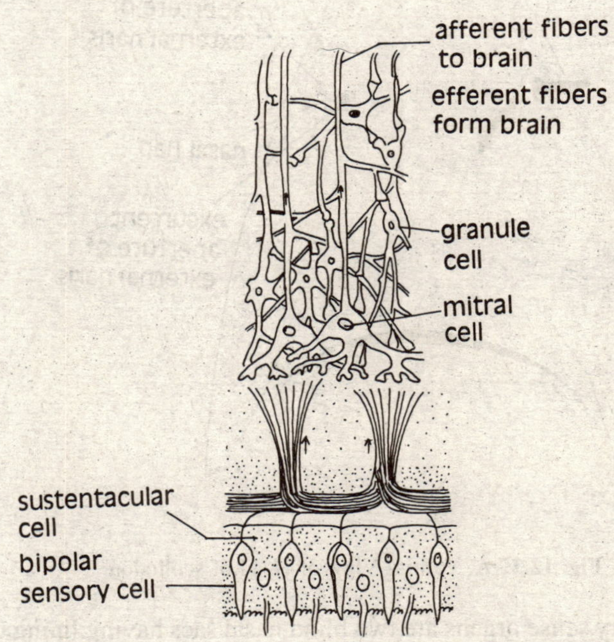

Fig. 12.24 Olfactory Epithelium in Shark

Special elongated *neurosensory cells* with hair-like processes at their free ends are also present in the mucous membrane. The neurosensory cells give fibres, which join to form olfactory nerves, which join olfactory lobes of the brain. The neurosensory cells are covered by mucus, which is protective substance and dissolves substances for smelling. The olfactory nerves are paired which indicate bilateral origin of the olfactory sac.

In *cyclostomes* a single median nasal aperture opens on the surface of the head and leads to a blind olfactory sac. The olfactory sac is lined with numerous folds covered with olfactory epithelium.

In elasmobranchs the olfactory organs are pairs structures. In *Scoliodon* each olfactory sac is situated in a cartilaginous capsule and does not communicate with buccal cavity (Fig. 12.25), the mucous membrane of the olfactory sac is produced in to a series of folds called *Schneiderian folds* containing olfactory cells and supporting cells.

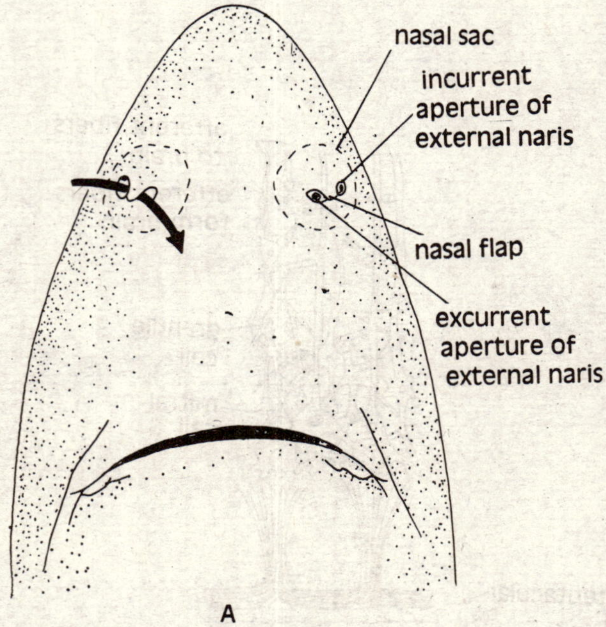

Fig. 12.25 A. Ventral View of Head of scoliodon

In *teleosts* the olfactory sense organs are two blind nasal sacs having lining of olfactory cells. The nasal sacs do not communicate with the buccal cavity and open to the exterior by openings.

In *Rana* the olfactory or nasal sacs are situated in the olfactory capsule of the skull separated from one another by nasal septum. Each sac communicates to the exterior by *external nares* and to the buccal cavity by *internal nares.* The mucous lining of the olfactory sac secretes mucus to keep the epithelium moist and its odour stimulates the hair of the neurosensory cells.

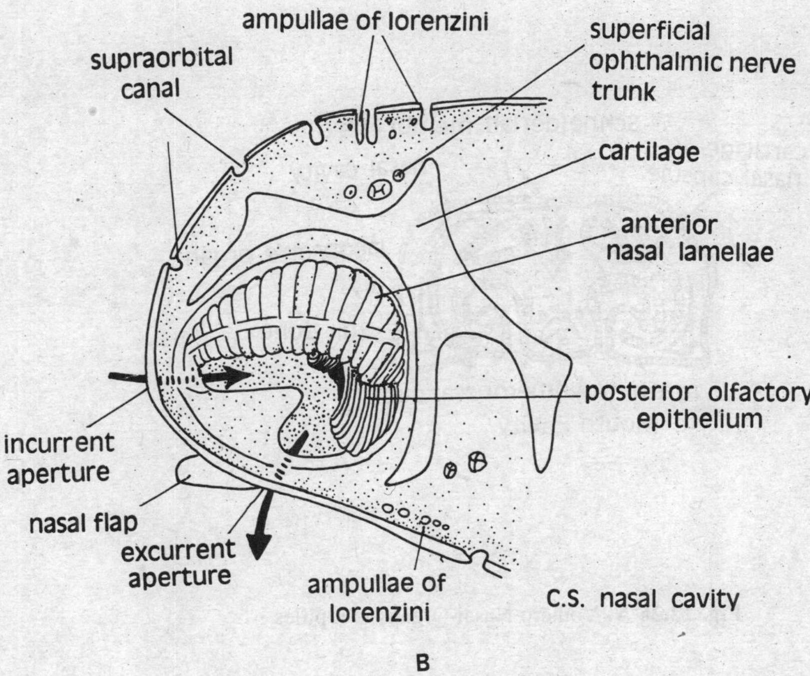

Fig. 12.25 B. C.S. Nasal Cavity of scoliodon

In *Uromastix* paired vomeronasal organs called the *Jacobson's organ* are present below the nasal cavity and above the buccal cavity (Fig. 12.26 A). It is believed to be an olfactory organ. The Jacobson's organs are derived embryonically from the olfactory cavity and consist of sensory epithelium innervated by olfactory and trigeminal nerves. This organ appears in amphibians but is best developed in lizards and snakes. Each organ opens in to the buccal cavity.

In *Columba* the olfactory organs are paired structures near the base of the upper beak. Each olfactory chamber is covered externally by ectoethmoid bone and two chambers are separated by a *mesethmoid bone*. This bone is produced in to three spiral processes called *turbinals* which increase the surface area of the mucous membrane. Each olfactory chamber is divided in to an anterior non-sensory vestibule and a posterior sensory part. The sensory part is having a single layered epithelium and the sense of smell is very poor.

In *Oryctolagus* and most mammals (Fig. 12.26 B). The sense of smell is highly developed. The elongated nasal passages contain folded turbinal bones, *ethmoturbinals, maxilloturbinals* and *nasoturbinals.* The olfactory epithelium covers these turbinals and contain neurosensory olfactory cells and supporting cells. Each nostril has an *ovoid sensory pad* with minute papillae and ridges and acts as distance receptory by testing the air. The epithelial lining of maxilloturbinals and nasoturbinals is not sensory but warms the inhaled air.

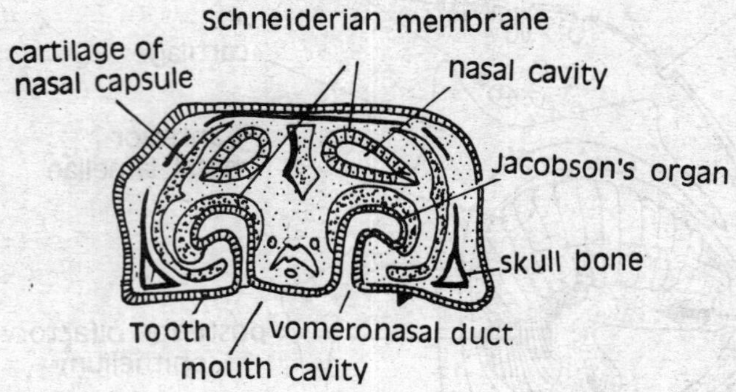

Fig. 12.26 A. Vomero Nasal Organs is reptiles

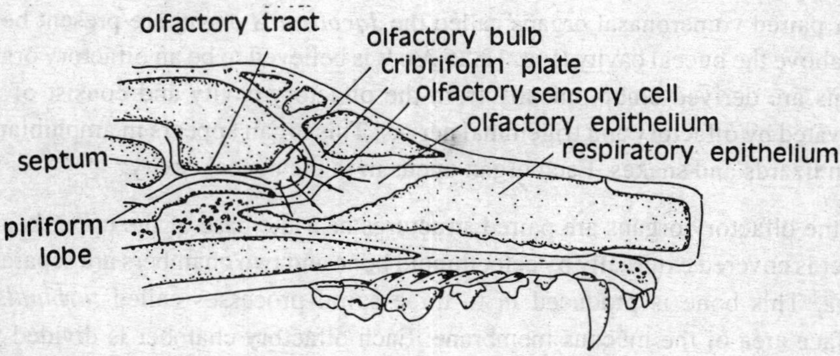

Fig. 12.26 B. Olfactory epithelium Nasal passages in mammals are lined with respiratory epithelium. The olfactory epithelium is a small region of this lining that contains specialized neuronal fibers that make contact with neurons of the olfactory tract. These processes relay impulses to the piriform lobe and septal area of the brain.

12.5.2. Gustatoreceptors

The organs of taste consist of taste buds present in all vertebrates. Their structure is fairly uniform. Each taste bud consists of a group of *neurosecretory cells* and *supporting cells* arranged

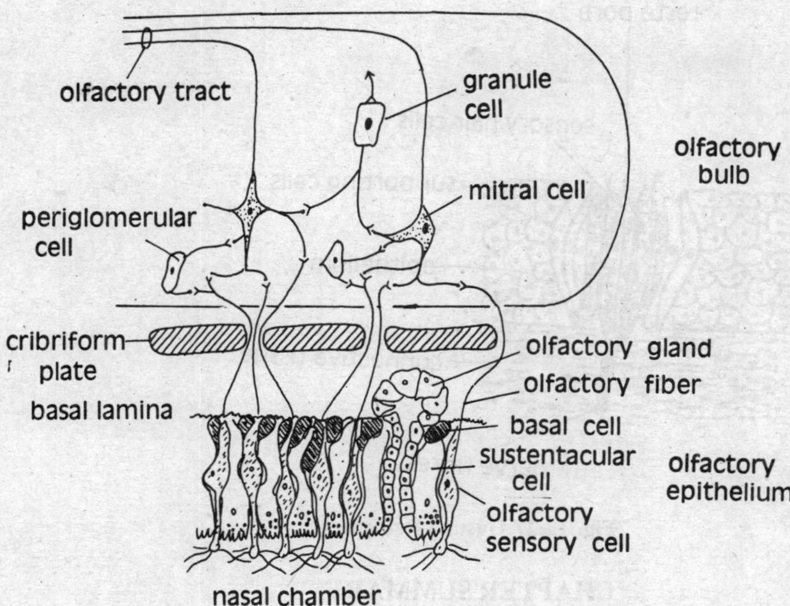

Fig. 12.26 C. Histology of olfactory epithelium. The olfactory epithelium includes supportive sustentacular cells, basal cells, and olfactory sensory cells. The apical surface of each olfactory cell develops cilia that project into the air passage. Its basal end consists of a nerve fiber that travels through the cribriform plate into the olfactory bulb, where it synapses with periglomerular mitral and granule cells. Fibers of the mitral cells constitute the olfactory tract, which goes to the brain.

to form barrel-shaped structures (Fig. 12.27). The taste buds are embedded in *stratified epithelium.* Substances in solution, which are sweet, salty, bitter and sour, may be tested by the taste buds. Each taste bud consists of outer covering of supporting cells and inner elongated neurosensory cells. Each neurosensory cell is narrow and long with a thin taste-hair at its tip and nerve fibre at its base. The taste hairs project above the surface or taste pore in mammals. They are innervated by V, VII, IX and X cranial nerves.

In *fishes* the taste buds are widely distributed on the roof, side walls and floor of the oral cavity and pharynx where they are situated in the path of respiratory water constantly passing though gills. In bottom feeder fishes the taste buds are scattered on the entire body surface to the tip of the tail.

In *tetrapods* the taste buds are having restricted distribution being confined to the tongue, oral cavity, and pharynx under the mucous covering of these structures. They are innervated by VII, IX and X cranial nerves. In *birds* the taste buds are very few. Four types of papillae have been recognized on the tongue of man. These are *filiform, foliate, fungiform* and *circumvallate.* The filiform, papillae do not have taste buds associated with them.

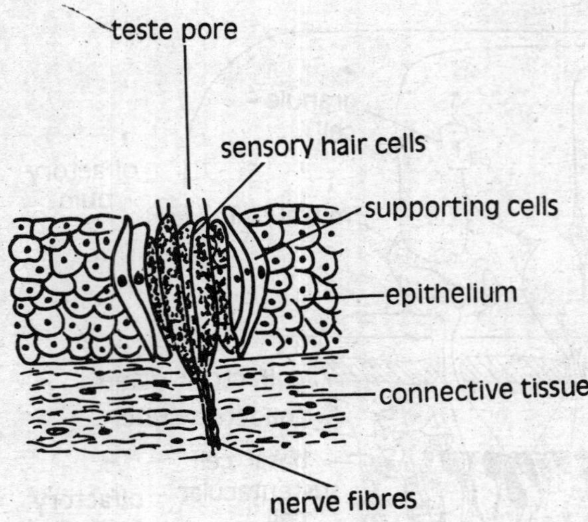

Fig. 12.27 Gustato-Receptors

CHAPTER SUMMARY

1. The receptor organs are closely associated with nervous system and yield conscious sensation to meet the environment. They transmit certain types of stimuli to the central nervous system.
2. The receptor organs have been classified into external or exteroceptors and internal receptors. The exteroceptors include photoreceptors (vision). olfactory receptors (smell), phonocreceptors (hearing), gustatoreceptors (taste), tangoreceptors (touch and presence), thermoreceptros (heat or cold), algesireceptors (pain) and rheoreceptors (currents of water). The internal receptors include proprioceptors (muscle position), statoreceptors (equilibrium) and interoceptors (hunger, thirst etc.). The receptor organs are also classified as general somatic (cutaneous, proprioceptors), special somatic (visual, neuromasts, membranous labyrinth), general visceral and special visceral (olfactory, gustatory, jacobsons organ) receptors.
3. The cutaneous receptors produce sensations of touch, heat, cold and pain. They include free nerve endings, encapsulated nerve endings (corpuscles) and Meissner's corpuscles.
4. The proprioceptors situated in striated muscle cells, in bursas of joints, tendinous endings of muscles, connective tissue and skeletal tissue help to corrdinate the position of limbs and assist in reflexy maintenance of posture.
5. The eye appears in cyclostomes for the first time in vertebrates. It originates from the embryonic forebrain as primary optic vesicle. The eyes of vertebrates show remarkable similarity in structure excepting minor variation to suit the particular needs of the animal concerned. The eye consists of three distinct layers sclerotic, choroids and a spherical lens infront to focus light on retina for the formation of image. The lens is

attached to the ciliary body by fiberous suspensory ligament. The sclerotic layer extends forwards to from iris diaphragm enclosing a slit-like pupil.

6. The neuromast organs are found in cyclostomes, fishes, aquatic larval stage of some amphibians and some adult amphibians. These are kinesthetic as proprioceptors are absent in aquatic animals. These receptors consist of neuromasts located on the surface of the skin of inside canals or grooves in the skin. The lateral line canal usually runs throughout canals or groves in the skin. The lateral line canal usually runs throughout the length of the body and terminates on the head in many branches. Pit organs, ampullae of Lorenzini and lateral line canal system are variation of the neuromasts.
7. The ear contains membranous labyrinth or internal car concerned with hearing in lower forms and hearing and equilibrium in higher forms. Sensory receptors are located in internal ear. In mammals the ear contains outer and middle ears for receiving and transmitting sound waves.
8. General visceral receptors are free sensory nerves ending to mucous membrane of the visceral organs and smooth and cardiac muscles. They play important role in homeostasis.
9. Special visceral receptors include chemoreceptors such as olfactory and gustatory organs. The olfactoreceptors are situated on the olfactory epithelium. The guststoreceptors consist of taste buds present in all vertebrates.

Bibliography

1. Alexander, R.M. (1975). *The Chordates,* Cambridge University Press London.
2. Colbert, E.H. (1955). *Evolution of the vertebrates.* John Wiley & Sons.
3. Das, S.M. (1936). *Herdmania. The Indian Zoological Memoirs.* Lucknow Publishing House. Lucknow. India.
4. De Beer, G.R. (1937). *The Development of Vestebrate Skull,* Oxford Claridon Press. Eaton, T.H. Jr. (1959), Comparative Anatomy of Vestebrates, 2nd ed. Harper and Row Publishers, New York.
5. Goodrich, E.S. (1958). *Studies on the Structure and Development of Vertebrate,* Reprinted by Dover Publications. New York.
6. Hyman, L.H. (1979). *Comparative Vertebrate Anatomy,* 3rd ed. edited by Marvale H. Wake. University of Chicago Press.
7. Jorden, E.L. and Verma. P.S. (1981). *Chordate Zoology,* 5th ed. S. Chand & Co. New Delhi.
8. Kardong, K.V. (2002). *Vertebrates: Comparative Anatomy, Function Evolution,* Tata Mc. Graw Hill Pub. Co. New Delhi.
9. Kent, G.C. Jr. (1961). *Comparative Anatomy of Vertebrates.* C. V. Mosby Company. St. Louis
10. Kingsley, J.S. (1974). *Outlines of Comparative Anatomy of Vertebrates.* P. Blakiston's Son & Co. 3rd ed, reprinted by Central Book Depot, Allahabad.
11. Mayr, E., Lindsley. E. G. and Usingcr, R.L. (1953). *Methods and Principles of Systemic Zoology.* Mc-Graw Hill Inc. New York.
12. Neal, H.V. and Rand. H. W. (1939). *Chordate Anatomy.* The Blakiston Company, Toronto.
13. Parker, T. J. and Haswell. W. A. (1972). *A Text Book of Zoology.* Vol. II. 7th ed. revised by A.J. Marshall The Mac Millan Co. London.
14. Prasad, S.N. (1967). *A Text Book of Vertebrate Zoology,* 6th ed. Kitab Mahal. Allahabad
15. Romer, A.S. (1959). *The Vertebrate Story* (Revision of Man and Vertebrates). The University of Chicago Press, Chicago.
16. Romer, A.S. (1966). *Vertebrate Palaeontology.* 3rd ed. University of Chicago Press, Chicago.
17. Romer, A.S. (1970). *The Vertebrate Body.* 4th ed. W.B. Saunders Company. Philadelphia.
18. Sarbahi, D.S. (1932). *The Endoskeloton of Labeo rohita.* J. Asiatic Soc. Bengal, 28.
19. Simpson, G.C. (1961). *Principles of Animals Taxonomy.* Columbia University Press. New York.

20. Storer, T.I. and Usmger. R.L. (1965). *General Zoology.* 4th ed. Mc. Graw Hill. Inc. reprint Tata Mc-Graw Hill Publishing Company. New Delhi.
21. Shrivastava, M.D. (1952). *An Introduction to the Comparative Anatomy of Vertebrate.* Pothishala Ltd. Allahabad.
22. Thillampalam, E.M. (1938). *Scoliodon, the Indian Zoological Memoirs,* Lucknow Publishing House. Lucknow.
23. Turner, C.D. and Bagnara. J.T. (1972). *General Endocrinology.* 5th W.B. Saunders Company, Philadelphia.
24. Weichert, C.K. and Presch. W. (1975). *Elements of Chordate Anatomy.* Mc. Graw Hill Co. New York.
25. Walter, H.E. and Sayles, L.P. (1965). *Biology of the Vertebrates.* The Mac Millan Company. New York.
26. Young, J.Z. (1981). *The Life of Vertebrates.* 3rd ed. Oxford University Press, London.
27. Young, L.Z. (1975). *The Life of Mammals their Anatomy and Physiology.* Clarendon Press, Oxford

Glossary of Technical Terms

Abdominal Cavity: Part of body cavity containing the digestive system.

Abducens Nerve: Sixth craniae nerve.

Abductor: Pertains to drawing a part away from the axis of the body.

Abomasum: fourth chamber of the stomach in ruminante containing glands

Acetabulum: cup shaped depression in pelvic girdle where head of the lemur fits.

Acanthodii: acanthodians a group of extinct armoured placoderm fishes which were completely covered with small bony plates or scales.

Acrania: contaning species of animals included in subphylum Cephalochordata.

Acrodont dentition: when teeth are attached on the rim of the jaw as in most fishes and smphibians.

Actinopterygii: a subclass of the class Osteichthyes including ray finned fishes.

Adductor: the one drawing a part towards the median line.

Adrenal: endocrine gland near kidney.

Agnatha: jaw less vertebrates.

Air bladder: gas bladder in some teleosts.

Air sacs in birds: distended blind ends of certain bronchi that project from the surface of the lungs.

Alisphenoid: one of the sphenoid bones in the skull of some vertebrates.

Allantois: an extraemryonic sac-like extension of the hind gut serving excretion and respiration.

Alveolus: small lobular chamber.

Amnion: inner foetal membrane surrounding the embryo.

Amniotes: animals possessing amnion in developing embryo such as reptiles, birds and mammals.

Ampulla: a saccular dilation of a canal.

Amniotes: animals such an fishes and amphibians which do not possess amnion.

Anapsida: a subclass of reptiles with solid skull without temporal fossa behind the eye.

Anastomosis: a connection between two blood vessels.

Anus: posterior opening of the alimentary canal.

Aortic arches: a series of arterial channels from ventral aorta which encircle the pharynx and empty into the dorsal aorta.

Appendicular skeleton: skeleton of appendages and girdles.

Archenteron: embryonic digestive tube.

Arcualia: arch shaped parts of the vertebrae

Atriopore: unpaired opening of the atrium in *Branchiostoma.*

Atrium: a chamber surrounding the pharynx where gill slits open in urochordates and cephalochordates; a part of the heart also called auricle.

Auditory: relating to hearing.

Auricle: receiving chamber of the heart, also called atrium; the external ear.

Autonomous nervous system: self controlling part of nervous system.

Autostylic jaw suspension: where jaw articulares directly to the cranium.

Axial skeleton: part of the endoskeleton present along the axis of the body.

Azygous: unpaired.

Basibranchial: median ventral component of a branchial arch.

Basihyoid: median ventral component of the hyoid arch.

Bicusped valve: valve in the heart contaning two flaps.

Blastocoel: cavity of the blastula.

Blastoderm: primitive cellular plate at the beginning of embryonic development.

Brachial: relating to the fore limb.

Branchial: relating to gills.

Branchimeric muscles: muscles of visceral arches.

Bronchus: branch of trachea leading to a lung.

Bronchioles: branches from bronchi which reach alveoli.

Buccal: relating to mouth.

Bunodont teeth: molars of some omnivorus animals such as man, bars, pigs with jaw rounded or blunt cusps.

Caecum: structure ending in a blind sac.

Canecllous bone: spongy bone.

Capillary: minute microscopic blood vessel which connects arteries and veins.

Capitulum: rounded extremity or head of a bone.

Carapace: dorsal part of the shell of tortoise.

Cardiac: relating to the heart.

Carpus: bone constituting the wrist.

Caudal: relating to the tail.

Cephalic: relating to the head.

Cerebellum: part derived from the metencephalon of the hind brain.

Cerebrum: part derived from the telencephalon of the fore brain.

Cervical: relating to the neck.

Chiasma: crossing of nerve fibers.

Chondro: relating to cartilaginous.

Chondrocranium: cartilaginous skull.

Chondrostei: a group of ray-finned fishes.

Chorda tendineae: numerous tendons given out by papillary muscles to the margins of valves to sustain them against the pressure of ventricular contraction.

Chorion: outer membrane surrounding the embryo.

Choroid: relating to a coat or membrane.

Chromatophore: pigment bearing cell.

Cloaca: a receptacle for urinogenital and rectal functions.

Coeliac: related to body cavity.

Coelom: body cavity lined by mesodermal epithelium.

Commissure: tissue mass connecting corresponding helves of spinal cord or brain.

Condyle: rounded articular surface at the end of a bone.

Constrictor: relating to binding or squeezing a part.

Corpus: any mass.

Cortex: outer portion of an organ.

Costal: relating to a rib.

Cranium: bones of the head.

Crista: structure present on the ampulla of semicircular canal of the internal car.

Crossopterygii: lobe finned fishes.

Ctenoid scale: scale with tooth-like projections.

Cuticle: horny epidermal layer of the skin.

Cycloid scale: scale heaving circular growth rings.

Cystic: related to a cyst or bladder.

Depressor: related to lowering or pulling down.

Dermatome: the foetal segment which gives rise to the skin.

Dermatocranium: part of skull made by membrane bones.

Diaphragm: the partition between the abdominal and thoracic cavities present in mammals.

Diapophysis: upper articular surface of the transveres process of a vertebra.

Diapsid: reptiles with two temporal openings on each side of the skull.

Diastema: a toothless interval between the last incisor and first cheek tooth.

Diencephalon: a part of the forebrain contaning thalamus and its appendages.

Digit: finger or too.

Diphycercal fin: the fin where the vertebral column extends upto the end of the fin and dorsal and ventral parts are eqwually developed; found in modern cyclostomes and dipnoi.

Distal: away from the median line or centre.

Dorsal: relating to the back or toe.

Ductus deferens: tubular structure carrying away any fluid.

Duodenum: first part of the small intestine.

Ecdysis: shedding of stratum corneum as a whole in amphibia and reptiles periodically.

Ectoderm: outermost primary germ layer.

Efferent: conducting certrifugally or outwards.

Elastics externa: outer sheath of the notochord.

Endocardium: lining of the heart cavities.

Endochondral: inside cartilaginous tissue.

Endoerine: relating to internal secration.

Endoderm: innermost primary germ layer.

Endolymph: fluid present in the membranous labyrinth of the internal ear

Endometrium: mucous membrane lining the uterus.

Endoskeleton: internal skeleton of the body.

Enterocoele: coelom in communication with the lumen of the gut.

Epicardium: visceral peritoneum covering the heart.

Epidermis: outer epithelium of the skin.

Epididymis: convoluted anterior portion of the exeretory duct of the testis.

Epiglottis: the structure covering over the glottis of the larynx.

Epimere: dorsal region of the mesoderm on lateral sides of the neural tube.

Epiphysis: terminal ossification developed separately on long bones.

Epithalamus: dorsal portion of thalamus of brain.

Epithelium: layer of cells covering a free surface.

Erythrocyte: red blood corpuscle.

Exoskeleton: skeleton present outside the body.

Extensor: relating to straightening a part.

Exteroceptors: external receptor organs.

Extrinsic: originating outside of the part upon which it works.

Fallopian tube: anterior part of the oviduct in mammals which is relatively short convoluted tube derived from the Mullerian duct.

Filum terminate: terminal delicate strand like ending of the spinal cord situated at the base of the tail.

Follicle: a crypt or circular or vesicular body present in some organ.